EXPLORING
ENVIRONMENTAL
ISSUES

AN INTEGRATED APPROACH

DAVID D. KEMP

Routledge
Taylor & Francis Group

LONDON AND NEW YORK

First published 2004
by Routledge
2 Park Square, Milton Park, Abingdon, Oxon OX14 4RN

Simultaneously published in the USA and Canada
by Routledge
711 Third Ave, New York, NY 10017

Reprinted 2007, 2008 (twice)

Transferred to digital printing 2008

Reprinted 2011

Routledge is an imprint of the Taylor & Francis Group, an informa business

© 2004 David D. Kemp

Typeset in Bell Gothic and Times New Roman
by RefineCatch Limited, Bungay, Suffolk

British Library Cataloguing in Publication Data
A catalogue record for this book is available from the
British Library

Library of Congress Cataloging-in-Publication Data
Kemp, David D., 1943
Exploring environmental issues: an integrated
approach / David D. Kemp.
p.   cm.
Includes bibliographical references and index.
ISBN 0–415–26863–X (hbk.: alk. paper) –
ISBN 0–415–26864–8 (pbk.: alk. paper)
1. Environmental degradation. 2. Nature – Effect
of human beings on. 3. Human ecology.   I. Title.
GE140.K48   2004
363.7 – dc22   2003016631

ISBN-10: 0–415–26863–X (hbk)
ISBN-10: 0–415–26864–8 (pbk)
ISBN-13: 978-0-415-26863-9 (hbk)
ISBN-13: 978-0-415-26864-6 (pbk)

TO MY WIFE, PAT

Contents

International co-operation
Is sustainable development the answer?

Acknowledgements

Although the author's name is the only one to appear on the title page of a book, the final product is always the culmination of the combined efforts of many people. This book is no exception.

At Routledge, Andrew Mould deserves thanks for his interest in the original concept, for his help throughout and his patience as the completion of the final product dragged on. Melanie Attridge was always ready with help and advice when I asked for it.

Lakehead University provided me with six months of administrative leave during which a major part of the research for the book was accomplished. I appreciate that opportunity very much. At Lakehead also I have to thank the students in my introductory Environmental Studies class over the past several years, who were unknowing guinea pigs for some of the ideas included in *Exploring Environmental Issues*. Their unsolicited comments at times made me rethink content and directions and I am sure the final product has benefited from that.

The reviewers of the original idea and those who commented on the draft have my thanks also, although I must admit I did not always embrace their suggestions completely. Any errors or omissions that arise from that are of course my responsibility.

After being through the book writing process several times, my wife, Pat, is perhaps used to the disruption it causes to home and family life, but I continue to be grateful to her for her patience and understanding. I also appreciate the contribution that my daughters, Susan and Heather, made to the completion of the book, from their suggestions on supplementary reading, to typing and photography.

My thanks are also due to the following for allowing me to reproduce copyright material: Figure 2.4, with permission from A.S. Goudie, *The Nature of the Environment* (2nd edn), Oxford: Blackwell Publishing (1989); Figure 2.16, with permission from P.D. Moore, B. Chaloner and P. Stott, *Global Environmental Change*, Oxford: Blackwell Science (1996); Figures 2.21 and 3.9, with permission of the McGraw-Hill Companies, from E.D. Enger and B.F. Smith, *Environmental Science* (8th edn), Dubuque IA: Wm. C. Brown (2002); Figure 5.2, with permission from C.C. Park, *The Environment: Principles and Applications*, London: Routledge (1997); Figure 5.7, with permission from A.S. Goudie, *The Nature of the Environment* (2nd edn), Oxford: Blackwell Publishing (1989); Figure 6.4, from Figure 3 in W. Simpson-Lewis, R. McKechnie and V. Neimanis (eds), *Stress on Land in Canada*, Environment Canada (1983), reproduced with permission of the Minister of Public Works and Government Services, 2003; Figure 7.1, with permission from M. McKloskey and H. Spalding, 'A reconnaissance level inventory of the amount of wilderness remaining in the world', *Ambio* 18 (1989): 221–7; Figure 7.6, with permission from J.H. Brown and M.V. Lomolino, *Biogeography* (2nd edn), Sunderland MA: Sinauer Associates Inc. (1998); Figure 7.13, with permission of the

McGraw-Hill Companies from G.W. Cox, *Conservation Ecology: Biosphere and Biosurvival*, Dubuque IA: Wm. C. Brown (1993); Figure 8.3, with permission from M.C. Healey and R.R. Wallace, 'Canadian aquatic resources; an introduction', *Canadian Bulletin of Fisheries and Aquatic Sciences* 215 (1987): 1–11; Figure 9.1, with permission from D.G. Thomas, 'Storm in a teacup? Understanding desertification', *Geographical Journal* 159 (1993): 318–31; Figure 9.4, with the permission of the Royal Meteorological Society, from S.E. Nicholson, 'Long-term changes in African rainfall', *Weather* 44 (1989): 47–56; Figure 9.5, with permission from R. Nowak, 'How the rich stole the rain', *New Scientist* 174, 2347 (2002): 4–5; Figure 9.7, with the permission of Pearson Education Ltd, from J.W. Watson, *North America: its Countries and Regions*, London: Longmans Green (1963); Figure 9.9, with permission from J. Jowett, 'People: demographic patterns and policies', in T. Cannon and A. Jenkins (eds) *The Geography of Contemporary China*, London/New York: Routledge (1990); Figure 9.10, with permission of Blackwell Publishing Ltd, from R.L. Heathcote, 'Images of a desert? Perceptions of arid Australia', *Australian Geographical Studies* 25 (1987): 3–25; Figure 10.6, from *SOE Report for British Columbia*, Environment Canada, reproduced with the permission of the Minister of Public Works and Government Services, 2003; Figure 10.7, with the permission of the American Meteorological Society, from E.W. Bierly and E.W. Hewson, 'Some restrictive meteorological conditions to be considered in the design of stacks', *Journal of Applied Meteorology* 1 (1962): 383–90; Figure 10.16, with the permission of Springer-Verlag, from H.H. Harvey, 'Effects of acid precipitation on lake ecosystems', in D.C. Adriano and A.H. Johnson (eds) *Acidic Precipitation II, Biological and Ecological Effects*, New York: Springer-Verlag (1989); Figure 10.18, from Indicator: Wet Sulphate Deposition, http://www.ec.gc.ca/soer-ree/English/Indicators/ Issues/AcidRain/Bulletin/arind3_e.cfm, Environment Canada (2002), reproduced with the permission of the Minister of Public Works and Government Services, 2003; Figures 11.6 and 11.7, from *Earth under Siege: Air Pollution and Global Change* by Richard P. Turco, copyright 1996 by Oxford University Press, Inc.; used by permission of Oxford University Press, Inc.; Figures 11.16 and 11.17, with the permission of the Intergovernmental Panel on Climate Change, from IPCC, *Climate Change 2001: Synthesis Report*, Cambridge: Cambridge University Press (2001); Figure 11.18, *Understanding Atmospheric Change*, Environment Canada, SOE Report 91-2, reproduced with the permission of the Minister of Public Works and Government Services, 2003.

David D. Kemp
Thunder Bay, 2003

Abbreviations

AAC	annual allowable cut
ABC	Asian Brown Cloud
AH	Arctic Haze
AIDS	acquired immune deficiency syndrome
ANWR	Arctic National Wildlife Refuge
APC	American Plastics Council
ASFR	age-specific fertility rate
ATP	adenosine triphosphate
ATV	all-terrain vehicle
BOD	biochemical oxygen demand
CBD	Convention on Biological Diversity
CBR	crude birth rate
CDM	Clean Development Mechanism
CDR	crude death rate
CFC	chlorofluorocarbon
CIDA	Canadian International Development Agency
CITES	Convention on International Trade in Endangered Species of Flora and Fauna
COP	Conference of the Parties
CPR	common-pool resources
CSCS	Comprehensive Soil Classification System
CSD	Commission on Sustainable Development
CSIRO	Commonwealth Scientific and Industrial Research Organization
DDT	dichlorodiphenyltrichloroethane
DNA	deoxyribonucleic acid
EASOE	European Arctic Stratospheric Ozone Experiment
EC	European Union
EIT	economies in transition
ENSO	El Niño–Southern Oscillation
EU	European Union
FAO	Food and Agriculture Organization
FBC	fluidized bed combustion
FCCC	(United Nations) Framework Convention on Climate Change

FEWS	Famine Early Warning System
FEWS NET	Famine Early Warning System Network
FGD	flue gas desulphurization
GAP	Global Program of Action for the Protection of the Marine Environment from Land-based Activities
GATS	General Agreement on Trade and Services
GATT	General Agreement on Tariffs and Trade
GEMS/Air	Global Emissions Monitoring System/Urban Air Pollution Assessment Programme
GEO	Global Environmental Outlook (Report)
GFR	general fertility rate
GHG	greenhouse gas
GM	genetically modified
GWP	global warming potential
HAP	hazardous air pollutant
HCFC	hydrochlorofluorocarbon
HDPE	high-density polyethylene
HFCs	hydrofluorocarbons
HIV	human immunodeficiency virus
IAEA	International Atomic Energy Agency
ICPQL	Independent Commission on Population and Quality of Life
IET	International Emissions Trading
IJC	International Joint Commission
IMF	International Monetary Fund
INDOEX	Indian Ocean Experiment
IPCC	Intergovernmental Panel on Climate Change
ITCZ	Intertropical Convergence Zone
ITTO	International Tropical Timber Organization
IUCN	World Conservation Union
JI	Joint Implementation
LIMB	limestone injection multistage burning
NAAQS	National Ambient Air Quality Standards
NAFTA	North American Free Trade Area
NAS	(US) National Academy of Science
NAWAPA	North American Water and Power Alliance
NCGCC	National Coordinating Group on Climate Change
NFAP	National Forestry Action Plan
NGOs	non-governmental organizations
NIMBY	Not In My Backyard
OECD	Organization for Economic Co-operation and Development
ODP	ozone depletion potential
ODS	ozone-depleting substances
OMA	Ontario Medical Association
OPEC	Organization of Petroleum Exporting Countries
ORV	off-road vehicle
PAN	peroxyacetyl nitrate
PCB	polychlorinated biphenyl
PET	polyethylene trephthalate
PFC	perfluorocarbon
POP	persistent organic pollutant
RAINS	Regional Acidification Information and Simulation
RNI	rate of natural increase
RSP	respirable suspended particulates

SCR	selective catalytic reduction
SPM	suspended particulate matter
SST	supersonic transport
SST	sea surface temperature
TDS	total dissolved solids
TFAP	Tropical Forestry Action Plan
TFR	total fertility rate
TNT	trinitrotoluene
TOGA	Tropical Ocean and Global Atmosphere Programme
TSP	total suspended particulates
UK DEFRA	United Kingdom Department of Environment, Food and Rural Affairs
UK DETR	United Kingdom Department of Environment, Transport and the Regions
UN	United Nations
UNCCD	United Nations Commission to Combat Desertification
UNCED	United Nations Conference on Environment and Development
UNCHE	United Nations Conference on the Human Environment
UNCOD	United Nations Commission on Desertification
UNDP	United Nations Development Program
UNECE	United Nations Economic Commission for Europe
UNEP	United Nations Environment Program
UNEP-WCMC	United Nations Environment Program–World Conservation Monitoring Centre
UNESCO	United Nations Educational, Scientific and Cultural Organization
UNICEF	United Nations Children's Fund
UNPD	United Nations Population Division
USAID	United States Agency for International Development
USEPA	United States Environmental Protection Agency
USGCRP	United States Global Change Research Program
USGS	United States Geological Survey
VLCC	very large crude carrier
VOCs	volatile organic compounds
WCED	World Commission on Environment and Development
WFP	World Food Program
WHO	World Health Organization
WMO	World Meteorological Organization
WSSCC	Water Supply and Sanitation Collaborative Council
WTO	World Trade Organization
WWF	World Wide Fund for Nature/World Wildlife Fund
ZPG	zero population growth

Introduction

Widespread popular interest in all things environmental grew rapidly in the second half of the twentieth century as knowledge of environmental issues that had previously been confined to the realm of natural science received greater exposure. Educators at all levels embraced the issues and numerous environmental interest groups emerged to highlight the various problems that society faced. Knowledge was spread through the millions of words of copy produced annually by the print media and the countless hours of audio and video coverage provided by their electronic counterparts. With technology creating an unprecedented ability to communicate rapidly and easily, in a society educated and interested enough to appreciate the issues, the rapid growth in interest and concern that took place was not particularly surprising. The result has been the creation of a society that is remarkably environmentally aware, with the preservation and rehabilitation of the environment not only of concern to individuals, but also an integral part of government policy.

Even with such attention, some environmental issues continue to be misunderstood, misinterpreted and misrepresented. The sheer volume of material and the complexity of the issues make this inevitable, but it may also be a reflection of the way in which the material is presented. As interest in global environmental issues expanded, the first indications of important discoveries or new initiatives often appeared in newspapers and popular magazines or on television, and that continues to be the case.

Because of limitations imposed by time and space, however, or even lack of specialist knowledge among the presenters, many of these articles or presentations cover only the salient points of an issue or oversimplify the elements involved. At the other end of the spectrum are the technical reports and academic volumes that remain unknown or inaccessible to those lacking scientific training or experience. There is a middle ground, however, in which readers or viewers with even a basic understanding of the workings of the environment and the relationships that exist among its components can access material at a level that meets their needs and expertise. As an introductory text *Exploring Environmental Issues* can provide the necessary environmental basics to allow readers to appreciate the implications of the information presented at the level of the print and electronic media, but also take them beyond that by providing a platform from which they can explore material available in academic and technical publications. It is set up to meet the needs of those who want no more than an introduction to the environmental issues that face society, but at the same time it provides a solid base for those who will go on to study the issues in greater detail and complexity.

To provide some perspective to current popular and academic views of the environment, the text begins with a review of the historical roots of the environmental movement, introducing the people and the events that contributed to its growth and development (Chapter 1). Much of the early interest grew out of an awakening appreciation of nature, and knowledge of the physical attributes of the environment continues to be a prerequisite for the study of the global environment. *Exploring Environmental Issues* sets the scene by examining the physical elements present in the lithosphere, atmosphere and hydrosphere and the ways in which they provide the necessary prerequisites for life in the biosphere (Chapters 2 and 3). Environmental issues include more than the physical conditions in the system, however. Most are multifaceted, involving a significant input from people, either through their numbers or technological development, and indeed, it might be argued that without human interference there would be no environmental problems. An examination of the role of human beings in the environment is therefore also essential in any study of environmental issues. The size, nature and structure of populations (Chapter 4) and their level of technological development (Chapter 5) are central to any consideration of the human input to environmental problems, but it extends beyond that. Society's perception of the environment and its relation to that environment is also important. For thousands of years the human animal lived according to the checks and balances of the natural environment and had a relationship to it that differed little from that of other animals. As populations expanded and technology advanced, that situation changed radically, leading to a conflict between what society wants from the environment and what the environment is able to provide, which is at the root of many of the problems considered in detail in Chapters 6 to 11.

Building on its physical and human base, the text examines various issues in turn, including those that threaten the land (Chapter 6), threaten plants and wildlife (Chapter 7), restrict the availability and quality of water (Chapters 8 and 9) and pose problems in the atmospheric environment (Chapters 10 and 11). Each chapter considers the nature of the issue, its development, its impact and the potential for mitigation. Specific concerns such as waste disposal, loss of biodiversity, drought, desertification, soil erosion, acid rain, ozone depletion and global warming are covered in such a way as to integrate the physical aspects of the problem with the technological, socio-economic and political contributions arising from human activities. An integrated approach is necessary to deal with the reality of a situation in which all environmental issues involve a series of interlinked and interrelated components that contribute individually and together to create the character of a specific issue.

The concept of sustainability and its extension into sustainable development has received increasing attention for its potential to contribute to the reduction or even the solution of a number of environmental problems. It appears as a participant in a number of the issues examined in Chapters 6 to 11. There is no doubt that it has much to offer, but its application also requires an appreciation of the integrated nature of environmental issues, without which the promise of ongoing development with no harm to the environment is unlikely to be fulfilled. Sustainable development is only one approach to solving the problems of environmental change and human vulnerability to it, and in the final chapter (Chapter 12) other directions are examined in terms of recent national and international approaches to the problems.

Each chapter includes a number of boxes, which may provide additional information on a specific topic, examine a case study that illustrates an issue introduced in the main body of the text, or offer technical information, allowing readers to decide on the level of detail and complexity they require or with which they are comfortable.

Within this format, the book is multi-disciplinary in its approach, particularly suitable for introductory courses in geography and environmental studies, but the ideas included in the various chapters have also been introduced to students in forestry, natural science and outdoor recreation programmes with some success. It contains a glossary to provide succinct definitions of a broad range of terms introduced in the book, allowing students in specific disciplines easy access to terms and concepts with which they are not initially familiar. In keeping with the multi-disciplinary approach, the books suggested for additional reading at the end of each chapter have been chosen to allow the reader to explore aspects of the environment represented by artists, historians, novelists and travellers as well as physical and natural scientists.

As a means of encouraging the reader to think beyond the material presented in the text, each chapter includes a number of questions for further study that can be tackled by individual students or developed for classroom or seminar activities. Some suggestions direct the readers to debate and discuss the global implications of environmental issues. Others ask the students to consider the local impacts of issues by identifying activities in their own communities that contribute to global environmental problems or are affected by them. Some of the projects are suitable for development as poster displays by a number of students. No introductory text can deal with every aspect of an issue in detail, but providing such activities, with the text as a base of information, allows readers and educators the opportunity to develop particular interests or pursue specific topics to a level beyond that of the book.

At the beginning of the twenty-first century, threats to the earth's environment are many and varied. Success in mitigating some existing environ-mental problems has shown that not all the threats are insurmountable, however. Given the complexity and global extent of many environmental issues it is clear that future success will require unprecedented co-operation at regional, national and international levels. That is likely to occur only if those in a position to make a difference – be they individuals or governments – understand why they have to co-operate and why they have to change habits that have been part of life for generations. Key to this is education, by governments and environmental groups, but particularly by schools, colleges and universities, and the aim of *Exploring Environmental Issues* is to contribute to that education.

1

The Environment and the History of Environmental Concerns

After reading this chapter you should be familiar with the following concepts and terms:

abiotic resources	ecology	new environmentalism
agricultural hearths	ecosystem	pollution
biotic resources	environment	preservation
conservation	environmental cycles	steady-state systems – open, closed
deep ecology	environmental interrelationships	technocentrism
dynamic equilibrium	environmental organizations	waves in the environmental
ecocentrism	Gaia hypothesis	movement

THE ENVIRONMENT AND ITS COMPONENTS

'Environment' is a term much used in modern society, but what does it mean? For many it is synonymous with nature; others see it as having a human element – as represented by the cultivated landscape of agricultural areas, for example, or the built environment of cities, perhaps. At its simplest, however, it is concerned with surroundings. The environment in which an object finds itself consists of all the other objects or elements that surround it (Figure 1.1). It involves more than that, however. One of the most important factors in any study of the environment is the idea of relationships. Objects do not exist in their environments in complete isolation. Each is affected by adjacent objects and in turn may influence them. In this, environmental studies, as a discipline, has much in common with the science of ecology. Both are concerned with the biotic (living) and abiotic (non-living) elements of the earth and their interactions with each other. In the past, ecologists tended to concentrate on the living elements in the system, investigating individual species of plants and animals or community patterns of interdependent organisms, which along with their immediate surroundings formed an ecosystem. Environmental scientists on the other hand approached their studies through the non-living elements of the system. Investigation of the atmosphere, hydrosphere and lithosphere, for example, provided the base upon which further studies of the biotic elements could be built. As the human element has assumed a greater and greater role in modern environmental and ecological studies, the difference between the disciplines has declined. Now, terms such as 'environment' and 'ecosystem' are often seen as interchangeable, particularly at the popular level, and although environmental scientists and ecologists continue to identify distinct differences between the disciplines, they often investigate similar issues using similar approaches and techniques. Both disciplines share the theme of 'interrelationships', a concept that recognizes the intimacy and potential impact of linkages in the natural system. When one element in an environment changes, others will be faced with the need to change also, a situation that has important implications for current and future environmental issues. Many current environmental problems, for example, have arisen because of ignorance of environmental interrelationships, or knowing disregard of them.

WHAT IS THE ENVIRONMENT?

Environment = Surroundings?

linkages/interrelationships ⌉
scale ⌉
change ⌋

Environment = Nature?

Environment = Nature
+
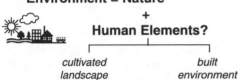
Human Elements?

cultivated landscape   *built environment*

Environment = Ecology?
|
Both study the links between living (biotic)
and non-living (abiotic) elements of the
earth/atmosphere system

Figure 1.1   What is the environment?

Plate 1.1 The natural environment – rocks and vegetation untouched by human activities.

There is also the question of scale. Although there is a tendency to talk about 'the' environment, there are, in fact, many environments, ranging from those at a microscopic scale to the whole earth environment itself. Like the individual elements in any one environment, the various environments are closely interlinked. As a result, the disruption of even a small-scale environment may ultimately be followed by change at a world scale.

Change is an integral part of any environment, and, under normal circumstances, environments are sufficiently dynamic or elastic to accommodate it. Indeed, most environments are in a continual state of flux, experiencing ongoing series of adjustments – a state known as dynamic equilibrium. When relative stability can be maintained with only minor adjustments, the environment is said to be in a steady state. These concepts have been central to the geographical approach to the environment for decades, but have received wider attention as part of the Gaia hypothesis promoted by James Lovelock (see Box 1.1, The Gaia hypothesis). Despite this ability to respond to change there are times when the amount of change exceeds the ability of the environment to accommodate it. The end result is environmental disruption. Although environmental change and environmental disruption occur naturally – climate change and associated environmental disruption appear regularly in the earth's history, for example – concern with these elements today is with those that involve some human input. In theory, human beings, as animals, are an integral part of the environmental scheme of things and subject to the controls and restraints that implies. In practice, the human element has become the main cause of change, disruption and deterioration in the environment. Why has this been so? It has much to do with the form that development has taken, along with society's attitude to the environment and knowledge of how it works.

In our modern, technology-based society the knowledge base is immense, but it is not limitless. In the environmental studies field many unknowns remain concerning the nature of the environment and the amount of change that it can accommodate.

Plate 1.2   The altered agricultural environment of the plain contrasts with the natural environment of the hills.

Plate 1.3   The built environment of a modern city. (Courtesy of Heather Kemp.)

BOX 1.1 THE GAIA HYPOTHESIS

First developed in 1972 by James Lovelock, and named after an ancient Greek earth goddess, the Gaia hypothesis views the earth as a single organism in which the individual elements coexist in a symbiotic relationship. Internal homeostatic control mechanisms, involving positive and negative feedbacks, maintain an appropriate level of stability. It has much in common with the concept of environmental equilibrium, but goes further in presenting the view that the living components of the environment are capable of working together actively to provide and retain optimum conditions for their own survival. In the simplest case, animals take up oxygen during respiration and return carbon dioxide to the atmosphere. The process is reversed in plants, the carbon dioxide being absorbed and oxygen being released. Thus the waste product from each group becomes a resource for the other. Working together over millions of years, these living organisms have combined to maintain oxygen and carbon dioxide at levels capable of supporting their particular forms of life and, through carbon dioxide, maintain the greenhouse effect at a level which can provide a temperature range appropriate for that life. This is one of the more controversial aspects of Gaia, flying in the face of conventional scientific opinion, which since at least the time of Darwin has seen life responding to environmental conditions rather than initiating them. Some interesting and possibly dangerous corollaries emerge from this. It would seem to follow, for example, that existing environmental problems which threaten current forms of life and life processes – global warming and ozone depletion, for example – are transitory, and will eventually be brought under control again by the environment itself. Some scientists view the acceptance of this aspect of Gaia as irresponsible, since it also requires acceptance of the efficacy of natural regulatory systems that are as yet unproven, particularly in their ability to deal with large-scale human interference. Lovelock himself has allowed that Gaia's regulatory mechanisms may well have been weakened by human activities, which have created so much stress on the environmental regulatory mechanisms that they may no longer be able to nullify the threats to balance in the system. The effects could even threaten the survival of the human species. Although the idea of the earth as a living organism is a basic concept in Gaia, the hypothesis is not anthropocentric. Humans are simply one of the many forms of life in the biosphere, and, whatever happens, life will continue to exist, but it may not be human life. For example, Gaia includes mechanisms capable of bringing about the extinction of those organisms that adversely affect the system. Since the human species is at present the source of most environmental deterioration, the partial or complete removal of mankind might be Gaia's natural answer to the earth's current problems.

For more information see:

Joseph, L.E. (1990) *Gaia: The Growth of an Idea*, New York: St. Martin's Press.

Lovelock, J. (1995) *Gaia; A New Look at Life on Earth* (2nd edn), Oxford: Oxford University Press.

Pearce, F. (2001) 'The Kingdoms of Gaia', *New Scientist* 170 (2295): 30–3.

Schneider, S.H. and Boston, P.J. (eds) (1991) *Scientists on Gaia*, San Francisco: Sierra Club Books.

Even when the knowledge is available, it is often ignored. For example, it is common knowledge that the earth/atmosphere system is a closed system in material terms and, as a result, resources are finite (Figure 1.2). This was expressed in a more popular form, in 1966, by the economist Kenneth Boulding through the concept of 'Spaceship Earth'. He likened the earth to a spaceship in which the occupants had to survive using the air, water and food available at lift-off, since – at that time, at least – there was no means of delivering additional material once the ship was in orbit. In the case of the earth, the concept applies most obviously to minerals and the other commodities upon which modern society has come to depend, but also to all those resources, including air, water, soil, vegetation and animals, that are normally considered part of the natural environment. The latter group of resources has been used for thousands of years, but are mostly still available because of very efficient recycling processes built into the earth/atmosphere system. Chemical elements such as carbon, nitrogen and sulphur are recycled continuously, as is water (Figure 1.3). Plants and animals are also recycled in a way, in as much as they are able to perpetuate their species through reproduction. Unfortunately, these cycles have been disrupted as a result of human activities, leading to a variety of environmental problems. The disruption of the carbon cycle, for example, is associated with global warming and the disruption of the sulphur cycle is associated with acid rain. Similarly, pressure on the hydrologic cycle

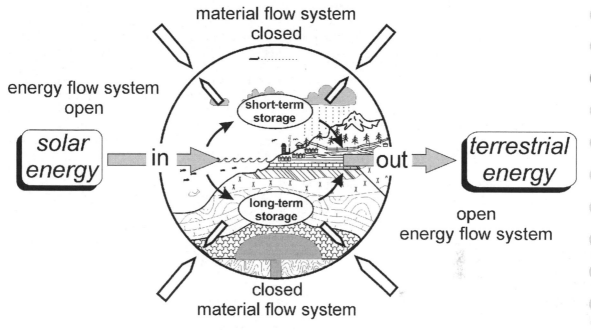

Figure 1.2   Energy and material flow in the earth/atmosphere system

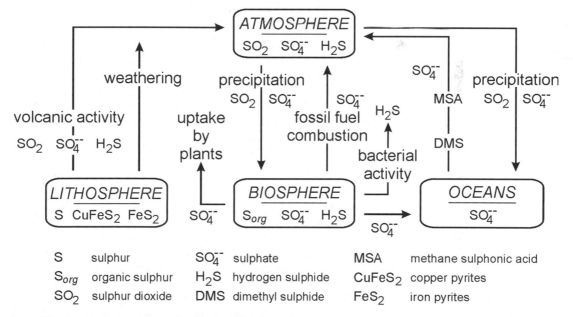

S	sulphur	SO_4^{--}	sulphate	MSA	methane sulphonic acid
S_{org}	organic sulphur	H_2S	hydrogen sulphide	$CuFeS_2$	copper pyrites
SO_2	sulphur dioxide	DMS	dimethyl sulphide	FeS_2	iron pyrites

Figure 1.3   The sulphur cycle: an example of a natural recycling system

in some areas has caused shortages of water and also serious water pollution.

In contrast, where energy is considered, the earth/atmosphere system is an open system, and therefore, in theory, much less restrictive than the closed material system (Figure 1.2). Solar radiation provides the energy required to allow the multitude of elements that make up the environment to function, being converted in the process into terrestrial energy, which ultimately passes out of the system into space. The smooth flow of energy is interrupted by storage mechanisms, which retain some of the solar energy in the system for periods ranging from a few months – in the case of the energy stored in plants during the growing season – to millions of years – in the case of the energy stored in fossil fuels in the earth's crust. Disruption of the flow of energy through the system by human activities is associated with a number of pressing environmental issues, including global warming, ozone depletion and increased atmospheric turbidity. The initial problems in these areas arose out of ignorance of the working of the environment, but even now, when a broad spectrum of society has been made aware of the causes, success in dealing with them has been limited at best.

SOCIETY AND ENVIRONMENT

Environmental studies are very much involved with the relationship between society and environment, but it is a relationship that is not always well understood and one that is always changing. There is widespread belief that, in the past, people enjoyed a much better rapport with their environment than they do today. To some extent that may have been true, but the relationship was not simple, and, from a human point of view, it was not always benign. At times the environment was restrictive and harmful to human activities. The early humans were subject to the same positive and negative environmental conditions as other animals. Such elements as predator/prey relationships and the carrying capacity of particular ecosystems applied and the early human populations must have experienced increases and decreases in numbers in much the same way, and probably for many of the same reasons, as other species. That being so, why did the human animal become the earth's dominant species? Little or no evidence to answer that question has survived from the earliest human societies, but it seems unlikely that they would have benefited from favourable environmental conditions any more than other species. Human success in becoming the dominant species did not result from a benign environment. It came about because society challenged the environment and did so successfully.

It is entirely possible that humans were in a better position to challenge the environment and make better use of the resources it provided than other animals. Their superior mental capacity allowed them to manipulate or stretch the relationships they had with their environment and a combination of skills that was broader than that of most other animals permitted them to survive under a wide range of environmental conditions. These attributes that allowed human beings to challenge the environment also ensured that they would change it, minimally at first, but to a greater and greater extent as time went by.

PREHISTORIC TIMES

As the last Ice Age was drawing to a close, some 13,000–15,000 years ago, the earth's human population survived by hunting and gathering. Such activities imply a reasonable balance between people and their environment to allow the relationship to be sustained and the hunters to survive. Whether or not this sustainability was the result of a deliberate strategy is not clear, but given low population densities, nomadic lifestyles and the absence of any mechanism other than human muscle by which the hunting and gathering groups could utilize the energy available to them, it is perhaps not surprising that it was achieved with minimal impact on the environment (Table 1.1). Other than the food they consumed, the main source of energy for the hunters was fire. It provided heat for warmth and cooking and on a larger scale for hunting, when strategically placed fires were used to drive game towards waiting hunters. Relatively large areas may have been burned during these hunting activities, and no doubt fires for heating and cooking sometimes got out of control and burned adjacent areas. However, fire is a natural element in many ecosystems and recovery would follow the fires. Similarly, when a decrease in the number of animals in an area reduced the success of the hunt or threatened the survival of a hunting group, their nomadic lifestyle allowed them to move on to new hunting grounds, leaving the animals in the original area to recover. Thus, when most of the earth's human population was involved in hunting and gathering, the environmental impact was local and short term, involving the temporary loss of vegetation and a reduction in the local animal population.

One possible exception has been postulated, however. During the late Pleistocene period the

TABLE 1.1 ENERGY USE, TECHNOLOGICAL DEVELOPMENT AND THE ENVIRONMENT

Time	Daily per capita energy consumption (kcal)	Main sources	Use	Environmental impacts
1,000,000 BC	2,000	Food; human muscle	Daily life	Minimal
100,000 BC	4,000–5,000	Food; fire; simple tools	Heating; cooking; hunting	Local and short term; changes to flora and fauna
5000 BC	12,000	Animals; agricultural produce	Cultivation; irrigation; construction; transport	Local and longer term; natural vegetation replaced by crops; aquatic environment altered; soil degraded
AD 1400	26,000	Wind; water; coal	Mechanical operations; pumping water; sawmilling; grinding grain; transport	Local and longer term or permanent; natural vegetation cleared; air and water pollution common in some areas
AD 1800	50,000	Coal; steam engine	Mechanical operations; industrial processes; transport	Local, regional and permanent; major landscape change begins; air and water pollution the norm in industrial areas
AD 1980–2000	300,000+ (developing nations about 30,000)	Fossil fuels; electricity; nuclear power; internal combustion engine	Mechanical operations; industrial processes; transport; social and cultural development	Local; regional and global; permanent and perhaps irreversible; air, water, soil deterioration on global scale; acid rain; global warming; ozone depletion; increased atmospheric turbidity

Source: compiled from data in A.K. Biswas, *Energy and the Environment*, Ottawa: Environment Canada (1974); M.H. Kleinbach and C.E. Salvagin, *Energy Technologies and Conversion Systems*, Englewood Cliffs NJ: Prentice Hall (1986).

extinction of megafauna species took place in areas as far apart as North America and Australia. Martin (1984) linked the disappearance of such animals as the mammoth and giant beaver in North America and giant kangaroos and wallabies in Australia some 10,000 years ago with the arrival of human hunting groups into areas previously uninhabited or at most sparsely inhabited. With no fear of humans, they were easy prey and their numbers were ultimately reduced below the level at which they could survive. Subsequent evidence from Australia and North America has provided strong support for Martin's proposal (Dayton 2001), but other potential causes have also been put forward. MacPhee and Marx (1997), for example, have suggested that the extinctions came about because the immune systems of the megafauna were unable to cope with the pathogens brought in by the migrating humans and they succumbed to disease rather than over-hunting. The late Pleistocene was also a period of major climate change and, although techniques for investigating these changes are becoming increasingly sophisticated, the role of climate in these extinctions is not clear (Barnosky 1994). It is possible that no single element was responsible. Changing climatic conditions, for example, may have caused the animals to become increasingly vulnerable to human predation or disease, and despite relatively primitive weaponry, the hunters were able to bring about their extinction.

In time some hunting and gathering communities left their nomadic lifestyle behind. Whether by accident or design, they discovered how to domesticate plants and animals and in so doing were able to pursue a more sedentary lifestyle, which led to the development of the first agrarian civilizations. Between 7000 and 3000 BP (c. 5000–1000 BC), these civilizations developed in Egypt, Mesopotamia, the Indus valley and the Yellow River basin (Hwang-He) in China. Towards the end of this period, the Mayan civilization grew up in Central America (Figure 1.4). Sedentary agriculture, permitted by the domestication of plants and animals, ultimately led to the development of permanent settlements and local urbanization. All of the Middle Eastern and Asian locations were on riverine plains in areas that experienced dry conditions for part of the year. Natural irrigation provided by seasonal over-bank flooding and artificial irrigation, using small dams, cisterns and ditches to redistribute the water, allowed year-round cropping and the accumulation of a food surplus. This in turn permitted a greater division of labour and the development of social, cultural and economic activities not possible in a migratory community or one dependent upon subsistence agriculture. Accompanying this was an increase in the level of human intervention in the environment, associated with accelerated population growth and a new technology based on agriculture. Natural vegetation was replaced by cultivated crops, the aquatic environment was altered, and the beginnings of soil degradation in the form of siltation and salinization became apparent in some areas (Jacobsen and Adams 1971).

THE AGRICULTURAL REVOLUTION

Significant as these developments were, they were limited in extent, and the level of human intervention in the environment increased only slowly over thousands of years. Agriculture gradually spread beyond the original hearths, sometimes encouraging permanent settlement, sometimes combining with hunting and gathering to perpetuate

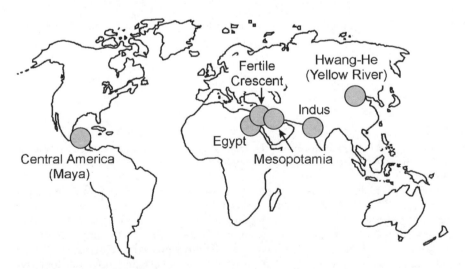

Figure 1.4 Distribution of the world's early civilizations. Those in Africa and Asia were established between 7000 and 3000 BP and the Mayans by about 3000 BP

nomadic lifestyles in shifting agriculture. Methods of converting the energy in falling water and wind were discovered and coal became the first of the fossil fuels to be used in any quantity. As late as the mid-eighteenth century, however, the environmental impact of human activities seldom extended beyond the local or regional level (Table 1.1). A global impact became possible only with the major developments in technology and the population increases that accompanied the so-called agricultural and industrial revolutions of the late eighteenth and early nineteenth centuries. This is not to imply that change had been absent prior to that time, but it was relatively slow – evolutionary rather than revolutionary. A period of more rapid change in agricultural activities began in Britain between about 1750 and 1850, with improvements introduced in all aspects of farming, leading to greater efficiency and allowing a substantial increase in food production. Greater attention was paid to maintaining and increasing the quality of the soil, by adding lime and manure. Land previously too wet to be used was brought into production by improving drainage, and soils that were too dry or light were treated with marl (clay rich in calcium carbonate) to improve their texture. New crops, such as turnips, potatoes and clover, were grown more frequently and crop rotation was introduced. Experiments with livestock breeding increased the quality and quantity of meat and wool. New mechanized or semi-mechanized implements were developed to deal with all aspects of cultivation, from ploughing and planting to harvesting. All of this reflected an improved knowledge of the science of agriculture, from soil improvement to plant and animal breeding and mechanization, but it also marked the beginning of a significant onslaught on the land. Soil composition and texture were changed, non-native plant and animal species were introduced and the new farm implements subjected the land to a much more intensive physical regime than it had experienced under simple manual labour.

In a few decades the agricultural revolution changed the landscape of Britain, replacing natural and existing cultivated vegetation with new crops and replacing open fields with enclosures surrounded by hedges and walls (Simmons 1996). In places it also contributed to environmental degradation in the form of soil erosion, where the enthusiasm for improvement brought land unsuitable for arable agriculture into production. As the new agriculture diffused throughout Europe and was carried to other continents through colonial expansion, it took with it a group of potentially serious environmental problems, ranging from the destruction of natural flora and fauna to the disruption of the hydrologic cycle and the initiation of soil erosion. This was not the result of any malicious intent. Indeed, improvements in agriculture were seen as natural and necessary, with a promise to enhance the quality of life for mankind. Initially, it seemed that the promise was being fulfilled and food production grew rapidly, but the situation was not sustainable. Ignorance of the impact of the new agricultural techniques on the environment ensured that mistakes would be made and the contribution of agricultural activities to environmental disruption and deterioration would grow.

THE INDUSTRIAL REVOLUTION

The changes in agriculture in the eighteenth and nineteenth centuries paralleled similarly innovative changes taking place in industry at that time. These brought about the industrial revolution, characterized by a major expansion in the use of coal as a fuel, in the steam engine and in the iron industry (Hudson 1992). Together they encouraged the growth of new industrial cities, incorporating heavy industries based on coal, iron and steel, as well as an expanding textile industry powered by the new steam engines. Railroads and steamships linked these cities with their sources of raw material and their markets. Population grew rapidly, fed by the food surpluses of the agricultural revolution, providing the necessary labour force and also creating a growing consumer demand. The exact relationship between population growth and technology remains a matter of controversy, but there can be no denying that, in combination, these two elements were responsible for the increasingly rapid environmental change, which began in the eighteenth century. The role of energy was particularly important, for it was the ability to concentrate and then expend larger and larger amounts of energy that made the earth's human population uniquely able to alter the environment (Table 1.1). Together, the rapidly growing population, new urbanization and industrialization created local and regional environmental stress through such elements as the inadequacy of sewage disposal techniques, mineral extraction, energy conversion and the sprawl of urban/industrial activities over the adjacent rural land. Since then the human impact on the environment has expanded from the local or regional level to the global and the results have become permanent or irreversible. Air pollution and water pollution are ubiquitous, natural vegetation has been used up faster than it can regenerate or has been replaced by cultivated crops, rivers have been dammed or diverted, natural resources have been dug from the earth in such quantity that people now rival geomorphological processes as agents of landscape change and, to meet the need for shelter, nature has been replaced by the built environment created by urbanization.

THE ENVIRONMENTAL MOVEMENT

As the impact of these changes became more and more obvious, and the magnitude of the problems involved became clear, concern for the environment grew, until today it is greater than it has ever been. One of the main forms in which this concern is expressed is in the environmental movement, a term which is widely and loosely used to include a variety of individuals and groups working through scientific, social or political agendas to achieve the common goal of defending the environment, conserving resources and generally protecting nature. In its modern form, the environmental movement dates from the 1960s and 1970s, a period which saw the creation of new environmental organizations such as Friends of the Earth, Pollution Probe and Greenpeace, and the celebration of the first 'Earth Day' on 22 April 1970. Prescient individuals such as Rachel Carson, with her exposure of the problems associated with pesticide use in *Silent Spring* (1962), and Paul Ehrlich, with his account of the potential threats from overpopulation in *The Population Bomb* (1968), influenced emerging environmental attitudes at the time, as did the publication of significant assessments of the earth's sustainability under human occupation such as *The Limits to Growth* (Meadows *et al.* 1972) and *Blueprint for Survival* (*Ecologist* 1972). Paehlke (1997) has identified this as the 'First Wave' in the environmental movement, extending from 1968 to 1976, but waves, whether natural or metaphorical, do not suddenly materialize, and this was no exception. It grew from a series of ripples that had first appeared some 200 years earlier (Figure 1.5).

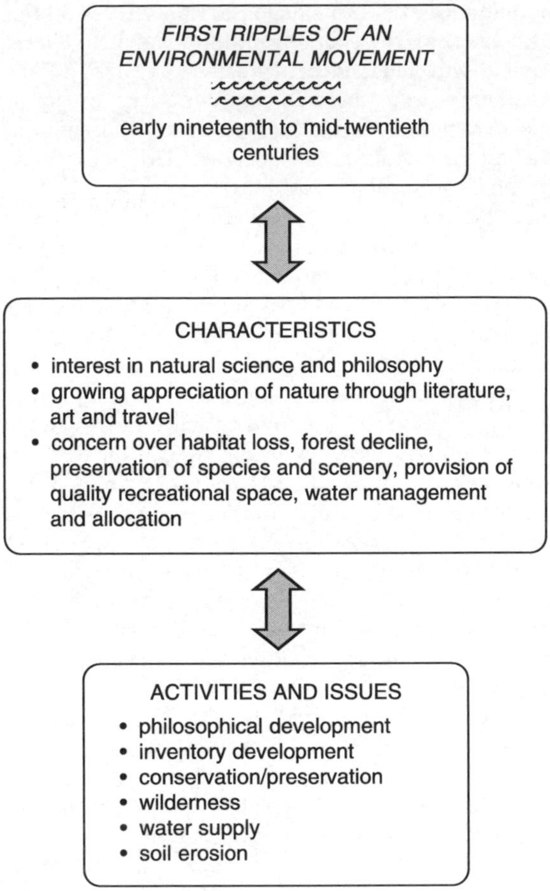

> **FIRST RIPPLES OF AN ENVIRONMENTAL MOVEMENT**
>
> early nineteenth to mid-twentieth centuries

> **CHARACTERISTICS**
> - interest in natural science and philosophy
> - growing appreciation of nature through literature, art and travel
> - concern over habitat loss, forest decline, preservation of species and scenery, provision of quality recreational space, water management and allocation

> **ACTIVITIES AND ISSUES**
> - philosophical development
> - inventory development
> - conservation/preservation
> - wilderness
> - water supply
> - soil erosion

Figure 1.5   The first ripples of an environmental movement

BRITAIN AND EUROPE

Humans have been curious about the earth and its physical attributes since at least the time of the ancient Greeks, but it was the rapid swelling of interest in natural science and philosophy in the eighteenth and nineteenth centuries that set environmentalism in motion (Figure 1.6). In Europe, the geological investigations of Playfair, Hutton and Lyell, for example, drew attention to the dynamic nature of the lithosphere (see Chapter 2) and its contribution to environmental change (Mannion 1997). In the biological sciences, the studies of Lamarck, Wallace and Darwin led to the recognition of the importance of gradual and cumulative change in plant and animal communities and produced the concept of evolution. Although Wallace shared very similar ideas, evolutionary concepts have come to be associated almost exclusively with Darwin. His classic *On the Origin of Species*, published in 1859, is popularly remembered for its development of the theory of evolution, but it was also a study of environmental change. In the concept of natural selection – commonly referred to as 'the survival of the fittest' – species able to adapt to a particular environment, or a change in the environment, survive, whereas those unable to adapt ultimately become extinct. In developing his evolutionary theories Darwin probably was influenced by the writings of the geologist Charles Lyell, who first recognized the role of gradual – as opposed to catastrophic – change in the physical environment, and by the work of Thomas Malthus on the relationship between population growth and food supply. This was also the era of the amateur naturalist. Spurred on by an interest in botany, zoology, geology and palaeontology, amateurs collected, dissected and catalogued, building an inventory of the natural

history of an area, which they shared, locally, nationally and internationally, with fellow enthusiasts. Some are recognized through the naming of a plant, insect, animal or fossil, and some, such as Gilbert White in *The Natural History of Selborne* (1789), published the results of their life's work, but most are long forgotten. Nevertheless, through their work they contributed to the data base upon which the scientific study of the environment would be built. As more and more knowledge accumulated some observers, such as Buffon, Von Humboldt and Woeikof, recognized that the environment was being changed by human activities (Thomas 1956). They noted, for example, the results of deforestation and the spread of urbanization. In Britain, in the mid-nineteenth century, Robert Smith first recognized the phenomenon of acid rain and its relationship to the burning of coal (Turco 1997). Such changes must also have been obvious to many others who did not put pen to paper, but the accumulation of scientific knowledge about the

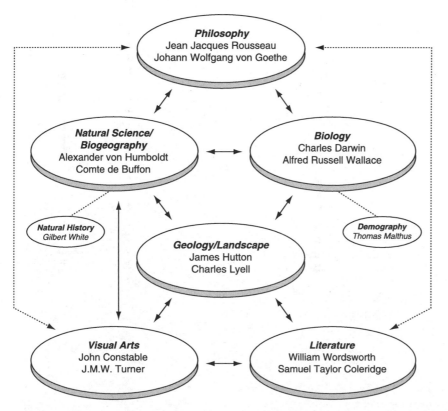

Figure 1.6   Precursors of environmentalism in Europe. The contribution of non-scientists is noteworthy

environment did little to slow the deterioration that was taking place during the growing industrialization that characterized much of Britain and Western Europe in the early nineteenth century.

Perhaps that is not surprising, since the study of current environmental problems has shown that progress is made only if knowledge reaches beyond the scientific community to give a broader audience an understanding of the issues involved. Providing that understanding has become the duty of a variety of governmental organizations and is one of the objectives of most modern environmental groups. In nineteenth-century Europe governments were generally uninterested and there were no environmental organizations as such, but there were individuals who, through the senses rather than the sciences, began to provide the opportunity for more and more people to appreciate nature or the natural environment. The philosophers Rousseau and Goethe, for example, explored the relationship between society and nature at a cerebral level, while poets such as Wordsworth and Coleridge considered the aesthetic pleasures and benefits of nature. Painters increasingly represented nature on canvas. Some such as Constable captured the realism of the landscape; others such as Turner incorporated a personal vision, which translated the elements of the environment into wild combinations of colour, light and shade (Clark 1969). These poets and painters were part of what came to be known as the Romantic movement. Whether or not it arose as a reaction to the rampant technology of the industrial revolution is debatable, but the Romantics certainly favoured feelings or sensation over reason (Ferkiss 1993). However it was done, all of this helped to foster a growing interest in nature. Prior to the end of the eighteenth century, few would have gone for a walk in the woods or climbed mountains for the sheer pleasure of it. By the middle of the next century, a tourist industry based on a desire to view 'wild' nature was well established, a favourite destination being the Swiss Alps. Although this touched only a few, mainly the wealthy, it was an essential part of the progress towards a better understanding of the environment and society's interrelationships with it.

Plate 1.4   The English Lake District – the inspiration for Wordsworth's nature poetry.

NORTH AMERICA

The growth of environmentalism and the environmental movement in North America had much in common with that in Europe, but also showed some important differences. When the Europeans came to North America, they met nature at a scale few had previously experienced. They found almost uninterrupted woodland from the tropical forests of what was to become Florida to the boreal forests of Canada in the north; they found coastal waters, rivers and lakes teeming with fish; they found wild game in abundance. In Spain, France and Britain, where most of the colonists originated, the landscape was not as domesticated as it would one day become, but it had been cultivated for centuries, with the result that the forests had been pushed back, wild animal populations had declined, crops had replaced natural vegetation and the landscape could be described as being more rustic than natural. The explorers, fur traders and fishermen, who were first to arrive used the resources of the New World for food and shelter – and profit, of course – and interacted with the indigenous population.

All that changed when the agricultural settlers arrived. These colonists viewed the land as wilderness that needed to be civilized to create a state similar to the one that they had left behind. Some of the timber was used for building settlements, for shipbuilding or the making of furniture and some was exported to the home countries in Europe, where even by the eighteenth century the forests had been decimated, but often the demand for new agricultural land was so great that the woodland was cleared by direct burning. The animals that lived in the forest lost their habitat and disappeared, while the Native Americans lost an environment upon which they had depended and like the animals were forced to move out. Treatment of the environment was extremely wasteful. Environmental resources were so abundant that they seemed never-ending, with new resources available simply by moving west away from the coast. That approach was central to the environmental history of North America in the 150–200 years following the arrival of the first colonists. In the south, for example, the land depleted of its nutrients by the heavy demands of the tobacco and cotton crops was simply abandoned and new land to the west was brought under cultivation. The land that was left was subject to major soil erosion, and only a few decades after the establishment of the Virginia colony soil eroded from the tobacco fields was already silting up Chesapeake Bay and the adjacent waterways. To the north, where the fur trade was a major staple, the problem of dwindling supplies of beaver pelts caused by the over-harvesting of the beaver in the eastern part of the continent was overcome by tapping into the abundant resources that remained in the west. The ongoing need for new supplies drove the fur trade beyond the Great Lakes into the far north-west and eventually to the Pacific coast. Later, the forest industry followed much the same pattern, harvesting trees with apparently no need for management in a land where the forest resources seemed limitless.

An awareness of nature and what was being done to it grew gradually, building through a combination of scientific or semi-scientific observations, artistic representation and philosophical debate, not unlike what had happened or was happening in Europe. In the eighteenth and nineteenth centuries, explorers such as Lewis and Clark recorded details of the country through which they passed, and the fur traders of the North West and Hudson's Bay Companies kept records not only of the furs they traded, but also of a wide range of environmental elements, such as the weather, the vegetation, the landscape and the waterways. A few years after arriving in the United States in 1803, the naturalist and artist John James Audubon began the task of painting the birds of North America and his illustrated books increased the awareness of the natural history of the continent. Other artists such as Cole, Doughty and Durand of the Hudson River school of landscape painters introduced the Romantic movement to the visual arts in North America with paintings that glorified nature, and on the literary scene James Fenimore Cooper, writing in the 1820s and 1830s, used the eastern wilderness as the setting for his novels about the early frontier. That wilderness was already disappearing from the eastern United States by that time and these paintings and novels exposed a wider public to its existence and former dominance in the landscape, albeit in a form coloured by the artistic licence necessary in such endeavours. Despite this, according to Ferkiss (1993), most Americans in the early nineteenth century continued to fear the wilderness, rather than revere it. They saw nature as something that had to be conquered if progress was to be made.

ENVIRONMENTAL PHILOSOPHERS AND ENVIRONMENTALISTS IN NORTH AMERICA

This attitude may even have applied to a group of pioneering environmental philosophers, including Emerson and Thoreau (see Box 1.2, Pioneering North American environmental philosophers and environmentalists), who are generally credited with sowing the seeds from which the environmental movement grew in the United States. Thoreau, particularly, has gained a reputation as the pre-eminent proto-environmentalist, with *Walden*, his account of the time he spent in the early 1850s communing with nature at Walden Pond near Concorde, Massachusetts, widely considered a classic of environmental literature (Buell 1995). Ferkiss (1993), however, has concluded that Thoreau was not the lover of nature or patron of the wilderness that later writers have made him out to be. He saw the natural landscape as one which was pastoral rather than wild; one that had been modified, even improved, by human intervention. Perhaps Thoreau's reputation is not entirely well deserved, but there can be no denying the environmental character of much of his writing. Cox (1993) has concluded that by his later years Thoreau was an ecologist in all but name and a perceptive observer of the changes taking place in the New England wilderness. Whatever interpretation is applied to their thoughts on nature and wilderness, by stimulating interest in the natural environment Emerson and Thoreau provided a philosophical base for American environmentalism upon which their contemporaries such as George Perkins Marsh and John Muir were able to build.

Marsh trained as a lawyer and became a diplomat, but he was also a self-trained physical geographer who was quick to appreciate the impact of society on nature and natural resources (Lowenthal 2000). His observations of the degradation caused by human activities on the landscape of the eastern Mediterranean and the rapid decline of the forests in eastern North America led him to believe that the environment could suffer irreparable damage at the hands of society. He was not against change, but saw that it had to be managed change (Buell 1995). He published his ideas in 1864 in *Man and Nature or Physical Geography as Modified by Human Action*, which included details of the impact of human activities such as mining, agriculture and forestry on the environment. Although his ideas are less widely known by environmentalists than those of Thoreau, they led to the first major environmental conference of modern times – 'Man's Role in Changing the Face of the Earth', held at Princeton University in 1956 – and gave an early glimpse of the catastrophic potential of human activities that was to become a central element in the environmental movement more than a century later. One of the first to put this early environmental thinking into practice was John Muir. Muir was born in Scotland, but spent most of his life in the United States, where his thoughts on nature and the wilderness were influenced by the writings of Emerson and Thoreau (Buell 1995). As a naturalist he was particularly concerned about the damage being done to the mountain and forest environments of the American west, and towards the end of the nineteenth century he turned to writing to promote their conservation. Within the conservation movement Muir was a preservationist, believing that nature had its own inherent value and should be preserved with little or no change, other than that which occurred naturally (Smith 2000). He advocated strong government participation and received the support of President Theodore Roosevelt, himself a strong believer in conservation. Muir was also a popular activist whose efforts helped to create the Sierra Club in 1892 and as its first president he was instrumental in having the Yosemite area of the Sierra Nevada designated as a national park. His legacy can be seen in the Sierra Club's involvement in the founding, preservation and expansion of parks and wilderness areas in the western United States, from Arizona to Alaska, over the century since its founding.

At odds with preservationists like Muir were those conservationists who saw no problem with the wise utilization of the economic resources of natural areas, while retaining as far as possible the environmental integrity of the areas so that they might remain available to future generations. In effect, their ideas foreshadowed the concepts of multiple land use and sustainable development that were to become central environmental issues in the latter part of the twentieth century. The main proponent of this utilitarian approach to conservation – as Ferkiss (1993) has termed it – was Gifford Pinchot, the first Director of the Division of Forestry when it was created in the US Department of Agriculture in 1898. He was a scientific forester who believed that the forests should be managed

BOX 1.2 PIONEERING NORTH AMERICAN ENVIRONMENTAL PHILOSOPHERS AND ENVIRONMENTALISTS

Ralph Waldo Emerson (1803–82) Philosopher and writer who embraced the transcendental philosophy of the divinity and unity of man and nature. Such beliefs foreshadowed later environmentalist concepts and ideas.

Henry David Thoreau (1817–62) A protégé of Emerson, Thoreau rejected materialism and sought to improve the quality and meaning of life by the contemplation and study of nature. Perhaps best known for his account of the time he spent at Walden Pond near Concord, Massachusetts, Thoreau also kept a journal in which for twenty-four years he recorded his philosophical and scientific observations. An ecologist in all but name, his observations made him aware of the concept of forest succession and as early as 1859 he advocated the creation of wilderness parks for the preservation of nature.

George Perkins Marsh (1801–82) A self-trained physical geographer, Marsh's main contribution to the early environmental movement was his appreciation of the human impact on nature and natural resources, which he detailed in his pioneering environmental studies text *Man and Nature*, published in 1864. Although less well known than his contemporary environmental pioneers, his ideas were revived in 1956 at the Princeton conference on 'Man's Role in Changing the Face of the Earth'.

John Muir (1838–1914) Appalled by the destruction of the environment in the California sierras, John Muir became one of the first environmental activists to use his writing and political contacts to promote the preservation of the western wilderness. In 1892 he was a founding member and first president of the Sierra Club. The original non-profit environmental conservation organization, after more than a century the Sierra Club remains a leader in the environmental movement.

Aldo Leopold (1887–1962) Although he lived and wrote more than a generation after the original environmental pioneers, Leopold's work in scientific wildlife management was ground-breaking. He appreciated the interrelationships among the various components of the environment and saw the concept of the ecosystem as central to the management of nature. Practising what he preached, Leopold supported the establishment of wilderness preserves and was a founding member of the Wilderness Society.

Rachel Carson (1907–64) Writer and naturalist who was the author of the best-selling book *Silent Spring* in which she drew attention to the impact of chemicals on the environment. When first published, the book was denounced as alarmist by the chemical industry and many biologists treated it with some scepticism. Carson's concerns were justified, however, and her book gave the environmental movement a major boost.

For more information see:

Carson, R. (1962) *Silent Spring*, New York: Houghton Mifflin.

Fleck, R.F. (1985) *Henry Thoreau and John Muir among the Indians*, Hamden CT: Archon Books.

Leopold, A. (1949) *A Sand County Almanac*, New York: Oxford University Press.

Lowenthal, D. (2000) *George Perkins Marsh: Prophet of Conservation*, Seattle WA: University of Washington Press.

Muir, J. (1894) *The Mountains of California*, New York: Century.

Thoreau, H.D. (1854) *Walden*, Boston MA: Ticknor and Fields.

commercially for their natural resources and did not support the preservation of national forest land for non-commercial purposes such as parks. Overall the utilitarian conservationists seem to have won the day, but environmental groups continue to press the preservationist approach. Aldo Leopold (see Box 1.2, Pioneering North American environmental philosophers and environmentalists), an American ecologist, regarded as the father of wildlife management and founding member of the Wilderness Society, incorporated aspects of both approaches in his teaching and writing, but the schism that was created more than a century ago still exists, being seen most recently in the debate over the development of oil and gas resources in the Arctic National Wildlife Refuge on the North Slope of Alaska.

ENVIRONMENTAL ISSUES AND THE ENVIRONMENTAL MOVEMENT IN THE TWENTIETH CENTURY

By the early decades of the twentieth century pressure from both industrial and agricultural development posed major threats to the environment, but the seriousness of the situation was not widely recognized. The atmosphere of large cities in Europe and North America was laden with smoke released by the industrial and domestic use of coal as a fuel. Habitat was lost to the spread of arable agriculture and forest exploitation worldwide, water was polluted by industry and the ecology of many areas was changed for ever. The migration of large numbers of Europeans to North and South America, Australasia and Africa in the second half of the nineteenth century and in the twentieth century up to the First World War ensured that the threats were worldwide. In Australia, for example, unique ecosystems were destroyed when the land was cleared for European-style farming. The environmental impact of introduced species, such as the rabbit, is well documented (Adamson and Fox 1982), but large herds of cattle and flocks of sheep that numbered in the millions also contributed to environmental deterioration, particularly in semi-arid areas where over-grazing led to serious soil erosion (Simmons 1996).

European colonization of Africa started later than in the Americas or Australasia, but by 1900, following the so-called 'Scramble for Africa' in the last decades of the nineteenth century, almost the entire continent was under the control of a handful of European powers. The late start, however, did not spare the African environment. In many areas the Europeans encountered an indigenous agriculture that was reasonably well suited to local environmental conditions. By altering agricultural practices, through the introduction of cash cropping, or livestock, for example, or by disrupting the social patterns that had grown up around traditional agricultural activities, the colonial powers paved the way for environmental deterioration. The clearing of marginal land, planting of environmentally inappropriate crops, imposition of permanent agriculture where shifting agriculture had been the norm and the resettlement of indigenous groups into areas that could not support traditional agriculture all contributed to habitat change, depletion of soil fertility, over-grazing and general degradation of the land (Mannion 1997).

Problems peaked in North America in the 1930s, with the drought that devastated the Great Plains. Cultivated crops could not withstand the drought in the same way that the natural grasses of the Plains could, and large areas became desert. Soil erosion was rampant, as the exposed, dry soil was lifted by the wind or washed away by any rain that did fall. As a result of the Dustbowl conditions, and their impact on the social and economic situation in both Canada and the United States, there was growing interest in dealing with soil erosion and the management of water supplies. This met with some success and although drought is still an integral part of the Plains environment, it has never again led to conditions that matched those of the Dustbowl.

By the middle of the century, the environmental cost of serious pollution was beginning to attract broader attention. Urban air pollution was particularly obvious. It was not a new phenomenon, but it had remained relatively localized in large cities that had high seasonal heating requirements, were heavily industrialized and had large volumes of vehicular traffic or a combination of all three. Paradoxically, it had often been seen as the price that had to be paid for a successful economy. Into the 1940s and 1950s, however, the economic and social costs of pollution were beginning to be recognized and attempts were being made to deal with it. Pittsburgh started in the late 1940s to deal with the pollution associated with the steel industry (Thackrey 1971) and, at about the same time, the state of California introduced laws in a first attempt to reduce the pollution associated with the increased use of the automobile (Leighton 1966). In London, England, a major smog episode in the winter of 1952 was so disastrous that it helped to bring about the introduction of a series of Clean Air Acts that were aimed at reducing air pollution (Brimblecombe 1987).

Pollution was every bit as bad – if not worse – in the waterways. The major rivers and lakes in both North America and Europe were choked with sewage or industrial waste, and the fish that had lived in them had either been killed off or become inedible. Eutrophication was rampant, leaving lakes covered by organic scum and rendering bathing beaches unusable because of the algae and weeds washed up on shore. Streams that appeared deceptively clean might, in reality, be polluted by invisible chemicals released from industrial plants or washed off agricultural land. Even in areas with limited amounts of industry, the waterways were no longer clean. In the relatively unpopulated north-western corner of the Canadian province of Ontario, for example, the presence of only one pulp and paper mill in the town of Dryden was enough to pollute a river many kilometres downstream. The mill spread mercury pollution north and west into the Winnipeg River system, contaminating fish and causing serious health problems for those who ate them.

Although air and water pollution received most attention, the land was not spared. There the problem was chemicals that had been introduced, with the best of intentions, to improve agricultural output or reduce damage to plants and animals by insects or disease. Rachel Carson, an American biologist, was the first to draw attention to the impact of these chemicals on the environment. The title of her best-selling book, *Silent Spring*, referred to the silence that fell over the land as birds succumbed to the chemical poisons released by the growing and often indiscriminate use of pesticides, herbicides and fertilizers. DDT, which up to that time had been viewed as almost a miracle pesticide, was identified as one of the main culprits. When it was published in 1962 the book was denounced by the chemical industry as alarmist, and many biologists treated it with some scepticism, but its concerns ultimately were justified (Cox 1993). It gave the environmental movement a major boost, and inspired an increasing amount of research over the next two decades into the problem of environmental pollution by chemicals.

Silent Spring may have been one of the triggers that led to the great upsurge of interest in the environment in the second half of the 1960s, which Paehlke (1997) has referred to as the First Wave in the modern environmental movement (Figure 1.7). Smith (2000) has identified other events in the United States that increased public awareness of environmental issues at that time. In early 1969 an oil drilling platform off the coast of southern California leaked tens of thousands of litres of crude oil over a period of eight months, leading to the pollution of beaches at Santa Barbara and neighbouring communities. Later that year, a cigarette discarded into the heavily polluted Cyahoga River in Cleveland set the river on fire. In Britain, the wreck of the *Torrey Canyon* off the Scilly Islands in 1967 played a similar role in alerting the public to the environmental consequences of increasing supertanker traffic. Although pollution

FIRST WAVE IN THE
ENVIRONMENTAL MOVEMENT

1968–1976

CHARACTERISTICS

- growing awareness of issues
- discontent with *status quo*
- detachment from existing social, political and economic order
- anti-technology, back-to-nature approach
- regulatory solutions favoured by decision makers

ACTIVITIES AND ISSUES

- air and water pollution
- deterioration in urban areas
- population growth
- resource depletion
- energy crisis
- nuclear power

Figure 1.7 The First Wave in the environmental movement, 1968–76

concerns such as these were central to Paehlke's First Wave, there were other factors, not necessarily unrelated to pollution, that also appeared. Energy issues emerged, for example, eventually surpassing pollution in terms of public concern in the mid-1970s, and the realization that population growth and resource depletion had to be addressed led to the serious reconsideration of approaches to development. The wave grew with the first Earth Day in 1970, the founding of Greenpeace in 1971, the creation of Green parties in Switzerland and New Zealand and the holding of the first conference to draw worldwide public attention to the immensity of environmental problems – the United Nations Conference on the Human Environment (UNCHE), Stockholm, 1972 (see Box 1.3, Development of environmental concern through international conferences). At the same time, attitudes to the environment were changing. Many environmental groups, such as Greenpeace and the Sierra Club, worked hard to build awareness of the problems, while the general alienation of young people from the existing social, political and economic order in the late 1960s and early 1970s was also reflected in environmental attitudes. Technology was seen as the main culprit in environmental deterioration, and the regulatory solutions favoured by decision makers were seen as useless by environmentalists unless they were accompanied by better education and greater appreciation of the environment.

Public pressure forced the political and industrial establishment to reassess its position on environmental quality. Oil companies, the forest products industry and even automobile manufacturers began to express concern for pollution abatement and the conservation of resources. Similar topics began to appear on political platforms, and although this increased interest was regarded with suspicion and viewed as a public relations exercise by some environmentalists, legislation was gradually introduced to alleviate some of the problems. By the early 1970s some degree of control seemed to be emerging. Despite this, the wave was beginning to break. The potential impact of Stockholm was not sustained and the environmental movement declined in the remaining years of the decade, pushed out of the limelight in part by growing fears of the impact of the energy crisis, which broke in 1973 (see Chapter 5). Memberships in environmental organizations – such as the Sierra Club and the Wilderness Society – which had increased rapidly in the 1960s, declined slowly, and by the late 1970s the environment was seen by many as a dead issue (Smith 2000).

Environmental deterioration did not disappear just because fewer people were concerned about it,

BOX 1.3 DEVELOPMENT OF ENVIRONMENTAL CONCERN THROUGH INTERNATIONAL CONFERENCES

Man's Role in Changing the Face of the Earth, 1956 Dedicated to George Perkins Marsh; a broad examination of human impact on the environment from earliest times.

Study of Critical Environmental Problems, 1970 An assessment of the issues characteristic of the 'Second Wave' (see Figure 1.8). This was the first major study to draw attention to the global extent of human-induced environmental issues.

United Nations Conference on the Human Environment, 1972 Held in Stockholm, it recognized the need to confront the growing threats to the environment. It formalized that recognition with the signing of the Declaration of the Human Environment and the creation of the UN Environment Programme.

World Commission on Environment and Development (the Brundtland Commission), 1987 Firmly combined economy and environment through its promotion of 'sustainable development', which requires development to be both economically and environmentally sound.

UN Conference on Environment and Development (the Earth Summit), 1992 Held in Rio de Janeiro, it produced a blueprint for sustainable development in the twenty-first century. Products included the 'Rio Declaration', 'Convention on Climate Change', 'Convention on Biodiversity', 'Statement of Forest Principles' and 'Agenda 21'.

Rio +5, 1997 A summit convened in New York, five years after the Earth Summit, to review progress on the issues raised in Rio. Particular attention was paid to Agenda 21, the blueprint for future environmental management. The general conclusion was that although some progress had been made in implementing sustainable development, few targets had been met in other areas.

Earth Summit +10, the World Summit on Sustainable Development, 2002 An international meeting held in Johannesburg to review progress towards sustainable development ten years after the Earth Summit. Despite attempts to encourage the implementation of Agenda 21, by identifying quantifiable targets, it is considered by many environmentalists to have done nothing to advance the solution of environmental issues.

In addition to these major international events, there have been hundreds of conferences and meetings that have dealt with individual environmental issues and the socio-economic, cultural and political concerns associated with them.

and by the mid-1980s there was a major resurgence of concern. Why it came about is not clear. It may simply have been a return to the natural progression started in the 1960s, but Smith (2000) has suggested that in the United States it was in part a public backlash against the perceived anti-environmentalism of the Reagan administration. Interest was broad, embracing all levels of society, and held the attention of the general public, plus a wide spectrum of academic, government and public-interest groups. The issues involved were part of a Second Wave in the environmental movement (Paehlke 1997) and most have continued on into the new century (Figure 1.8). They are global in scale and although they appear different from issues of earlier years, in fact they share the same roots. Topics such as acid rain and global warming are linked with the sulphurous urban smogs of two or three decades ago by society's continuing dependence on fossil fuels. Societal pressures on land of limited carrying capacity contribute to famine and desertification much as they did in the past. The depletion of the ozone layer, associated with modern chemical and industrial technology, may be considered as only the most recent result of society's continuing search to improve its quality of life, all the while acting in ignorance of the environmental consequences. This Second Wave is characterized by a new environmentalism, in which there is growing awareness of the breadth and complexity of the issues. One of the results is that the economic and political components of the issues are better understood and better addressed than in the past. Reporting in 1987, the World Commission on Environment and Development – commonly called the Brundtland Commission after its chairwoman, Gro Harlem Brundtland – firmly combined economy and environment through its promotion of sustainable development, a concept that required development to be both economically and environmentally sound so that the needs of the world's current population could be met without jeopardizing those of future generations. Part of the commission's mandate was to explore new methods of international co-operation that would foster understanding of the concept and allow it to be developed further. To that end, it promoted a major international conference, held in Rio de Janeiro in 1992 as the Earth Summit or the UN Conference on Environment and Development. The theme of economically and environmentally sound development was carried through the conference and was central

> ### SECOND WAVE IN THE ENVIRONMENTAL MOVEMENT
>
> 1986–1994

> ### CHARACTERISTICS
>
> - concerns increasingly global in nature
> - re-emergence of nineteenth-century preservationist approach
> - environmental groups more professional but also more radical and aggressive
> - greater appreciation of environmental issues by politicians and economists
> - multiple-tools approach to solutions

> ### ACTIVITIES AND ISSUES
>
> - global warming, ozone depletion, acid rain
> - wilderness and habitat preservation
> - biodiversity
> - water quality and availability
> - urban planning and land use
> - waste disposal
> - indoor air quality
> - sustainability and the environment

Figure 1.8 The Second Wave in the environmental movement, 1986–94

to most of the treaties and conventions signed at the summit. It was also included in many of the agreements reached at the Global Forum, a conference of non-governmental organizations held in Rio at the same time as the Earth Summit (Box 1.4). Rhetoric often exceeds commitment at such wide-ranging international conferences and concerns were expressed at the time regarding the effectiveness of the summit (see, for example, Pearce 1992a). One initiative – the Framework Convention on Climate Change – has retained a very high profile, with the scientists and politicians meeting regularly to wrestle with the environmental issues associated with global warming. Most of the others, however, are progressing much more slowly. In one respect the Earth Summit was successful. By bringing

BOX 1.4 TREATIES SIGNED AT THE WORLD ENVIRONMENT MEETINGS IN RIO DE JANEIRO, JUNE 1992

UNITED NATIONS CONFERENCE ON ENVIRONMENT AND DEVELOPMENT (UNCED)

Government treaties and other documents

The Rio Declaration   Including twenty-seven principles – key elements of the political agendas of both industrialized and developing nations

Convention on Climate Change   Included as an objective the stabilization of greenhouse gas concentrations, but no agreement on specific emission targets or dates – led ultimately to the Kyoto Protocol

Convention on Biodiversity   Goals included conservation and the sustainable use of biological diversity, plus fair sharing of products made from genestocks

Statement of Forest Principles   Was not a treaty but a statement of seventeen non-binding principles for the protection and sustainable development of all forests – tropical, temperate or boreal

Agenda 21   Attempted to embrace the entire environment and development agenda. It consists of four sections – social and economic dimensions, conservation and management of resources for development, strengthening the role of major groups, means of implementation – and forty chapters covering all aspects of the environment, including issues such as climate change, ozone depletion, transboundary air pollution, drought and desertification

GLOBAL FORUM

Non-governmental organization (NGO) treaties and other documents

Earth Charter   A short statement of eight principles for sustainable development intended to parallel the Rio Declaration

TREATY GROUPINGS

NGO co-operation and institution building cluster   Included treaties on technology, sharing of resources, poverty, communications, global decision making and proposals for NGO action

Alternative economy issues cluster   Included treaties on alternative economic models, trade, debt, consumption and lifestyles

Major environmental issues cluster   Included treaties on climate, forests, biodiversity, energy, oceans, toxic waste and nuclear waste

Food production cluster   Included treaties on sustainable agriculture, food security and fisheries

Cross-sectorial issues cluster   Included treaties on racism, militarism, women's issues, population, youth, environmental education, urbanization and indigenous peoples

Source: after D.D. Kemp, *Global Environmental Issues: A Climatological Approach* (2nd edn), London and New York: Routledge (1994).

politicians, non-governmental organizations and a wide range of scientists together, and publicizing their efforts by way of thousands of journalists, it ensured that knowledge of the perilous state of the environment was widely disseminated and through that it added momentum to growing concern over the issues. Progress in dealing with environmental problems often seems to be minimal, but without these two elements – knowledge and concern – it would be even slower than it has been.

The concept of sustainable development has not been embraced by all environmentalists (Figure 1.9). Those who support it are seen as technocentric in their approach, using technology and managerial techniques to allow the environment to be administered for the benefit of society. In opposition are those who hold the ecocentric view that the human species is not necessarily the most important species in the natural environment, and as a result priority should not routinely be given to human needs when dealing with environmental issues. The difference is not unlike that between the proponents of utilitarian conservation advocated by Pinchot and the preservationist views of Muir at the beginning of the twentieth century. Ecocentrism has been embraced directly and indirectly by a number of environmental philosophies, from social ecology, in which the human domination of nature is viewed as an extension of the hierarchical nature of society,

MODERN ENVIRONMENTALISM?

Technocentric Environmentalism

using technology and managerial techniques to allow the environment to be administered for the benefit of society

sustainable development?　　　conservation?

Ecocentric Environmentalism

since the human species is not necessarily the most important species in the natural environment priority should not be given routinely to human needs when dealing with environmental issues

deep ecology?　　　preservation?

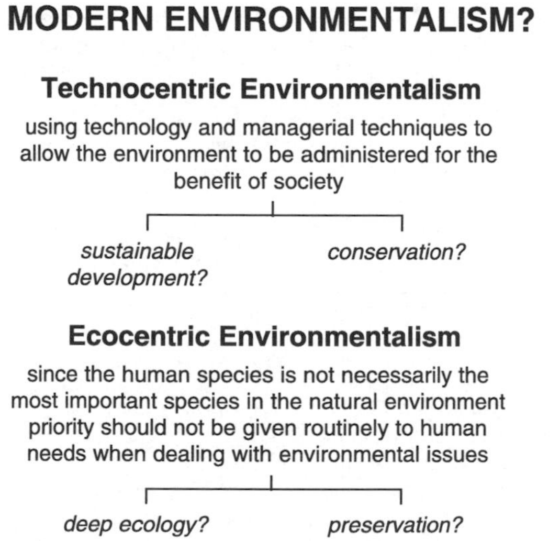

Figure 1.9　The nature of modern environmentalism. The techno-centric/ecocentric split is not universally accepted, but it has a wide following

with its emphasis on profit and group dominance, to eco-feminism, which claims particular ties between women and the environment as a result of such shared elements as productive and reproductive functions (Hessing 1997). An approach to ecocentrism that has been particularly widely promoted is deep ecology. First proposed by Arne Naess in 1973 and developed over the following decade, deep ecology involves a holistic approach, which recognizes the importance of individual elements in the environment and their relationship to each other (Hessing 1997). It also recognizes that there will be times when the intrinsic natural value of an environment or some component of the environment will have to be judged against the economic or social value that society places on it. In such cases, it is invariably assumed that human need takes precedence, but deep ecologists, being strong advocates for the environment, regularly challenge that assumption.

Attempts at translating these different philosophical concepts into reality have been accompanied by an ongoing clash of views on the best or most effective approaches to the issues involved (Goldfarb 2001). For some in the environmental movement the technocentric approach, even with the inclusion of sustainable development and conservation techniques, is insufficient to deal with the problems of technology, economics and politics that are central to modern environmental issues, and may at best only slow environmental deterioration. In turn, those who embrace ecocentrism are often seen as unrealistic in their demands, creating a false equality among the components of the biosphere, and ignoring the fact that human beings have technical and intellectual attributes that make them different from other living creatures. Not all environmentalists belong in these broad groups, of course. Each philosophy has its central core of proponents, but at the individual level more mundane factors such as personal values, political beliefs and self-interest play a part in the development of attitudes to the environment and the ways in which it should be managed or protected.

Modern environmentalism includes an aggressive element, with environmental groups much more militant and ready to take direct action. That action may include direct legal challenges to perceived environmental destruction, or the non-violent, confrontational approach pioneered by Greenpeace, carried out in a well planned, professional manner. Several degrees more radical are groups such as Earth First and the Sea Shepherd Society, which have engaged in or planned illegal activities such as the spiking of trees to make them difficult and dangerous to harvest, the blockading or occupancy of threatened areas and the sinking of whaling ships. Many radical environmentalists support the ecocentric philosophy of deep ecology. As a result, some of the issues that were central to the environmental movement some 100–50 years ago – wilderness preservation, for example – have resurfaced. The modern approach appears much more drastic than its predecessor, but Muir and his fellow members of the Sierra Club were viewed as radical in their day, and future generations may see the current environmental confrontation as acceptable, perhaps even necessary.

One major advantage that modern environmentalists have over previous generations is the ability to collect and analyse data. Although an element of ignorance remains in many areas, the knowledge base that would allow some of the most serious environmental issues to be tackled is already phenomenally large and growing daily. Until that fund of knowledge has been translated into action, it will not be possible to slow down and eventually reverse the environmental deterioration that threatens the world.

SUMMARY AND CONCLUSION

The environment in which an object finds itself is a combination of the various physical and biological elements that surround it and with which it interacts. Although it is common to refer to 'the' environment, there are in fact many environments, all capable of change in time and place, but all intimately linked and in combination constituting the whole earth/atmosphere system. They vary in scale from the microscopic to the global and may be subdivided according to their attributes. The aquatic environment, for example, is that of rivers, lakes and oceans, the terrestrial environment that of the land surface. The term 'built environment' has been applied to areas such as cities, created by human activity. The human element has a dominant role in modern environmental studies, a situation that has developed in a series of waves over the past 150–200 years. The environmental movement has its roots in the growing concern for nature which characterized all sectors of society – from literature to science – in the nineteenth century. Interest was mainly in the cataloguing and conservation of flora and fauna and their natural habitats, leading to the creation of national parks, forest reserves and game preserves. Between the world wars, particularly where drought devastated large areas of agricultural land in the 1930s, more attention was paid to soil conservation. By the 1950s and 1960s pollution had become the central environmental issue. After a decline in the 1970s, when concern over energy replaced the environment in public interest, the environmental movement rebounded, reflecting an increased level of concern with society's ever-increasing ability to disrupt environmental systems on a large scale. A new environmentalism emerged, characterized by a broad global outlook, increased politicization and a growing environmental consciousness that took the form of waste reduction, prudent use of resources and the development of environmentally safe products. There is also growing appreciation of the economic and political components in environmental issues, particularly as they apply to the problems arising out of the economic disparity between rich and poor nations. The modern environmental movement is aggressive, with certain organizations using direct action in addition to debate and discussion to draw attention to the issues. Rising above all of this is the recognition that education in environmental issues is essential if the earth's environmental problems are to be resolved.

SUGGESTED READING

Clark, K. (1969) *Civilization*, London: BBC/John Murray. Clark's personal view of the development of Western civilization, with a chapter on 'The worship of nature' that illustrates the role of European poets, artists and philosophers in setting the stage for the growth of the environmental movement.

O'Riordan, T. (1976) *Environmentalism*, London: Pion. Written at a time when the oil crisis of the mid-1970s had the world in its grip, it focuses on a resource use and management approach to the environment.

Thomas, W.L. (1956) *Man's Role in Changing the Face of the Earth*, Chicago: University of Chicago Press. A report on the first major international conference to address the impact of human activities on the environment, held at Princeton University in 1956 and attended by a variety of academics including geographers, biologists, economists and historians.

QUESTIONS FOR REVISION OR FURTHER STUDY

1   In its broadest sense, the environmental movement works to reduce and prevent damage to the environment. There are different approaches to these goals, however. In the nineteenth century the debate was between conservation and preservation, in the twenty-first there is a similar debate between technocentrism and eco-centrism. Examine current environmental issues and try to decide which of these approaches is likely to be most effective in reducing or solving the problems you can identify. (It may be worth while to consider this question at the beginning of your course and again at the end, and compare your responses.)

2   List by-laws, statutes and other ordinances that have been passed in your community in an attempt to improve or maintain environmental quality. How successful have they been? What are the main reasons for success or failure?

3   Read the poems of Wordsworth or other nature poets, look at the paintings of Turner, Constable or the Hudson River school and consider how the view of the environment represented there compares with your own.

2

The Physical Environment

After reading this chapter you should be familiar with the following concepts and terms:

ablation
abyssal zone
aerosols
albedo
allotrope
aquifer
Arctic Haze
artesian well
atmosphere
atmospheric circulation
atmospheric turbidity
base flow
biosphere
CFCs
carbon dioxide
continental drift
continental shelf
convection cell
Coriolis effect
craton
delta
earthquakes
ecosphere
environmental lapse rate
exosphere

flood plain
fold mountains
fossil fuels
fuelwood
global warming
greenhouse effect
greenhouse gases
ground water
groundwater mining
gyre
hydrologic cycle
hydrology
hydrosphere
index cycle
infrared radiation
jet streams
lithosphere
mesopause
mesophere
methane
mobile belts
neritic zone
nitrogen
ocean currents
ocean trenches

orogenesis
oxygen
ozone
permafrost
photochemical smog
plains
plate tectonics
Rossby waves
shield
solar radiation
stratopause
stratosphere
tectonic activity
terrestrial radiation
thermohaline circulation
thermosphere
total dissolved solids
tropopause
troposphere
ultraviolet radiation
volcanoes
water table
water vapour

There are many different approaches to the study of the earth/atmosphere system, but from the environmental viewpoint there are advantages in considering the system as a group of interlocking components. The solid inorganic portion of the earth's surface is the lithosphere, consisting of the rocks of the crust, plus the broken and unconsolidated particles of mineral matter that rest on it. The vast gaseous envelope of air that surrounds the earth is the atmosphere. The waters on and in the earth's surface make up the hydrosphere and all living things comprise the biosphere or ecosphere.

These four spheres are not discrete entities. They are strongly interlinked with the combination of the elements in the lithosphere, hydrosphere and atmosphere providing the conditions appropriate for the life that is characteristic of the biosphere (Figure 2.1). Each contains elements of the others. The oceans are a major part of the hydrosphere, but they receive sediments from the lithosphere and the

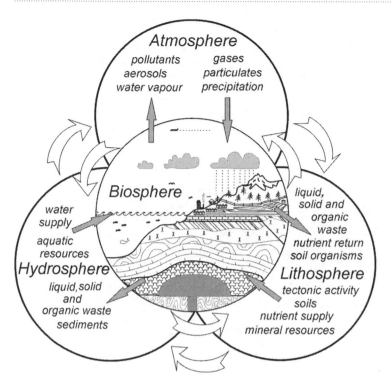

Figure 2.1   A schematic representation of the relationships among the various elements that make up the environment

fish and other organisms they contain are part of the biosphere. The atmosphere contains water from the hydrosphere and particulate matter from the lithosphere. Similarly, soil consists mainly of mineral matter from the lithosphere, but also contains organisms that are part of the biosphere, soil moisture from the hydrosphere and air from the atmosphere in its pore spaces. Many modern environmental issues arise as a result of society's failure to appreciate the complexities of these inter-relationships. The components of the physical environment are considered in this chapter, followed by an examination of the living environment in Chapter 3.

LITHOSPHERE

When environmentalists deal with the primary environmental elements, the atmosphere, hydro-sphere and biosphere usually receive most of the attention. The lithosphere seems relatively less important, perhaps because it appears to have a stability that the others lack. It is not subject to the

short-term changes associated with the weather systems in the atmosphere, for example, nor does it experience the daily tidal variations of the oceans or the seasonal changes in river flow characteristic of the hydrosphere. It appears to endure, unchanging, for decades or even centuries. Yet, if modern environmental problems are viewed in terms of relationships between people and their environment, then the lithosphere deserves greater consideration, since that is where most people are located and that is where most of the problems are initiated.

In geological terms, the lithosphere consists of the earth's crust and the underlying rigid section of the mantle. It is thinnest beneath the ocean basins, but thickens under the continental blocks to as much as 300 km in places. It is these continental blocks that make up the earth's land surface and are most directly involved in environmental issues (Press and Siever 1994). The land surface is very unevenly distributed into a number of continents separated from each other by large expanses of ocean. Most of the land is in the northern hemisphere, whereas the southern continents such

Plate 2.1   The subdued forest-covered landscape of the Canadian Shield.

as Africa, South America and Asia taper quite markedly to the south. What does that have to do with modern environmental issues? Consider where most of these issues have originated – in the northern hemisphere. Basically the sequence has been – more land, more resources: more resources, more people: more people, more development: more development, greater stress on the environment. All of that probably oversimplifies the situation to some extent, but it is true that all of the major industrial nations are in the northern hemisphere and it is there that problems such as air pollution, water pollution and waste disposal have arisen.

STRUCTURAL FEATURES OF THE CONTINENTS

Shield

When viewed at a large scale, all of the continents have structural features in common, namely shield areas, fold mountains and plains (Klein 2002). The shield areas are the oldest geologically, with ages ranging through 3 billion years in the Baltic, 3.8 billion years in Greenland and 4 billion in South Africa to 4.2 billion in Australia. They are the exposed areas of ancient, stable continental rocks or cratons, buried in places by younger sediments. The Canadian Shield, for example, is a large area of exposed rock of Precambrian age (>2.5 billion to 3 billion years ago), which extends westwards beneath the younger rocks of the Great Plains and southwards into Arizona, New Mexico and Texas. Together, the shield and its buried platform comprise the North American craton. Similarly, in Europe, the Fenno-Scandian or Baltic Shield is exposed in Scandinavia in the north and Ukraine in the south, but exists as a buried platform stretching eastwards into Russia and westwards into Western Europe and Britain. Elsewhere in the northern hemisphere, shield rocks are exposed in Siberia and in India. In the southern hemisphere, most of central and western Australia are underlain by cratonic rocks and in South America the principal shield areas are in Brazil and Guyana, separated by the Amazon basin. Almost all of Africa is shield, with large blocks of ancient rocks in West Africa, the Congo and Southern Africa, covered in places by younger strata. Although Antarctica is almost completely covered by ice, geophysical surveys have shown that the eastern part of the continent is composed of a thick cratonic block, some 2 billion years old (Figure 2.2).

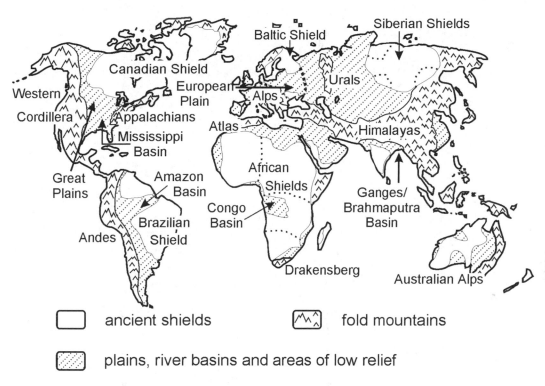

legend:
☐ ancient shields   ⌖ fold mountains

▨ plains, river basins and areas of low relief

Figure 2.2   The structure and morphology of the lithosphere

All of these shield areas have a subdued topography, as a result of ongoing erosion over millions of years. In North America and Europe most of the soil covering was scraped away by glaciers during the last Ice Age, and since soil-forming processes work only slowly in cooler latitudes, large areas of bare rock are common. In such areas as Australia and parts of Africa, soils on the shield tend to be thin because of the limited moisture available, but elsewhere in the tropics, as in Brazil, high temperatures and abundant moisture encourage the creation of deep soils, although they may not be particularly fertile. The combination of inhospitable climate and poor soil has ensured that few shield areas support a large population (see Box 2.1, Environmental issues in shield areas).

Fold mountains

Along the margins of some of the continental cratonic blocks are mobile belts which have been recently tectonically active, and in some cases remain active. These are the areas in which the world's fold mountains have been formed (Figure 2.2). Sediments eroded from the shield areas were deposited in adjacent depressions before being crumpled and folded as the earth's crustal plates collided and moved over or under each other (Goudie 2001). During this period of mountain building or orogenesis, volcanic material was injected into cracks in the sedimentary rocks or pushed through to the surface to create volcanoes (see Box 2.4, Plate tectonics). All current fold mountains are geologically younger than the shield areas, but there is evidence that the eroded roots of ancient fold mountains have been incorporated into some cratons. Although the various fold mountain belts were formed by the same processes, they differ in age. The mountains of Scandinavia and Scotland, for example, are several million years older than the Alps, and in North America the Appalachians in the east are older than the Cordilleran ranges of the west. Even in the Cordillera there is a considerable age

BOX 2.1 ENVIRONMENTAL ISSUES IN SHIELD AREAS

One of the characteristics that all shield areas share is their mineral potential. Because the rocks are old they have undergone the heating and pressure changes associated with metamorphosis or, where they include the roots of ancient mountain chains, they have been injected with magma from the molten material that lies beneath them. Together these have caused minerals to accumulate in concentrations that exceed normal levels in the earth's crust, and as a result the world's shield areas have become major mineral storehouses. Obtaining these minerals has led to many environmental problems, from the landscape and hydrological changes associated with extraction to the disposal of mine waste. Often these are local problems, as in the mining of gold in Brazil, but sometimes they have wider implications. The release of acid gases from the smelting of nickel ores found in the Canadian Shield at Sudbury in Ontario, for example, has caused acid rain hundreds of kilometres downwind from the smelters.

The natural vegetation of the northern shield areas in North America, Europe and Russia is coniferous forest, which for many years has provided timber for construction and for the making of pulp and paper. So great is the demand in some areas that the forest is being cut much faster than it can reproduce naturally or be replaced by reforestation. The landscape has been changed, the hydrologic cycle has been altered and the animals that live in the forest have lost the habitat to which they had become adapted. Where the management of the forest resources has allowed the rate of harvesting to exceed the maximum sustainable yield, as has been the case in parts of Canada and Russia, there has been a major decline in the area available for commercial forestry.

Extractive industries such as mining and forestry have encouraged local concentrations of population but, overall, shield areas are not densely populated. Despite this, it is clear that they are not spared the effects of environmental stress and degradation.

For more information see:

Evans, A.M. (1993) *Ore Geology and Industrial Minerals: An Introduction* (3rd edn), Oxford/Boston MA: Blackwell.

Plate 2.2   An isolated mine for extracting the mineral riches of the Canadian Shield.

Plate 2.3(a)   Young rugged fold mountains in the Western Cordillera of North America.

Plate 2.3(b)   The rounded fold mountains of the Scottish Highlands, worn down by millions of years of erosion.

difference between the Pacific coastal ranges and the Rocky Mountains in the interior.

Like the shield areas, the fold mountains have been subject to prolonged erosion, and since some mountain ranges are older than others, the impact of that erosion varies. Younger fold mountains such as the Alps, Rockies and Himalaya remain high and rugged, whereas older ranges such as the Appalachians, the mountains of north-west Europe or the Eastern Highlands of Australia, subject to erosion for a longer period of time, tend to be lower and more rounded. The younger fold mountains provide the greatest relief on the earth's surface, and the greatest elevation in Mount Everest, at nearly 9000 m above sea level, in the Himalayas. The conditions that produce folding are also responsible for creating deep ocean trenches such as the Marianas Trench, where the ocean is more than 11,000 m deep. On land, the greatest depression is beneath the ice in Antarctica at almost 2500 m below sea level. The floor of Lake Baykal, the world's deepest lake, is at 1500 m below sea level and the Dead Sea between Jordan and Israel is some 400 m below sea level. In contrast, Death Valley, the lowest point in the western hemisphere, is less than 100 m below sea level. On a human scale of perception the maximum relief of 20 km appears significant, but on a planetary scale it is much less so. If the earth were reduced to the size of a basket-ball or soccer ball, even a difference that large would be lost in the normal surface roughness of the ball.

BOX 2.2 ENVIRONMENTAL ISSUES IN THE MOUNTAINS

Like the shield areas, fold mountains in the past did not support large populations. The higher mountains, particularly, were barriers to movement, with harsh climatic conditions and steep slopes thinly covered with soil that limited agriculture. Where climatic conditions were less inhospitable, as in some of the tropical areas of Africa and South America, or in the older mountain areas, which were lower, with slopes that were less steep, population numbers were higher. The population was not evenly spread, however, being concentrated in the valleys, where flat land suitable for settlement and cultivation was available. These concentrations created local environmental stress, but, for the most part, natural restrictions on population growth and the general inaccessibility of mountainous regions ensured that they were spared major environmental problems. Developing technology, growing demand for resources and changing socio-economic conditions that encouraged rapid population growth changed the situation. Activities ranging from mining, forestry and hydro-electric power production to tourism and recreation have now created serious and growing environmental deterioration in many mountainous areas.

The tectonic processes that created the fold mountains also caused mineralization. As a result, the world's mountains are a major source of valuable minerals, particularly non-ferrous metals such as gold, silver, copper and zinc. Mining of these metals has been part of the mountain economy in areas such as the Urals, the Alps and the Andes for centuries, but the scale of the operations was small and it is the modern exploitation of these minerals that is causing serious environmental problems, such as slope destabilization, pollution and waste disposal, in many areas. The process of extraction alters the environment directly and since the amount of rock excavated always exceeds the amount of ore extracted, disposing of waste becomes a problem, aggravated by the steep slopes and narrow valleys common in mountainous areas. Modern mining operations can handle ores in which the metal content is less than 1 per cent. To do so economically, however, it is necessary to concentrate the ores at the mine site.

Concentrating processes vary with the kind of ore involved, but include crushing, washing and roasting plus a variety of chemical treatments that without appropriate precautions can have serious environmental consequences. Crushing and washing produce large volumes of fine rock debris, which may cause atmospheric pollution or contribute to increased turbidity and silting in streams. Roasting releases acidic sulphur gases into the atmosphere, while chemical processing can raise the levels of mercury and cyanide in the environment adjacent to mine sites. The direct environmental problems of mining in mountainous regions are obviously limited to those areas where the minerals are available, but they are often serious, particularly in the developing world, where environmental regulations tend to be less stringent than elsewhere.

Although many of the higher fold mountains extend above the tree line, the well forested lower slopes, particularly in mid-latitudes, have been a source of timber for centuries. At the low levels of technology that prevailed in the past, with few

BOX 2.2 – continued

exceptions, the rate of forest renewal kept pace with the rate of harvesting and only limited environmental deterioration occurred. More recently the growing demand for wood and the development of commercial harvesting techniques that can remove timber from even the steepest slopes have led to serious environmental damage. Hill slopes that have been cleared of vegetation experience more rapid run-off, which leads to soil erosion and flooding problems in the land immediately downstream from the harvested areas. Even where there is no commercial logging, wooded mountain slopes are often laid bare for other reasons. The demand for fuelwood is so high in the foothills of the Himalayas in India and Nepal that many hillsides have been stripped of wood, leaving them exposed to excess run-off and soil erosion.

As the world's population has grown and, in many areas, become more affluent, the demand for certain types of resources has increased. Energy has led that growing demand. Although energy resources take many forms, they are often converted from one form to another and in many cases the final conversion is to electricity. One of the most popular ways of providing electricity is by the conversion of the power of flowing or falling water through hydro-electric generation. Hydro-electricity is often seen as an environmentally friendly form of energy, but it is not problem-free. The water used to produce the electricity is returned to the system virtually unaltered and the process produces no air pollution – both serious problems associated with thermally produced electricity. However, in setting up the infrastructure to produce hydro-electricity the hydrologic cycle is disturbed by the creation of artificial lakes and the flow regime of rivers and streams is altered. Together, these have significant environmental consequences. Since the gradients that produce rapidly flowing or falling water and the deep steep-sided valleys suitable for dam and reservoir construction are characteristic of mountainous areas, it is there that the immediate environmental impact is experienced.

Another source of environmental stress in recent years is the growth of recreation and tourism in mountainous areas. This use of the mountains is not new. The multitude of resorts and other recreational facilities that have sprung up in mountainous areas around the world were preceded by Alpine communities like St Moritz, first visited by British tourists

in the nineteenth century, and the hill resorts of theorient, where the colonial administrators of India, Malaya and Indonesia escaped the tropical heat of the lowlands. What is new is the intensity of the use. The solitude, scenery, steep slopes and deep snow now attract visitors year round, and as their numbers increase, the impact on the environment increases also. Even relatively small resorts create problems of water supply and waste disposal and alter the environment beyond the boundaries of the resort itself. To build ski runs, for example, trees have to be removed, while the making of snow, plus the packing and grooming of the hills, creates an artificial moisture and temperature regime to which the surviving vegetation and wildlife must adjust. During the summer months, hiking, trekking and backpacking tempt thousands of people into the mountains. The stress that they create has reached crisis proportions in some parts of the Alps, where the destruction of vegetation and the compaction of soil along frequently used trails have contributed to severe slope erosion. Even in areas such as Nepal, relatively inaccessible by European standards, the volume of tourists has reached such a level that in some parts of the Himalaya garbage and human waste left by hiking and climbing parties has become a serious problem.

These problems are recognized and attempts have been made to deal with them, through the creation of parks, where development can be controlled, or through legislation that restricts certain activities in susceptible areas. Despite this, the growing demand for outdoor recreation, the improved accessibility provided by new access roads and the increasing number of off-road vehicles, will ensure that pressure on the environment in mountainous areas will become even more intense.

For more information see:

Fox, D.J. (1997) 'Mining in mountains', in B. Messerli and J.D. Ives (eds) *Mountains of the World: A Global Priority*, New York: Parthenon.

Ives, J.D. and Messerli, B. (1989) *The Himalayan Dilemma: Reconciling Development and Conservation*, London: Routledge.

Price, M.F., Moss, L.A.G. and Williams, P.W. (1997) 'Tourism and amenity migration', in B. Messerli and J.D. Ives (eds) *Mountains of the World: A Global Priority*, New York: Parthenon.

Plate 2.4 A mountain ski resort. Popular, expensive and often crowded, ski resorts threaten the environment in mountainous areas.

Plains

The plains are large areas of low relief composed of sediments eroded from the adjacent shield or fold mountains and deposited in major sedimentary basins. Over millions of years these sediments have become consolidated into sandstones or shales and have been augmented by limestones, formed from the skeletons and shells of organisms that lived in the basins, or by chemical processes that led to the precipitation of carbonate compounds. For the most part, these sediments have remained horizontal or only gently tilted, which contributes to their low relief. Close to their boundaries with the fold mountains some of the older sediments have been uplifted or even buckled by the ongoing tectonic activities associated with the mountains. Commercially valuable resources such as coal, petroleum and building stone are often present in

these sediments. The plains also include more recent sedimentary deposits in the form of unconsolidated fluvial, lacustrine and (in higher latitudes) glacial deposits, which have been deposited relatively recently and tend to maintain the low relief of the plains. Lacustrine deposits, such as those left behind by glacial Lake Agassiz in the Canadian province of Manitoba, produce a very flat landscape. In contrast, the more complex depositional environment associated with fluvial and fluvioglacial processes introduces minor relief into the plains, while glacial activity can create a rolling landscape as a result of the irregular deposition of glacial drift (Strahler and Strahler 2003).

Plains are also found in coastal areas, formed by marine erosion and deposition, or caused by changing sea level. The continental shelves that surround the continental land masses represent the continuation of these coastal plains beneath the oceans. In the past, when sea level was lower than it is now, these shelves were exposed as dry land and the existing coastal plains increased in extent. Such was the case during the last Ice Age, when so much water was incorporated in the land ice sheets that sea level fell by as much as 80 m, causing large sections of the continental shelf to be exposed as coastal plains. Coastal and fluvial plains are often contiguous, and the creation of deltas can be viewed as the extension of the river flood plains into the coastal zone.

The world's major plains include the Great Plains of North America and the Northern European Plain with its extension into Russia and Ukraine (Figure 2.2). The larger river basins of Asia fit this category, as do the Amazon basin and the pampas of Argentina in the southern hemisphere. It is only in Africa that they are limited. The basins of the Congo and the Nile would be included, but the high plains of East and South Africa are geologically part of the African Shield areas, which have been worn down by erosion to a relatively subdued surface, but are structural rather than depositional in nature.

TECTONIC AND GEOMORPHOLOGICAL PROCESSES IN THE LITHOSPHERE

The lithosphere as a whole is usually considered to be much more stable and less subject to rapid change than the hydrosphere and atmosphere, and

BOX 2.3 ENVIRONMENTAL ISSUES ON THE PLAINS

Much of the environmental stress in the shield areas and in the mountains can be linked directly with the needs and wants of the more densely populated plains. However, the human impact on the plains has, in many ways, been more significant and has been growing for several thousand years. It was on the plains that the world's first agricultural revolution took place, when society began to domesticate plants and animals (Figure 1.4). This was a very important stage in human development, but it was also a very important stage in society's relationship with the environment. It was the first time that human activity had altered the environment to any extent. Domesticated crops replaced natural plants; domesticated flocks and herds replaced wild animals; the aquatic environment was altered by irrigation and the inexorable deterioration of the soil began.

Initially the stress on the environment was only local, but it was real enough and some of the results of early agricultural activities in Mesopotamia are still evident to this day in the salinization of irrigated soil in southern Iraq. With time, technology developed, population increased and along with these the disruption of the natural environment grew also. The relatively flat, fertile, well watered land of the plains encouraged the development of new agricultural techniques that increased food production. The plains of North America and Europe and, to a lesser extent, those of Argentina and Australia became massive producers of grain and beef, which were transported around the world to feed a burgeoning population. All this came at a cost to the soil, however. Except in the most fertile areas along river flood plains, where fertility was maintained by annual flooding, nutrient levels fell and productivity declined. Where it was available reduced productivity was countered by bringing new land under cultivation, but where that was not possible attempts were made to return the nutrients to the soil by adding increasingly large amounts of fertilizer. While that improved production, it eventually led to pollution and eutrophication in rivers and lakes when the excess fertilizers were washed off the land. Herbicides and pesticides, seemingly essential to modern commercial farming, also entered the waterways along the same paths. In places, inappropriate land use led to serious erosion, perhaps best exemplified by the Dustbowl conditions on the Great Plains in the 1930s.

Environmental problems also arose from the growth of industry on the plains from the eighteenth century onwards. Economically valuable resources lie beneath the plains, often easily accessible, but not without environmental disruption (see Chapter 6).

Limestone and sandstone can be quarried for building material, clay is available for the making of bricks, fluvial or fluvioglacial sands and gravel provide for the needs of the construction industry. Extracting and using these materials changes the landscape, at the very least on a local scale. Large-scale change is associated with the mining of energy resources such as coal. In the Great Plains, in the United States, large reserves of coal lie close to the surface west of the Mississippi. The strata are horizontal, allowing them to be mined relatively easily by stripping the overburden and digging up the coal. In the past, strip mining of this type was a major source of environmental degradation. The overburden was deposited in ridges parallel to the workings, where, being unconsolidated, with little or no vegetation, they were subject to erosion by wind and heavy rain. Erosion by running water carried fine sediments into adjacent streams and sulphur-rich minerals associated with the coal contributed to acid mine drainage. Although laws now require mining companies to rehabilitate the land once all the coal has been extracted, it is impossible to repair all the environmental damage. Oil and natural gas are also found beneath the plains, but their extraction is generally less directly damaging to the environment than coal mining. Exceptions are found in the mining of tar sands and oil shale in which the oil is either too thick to flow (tar sands) or incorporated in the rock (oil shale). Major deposits of tar sands are located in Venezuela, the United States and Russia, but by far the largest are those being worked in Alberta, Canada, where there are estimated reserves equivalent to 800 billion to 900 billion barrels of crude oil. Since approximately 2.5 tonnes of sand are required to produce one barrel of oil, the initial extraction of the tar sands causes major damage to the landscape, while the separation of the oil from the sand particles leaves behind large volumes of contaminated water and waste sand. Prior to the development of liquid petroleum resources, oil shale was an important source of oil in nineteenth-century Europe. Although significant deposits exist in Baltic Europe, Brazil and the United States, none is being worked at present. On average, nearly 1.3 tonnes of shale must be processed to release one barrel of oil, and the process is accompanied by serious threats to the environment. When processed to release the oil, the waste rock increases in volume several times, creating serious disposal problems. Leachate contamination of the adjacent water pollution from the refining process are also potential hazards.

BOX 2.3 – continued

The agricultural and mineral resources of the plains attracted a growing population, and towns and cities sprang up. The presence of the rivers and the relatively flat land allowed easy transport, initially by water and later overland by road and railway. As transport systems improved, the resources of the shield and mountain areas were brought to the plains to be processed. Processing and manufacturing led to the growth of industrial towns with their accompanying pollution. The rivers became polluted with human and industrial waste or dried up as more and more water was withdrawn for growing urban and industrial needs. Pollutants darkened the skies and the precipitation that fell from the clouds became acidic. The plains of North America and Europe suffered most, but Asia was not immune. In the southern hemisphere, industrial exploitation was less of a problem, but there too the destruction of vegetation and soil erosion, as a result of pressure from rapidly growing populations, placed the environment under increasing threat.

For more information see:

Jacobsen, T. and Adams, R.M. (1971) 'Salt and silt in Ancient Mesopotamian agriculture', in T.R. Detwyler (ed.) *Man's Impact on Environment*, New York: McGraw-Hill.

Schumacher, M.M. (ed.) (1982) *Heavy Oil and Tar Sands Recovery and Upgrading: International Technology*, Park Ridge NJ: Noyes Data Corp.

to a large extent that is true. The lithosphere is not static, however. The processes of weathering, erosion, transport and deposition continue slowly and relentlessly to change the shape of the landscape (Figure 2.3). Mass movement on slopes, flowing water, waves and ice remove some 10 million tonnes of sediments from the continental land masses every year, with as much as 95 per cent being eroded and transported by rivers (Park 1997). During the Pleistocene epoch, ice replaced running water as the dominant form of erosion in many northern areas, which retain the typically glaciated landscapes formed at that time. Wave action in coastal areas creates cliffs and beaches and under stormy conditions can cause very rapid change. As sea level fluctuates, the pace of marine erosion and deposition changes also. With a rising sea level, the amount of land lost to the sea increases, and if the sea-level rise expected to accompany global warming does occur, many coasts, particularly those formed in unconsolidated sediments, will suffer serious and increasing erosion. Although weathering, erosion, transportation and deposition are entirely natural in origin, they can be deliberately or inadvertently enhanced or decreased by human activities.

TECTONIC PROCESSES

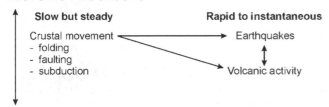

GEOMORPHOLOGICAL PROCESSES

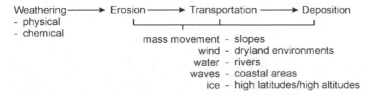

Figure 2.3   Tectonic and geomorphological processes in the lithosphere

In addition to these relatively slow, but steady, elements of landscape change, there are various catastrophic events that bring about very rapid change (Figure 2.3). The crustal plates upon which the continents rest are always moving, slipping over and under each other to contribute to slow change (Harrison 2002), but also occasionally causing rapid change through such activities as earthquakes and volcanic eruptions (Figure 2.4) (see Box 2.4, Plate tectonics). Earthquakes, for example, can change the landscape in minutes by altering the courses of rivers, creating lakes and causing land-slides. Volcanoes can create new land in the form of hills or mountains or cause islands to appear and disappear. Lava flows have been responsible for major physiographic features such as the Columbia-Snake plateau of the north-western United States and the Deccan plateau of India, and they continue to be a significant element of landscape change in Hawaii and Iceland. Also important from an environmental point of view is the increased atmospheric turbidity caused when particulate matter is thrown up into the atmosphere during volcanic eruptions.

BOX 2.4 PLATE TECTONICS

Once the general shape and distribution of the earth's land masses and oceans had been established by exploration and survey, it became clear to some observers that the shapes of the various continents were not random or unrelated to each other. Francis Bacon noted in the seventeenth century, for example, that the eastern shoulder of South America would fit remarkably well into the hollow of the West African coast on the opposite side of the Atlantic, implying that they were one in the past, but was unable to explain how they had come to be separated by a wide stretch of ocean. Although later scientists and natural philosophers also examined such issues, it was not until the early twentieth century that the distribution and structure of the continents and oceans received serious scientific attention.

CONTINENTAL DRIFT

In 1912, Alfred Wegener proposed the theory of continental drift, in which he used a combination of geological, palaeontological and biogeographical evidence to show that the earth's continents had not always occupied the same location on the earth's surface. He concluded that some 2 billion to 2.5 billion years ago the earth's continental blocks were combined in a super-continent, that he called Pangea, surrounded by an ocean – Panthalassa. Pangea was composed of lighter crustal materials floating on the denser crust of the ocean basins. With time, Pangea split apart into a northern continent – Laurasia – which included modern North America, Europe and Asia and a southern continent – Gondwanaland – which would eventually become South America, Africa, Australia and Antarctica. As Laurasia and Gondwanaland split further, mountain building took place along the edges of the continental blocks as a result of movement through the oceanic crust or collisions between blocks. The Western Cordillera were formed almost as a bow wave, for example, as the North American block drifted westwards across the oceanic crust, while the Alps and Himalayas were created as a result of collisions between the African and European blocks and Indian and Asian blocks respectively.

Although many of Wegener's ideas were subsequently shown to be valid, he received little support at the time, and it took a quarter of a century before the concept of continental drift appeared again with the publication of Alexander du Toit's *Our Wandering Continents* in 1937. One of the problems with the concept was the absence of an obvious mechanism by which the mobility of the continents could be initiated and sustained. In 1944 the eminent geologist Arthur Holmes suggested convection currents in the mantle as a potential mechanism by which the crustal blocks could be moved, and although the processes involved are now known to be more complex, the mobility of the upper part of the mantle as a result of heat transfer from below is an integral part of the theory of plate tectonics, which ultimately evolved from Wegener's ideas of continental drift.

THE MODERN THEORY OF PLATE TECTONICS

In the theory of plate tectonics, the lithosphere is seen to consist of a number of rigid plates of various sizes that are capable of movement over the asthenosphere, the most easily deformed part of the underlying mantle (Figure 2.4). Individual plates consist of continental or oceanic crustal material or a combination of both, and although the central sections of the plates tend to be stable, around their edges

BOX 2.4 – continued

interaction with adjacent plates may cause major tectonic activity. As oceanic plates move away from each other, for example, sea-floor spreading takes place and the upwelling of magma forms new crustal material. Over millions of years, this has created mountainous ridges on the ocean floor between the separating plates. The mid-oceanic ridges that lie beneath the Atlantic and south-east Pacific Oceans, for example were formed by submarine volcanic activity associated with sea-floor spreading. Tectonic activity in the form of earthquakes and volcanic eruptions continues in these areas, most obviously in volcanic islands like Iceland, where the ridges emerge from below the surface of the ocean. Continental plates, in which the crust is more brittle, may crack or split as they separate. The rift valleys that stretch from East Africa through the Red Sea to the Middle East were formed in this way, with the rifting accompanied by significant seismic activity and the formation of volcanic mountains such as Mount Kenya and Mount Kilimanjaro.

When crustal plates move together or collide, the results depend upon the extent and nature of the contact. If a collision takes place between two oceanic plates, one plate will slide beneath the other in a process referred to as subduction. Similarly, in a collision between an oceanic plate and a continental plate, the former usually subducts. The Pacific plate subducts beneath the North American plate, for example, as does the Nazca plate beneath the South American plate. Such subduction zones, where the descending parts of the lithosphere are reabsorbed into the mantle, are the most tectonically active areas of the earth's crust, characterized by frequent seismic activity and volcanic eruptions. They are often marked by deep oceanic trenches – the 11,000 m deep Mariana Trench, for example – and their associated volcanic island arcs. Subduction zones are also sites of mountain building, or orogenesis. Sediments eroded from the continental plates are deposited in the trenches, where as a result of continued movement at the plate boundaries they are crumpled, folded and uplifted to form mountain belts. In addition, the intense volcanic activity associated with subduction zones causes igneous material to be injected into the sedimentary rocks or produces ranges of volcanic mountains such as the Cascades in western North America. The great ranges of mountains that stretch from the southern tip of South America to Alaska mark the location of a major group of subduction zones caused by the collision of the continental plates of the Americas with the oceanic plates of the Pacific.

Subduction does not normally take place when two continental plates collide. As the plates approach each other, downwarping of the crust creates a basin into which sedimentary materials are deposited. Ultimately, when the plates collide, the sediments are folded, deformed and uplifted to form mountain belts. The Himalayas were created, for example, through the collision of the Eurasian and Indian plates while the Alps were produced when the African and Eurasian plates collided. In both places, sediments containing marine fossils are now found in the mountains several thousand metres above sea level.

In places, crustal plates do not collide or move apart, but move past each other along transform faults. Such faults are responsible for the offsets that occur in the mid-oceanic ridges as adjacent parts of the sea floor spread at different rates. In most cases movement along these faults is not continuous or smooth. Sometimes the plates lock together until sufficient energy builds up in the fault to break the lock. The sudden release of energy that occurs causes an earthquake. The best known of these transform faults is the San Andreas fault in California. There the Pacific plate is moving north-westwards past the North American plate, carrying the western coast of California – and the cities of Los Angeles and San Francisco – with it. Movement along the San Andreas fault caused the disastrous San Francisco earthquake of 1906, and since then a continuing series of minor to moderate-scale earthquakes has occurred along the main fault and adjacent fault lines. Because of its potential to cause serious physical and economic damage to the region, as well as major loss of life, the San Andreas fault has been studied intensely. It is known that the main fault is locked in several places, allowing pressure to build up in the crust, and the consensus among seismologists is that a major earthquake is long overdue. The tectonic stage appears to be set for a catastrophic earthquake when the locks break and the energy stored in the crust is released.

The influence of plate tectonics extends beyond the lithosphere to the hydrosphere and the atmosphere. Sea-floor spreading and the movement of the crustal blocks gradually change the shape of the ocean basins, for example, leading to changes in sea level and oceanic circulation patterns. Ocean chemistry is influenced by the release of hydrothermal fluids from vents in the ocean crust along the mid-oceanic ridges. In the past, tectonic events such as orogenesis must have had a major impact on climate, through the disruption of global atmospheric circulation patterns. The building of the western

BOX 2.4 – continued

Cordillera and the Andes, for example, placed barriers across the westerly air flow in mid-latitudes in both hemispheres, which must have caused changes in climate in the areas downwind from the mountains. Similarly, the current monsoonal circulation in East Asia would not have existed prior to the elevation of the Himalayas and the Tibetan plateau.

In human terms, these changes involve such long time periods that their impact is of limited interest beyond the scientific community. However, there are climate impacts of plate tectonic activity that take place over a much shorter term and have the potential to contribute to environmental change. The volcanic activity associated with subduction zones and mid-oceanic ridges, for example, has always been an important source of particulate matter and gases such as carbon dioxide and sulphur dioxide for the atmosphere. The impact of these aerosols on climate is well known, and any

increase or decrease in volcanic activity will contribute to climate change, either through variations in atmospheric turbidity or through changes in the greenhouse effect.

The activities associated with plate tectonics have determined the current nature and distribution of the earth's continents and oceans. Their influence continues through such geophysical phenomena as volcanic activity and earthquakes and the effects can extend beyond the lithosphere to the hydrosphere and atmosphere.

For more information see:

Cattermole, P. (2000) *Building Planet Earth*, Cambridge: Cambridge University Press.

Du Toit, A.L. (1937) *Our Wandering Continents*, Edinburgh: Oliver and Boyd.

Harrison, C.G.A. (2002) 'Plate tectonics', in A.S. Goudie (ed.) *The Encyclopedia of Global Change*, New York: Oxford University Press.

Thus the lithosphere participates actively in the various processes that are characteristic of a dynamic environment (Press and Siever 1994). Through the various geomorphological processes it experiences continuing change and because of the effects of these changes on the other environmental elements it plays a very active role in the environmental system. It is linked to the hydrosphere through its influence on the hydrologic cycle (see Box 2.5, The hydrologic cycle), to the atmosphere through the processes that produce gases and particulate matter and to the biosphere through soils, vegetation and animals. All of these links are two-way, of course. Water moving through the hydrologic cycle helps to produce the erosion and deposition that change the face of the earth, gaseous exchange and the provision of water from the clouds link the atmosphere and lithosphere, while the growth of vegetation and the activities of animals also contribute to landscape change.

ATMOSPHERE

The atmosphere is a thick blanket of gases which completely envelops the earth and is held in place by gravity. The mix of gases in the atmosphere is remarkably uniform throughout, particularly in its

lower layers, but it also contains suspended liquid and solid particles, which can vary considerably in type and concentration from time to time and from place to place. It has no obvious outer limit. Although traces of atmospheric gases have been detected far out into space, nearly 99 per cent of the atmosphere lies within 30 km of the earth's surface, with the greatest concentration, 50 per cent of the total, being in the lowest 5 km (Kemp 1994).

ATMOSPHERIC GASES

Oxygen and nitrogen

Oxygen and nitrogen account for 99 per cent of the total volume of gases in the atmosphere. The other 1 per cent is a mixture of a group of minor gases in which argon accounts for 0.93 per cent (Table 2.1). Argon is inert and contributes little or nothing to atmospheric activities, but some of the other so-called minor gases have an impact that is far from minor.

Oxygen (21 per cent of the total) is chemically very reactive. Through oxidation it combines readily with other elements and contributes to such processes as combustion, corrosion, metabolism and respiration. Although it is continuously consumed

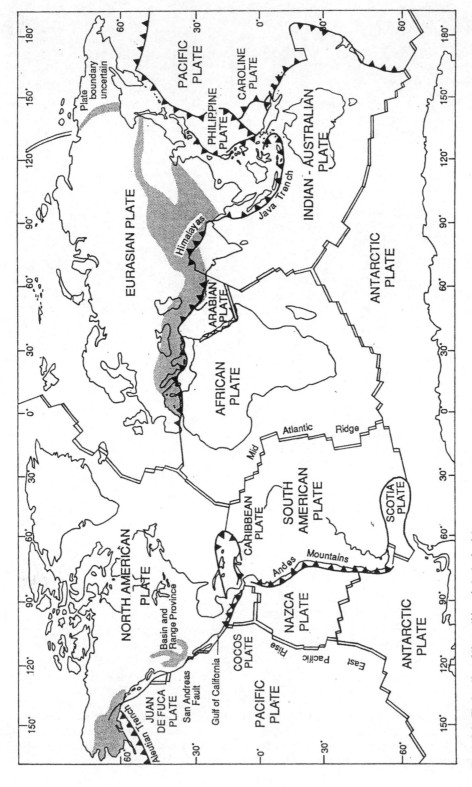

Figure 2.4   The location of the earth's main tectonic plates
Source: Goudie (1989) with permission.

BOX 2.5 THE HYDROLOGIC CYCLE

The hydrologic cycle is a complex group of processes by which water in its various forms is circulated through the earth/atmosphere system.

THE MECHANICS OF THE HYDROLOGIC CYCLE

The energy to drive the cycle is provided by solar radiation. Although the amount of water in the system does not change, it can exist as a gas (water vapour), a liquid (water) or a solid (ice), and as it moves through the cycle it also moves through these different states. The basic processes involved in the cycle are evaporation, condensation and precipitation. Where sufficient energy is available, liquid water on the earth's surface is converted into water vapour by evaporation and taken up into the atmosphere. Water vapour is also carried into the atmosphere by transpiration, when plants absorb liquid water through their root systems and release it through their leaves as water vapour, and by sublimation, in which solid water in the form of ice or snow is converted directly into a vapour. As long as temperature and pressure conditions remain favourable, the water will remain in its vapour state, but eventually – when cooling takes place, for example – it will be converted back into water again through condensation or, if temperatures are low enough to cause freezing, even into ice. After conversion, it may still remain in the atmosphere in the form of water droplets or ice crystals, which can be seen as clouds when quantities are sufficiently large. When conditions are right for precipitation, the water falls out of the clouds as rain or snow, to return to the surface, where it is again available for evaporation. The precipitation falling on the oceans is directly incorporated into the ocean reservoir, but precipitation over the land follows a number of different paths. It may be intercepted by vegetation, infiltrate through the soil and rock into the groundwater sector or flow across the surface into rivers and lakes to become part of the run-off sector and eventually return to the ocean.

Since oceans cover 71 per cent of the earth's surface, they account for the bulk of the evaporation that takes place and also receive the greatest return from precipitation (Figure 2.7). Evaporation does exceed precipitation, however, and the excess is advected to provide precipitation over the land surface, where evaporation is less. The redistribution of water eventually produces a balance, but the smooth flow of water through the system is disrupted by short and long-term storage in lakes, the ocean deeps, ice sheets and the groundwater reservoir. Although it is usual to describe the hydrologic cycle as one all-encompassing system, there are in fact many regional and seasonal variants that may have environmental or societal implications. The relationship between the various elements in the cycle in a hot, semi-arid area, for example, is very different from that in a cool, moist location. Similarly, in higher latitudes the difference in the cycle between summer and winter is quite distinct.

ENERGY FLOW IN THE HYDROLOGIC CYCLE

Changes in the state of water and the redistribution of water in the earth/atmosphere system, which are an integral part of the hydrologic cycle, have implications for the earth's energy budget. The evaporation of water requires the uptake of energy to bring about the conversion from a liquid to a vapour. As long as the water remains in the form of a vapour, it retains that energy in the form of latent heat. That energy is released into the environment when the conversion back to a liquid takes place. Vapour in the atmosphere is easily moved, however, and the release of the latent heat may take place thousands of kilometres from where it was originally absorbed. During the summer monsoon in eastern Asia, for example, the northward transfer of significant amounts of water vapour from the southern oceans brings with it energy that is released above the northern land masses. Similar energy transfer takes place in the liquid/solid/liquid conversion, but the amounts involved are much less than those in the liquid/vapour/liquid sequence and much more restricted in distribution.

WATER USE AND THE HYDROLOGIC CYCLE

The recycling of water through the hydrologic cycle is essential for life on earth. It provides plants and animals with a continuing supply of fresh water from a reservoir that amounts to less than 1 per cent of the total water in the hydrosphere. The impact is direct, through the provision of the water, but is also indirect through the influence of the hydrologic cycle on such elements as the circulation of the atmosphere

BOX 2.5 – continued

and the redistribution of energy in the earth/atmosphere system.

The greatest pressure from human activities is on the run-off sector of the cycle, from which water is diverted for domestic, agricultural and industrial uses, but modern society interferes with almost all aspects of the cycle. Evaporation and transpiration are disrupted by agricultural and forestry practices, boreholes and wells allow access to the groundwater supply and the construction of dams and reservoirs places additional water in storage. Since the cycle is a closed system in material terms, human activities do not deplete the entire system, but excess withdrawal from the run-off or groundwater sectors can create local shortages of water. Most human uses involve only short-term withdrawal from the system, but when the water is returned its quality is often much impaired by a variety of pollutants. The cleansing of such polluted water, mainly through evaporation, condensation and precipitation, is an important but less well acknowledged feature of the hydrologic cycle.

For more information see:

Speidel, D.H. and Agnew, A.F. (1988) 'The world water budget', in D.H. Speidel, L.C. Ruedisili and A.F. Agnew (eds) *Perspectives on Water*, New York and Oxford: Oxford University Press.

Ward, R.C. and Robinson, M. (2000) *Principles of Hydrology* (4th edn), Maidenhead: McGraw-Hill.

TABLE 2.1 THE AVERAGE GASEOUS COMPOSITION OF AMBIENT AIR

Gas	% by volume	Parts per million
Nitrogen	78.08	780,840.00
Oxygen	20.95	209,500.00
Argon	0.93	9,300.00
Carbon dioxide	0.0345	345.00
Neon	0.0018	18.00
Helium	0.00052	5.20
Methane	0.00014	1.40
Krypton	0.00010	1.00
Hydrogen	0.00005	0.50
Xenon	0.000009	0.09
Ozone	Variable	Variable

by these reactions, a very effective recycling process ensures that it is not under any serious threat of depletion. Oxygen takes several forms or allotropes, the most common being diatomic or molecular oxygen (O_2). In addition it may be present in the atmosphere as atomic oxygen (O) or as its triatomic allotrope, ozone (O_3). Both normally exist in only small quantities subject to rapid change, but ozone has an important role in several environmental issues. It is present in the lower atmosphere, where it is a common constituent of photochemical smog, and in the upper atmosphere, where it filters out ultraviolet solar radiation, which can be harmful to living organisms. Oxygen is essential to life on earth, being absorbed by animals during respiration, releasing energy through combustion and meta-bolism, and, in the form of ozone, protecting the biosphere from excess ultraviolet radiation.

Nitrogen, like oxygen a colourless, odourless gas, accounts for 78 per cent of the volume of the atmosphere. It is a fundamental element in all living organisms, being present in proteins and nucleic acids, for example, yet molecular nitrogen (N_2) is chemically relatively inactive. It seldom becomes involved in biospheric chemical or biological processes except under special circumstances. During thunderstorms, for example, the enormous energy flow in a lightning stroke causes nitrogen to combine with oxygen to form oxides of nitrogen (NO_x), which may be carried by precipitation into the soil. On a much less spectacular scale, but capable annually of processing about fifty times the amount of nitrogen fixed in thunderstorms, there are bacteria found in the soil and in the roots of leguminous plants that are able to fix the nitrogen required for the synthesis of the complex nitrogen compounds found in all forms of life on earth. A limited amount of nitrogen is also fixed by biota in the oceans.

The level of nitrogen in the atmosphere is maintained by a very complex, but efficient, nitrogen cycle in which the nitrogen obtained from the atmosphere is ultimately returned there (Figure 2.5). When plants and animals die, for example, the complex nitrogen compounds that they contain are broken down by denitrifying bacteria to release the nitrogen back into the environment. Nitrogen in the nitrogenous waste excreted by animals is also recycled back into the atmosphere through the activities of these bacteria. Human activity has

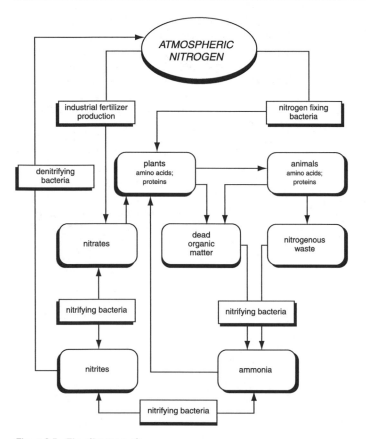

Figure 2.5   The nitrogen cycle

disrupted the cycle through the introduction of agricultural and industrial activities involving nitrogen compounds. Nitrogen fertilizers add to the quantities of nitrate available in the soil; the growing number of animals and people increases the volume of nitrogenous waste produced; industrial activities cause the formation of NO_x and other gases that disrupt the flow of nitrogen in the system. This has resulted in the creation or augmentation of a number of environmental issues – from eutrophication to ozone depletion – involving nitrogen compounds.

The greenhouse gases

After oxygen, nitrogen and argon, the remaining gases account for only about 0.04 per cent of the atmospheric gases by volume. They have an influence quite out of proportion to their volume, however. The most abundant of them is carbon dioxide

(CO_2), only 0.03 per cent by volume, but important to life on earth because of its participation in photosynthesis and its contribution to the greenhouse effect.

The atmosphere is selective in its response to radiation – some it allows to pass through, some it traps (Figure 2.6). It is transparent to high-energy short-wave radiation, such as that from the sun, with the exception of the ultraviolet radiation absorbed in the ozone layer. In contrast, it is partially opaque to lower-energy, long-wave radiation, such as that emanating from the earth's surface. A high proportion of the solar energy arriving at the outer edge of the atmosphere is therefore transmitted through to the surface, retaining its high energy content as it does so. Once it arrives, some of it is absorbed, the earth's surface heats up and it begins to send terrestrial radiation back into the atmosphere. This longer-wave radiation, from the infrared end of the spectrum, is captured by the atmosphere, causing the temperature to rise. The

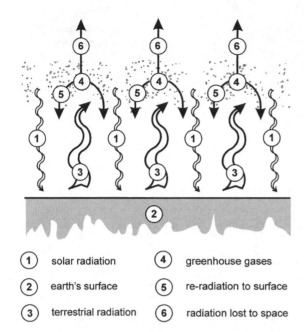

(1)	solar radiation	(4)	greenhouse gases
(2)	earth's surface	(5)	re-radiation to surface
(3)	terrestrial radiation	(6)	radiation lost to space

Figure 2.6   A schematic representation of the greenhouse effect

process has been likened to the way a greenhouse works – allowing sunlight in, but trapping the resulting heat inside – hence the name 'greenhouse effect' that has been applied to it. The analogy is not perfect, but the term continues to be used universally to describe the process. It is the glass in the greenhouse that prevents the heat from escaping by disrupting convection. There is no such physical barrier in the atmosphere, where the capture of the outgoing terrestrial radiation is effected largely by the carbon dioxide, along with about twenty other minor gases, including methane (CH_4), nitrous oxide (N_2O), tropospheric ozone (O_3) and chlorofluorocarbons (CFCs), which together are called the greenhouse gases. Water vapour also exhibits greenhouse properties, but has received less attention than the other gases. Without these gases, global temperatures would be much lower than they are at present – perhaps averaging only $-17°C$ compared to the existing average of $+15°C$ – and the nature of the environment would be quite different.

Since the greenhouse effect depends upon carbon dioxide and the other greenhouse gases, it follows that change in the relative concentration of these gases will impact the greenhouse effect (see Chapter 11). There is evidence that the proportion of carbon dioxide in the atmosphere has varied considerably

in the past, being greatest perhaps during the Carboniferous period and lowest during the Ice Ages. Such changes were brought about by natural processes, presumably, but since the middle of the nineteenth century, human activities have had a major role in increasing the intensity of the greenhouse effect through the anthropogenic production of carbon dioxide. Levels of methane, an even more effective greenhouse gas than carbon dioxide, have risen also, particularly in the last two decades of the twentieth century. There is evidence that this has caused the mean global temperature to rise, and that the increase is likely to continue. Concern over the climatological, environmental and economic consequences led to the emergence of global warming as the leading environmental issue of the late twentieth century.

Oxides of sulphur and nitrogen and other trace gases

There are many other gases that from time to time become constituents of the atmosphere. These include sulphur dioxide (SO_2), oxides of nitrogen (NO_x), hydrogen sulphide (H_2S) and carbon monoxide (CO), along with a variety of more exotic hydrocarbons, which even in small quantities can be harmful to humans. In the past, these have given rise to local problems, following volcanic eruptions or forest fires, for example, or associated with urban pollution from industrial or vehicular sources, but in recent years concern has increased over the widespread dissemination of some of these gases and the consequent larger-scale environmental impact. Increasing industrial activity during the twentieth century and the continued reliance on fossil fuels as energy sources caused a gradual but steady growth in the proportion of sulphur and nitrogen oxides in the atmosphere, particularly in the developed countries. The growth of industry in developing countries, accompanied by the burgeoning use of cars and trucks worldwide, has led to the much more widespread production of these gases and the problems they present have become ubiquitous. Few of the world's major cities are now free of the photochemical smog produced from the exhaust gases of the internal combustion engine. In combination with atmospheric water, gases, such as the oxides of sulphur and nitrogen, create acid precipitation, which caused serious damage to aquatic and terrestrial ecosystems in Europe and

North America in the 1960s, 1970s and 1980s (see Chapter 10). Since then there has been some success in slowing and even reversing the damage in these areas, but in developing countries such as India and China, where development is being sought through the use of fossil fuel-based energy, acid rain is a growing problem.

WATER IN THE ATMOSPHERE

The production and distribution of acid rain are related to a number of factors, but its existence would not have been possible without the presence of atmospheric water. In its pure state water is a combination of hydrogen and oxygen (H_2O), but it is seldom completely pure even in those areas where atmospheric pollution is limited. In combination with carbon dioxide, for example, it forms weak carbonic acid, which contributes to chemical erosion in areas of chalk and limestone bedrock. Lists of the principal atmospheric gases are commonly based on the composition of dry air, but the atmosphere is never completely dry. The actual amount of water in the atmosphere varies from

fractional amounts in some cold, continental areas to as much as 4 per cent by volume above the humid tropics (Barry and Chorley 1998). Although the total amount of water in the atmosphere at any given time is relatively small it is regularly replenished through the hydrologic cycle, which circulates water through the biosphere (see Box 2.5, The hydrologic cycle) (Figure 2.7).

The role of atmospheric water in the earth/atmosphere system is twofold. First, as part of the hydrologic cycle, it is directly involved in the distribution of moisture across the earth's surface. Rain and snow, and the clouds that produce them, are the most obvious indicators of this, but even in the clear skies above the world's driest deserts there is water present. Second, because of its ability to reflect and absorb radiation and to change state between solid, liquid and gas, it makes a major contribution to the earth's energy budget at both regional and global scale. Atmospheric water is an integral part of many environmental issues such as drought, desertification and acid precipitation. In addition, because of its role in the earth's energy budget, any change in the distribution of water in the system can augment or diminish the impact of

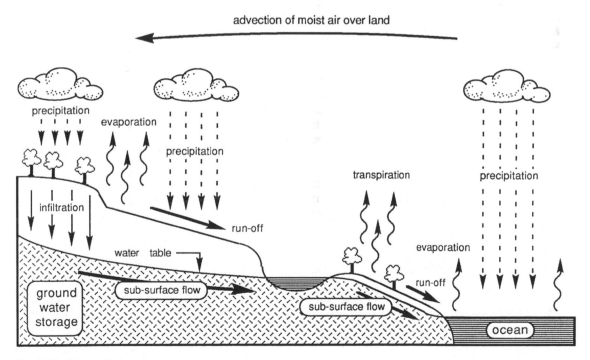

Figure 2.7 The hydrologic cycle

other elements, including the greenhouse effect and ozone depletion, which, in whole or in part, make their presence felt through that budget.

ATMOSPHERIC AEROSOLS

In addition to the gaseous components of the atmosphere, there are also non-gaseous materials present in the form of solid or liquid particles dispersed in the air (Kemp 1994). These are known as aerosols or particulate matter. Collectively, they are often regarded as synonymous with air pollution, although many of the elements involved – dust, soot, smoke and salt particles, for example – are regularly produced by natural processes such as volcanic activity, forest or grass fires, evaporation and air movement, while normal biological processes create and release spores, pollen grains, bacteria, viruses and a variety of other microscopic particles (Figure 2.8). From time to time in the past the levels of such aerosols have increased, sometimes dramatically, but in most cases the atmosphere's built-in cleaning mechanisms were able to react to the changes and the overall impact was limited in extent and duration. For example, when the island of Krakatoa exploded in 1883, it threw several cubic kilometres of volcanic dust into the atmosphere. Through processes such as co-agulation, dry sedimentation and wash-out by

precipitation, almost all of the dust was returned to the earth's surface in less than five years. The 'red rain' that occasionally falls in northern Europe is a manifestation of this cleansing process, being caused when dust from the Sahara Desert is carried up into the atmosphere by turbulence over the desert and washed out by precipitation in more northerly latitudes. Thus the atmosphere can normally cope with the introduction of aerosols by natural processes. Cleansing is never complete, however. There is always a global background level of atmospheric aerosols, which reflects a dynamic balance between their output from natural processes and the efficiency of the cleansing mechanisms. It is now thought that the proper working of the earth/atmosphere system may require a certain proportion of non-gaseous material to be present in the atmosphere – to provide condensation nuclei for the precipitation process, for example – but it is also clear that the existing level is higher than naturally normal and increasing. The increase has been linked with human industrial and agricultural activities, plus rising levels of energy consumption. Although the output from such sources is only 10–15 per cent of the annual aerosol production, the chemical content of the anthropogenic emissions and their distribution can lead to serious environmental problems. Redistribution of the aerosols by wind and pressure patterns ensures that they do not always remain in the atmosphere close to their

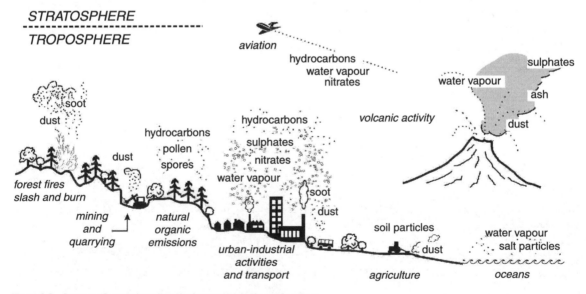

Figure 2.8   Sources of aerosols and particulate matter in the environment

source, but may be carried into non-industrial areas with little human habitation. This is the case with the Arctic, a region of low population density and no industry, which suffers regularly from Arctic Haze, brought about by aerosols carried into the area from industrial mid-latitudes in Eurasia and North America. As a result, the environment in the Arctic has levels of heavy metals, pesticides and a variety of toxic chemicals that match those in industrial areas to the south (Barrie 1986).

The presence of aerosols disrupts the flow of radiation through the atmosphere, and by observing the degree of disruption it is possible to obtain a measure of the level of atmospheric turbidity – the dirtiness or dustiness of the atmosphere. Some studies suggest that high aerosol levels contribute to global cooling through the disruption of the flow of solar radiation, but there is also evidence that in some cases they actually produce a slight warming through their ability to absorb and re-radiate outgoing terrestrial radiation. Resolution of this apparent contradiction will be possible only through systematic observation and monitoring of atmospheric aerosol levels and temperatures.

THE STRUCTURE OF THE ATMOSPHERE

Although its gaseous constituents are well mixed and spread relatively evenly around the earth, the atmosphere is not physically uniform throughout. Temperature and air pressure differences cause the formation of distinct zones or layers, which provide form and structure (Ahrens 2002). Temperature trends and differences, for example, are the basis for the delineation of the atmosphere into a series of layers stretching from the earth's surface to 500 km out into space (Figure 2.9).

The lowest layer is the troposphere, reaching from the surface up to about 8 km above the poles and 16 km at the equator, but with variations brought about by changes in seasonal energy budgets. Within the troposphere, temperatures characteristically decrease with altitude at a rate of 6.5°C per kilometre – the tropospheric or environmental lapse rate – but close to the surface the lapse rate is often quite variable, which helps to produce instability in the system and makes the troposphere the most turbulent of the atmospheric layers.

The upper limit of the troposphere is the tropopause, where temperatures have declined on average

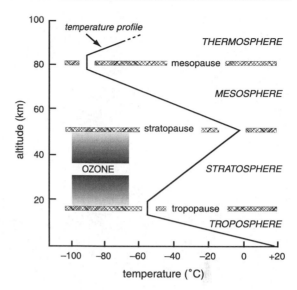

Figure 2.9 The vertical structure of the atmosphere and its associated temperature profile

to between −50°C and −60°C. Above the tropopause lies the stratosphere, a layer in which the decline in temperature is arrested and then turned around so that the temperature begins to increase with altitude. At the stratopause – the upper limit of the stratosphere – some 50 km above the surface, temperatures close to or even slightly above 0°C are not uncommon. Within the stratosphere there is a layer of ozone gas, which absorbs short-wave solar radiation and is therefore directly responsible for the warming. Rising temperatures in the stratosphere tend to dampen any turbulence and as a result conditions in the stratosphere are remarkably stable. Beyond the stratopause – the upper limit of the stratosphere – there are two other atmospheric layers, the mesosphere and the thermosphere, separated by the mesopause. Temperatures decline in the mesosphere and begin to rise again in the thermosphere, where they reach values as high as 1000°C.

In some cases the atmospheric layers are grouped as: the lower atmosphere (troposphere), the middle atmosphere (stratosphere and mesosphere) and the upper atmosphere (thermosphere) or the homosphere (troposphere, stratosphere and mesosphere), where the atmospheric gases are well mixed, and the heterosphere (thermosphere and beyond), where the atmosphere is thin, the gases are less well mixed and diffusion of the heavier and lighter gases takes place. At the outer edge of the thermosphere, some

Plate 2.5 A roll cloud. A low, horizontal tube-shaped cloud associated with thunderstorms, which produce stormy, turbulent conditions in the troposphere. (Courtesy of Heather Kemp.)

350–500 km from the earth's surface, lies the exosphere, where gaseous molecules with sufficient velocity may escape the earth's gravity.

The direct contribution of the outermost atmospheric layers to the biosphere is limited, but conditions in the troposphere and stratosphere have an integral role in many environmental issues. This is particularly true of the lowest 1–2 km of the troposphere, commonly referred to as the planetary boundary layer, where material and energy fluxes between the earth's surface (lithosphere and hydrosphere) and the atmosphere take place. These exchanges are complex, and human interference with the mechanisms involved has contributed to the creation and intensification of a variety of environmental issues.

THE EARTH'S ENERGY BUDGET

Virtually all the earth's energy is received from the sun in the form of solar radiation, and balancing that inflow is an equivalent flow of energy returning to space as terrestrial radiation (Ahrens 2002). Without that balance between these energy fluxes the earth would become increasingly warm or increasingly cool. In practice, the concept does seem to apply in general terms to the earth as a whole over an extended time period, but it is not applic-

able to any specific area over a short period of time. Seasonal differences in land cover, seasonal and annual variations in the hydrologic cycle or changes in atmospheric turbidity all contribute to local and short-term imbalance in the energy budget. Human interference through land clearing, water use and atmospheric pollution has a similar impact (see Box 2.6, Radiative forcing). In most cases, however, the earth/atmosphere system is sufficiently dynamic to accommodate the differences and over the long term the global balance is retained. This is not to imply that the budget is never unbalanced. The cooling of the earth that caused the Ice Ages, for example, probably represents the disruption of the global budget by natural processes. Current concern is centred on the potential impact of human interference, with the enhancement of the greenhouse effect and subsequent global warming receiving most attention.

Only about 50 per cent of the solar radiation that reaches the outer edge of the atmosphere is absorbed by the earth's surface (Kemp 1994). Half of that is received as direct radiation and half as diffuse radiation that has been scattered by water vapour, dust and other aerosols before eventually reaching the surface. Some 30 per cent of the original incoming radiation is reflected back into space either from the surface of the land and sea or from clouds and aerosols in the atmosphere. This

BOX 2.6 RADIATIVE FORCING

The earth's energy budget is considered to be in equilibrium, with incoming solar radiation being balanced by outgoing terrestrial radiation over the long term. Any factor capable of disturbing that energy balance is called a radiative forcing agent. Natural forcing agents include changes in solar radiation – either from variations in the output of energy from the sun or variations in the earth's orbit – changes in planetary albedo and changes in atmospheric aerosol concentration. Individually or in combination, these have altered the flow of energy into and out of the earth/atmosphere system for millions of years (Table 2.2).

These natural forcing agents continue to disrupt the energy budget, but current concern is with anthropogenic radiative forcing and the climate change it is likely to produce. Ozone depletion, increased atmospheric turbidity and the enhancement of the greenhouse effect caused by human activities have already disturbed the earth's energy budget and have the potential to do much more.

It is estimated, for example, that the effect of rising levels of greenhouse gases on the radiative balance of the earth will be greater than that caused by any other forcing agent, natural or anthropogenic.

Radiative forcing may be positive or negative, as in the case of solar variability, which created radiative forcing of between +0.1 and −0.1 W m^{-2} (watts per square metre) between 1990 and 2000. Aerosols generally have a negative impact. When Mount Pinatubo erupted in 1991 it reduced global net radiation by between 3 W m^{-2} and 4 W m^{-2}. In contrast, greenhouse gases are positive radiative forcing agents, and a forcing of +2.45 W m^{-2} has been attributed to the increase in greenhouse gas levels since pre-industrial times.

For more information see:

Harvey, L.D.D. (2000) *Global Warming: The Hard Science*, Harlow: Pearson.
IPCC (2001) *Climate Change 2001: Synthesis Report*, Cambridge: Cambridge University Press.

TABLE 2.2 RADIATIVE FORCING AGENTS

Forcing agents	Radiative forcing (W m^{-2})	Comments
Greenhouse gases	+0.56	Business-as-usual
	+0.41	Major emission controls (see Chapter 11)
Solar variability	+0.1	E.g. orbital changes and changes in solar
	−0.1	irradiance – sunspot cycles
Large volcanic eruption	−0.2	E.g. El Chichon, Mt Pinatubo
Anthropogenic sulphur emissions	+0.15	Difficult to estimate – total emissions are
	−0.15	declining, regional differences remain
Stratospheric H$_2$O	+0.02	

Source: based on data in IPCC *Climate Change 1994: Radiative Forcing of Climate Change and an Evaluation of the IPCC 1992 Emissions Scenarios*, Cambridge: Cambridge University Press (1994).

is a measure of the earth's reflectivity, or albedo, and the radiation bounced back into space in this way has no direct role in the earth/atmosphere system. The remaining 20 per cent is absorbed in the atmosphere, mainly by oxygen, ozone and water vapour, and some of the resulting long-wave energy is re-radiated towards the earth's surface (Figure 2.10).

Most of the solar energy absorbed by the earth's surface is re-radiated as long-wave terrestrial radiation, but some is transferred into the atmosphere by convective and evaporative processes. The greenhouse gases trap the bulk of this outgoing terrestrial radiation, except for small amounts (about 5 per cent) which escape directly to space through the so-called atmospheric window. Some of the energy absorbed by the atmosphere is emitted to space also, but most accumulates to be re-radiated back

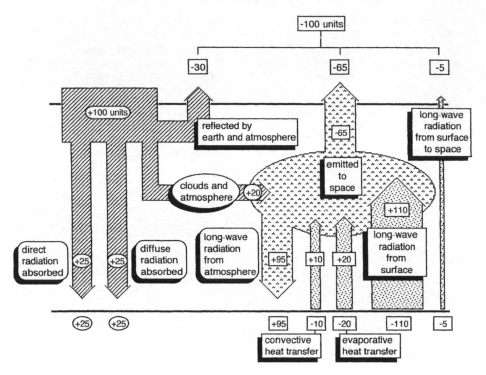

Figure 2.10   The earth's energy budget

towards the surface. The exchange of energy between the atmosphere and the earth's surface involves amounts apparently in excess of that provided by solar radiation. This situation is a direct function of the greenhouse effect, and its ability to retain energy in the troposphere. Eventually all the long-wave radiation passes out of the system, but not before it has provided the energy necessary to allow various atmospheric and terrestrial processes to function.

Superimposed on this general picture are a series of smaller-scale budgets, which vary in time and place. There is a definite latitudinal imbalance in the budget, for example (Figure 2.11). The equator annually receives about five times the amount of solar radiation reaching the poles, and those areas between 35°N and 35°S latitude receive more energy than is returned to space. Poleward of these latitudes, there is an excess of outgoing radiation over incoming, which creates a radiation deficit in higher latitudes. At first sight, such an imbalance would lead to higher latitudes becoming progressively colder and equatorial latitudes becoming progressively warmer. In reality, as soon as these latitudinal differences develop they initiate circulation patterns in the atmosphere and in the oceans which combine to transfer heat from the tropics towards the poles, and in so doing to serve to counteract the imbalance (Barry and Chorley 1998).

THE CIRCULATION OF THE ATMOSPHERE

In the first attempt to explain the circulation of the atmosphere in 1735, George Hadley visualized a

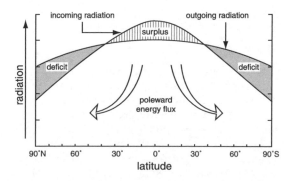

Figure 2.11   Latitudinal variation in the earth's energy budget

simple convective system, based on a non-rotating earth with a uniform surface that was warm at the equator and cold at the poles. Warm buoyant air rising at the equator spread north and south in the upper atmosphere, eventually returning to the surface in high or polar latitudes. From there it flowed back across the surface towards the equator to complete the circulation.

Hadley's original model, with its single convection cell in each hemisphere, was eventually replaced by a three-cell model as technology advanced and additional information became available, but his contribution was recognized in the naming of the tropical cell as the Hadley cell. The three-cell model initially assumed a uniform surface, but the rotation of the earth was introduced, and with it the Coriolis effect, which causes moving objects to swing to the right in the northern hemisphere and to the left in the southern. Thus the winds became westerly and easterly in

the new model, rather than blowing north or south as in the one-cell version. The three cells plus the Coriolis effect in combination produced alternating bands of high and low pressure, separated by wind belts that were easterly in equatorial and polar regions and westerly in mid-latitudes. Elements of this model can be recognized in existing global wind and pressure patterns, particularly in the southern hemisphere, where the great expanse of ocean in high latitudes more closely resembles the uniform surface of the model (Figure 2.12).

In the late 1940s and 1950s, as knowledge of the atmosphere improved, it became increasingly evident that the three-cell model oversimplified the general circulation. The main problem arose with the mid-latitude cell. Observations indicated that most energy transfer in mid-latitudes was accomplished by horizontal cells – such as travelling high and low-pressure systems – rather than by the

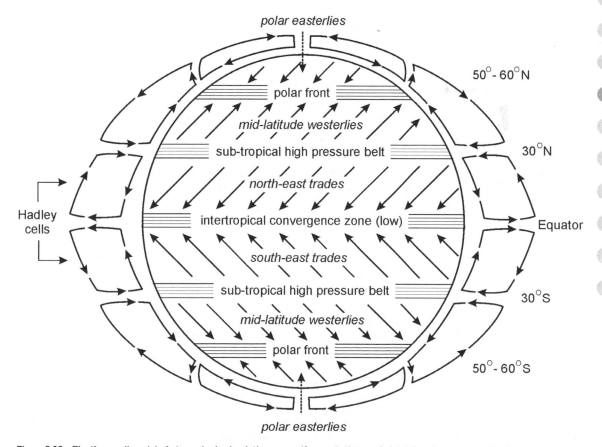

Figure 2.12 The three-cell model of atmospheric circulation on a uniform, rotating earth, heated at the equator and cooled at the poles

vertical cell indicated in the model. Modern interpretations of the general circulation of the atmosphere retain the tropical Hadley cell, but horizontal eddies have come to dominate mid-latitudes, and have even replaced the simple thermal cell of polar latitudes (Barry and Chorley 1998).

Conditions in the upper atmosphere are an integral part of modern studies of the atmospheric circulation. The upper atmospheric circulation is quite complex in detail, but in general terms it is characterized by an easterly flow in the tropics and a westerly flow in mid to high latitudes. The upper westerlies include a pattern of waves, called Rossby waves, which vary in amplitude in a quasi-regular sequence represented by an index cycle. When the amplitude of the waves is greatest, cold air is carried equatorwards and warm air polewards, resulting in a significant latitudinal energy transfer during the run of any one cycle. Within these broad air flows, at the tropopause, there are relatively narrow bands of rapidly moving air called jet streams, in which wind speeds may average 125–130 km per hour, although much higher speeds may occur

(Figure 2.13). Modern representations of the general circulation of the atmosphere take into account the non-uniform nature of the earth's surface, with its mixture of land and water, and include consideration of seasonal variations in energy flow. Differences in their physical properties cause land and sea to warm up and cool down at different rates. This creates significant temperature differences between land and water, and produces a series of pressure cells rather than the simple belts of the original models. By altering the regional air flow, such pressure differences cause disruption of the theoretical wind patterns. The changing location of the zone of maximum insolation with the seasons also causes variations in the location, extent and intensity of the pressure cells and wind belts. Although such changes are repeated year after year, they are not completely reliable, and this adds an additional degree of variability to the atmospheric circulation. Other elements that contribute to that variability include variations in solar output and changes in atmospheric turbidity. These elements have always been part of the natural system, but

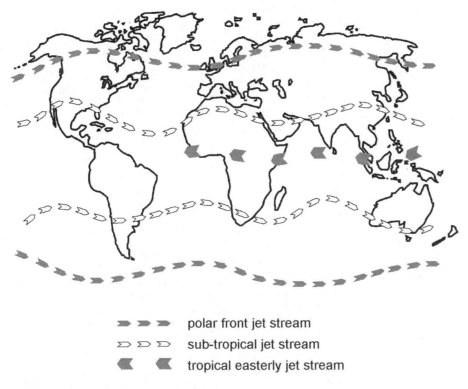

> ➤ ➤ ➤   polar front jet stream
> ᗡ ᗡ ᗡ   sub-tropical jet stream
> ◀ ◀   tropical easterly jet stream

Figure 2.13 The jet streams

increasingly they are being augmented by a range of human activities that have the potential to alter the dynamic energy balance in the system, and through that cause changes in the atmospheric circulation.

ENVIRONMENTAL ISSUES IN THE ATMOSPHERE

Without the atmosphere, life on earth, if it existed at all, would be quite different from what it is today. The atmosphere performs several functions that have allowed humans to survive and develop almost anywhere in the earth's surface. First, it provides and maintains the supply of oxygen required for life itself. Second, it controls the earth's energy budget through such elements as the ozone layer and the greenhouse effect and, by means of its internal circulation, distributes heat and moisture across the earth's surface. Thirdly, it has the capacity to dispose of waste material or pollutants generated by natural or human activity. Through ignorance of the mechanisms involved, or lack of concern for the consequences, society has interfered with all of these elements and created or intensified problems, many of which are now recognized as being of global concern (Kemp 1994). These include atmospheric pollution, acid precipitation, global warming and ozone depletion (see Chapters 10 and 11). Most of these issues have their roots in changes in the composition of the atmosphere, interference with the global energy budget or lack of understanding of the elements of the general circulation, either individually or in combination.

HYDROSPHERE

The hydrosphere is that part of the environment that includes all of the water in and on the lithosphere and above it in the atmosphere. It is most obvious in oceans, seas, rivers and lakes, but the ground water in the rocks beneath the surface, the water vapour and water droplets in the atmosphere and the water contained in the living elements of the biosphere are also part of the hydrosphere. The water in the hydrosphere is very unevenly distributed. In some places it is abundant, in other places it is absent, and society spends much time, money and energy redistributing it. Some 97 per cent of the world's water is in the oceans, while a further 2 per cent is in the form of ice and snow, and

TABLE 2.3 THE PHYSICAL DISTRIBUTION OF THE WORLD'S WATER (%)

Oceans	97.00	
Fresh water	3.00	
of which		
Snow and ice		77.09
Ground water		22.3
Lakes and rivers		0.53
Atmosphere		0.03
Living organisms		0.0016

Source: based on data in J.P. Peixoto and M. Ali Kettani, 'The control of the water cycle', in F. Press and R. Seiver (eds) *Planet Earth*, San Francisco: Freeman (1976).

1 per cent is on land as fresh water in rivers, lakes and the groundwater system, with small amounts incorporated in the plants and animals of the terrestrial biosphere (Table 2.3). The oceans support large populations of plants and animals that have adapted to the chemical and physical conditions in the marine environment. The survival of terrestrial flora and fauna despite their access to only a small proportion of the total available water is made possible by the natural recycling of the water through the hydrologic cycle (see Box 2.5, The hydrologic cycle). Hydrology is the study of all the water in the system. Physical hydrology focuses on the distribution and circulation of water, whereas applied hydrology is more concerned with water and human activities, and includes consideration of such elements as water availability, water quality, irrigation, drainage, erosion and flood control (Ward and Robinson 2000).

OCEANS

The oceans cover about 71 per cent of the earth's surface and contain 97 per cent of the world's total water supply. Beneath the ocean surface, the morphology of the ocean floor includes a complex combination of physical features, from the extensive plains of the continental shelf to deep troughs and high mountain ridges (Figure 2.14). Beyond the continental shelf, at the base of the continental slope, lying more than 2000 m beneath the surface, is the abyssal zone. The main feature of this zone is the abyssal plain, a region of low relief beyond the

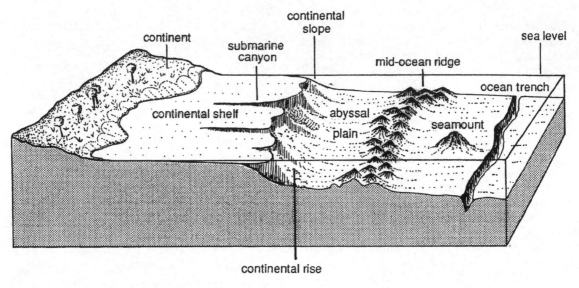

Figure 2.14   The morphology of the ocean basins

continental margins. Its surface is broken in places by ridges of volcanic mountains associated with sea-floor spreading or by individual sea mounts rising out of the plain, and, particularly in the Pacific Ocean, by linear or curvilinear trenches that reach depths in excess of 10,000 m beneath the sea surface and represent the hadal zone, the deepest part of the oceans. Sediments deposited in the upper levels of the oceans ultimately reach the abyssal zone, but little light penetrates, energy levels are low and plant and animal life is sparse (Duxbury and Duxbury 1997).

That part of the ocean in coastal areas and above the continental shelf is the neritic zone. Conditions such as temperature, salinity and sedimentation are much more variable in the neritic zone than in the ocean deeps, and because it is penetrated by solar radiation it also has a much richer plant and animal community.

The circulation of the oceans

The water in the ocean basins is in constant motion, particularly at the surface, in a circulation that is intimately linked with the circulation of the atmosphere (Segar and Segar 1998). The prevailing winds in the atmosphere, for example, drive water across the ocean surface at speeds of up to 5 km per hour, in the form of broad, relatively shallow currents or drifts (Figure 2.15). Depending on the direction of flow, some currents carry warm water polewards, while others carry cooler water into lower latitudes. The major currents assume roughly circular forms, called gyres, centred on the subtropical high pressure cells and constrained by the shape of the ocean basins. The water in these gyres circulates clockwise in the northern hemisphere and anticlockwise in the south under the influence of the Coriolis effect and the overall atmospheric circulation. It is only in high latitudes in the southern hemisphere that the absence of land allows the West Wind Drift or the Antarctic Circumpolar Current to travel unobstructed around the earth.

In addition to these surface flows, density differences, in part thermally induced, cause horizontal and vertical movements within the oceans. This is the thermohaline circulation. For example, cold, dense water, sinking in the Antarctic, flows along the bottom of the ocean basins and spreads northwards into the southern Pacific, Atlantic and Indian Oceans. The North Atlantic off Greenland and Labrador is also a source of cold water that sinks and flows south, where it mixes with the Antarctic water and is carried into the Indian and Pacific Oceans. The deep water is gradually warmed and mixed upwards, to become part of the surface or near surface circulation, eventually returning to the Antarctic and North Atlantic to complete the circulation (Figure 2.16). The complete journey

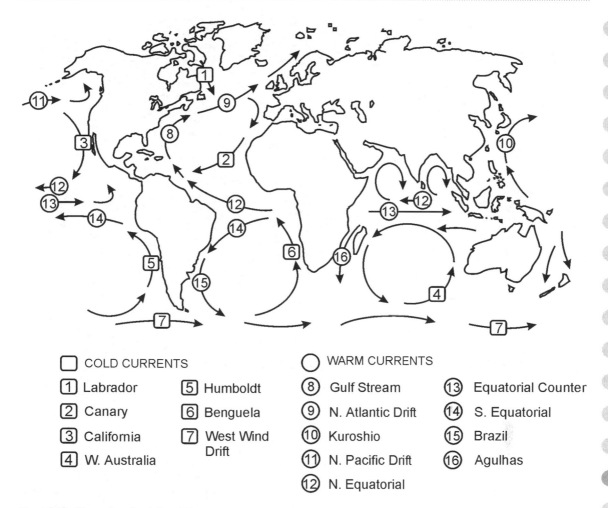

COLD CURRENTS

1 Labrador	5 Humboldt	
2 Canary	6 Benguela	
3 California	7 West Wind Drift	
4 W. Australia		

WARM CURRENTS

8 Gulf Stream	13 Equatorial Counter
9 N. Atlantic Drift	14 S. Equatorial
10 Kuroshio	15 Brazil
11 N. Pacific Drift	16 Agulhas
12 N. Equatorial	

Figure 2.15   The surface circulation of the oceans

of water on this so-called 'Great Oceanic Conveyor Belt' is thought to take at least 1000 years (Broecker 1991).

All of these processes help to offset the imbalance of energy that develops between equatorial regions and the poles. This is illustrated particularly well in the North Atlantic, where the warm waters of the Gulf Stream Drift ensure that areas as far north as the Arctic circle are anomalously mild during the winter months. Currents such as the California and Canary currents in the northern hemisphere and the Benguela and Humboldt in the southern return the cooler water to tropical latitudes, where it is warmed again. One ocean current which does not appear as part of the well established pattern, but which makes a major contribution to energy transfer, is El Niño. This is a flow of warm surface water which appears with some frequency in the equatorial regions of the eastern Pacific and along with its cool water counterpart, La Niña, has an influence that extends well beyond that area (see Box 2.7, El Niño–La Niña; Figure 2.17).

Through the flow of these currents across the ocean surface, the oceanic circulation contributes to the transfer of energy from low latitudes towards the poles. On average, oceanic transport accounts for 40 per cent of the total poleward transfer of energy in the ocean/atmosphere system, while the atmospheric circulation accounts for the remaining 60 per cent.

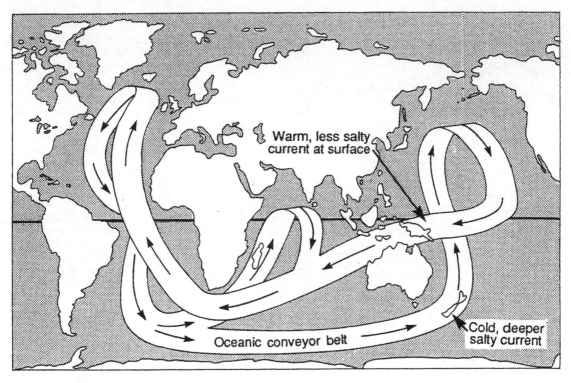

Figure 2.16 The Great Oceanic Conveyor Belt
Source: Moore *et al.* (1996) with permission.

ICE AND SNOW: THE CRYOSPHERE

The cryosphere is that part of the hydrosphere in which water is present as ice or snow (Wadhams 2002). It accounts for about 2 per cent of the total water in the environment and more than 70 per cent of the fresh water. The cryosphere includes the permanent or semi-permanent snow and ice in polar ice sheets, mountain glaciers and permafrost, as well as seasonal snow cover and river and lake ice (Figure 2.18). At various times in the past it has incorporated more or less water depending upon overall climatic conditions. During the Ice Ages, for example, the amount of water held in snow and ice was much greater than at present, so much so that ice several kilometres thick covered large areas of North America and Europe and sea level was tens of metres below its current level. Some of the ice formed at that time remains as part of the modern cryosphere. At other times, during interglacials, for example, the volume of ice and snow was much diminished.

The actual amount of water held in a solid state in the cryosphere is quite changeable over a range of time scales. Seasonal snow and ice cover may exist for only a few months before reverting to its liquid state, but given appropriate weather conditions changes between liquid and solid or solid and liquid may occur over time scales of weeks or even days. Mountain glaciers respond at varying rates depending upon such factors as latitude and local climate conditions. Rates of ablation (melting) and accumulation in alpine glaciers tend to change more rapidly than in polar glaciers, for example, with time scales ranging from a few years to centuries. Although permafrost is a contraction of 'permanently frozen ground', in most areas it is subject to some degree of seasonal melting in the upper or active layer, which varies in depth according to such factors as the nature of the surface (sand, gravel, clay, rock, for example), water and ice content, vegetation cover and the aspect of the surface (flat, sloping north or south, for example). Beneath the active layer the permafrost may change little from year to

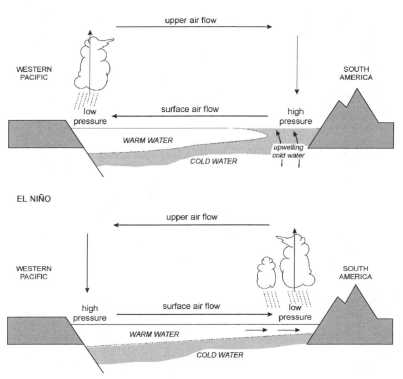

Figure 2.17   The El Niño circulation in the Pacific Ocean

BOX 2.7   EL NIÑO–LA NIÑA

The normal atmospheric circulation pattern in the equatorial Pacific Ocean is characterized by the trade winds, blowing from the north-east in the northern hemisphere and the south-east in the southern hemisphere. Together these winds push the warm surface water of the ocean to the west, allowing it to be replaced by cold water welling up in the eastern equatorial Pacific off the coast of South America. On occasion, the trade winds weaken and the warm water from the western Pacific returns eastwards as an anomalously warm flow of surface water known as El Niño (Figure 2.17). The name originally referred to a warm current that appeared off the coast of Peru close to Christmas – hence El Niño, the (Christ) Child – but now it is applied to a larger-scale phenomenon, capable of influencing global climatic conditions. It owes its development to the Southern Oscillation, a periodic fluctuation in atmospheric pressure in the southern Pacific first recognized in the 1920s by Sir Gilbert Walker as he sought to develop methods of

forecasting rainfall in the Indian monsoon. The term ENSO is commonly used to refer to the combination of El Niño and the Southern Oscillation.

An indication of the state of the Southern Oscillation can be obtained by comparing barometric pressure differences between Tahiti in the eastern Pacific and Darwin in northern Australia. Pressure at these two stations is negatively correlated, high pressure over Tahiti normally being accompanied by low pressure over Darwin, for example. In contrast, low pressure at Tahiti is matched by high pressure at Darwin. These pressure differences induce a strong latitudinal circulation in the equatorial atmosphere – named the Walker circulation. Periodically the regional pressure patterns reverse, and it is that phenomenon that is referred to as the Southern Oscillation. It has a periodicity of one to five years, and in its wake it brings changes in windfields, sea surface temperatures and ocean circulation patterns.

BOX 2.7 – continued

When an El Niño event ends, the circulation patterns tend to return to normal, but in some years – when the difference between the high pressure over Tahiti and the low pressure over Darwin is particularly well marked, for example – the equatorial easterlies return stronger than normal to push the cold water welling up off the South American coast far out into the Pacific. This creates a cold current, flowing east to west, which has been given the name La Niña. Like El Niño, its effects are felt beyond the equatorial Pacific Ocean.

The climatological impacts of El Niño and La Niña were first recognized in South America. During normal years, with easterly winds and cold water offshore, Peru and Ecuador experienced drier cooler conditions. In contrast, during El Niño years the warm water offshore brought with it atmospheric instability, which led to increased precipitation, often in such quantities that it caused serious soil erosion and massive flooding. The impact was also felt in other elements of the environment. In normal or La Niña years, for example, the nutrients brought to the surface by the upwelling water provide an important food source for fish, which in turn support sea mammals and large populations of seabirds. In contrast, during El Niño events, nutrient supply is reduced, fish populations decline and, deprived of their food source, larger sea mammals and birds move away. El Niño events also disrupt the lives of people depending upon the resources of the sea, and through the heavy rains accompanying them damage soils and crops and reduce the local food supply.

Beyond the Pacific, El Niño years are also years of drought in Australia and Brazil, and in India the monsoon rains are often less than normal. They are also associated with increased precipitation in normally dry southern California, and to the north, on the Canadian prairies, El Niño events bring warmer, drier winters. Although La Niña has received less attention than El Niño, it also appears to cause changes in weather patterns beyond its source region. It is, for example, associated with increased precipitation in the Sahel and in India, and below-normal temperatures in central Canada.

For more information see:

Glantz, M.H. (1996) *Currents of Change: El Niño's Impact on Climate and Society*, Cambridge/New York: Cambridge University Press.
Philander, G. (1989) *El Niño, La Niña and the Southern Oscillation*, Orlando FL: Academic Press.

year, but it does respond to changing climatic conditions over time scales of a century or more. In northern Canada and Eurasia the southern limit of the permafrost is retreating northwards, presumably as a result of the rising temperatures in the twentieth century. If global warming continues as predicted, increasingly large areas in high latitudes will become free of permafrost as the ice in the ground reverts to water. Polar ice sheets, such as those in Greenland and Antarctica, are normally considered permanent features of the cryosphere, but they do respond to environmental change, if only slowly, at time scales of several thousand years. Despite these slow response times, late twentieth-century warming appears to have had a measurable impact on the ice sheets and there is growing concern that continued warming will cause significant changes in sea level as the water from the melting snow and ice runs off into the oceans. According to IPPC estimates, sea level may rise by 9–88 cm by the year 2100, with most of the increase attributable to the melting of snow and ice (IPCC 2001a).

Changes in the amount of snow and ice in the hydrosphere would have environmental impacts extending beyond the more obvious hydrological changes. The disappearance of ice and snow would reduce the albedo of the earth's surface, for example, allowing the absorption of additional solar radiation and altering the earth's energy budget. In some areas, the melting of permafrost would release methane currently trapped in the ice to add to the enhancement of the greenhouse effect. Increased run-off of fresh, cold water into the oceans from melting snow and ice could bring about climate change through the alteration of the thermohaline circulation and the subsequent disruption of existing patterns of energy exchange. Thus, although at first sight the cryosphere may appear to be the least active part of the hydrosphere, it is quite variable at a wide range of time

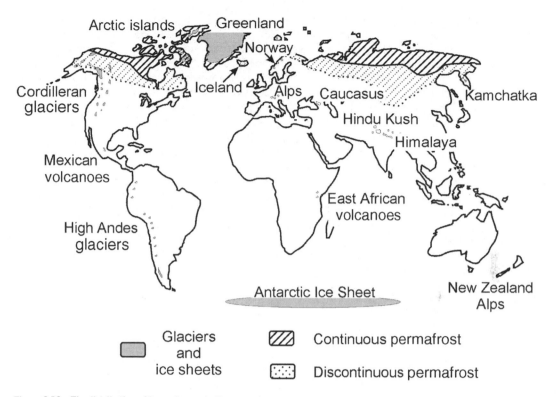

Figure 2.18   The distribution of ice and snow in the cryosphere

scales and as a result has the potential to contribute significantly to environmental change.

GROUND WATER, SURFACE WATER AND WATER IN THE ATMOSPHERE

The water in, on and above the earth's surface accounts for only about 1 per cent of the water in the hydrosphere and is about 20–5 per cent of the earth's fresh water. Despite the small amounts involved, it is upon this sector of the hydrosphere that living terrestrial organisms, including humans, have to depend for the water they need. Water is the main constituent of all living organisms – human bodies comprise about 65 per cent water. It helps cells to maintain their form, and the chemical processes that are involved in metabolism take place in a watery solution. The digestion of food in mammals, the transport of bodily wastes and the maintenance of a stable body temperature through perspiration and evaporation all require water. Without a regular supply of water, organisms are unable to survive, as is evident during prolonged drought, when plants and animals become dehydrated and die.

Water is present in the atmosphere in only miniscule amounts, but it plays an important role in the aquatic environment by providing the precipitation to replenish the groundwater and surface water reservoirs. If precipitated completely and evenly across the earth's surface, the amount of water in the atmosphere at any one time would produce the equivalent of no more than 25 mm of rainfall (Barry and Chorley 1998). In reality the distribution is very uneven, as a result of regional variability in the dynamic processes that produce precipitation. Intense thunderstorms can yield 25 mm of rain in a matter of minutes, whereas the same amount may take several months, or even years, to accumulate under more stable atmospheric conditions. Variations such as these account for annual precipitation totals that range from virtually nothing in some of the world's deserts to more than 10,000 mm in the monsoon lands of the tropics. These differences are reflected in the availability of ground water and surface water in these areas.

The largest volume of liquid fresh water – about

22 per cent of the total – is in the groundwater sector. Ground water originates in the precipitation that percolates down into the cracks or pore spaces in the sediments and rocks beneath the earth's surface. Rocks that are sufficiently porous and permeable to store significant quantities of ground water are called aquifers (Figure 2.19). The upper limit of groundwater saturation in an aquifer is the water table. Ground water moves under the influence of gravity, although usually only slowly, and may return to the surface naturally – through springs, for example, or by seepage into rivers and lakes. The water that moves out of the groundwater system is replaced by precipitation and the level of the water table is determined by the difference between the rate of outflow and the recharge.

Aquifers provide water for human settlement in those areas where the surface water is absent or inadequate. Where the aquifer is present at the surface, ground water is easily accessible through shallow wells, but in places, where the aquifer lies beneath a less permeable layer, the ground water can be reached only by digging or boring through the overlying layer. In such cases, the aquifer may contain sufficient hydraulic pressure to cause the ground water to flow freely to the surface as an artesian well.

Groundwater reservoirs vary in size and capacity from small local aquifers to systems, such as the Ogallala aquifer beneath the Great Plains in the United States or the Great Artesian Basin in Australia, that extend over several thousand square kilometres. The scale of groundwater use ranges from provision of water for single-family rural properties to public water supply in major cities such as Beijing, Dhaka and Mexico City. It is estimated that in Asia alone 1 billion to 1.2 billion

people depend upon ground water to meet their needs, and even in developed areas groundwater use is high. More than 100 million people in Western Europe and about the same in North America depend on aquifers for their water supply (Foster *et al.* 1998). In rural areas, where precipitation and surface water supplies cannot meet the needs of agriculture, ground water is being used in increasing quantities for irrigation.

As a result of increased demands on the groundwater system, in many areas the rate at which water is being withdrawn exceeds the rate of recharge, the water table is falling and the groundwater supply is declining rapidly (see Chapter 8). In parts of the Ogallala aquifer, for example, water is being withdrawn at a rate that is twenty times greater than it is being replaced, and the water table has declined by between 30 m and 60 m since the 1940s. Such an approach to the use of ground water is sometimes referred to as groundwater mining. Similar problems of aquifer depletion are occurring in Saudi Arabia, China and India, where surface water is in short supply and ground water is essential to meet local needs. The removal of ground water without recharge can also lead to land subsidence, as in Mexico City, where depletion of the underlying aquifer has allowed the land to sink, causing structural damage to buildings, roads and sewers. In contrast, in London, England, reduced groundwater use has allowed the water table to recover by as much as 20 m (Goudie 2002).

Another threat to groundwater supplies comes from contamination. Leakage from sanitary sewers, landfill sites, septic systems and livestock feedlots introduces faecal pathogens and other toxins into groundwater systems. Petroleum and chemical spills frequently seep into aquifers before they

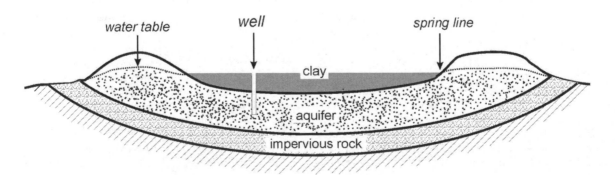

Figure 2.19 The characteristics of a simple aquifer

can be cleaned up and excess use of fertilizers and pesticides also contributes to pollution in some areas. The slow through-flow in most aquifers ensures that the impact of those pollutants continues for many years. In coastal areas the removal of fresh water from aquifers may allow the incursion of salt water, which effectively prevents further use of the system for most domestic and industrial purposes.

Although the largest amount of the earth's liquid fresh water is contained in groundwater reservoirs, society's perception of the world's freshwater supply is obtained from the visible water in the lakes and rivers on the earth's surface. This surface water accounts for less than 0.5 per cent of the total fresh water and is very unevenly distributed (Figure 2.20). Most of it is held in large lakes such as the Great Lakes in North America and the rift valley lakes of East Africa, the greatest volume in any one lake being in Lake Baykal in Russia. (The Caspian Sea is actually the largest lake by far, but it is saline.) Lake volumes are maintained when inflow from surface streams or ground water matches outflow. From a human point of view lakes appear permanent, but in geological terms they are ephemeral features, subject to filling by sedimentation or drainage through the deepening of overflow channels.

Society uses lakes for a variety of purposes, including the supply of water for domestic, agricultural or industrial purposes, transport, recreation and the provision of food. These uses are not always well planned. In 1960, for example, the Aral Sea in Uzbekistan and Kazakhstan was the fourth largest lake in the world, with an area of 66,900 km^2. By the end of the twentieth century, major water diversion from the rivers flowing into it had reduced its area by more than half, with disastrous environmental results (see Chapter 8). In contrast, in many parts of the world reservoirs have been created, to provide some or all of the attributes of natural lakes. Most reservoirs are relatively small, but some, such as Lake Kariba (5500 km^2) on the Zambezi River in Africa, the Altimera/Xingu project (18,000 km^2) in the Brazilian Amazon and the Three Gorges dam (632 km^2) in China compare in area and volume with large natural lakes. The creation of such features has significant environmental consequences, from habitat destruction to the addition of greenhouse gases such as methane to the atmosphere.

The smallest, most variable but in some ways the most dynamic contributors to surface fresh water are rivers and streams. Stream flow varies both regionally and seasonally, but accurate measurements are not readily available because of the poor network of gauging stations in many river basins.

Figure 2.20   The world's major rivers and lakes

To overcome the limited availability of direct measurements, estimates can be made by comparing precipitation and evaporation, with the difference between the two representing run-off. As a result, most data on stream flow or run-off are considered to represent best estimates rather than accurate measurements. The best estimates of the annual volume of run-off indicate that the greatest flow is in Asia from major rivers such as the Huang He, Yangtze, Ganges, Brahmaputra and Mekong, followed by South America, where the Amazon alone accounts for almost half of the volume of surface water flowing from that continent into the oceans. Elsewhere, major rivers such as the Nile, Zambezi and Zaire (Congo) in Africa and the Missouri/Mississippi system and the St Lawrence in North America account for the bulk of the run-off on these continents. The lowest continental stream flow is in Australia.

The amount of water carried by a river depends upon a number of elements. Water is supplied initially by precipitation and groundwater inflow within the catchment area or drainage basin of the river. Thus there is a strong correlation between the area of a river's drainage basin and its discharge – a larger area intercepts more precipitation and ground water. The latter provides a relatively steady input, called the base flow, while variability in discharge rates depends upon the amount, timing and nature of the precipitation. During the northern winter, for example, the bulk of the precipitation falls as snow, rivers and streams in North America and Eurasia freeze over and run-off is reduced. The flow increases rapidly during the spring melt, when several months' accumulation of precipitation is channelled through the waterways in only a few weeks. Thus, although rivers such as the Ob and Yenesei, which flow north into the Arctic Ocean, have annual discharge rates comparable to the Mississippi, a much greater proportion of their annual discharge occurs in the spring. At that time, they are often prone to flooding when the volume of water exceeds the capacity of the channel and the water flows over their banks. Intense precipitation, even over such a short time span as a few hours, can cause flooding on some rivers at any time of the year. Variations in flow may be greater yet in areas that have distinct wet and dry seasons, where even major rivers experience very low water during the dry season. In arid areas, stream flow occurs only when there is an external source of water, as in the case of the Nile, for example, rising in Lake Victoria

and being fed by the rains in the mountains of Ethiopia. Otherwise, streams will dry up completely, particularly if the water table is also low and there is no contribution from groundwater inflow.

In addition to discharging water into lakes, seas and oceans, rivers and streams also carry solids suspended or dissolved in the water (Figure 2.21). The volume of suspended sediment moved through river basins annually is large, measuring some 16 billion tonnes per annum. It varies according to geology, climate, topography, vegetation cover and land use, and has been augmented by soil erosion associated with such human activities as forestry and agriculture (Strahler and Strahler 1992). The average annual yield is 125 tonnes per square kilometre (t km^{-2}) per annum, but some rivers in India and China carry more than 2000 t km^{-2} per annum and between them the Brahmaputra, Ganges and Huang He annually carry about 20 per cent of the world's total sediment load. Not all the sediment reaches the oceans. Some is deposited in lakes and reservoirs or in stream beds when flow is reduced. The sediment deposited in the stream beds may be picked up again when the stream flow increases and eventually make its way to the ocean, but the sediment in the lakes and reservoirs remains, gradually filling them up. While this is a natural process for lakes, it is being accelerated by human activities and the shoaling that occurs creates problems for transport and damages fish habitat. The accumulation of sediments in reservoirs gradually reduces their ability to function as producers of electricity or providers of irrigation water.

Although the water in rivers and streams is much less saline than that in the oceans, it does contain dissolved solids (Figure 2.22). The concentration of total dissolved solids (TDS) in sea water is about 35,000 parts per million (ppm), whereas the average for river water is about 100 ppm. They have their origins in a variety of processes, including the chemical weathering of rocks, leaching of salts from soils, atmospheric loading from polluted rain and organic processes that lead to the release of salts from plants and animals (Maybeck 2002). TDS levels vary from river to river, but even the highest river values are well below those in the oceans. Rivers with larger discharges, such as the Amazon and the Zaire, tend to have lower TDS values – about 50 ppm on average – while those flowing through arid zones – such as the Colorado, with a TDS of 700 ppm – tend to have higher values. Among the most common dissolved solids are the

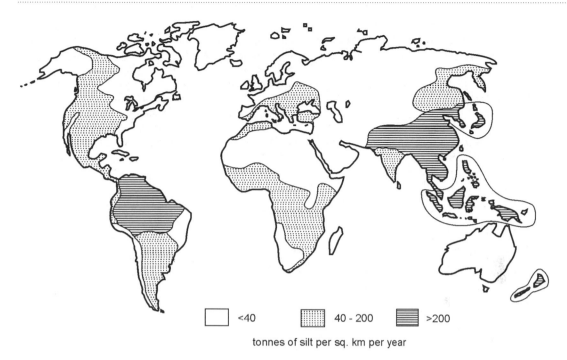

Figure 2.21   Suspended sediment transport in some of the world's major river basins. See Figure 2.20 for the names of the major rivers
Source: E.D. Enger and B.F. Smith, *Environmental Science* (8th edn), Dubuque IA: Wm. C. Brown (2002), with permission of the McGraw-Hill Companies.

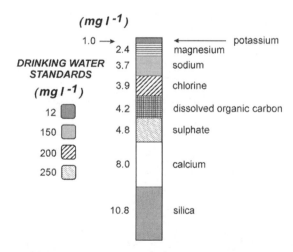

Figure 2.22   The average concentration of dissolved solids in river water
Source: based on data in Maybeck (2002).

salts of calcium and magnesium, derived from the chemical weathering of limestone rocks, but nitrates, sulphates and chlorides are also found, along with metallic ions, such as iron, manganese and sodium. Although all of these occur naturally,

human activities have added to the level of TDS in many rivers. Waste water from industrial activities often contains high levels of dissolved salts; the leaching of fertilizer added to soils carries nitrates into rivers; the removal of irrigation water from rivers in hot, arid areas raises salinity levels and the return flow of the water brings with it nitrate and chloride salts. Because TDS levels can have an impact on water use, they are used in the establishment of water quality standards. Although they are considered mainly in human terms – suitability of water for drinking, for example – water quality standards involving TDS levels are also applied to agriculture, fisheries and certain industries.

ENVIRONMENTAL ISSUES IN THE HYDROSPHERE

Water is a renewable resource, with its supply being maintained by the hydrologic cycle, which not only replaces the water once it has been used, but also cleans it. The use and abuse of water are growing so rapidly, however, that the hydrologic cycle cannot keep up, and serious problems of quality and quantity have arisen worldwide. The causes and nature of these problems are examined in detail in Chapter 8.

SUMMARY AND CONCLUSION

The abiotic components of the physical environment in the lithosphere, atmosphere and hydrosphere provide the physical framework that supports the various forms of life in the biosphere. Although they can be described as separate entities consisting of land, air and water, these components are intimately linked through a series of complex, dynamic interrelationships that together define the character of the environment in a specific area. Variations in the relationships in time and place create short-term and long-term changes in the environment and contribute to differences in regional physical geography. Human interference in the physical environment has disrupted many of the relationships, creating or aggravating environmental problems that threaten life in the biosphere.

SUGGESTED READING

Cattermole, P. (2000) *Building Planet Earth*, Cambridge: Cambridge University Press. A well illustrated account of the lithosphere and the processes that gave the earth its present form.

Barry, R.G. and Chorley, R. (2000) *Atmosphere, Weather and Climate* (7th edn), London: Routledge. Latest edition of a classic climatology text, it introduces the complexity of the atmosphere at a moderately technical but readable level.

Dingman, S.L. (2002) *Physical Hydrology* (2nd edn), Englewood Cliffs NJ: Prentice Hall. Comprehensive, advanced text, combining qualitative and quantitative consideration of the various sectors of the hydrological cycle.

Thurman, H.V. and Trujilo, A.P. (2001) *Essentials of Oceanography* (7th edn), Englewood Cliffs NJ: Prentice Hall. A relatively non-technical introduction to the physical and biological aspects of oceanography.

QUESTIONS FOR REVISION AND FURTHER STUDY

1   On a blank map of the world plot the distribution of earthquakes and active volcanoes. Compare your results with the distribution of the earth's tectonic plates as indicated in Figure 2.4. What relationships do you see between the plates, earthquakes and volcanoes? How do you explain them?

2   Explain the contribution of (a) carbon dioxide, (b) aerosols, (c) water vapour to the earth's energy budget.

3   Consider the components of the hydrologic cycle – evaporation, precipitation, run-off, groundwater storage – in the area in which you live. How have human activities disrupted the cycle?

3

The Biosphere:
The living environment

After reading this chapter you should be familiar with the following concepts and terms:

abiotic
allelopathy
animal
azonal soil
benthic ecosystem
biological weathering
biome
biosphere
biotic
black earth (chernozem)
broad-leaved trees
brown earth
carnivore
chemical weathering
climax community
coniferous trees
cryosols
deciduous trees
ecosphere
ecosystem
ecotone
ecozone

epiphyte
eutrophication
eutrophic lake
evapotranspiration
evergreen trees
fire climax
food chain
habitat
herbivore
humus
hydrophyte
intrazonal soils
laterite
leaching
mesophyte
needle-leaved trees
nekton
niche
oligotrophic lake
omnivore
osmosis
pampas

pelagic ecosystem
permafrost
photosynthesis
physical weathering
phytoplankton
plankton
plant
plant succession
podzol
prairie
regolith
soil – classification, horizon,
    profile, structure, texture
steppe
stomata
taiga
transpiration
trophic level
xerophyte
zonal soil
zooplankton

The biosphere, or ecosphere as it is sometimes called, is the living environment. It is not a discrete entity in the way that the other spheres are, rather it is an interactive layer, incorporating elements of the lithosphere, atmosphere and hydrosphere, which combine in a variety of complex ways to maintain life in the system (Figure 2.1). Thus the plants and animals and other less visible constituents of the biosphere are secondary environmental elements dependent upon interaction with the primary components – lithosphere, atmosphere and hydrosphere – for their continued existence (Box 3.1). Without the appropriate mix of gases in the atmosphere or the water in the hydrosphere, the earth's flora and fauna could not survive. Because it represents the results of a wide range of interactions, the biosphere is the most complex of all the spheres that make up the global environment, and perhaps as a result it often provides the first indications of environmental change.

SOIL

Supplying the essential link between the abiotic (non-living) and biotic (living) groups in the

BOX 3.1 BIOGEOCHEMICAL CYCLES

Essential to the successful operation of a closed material flow system such as the earth/atmosphere system are procedures that allow an element to be used and reused an infinite number of times. This is accomplished through a biogeochemical cycle in which an element moves between sources and sinks along well established pathways. To facilitate progress through the cycle, it may change state – from a liquid to gas, for example – often in combination with other chemicals and involve both organic and inorganic phases. Although any given cycle is ultimately balanced, flow-through is normally uneven. Quantities of an element may be shunted out of the cycle for periods of time. Carbon and phosphorus, for example, are regularly stored in the oceans, perhaps for years at a time, before becoming active in their respective cycles again. Similarly, fossil fuel deposits represent carbon that has not been directly involved in the carbon cycle for millions of years. Several current global issues involve disruption of biogeochemical cycles. Enhancement of the greenhouse effect has come about as a result of inadvertent human interference in the carbon cycle (see Chapter 11), and problems with acid rain can be traced to the loading of the atmospheric sector of the sulphur cycle with additional sulphur dioxide (see Chapter 10).

THE NITROGEN CYCLE

Nitrogen is necessary for the synthesis of the complex nitrogen compounds found in all forms of life on earth. The main nitrogen reservoir is the atmosphere, but gaseous nitrogen is relatively inactive, and the bulk of the nitrogen being cycled through the earth/atmosphere system is held in the organic and inorganic nitrogen compounds present in the soil and in living organisms (Figure 2.5). Free nitrogen is brought into the system from time to time during thunderstorms or through the activities of nitrogen-fixing bacteria in the soil. Plants absorb nitrogen mainly through inorganic compounds such as nitrates, and convert them into more complex compounds such as amino-acids or proteins. When the plants die, bacteria change the organic compounds in the vegetable matter back into nitrates that can enter the loop again, although recent studies have shown that some forests lose nitrogen as dissolved organic nitrogen rather than in an inorganic form. Some of the nitrates may be subject to denitrification, resulting in the release of nitrogen gas or nitrous oxide (N_2O) into the atmosphere. When animals eat plants they absorb nitrogen compounds which remain in their bodies until they die. The compounds are then released into the environment again to be converted into inorganic products such as nitrates and nitrites or perhaps returned to the atmosphere as nitrogen gas. Nitrogenous waste, containing ammonia (NH_3) and urea ($CO(NH_2)_2$) excreted by animals, is also involved in the cycle. Gaseous nitrogen is reintroduced into the main part of the cycle through the activities of nitrogen-fixing bacteria. Thus the nitrogen cycle is not a simple cycle, but rather a complex group of integrated loops.

Human activity has disrupted the volumes involved in the cycle through agricultural and industrial practices. Nitrogen fertilizers, for example, add to the quantities of nitrate available in the soil; the growing number of people and animals in the world increases the volume of nitrogenous waste produced; industrial activities cause the formation of oxides of nitrogen (NO_x), that bring additional atmospheric nitrogen into the system. This has resulted in the creation or augmentation of a range of environmental problems – from eutrophication to air pollution and ozone depletion – that involve nitrogen and its compounds.

For more information see:

Bacon, P.E. (ed.) (1995) *Nitrogen Fertilization in the Environment*, New York: Dekker.

Bolin, B. and Cook, R.B. (1983) *The Major Biochemical Cycles and their Interactions*, New York: Wiley.

Perakis, S.S. and Hedin, L.O. (2002) 'Nitrogen loss from unpolluted South American forests mainly via dissolved organic compounds', *Nature* 415 (6870): 416–19.

environment is the soil. It depends for its existence on the various landscape, atmospheric and hydrologic processes that bring about the breakdown of solid rock and in turn it provides suitable conditions for the growth of vegetation. The vegetation then provides the necessary food and shelter for the animal community. Plants and animals return organic material to the soil, firmly establishing its role in the biosphere.

Soil is basic to the survival of society. The food supply of all animals – including humans – comes ultimately from green plants, and these plants draw their life from the soil. When measured on a human scale, soil-forming processes are slow and the earth's supply of soil is therefore essentially finite. Demands on the soil are increasing constantly, with the result that in some areas its ability to provide an adequate food supply has been reduced or even lost completely. Concern for the survival of the soil should be high, for, in the simplest possible terms, if the soil fails, the links between the biosphere and the other spheres will be lost, plant and animal life will be diminished and the consequences for society will be serious.

Soil is a mixture of mineral matter, organic matter, air and water capable of supporting plant growth (Figure 3.1). Its composition varies with time, place and use. The mineral particles differ in size and chemical composition, depending upon the original bedrock source or the nature and extent of the breakdown or weathering of the rock. Organic matter may be dead or decaying plant and animal remains, but it also includes the roots of growing plants, macro-organisms such as earthworms and micro-organisms such as bacteria. The air in the pore spaces between the mineral particles includes the usual atmospheric gases, but it may also contain gases produced during biological and chemical activity. Similarly, soil water is never pure, but contains a variety of chemicals in solution. A mature soil is a complex, dynamic medium with characteristics that reflect the relative proportions of these various elements and the interactions that take place among them (Coleman and Crossley 1996).

SOIL FORMATION

Physical weathering

The formation of soil is initiated by physical, chemical and biological processes that work together to cause the disintegration or weathering of exposed rock (Figure 3.2) (Paton *et al.* 1995). Physical weathering includes freeze/thaw action, salt wedging and thermal disintegration. When water freezes it expands and, if it is confined in some way, that expansion can exert considerable pressure on its surroundings. Water that freezes in a glass bottle or a pipe, for example, can exert sufficient pressure to burst the bottle or crack the pipe. Similarly, in the natural environment, water freezing in cracks in a rock can cause the cracks to widen or cause pieces of the rock to break off. Over time, after many cycles of freezing and thawing, the original surface of the rock becomes a layer of debris ranging in size from large boulders to small rock particles. Also referred to as frost wedging or frost shattering, this form of weathering is common in mid to high latitudes or in mountainous areas

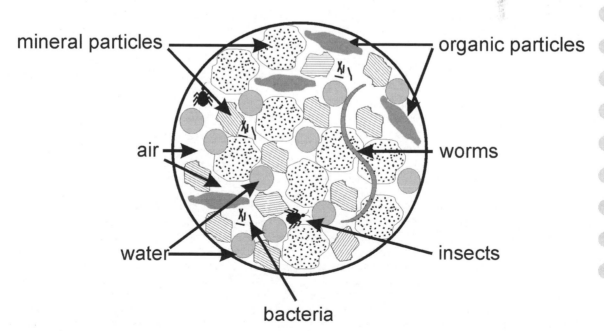

Figure 3.1 The common constituents of soil

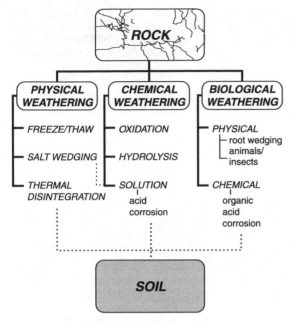

Figure 3.2   The processes involved in the creation of soil

integrity of the rock. This type of weathering is usually associated with desert regions or high altitudes where solar radiation levels are high, but wherever it occurs it is a very slow process.

Chemical weathering

Chemical weathering occurs as a result of chemical processes such as oxidation and hydrolysis, in which rock-forming minerals react with oxygen and water to create new compounds that are usually softer, more bulky and less stable than the originals (Kittrick 1986). Iron compounds, which are very common in the earth's crust, are particularly prone to oxidation. The red or brown iron oxides that result are softer than the original compounds, weakening the rocks and leading to their breakdown. Similar effects occur with the oxidation of aluminium compounds, also a common constituent of the earth's rocks. The red, yellow and orange soils of tropical regions owe their coloration to the oxidation of iron and aluminium compounds.

Hydrolysis is also a major contributor to chemical weathering. It occurs when one or more of the elements in a compound combines with water to form a new compound, causing the original to be destroyed. Igneous rocks contain a range of minerals, some of which are susceptible to hydrolysis and others that are more resistant to the process. Silicate minerals such as feldspars, for example, are hydrolysed to clay, whereas quartz remains relatively unchanged. The net result is the destruction of the rock and the formation of fine clay particles and coarser quartz particles that will ultimately contribute to the texture of the soil.

Other chemical processes work on specific minerals or compounds to cause weathering. Minerals such as rock salt can be dissolved directly by water, but more frequently solution comes about as a result of acid in the water. Rain is frequently a weak solution of carbonic acid, which reacts with such elements as calcium, magnesium and potassium to form carbonates. They take up more space than the original constituents of the rock, and therefore exert pressure on adjacent minerals. These carbonates are also more susceptible to solution weathering. Calcium carbonate, for example, reacts with the carbonic acid in water to form calcium bicarbonate, which is readily soluble in water. Its removal contributes to further deterioration of the rock.

where daily and seasonal temperature fluctuations initiate the freeze/thaw cycle.

Salt wedging works in a similar fashion, with the formation of salt crystals during the evaporation of water containing dissolved salts. The salt crystals have the same impact as ice, widening cracks in the rock and causing particles to split apart, although the process is less effective than frost wedging. It is most common in arid or semi-arid areas when capillary action draws water containing relatively high levels of dissolved salts to the surface, where it evaporates, leaving the salt crystals behind. It is not effective in humid regions where the net movement of water is downward from the surface and any crystals that do form are readily dissolved again before they can damage the rock.

Direct thermal disintegration may occur in areas where the rocks are exposed to high levels of solar radiation and where the diurnal temperature range is considerable. Daytime heating of the rock will cause the surface to expand while overnight cooling will cause it to contract again. Expansion and contraction take place in a thin layer close to the surface of the rock and eventually the stresses created will cause the layer to crack and separate from the rock beneath it. In rocks that are composed of a variety of minerals, differential expansion and contraction between minerals will also destroy the

The form and rate of chemical weathering vary with such factors as rock type and climatic conditions. Igneous rocks rich in silicates may weather relatively rapidly, whereas sandstones in which the main constituent is quartz will be much more resistant. Whatever the rock type, however, chemical weathering takes place more rapidly under hot conditions than under cool and more rapidly under moist conditions than under dry. Thus chemical weathering is always more effective in the hot, wet tropics than in cold, dry regions such as the Arctic.

Biological weathering

Biological weathering includes processes that may be either physical or chemical in nature. Plant roots growing in cracks help to wedge the rock apart. Tree roots provide the most obvious example, but even small and relatively primitive plants such as lichens can make a contribution through their growing rootlets. As weathering proceeds and the rock particles become finer, the contribution of growing plants increases and burrowing animals and insects become involved. During their life cycle plants also produce a variety of chemicals that promote chemical weathering. Lichens, for example, secrete chemicals that allow them to absorb chemicals from the rock surfaces on which they grow. When plants die the decay that follows often liberates organic acids that also contribute to chemical weathering.

The basic result of weathering is the breakdown of solid rock into smaller individual pieces, creating a layer of loose, inorganic material called regolith overlying the unaltered rock underneath. Given time, the layer of weathered material thickens and the particles in the upper part of the regolith become gradually finer, providing the physical base for the formation of the soil. Additional ingredients are needed before that can happen, however. Nutrients released during weathering support the growth of plants and they in turn provide the necessary organic material for the developing soil. Animals contribute organic waste, while earthworms and insects help to mix the components. The product of all this activity is a mixed layer of organic and inorganic material, air and water that has developed a characteristic texture and structure (see Box 3.2, Soil texture and structure) (Figures 3.3–4) and is

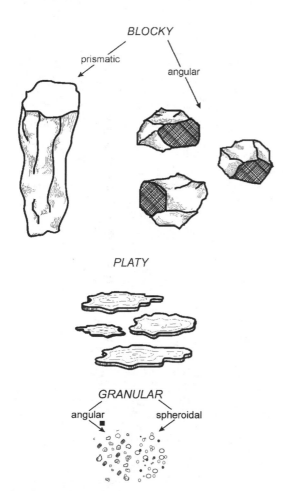

Figure 3.3   Common structural forms in soil

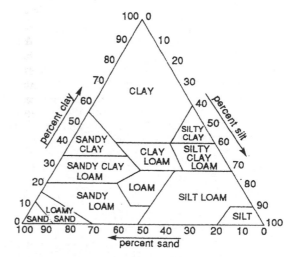

Figure 3.4   A soil texture diagram

BOX 3.2   SOIL TEXTURE AND STRUCTURE

Soil texture and structure describe the physical characteristics of individual soils, according to the size and type of mineral particles they contain and the manner in which they are arranged in the soil.

SOIL TEXTURE

Soil texture is a measure of the proportion of sand, silt and clay particles in the soil (Figure 3.4). Sand particles are the largest, ranging between 2.0 mm and 0.06 mm in diameter, silt particles are between 0.06 mm and 0.002 mm in diameter, and clay particles are smallest, being less than 0.002 mm in diameter. The mixture may be heavily weighted towards one particular size fraction – almost all sand or all clay, for example – or evenly proportioned with a balance of each size. The proportions of the various particles in a specific soil can provide information on other characteristics of the soil. A sandy soil, for example, has abundant interconnected pore spaces, which means that it is highly permeable and readily transmits any water that enters it. As a result, sandy soils are susceptible to leaching, particularly in humid regions, where the net downward flow of water through the soil profile carries nutrients with it beyond the reach of the plants that need them. In contrast, clay particles are smaller and have a platy structure that allows them to pack closely together, creating smaller pore spaces. Since the pores are poorly interconnected, permeability is low and once the water gets into clay soil it tends to remain. Although this helps to keep nutrients in the soil, it can also contribute to poor drainage, which may need to be dealt with before the agricultural potential of the soil can be realized. Important constituents of the clay sector are the clay colloids. These are small flakes of clay, less than 0.001 mm in diameter, that carry negative electrical charges on their surface, allowing them to attract nutrients such as calcium, potassium and ammonium, which are present in the soil in the form of positively charged particles called ions. Attached to the colloids, they are not leached out of the soil profile by percolating water and remain available as nutrients for growing plants. In this, the clay colloids work with small particles of humus, which have similar properties. Together they form the clay–humus complex. The clay and humic colloids help to maintain the fertility of the soil and improve its water capacity, through their ability to absorb moisture.

From an agricultural point of view, the best soils are those with a balanced texture, containing a mix of all three fractions, so that they are easy to cultivate but retain sufficient moisture for plant growth and are not subject to undue leaching. Such soils are called loams. In working any soil, considerable time and effort must be invested in maintaining the benefits of a good textural mix and reducing the impact of a poor one. If that is not done the quality of the soil is reduced and it may eventually deteriorate to such an extent that it becomes subject to soil erosion.

SOIL STRUCTURE

Individual soil particles seldom exist separately in the soil. Sand, silt and clay combine to form aggregates, which vary in size, shape and composition and together give structure to the soil (Figure 3.3). Although there is no widely accepted classification of structural types, there are a small number of common forms. Blocky structures, with aggregates that have flat faces and relatively sharp edges or corners, are common in soils that have high proportions of clay. Platy structures are also common in clay-rich soils. They contain aggregates that are relatively thin and flat, elongated horizontally. The aggregates in soils that have a prismatic structure share the relatively sharp edges and flat faces of blocky aggregates, but they are longer, reaching vertically down into the soil profile. Aggregates that are more or less rounded are said to have a granular or crumb structure. These are common in loams and often seen as providing a good balance between aeration and drainage as well as being easier to cultivate than the other forms. The durability of the aggregates is important in the maintenance of a good structure. Aggregation is usually better developed, and the aggregates are more durable in moist soils than in dry, and in soils that have a good supply of organic matter, such as humus, and nutrients. Humus helps to retain moisture in the soil while the nutrients by bonding with clay and humus colloids help to hold the aggregates together. Many common methods of cultivation are designed to produce a structure that will improve soil drainage, aeration and availability of plant nutrients. Harrowing freshly ploughed fields, for example, will break down large aggregates into smaller pieces, increasing pore space for improved water retention. Autumn ploughing allows large clods of soil to be exposed to frost to be broken down naturally. The addition of lime to clay soils causes individual clay particles to flocculate, or combine into larger units, with pore spaces that improve

BOX 3.2 – continued

moisture percolation and allow easier root penetration for plants. If cultivation techniques are poor, however, particularly if they lead to low levels of organic matter and nutrients, the structure of a soil is more easily broken down and the soil becomes susceptible to erosion.

For more information see:

Brady, N.C. (1990) *The Nature and Properties of Soil*, New York/London: Macmillan/Collier Macmillan.
Strahler, A.H. and Strahler, A.N. (2003) *Introducing Physical Geography*, New York: Wiley.

capable of providing ongoing support for a community of flora and fauna – in short, a soil. The physical, chemical and biotic activities do not end when a recognizable soil has been formed. They continue to renew and revitalize the soil, maintaining nutrient levels and allowing it to adjust to natural environmental change. It is this dynamism that makes the soil strong, and when it is lost – through human interference, for example – the soil becomes vulnerable to decay and destruction.

The time that it takes for a soil to form varies according to the rate at which the various soil-forming processes work and the way in which they interact with each other. Overall the rate is slow, however, with little soil formation apparent in a human life span. It may take several centuries for a few centimetres of soil to develop on a newly exposed surface. In contrast, it is possible for a soil to be destroyed or degraded in only a few years.

SOIL CHARACTERISTICS

Soils reflect the geological, climatological, topographical and biotic conditions under which they were formed (see Box 3.3, Soil profiles) (Figure 3.5). Characteristics such as colour, texture, structure, depth and profile, which arise out of the interaction of these factors, are used by soil scientists in the classification of soil types, but, individually and together, they can also provide a considerable amount of information about the ability of a particular soil to perform its role as a basic resource – in agriculture or forestry, for example. They can give an insight into the inherent fertility of the soil, the use to which it is most suited, the methods required to maintain or improve fertility and any misuse to which the soil may have been subjected (Singer and Munns 1991).

Geological conditions play an important part in developing the original character of the soil. The colour of a soil is often that of the parent rock or indicative of its chemical composition. Iron-rich rocks tend to produce red or brown soils, for example. The chemical composition of the rock is also carried through into the soil in other ways. The rocks of the Canadian Shield are chemically acidic and therefore produce acid soils whereas in areas such as southern England, underlain by limestone and chalk rich in calcium, the soils are low in acidity. This in turn will influence the vegetation that can grow in these areas and ultimately the type of agriculture possible. The parent rock can also influence soil texture – soils derived from sandstone will have a coarse texture, whereas those derived from shale will be much finer. Where the parent material from which the soil is derived is unconsolidated sediment the texture may be quite homogeneous. Lacustrine sediments or those deposited

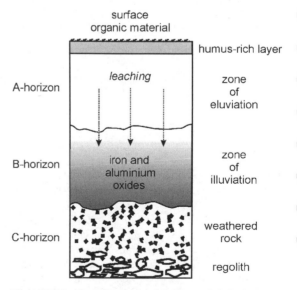

Figure 3.5 A soil profile showing the characteristic horizons of a podzol

BOX 3.3 SOIL PROFILES

Mature soils have a profile that consists of a series of distinct layers or horizons reaching from the surface to the solid rock beneath (Figure 3.5). Soil profiles are used extensively in the development of soil classifications. Soils are differentiated according to physical, chemical and biological properties, which give each horizon a characteristic colour, texture or structure. Soil horizons are commonly designated A, B or C. The A horizon is the uppermost, containing humus and other organic matter, but usually showing evidence of leaching or eluviation, particularly in its lower levels. Beneath it, the B horizon is a zone of illuviation or deposition for the minerals, such as iron and aluminium oxides, and clay particles carried down from the A horizon. The C horizon consists of the parent material of the soil, usually in the form of weathered rock particles that grade down into the regolith. In soils where there is a distinct organic layer at the surface, the letter O is used to designate an organic horizon.

The different horizons vary from soil to soil, and the full sequence may not always be present. Profiles with distinct horizons are typically well developed in humid mid-latitude soils such as Podzols or Spodosols, but may be completely absent in tropical and subtropical Oxisols. Differences occur within horizons also. The leaching that is characteristic of the A horizon, for example, increases with depth, and the upper part of the horizon will contain more organic matter than the lower. Where there is a distinct organic layer in the A horizon, including partially altered vegetable matter, it is designated as A_0 whereas the lower layer or layers will be given other designations – A_1 or A_2, for example. In soils where the lower part of the A horizon is particularly strongly leached, a separate E (for eluviation) horizon may be designated as part of the profile. Similarly in the B horizon the level of illuviation can

vary. In some cases the iron and aluminium oxides may be evenly distributed through the horizon, but they can also form a distinct hardpan, which is increasingly impenetrable by water and therefore disrupts drainage.

Human activities involving agriculture, forestry and industry can alter soil profiles. Although slow but steady change is a natural part of the evolution and maintenance of a soil profile, changes introduced by human activities take place more rapidly and tend to be negative. Cropping, which involves the removal of vegetation, for example, reduces the amount of organic material being returned to the soil and changes the nature of the A horizon. Changes in drainage associated with agriculture or industry can alter the amount and movement of water in the soil profile. The felling of trees will reduce the amount of organic material normally returned to the soil, while the greater exposure of the soil surface to precipitation will encourage increased leaching. The greatest impact is on the A horizon, which may be changed to such an extent that it can no longer support crops or even the natural vegetation characteristic of a particular profile. In severe cases where the integrity of the horizon is lost it becomes subject to erosion. Such disruption of natural soil profiles has become so prevalent that some soil classifications include a designation – anthroposols – in which the soil horizons owe their characteristics to human intervention in the natural soil profile.

For more information see:

Paton, T.R., Humphreys, G.S. and Mitchell, P.B. (1995) *Soils: A New Global View*, London: University College of London Press.
Strahler, A.H. and Strahler, A.N. (2003) *Introducing Physical Geography*, New York: Wiley.

by the wind, for example, are composed of fine silts or clays with very little from the coarser fractions. Where soils have been cultivated for a long time, and have been subject to erosion or have been treated with chemical or organic fertilizers, the relative importance of the geological factor is reduced.

The climatology of the region in which a soil is being formed has an important impact on the processes involved and on the final product (McKnight and Hess 2000). Temperature and moisture are the most significant elements, together controlling the nature of the processes and the rate at which

they take place. Soils form most rapidly when temperatures are high and moisture is abundant, mainly because chemical weathering, organic decomposition and plant growth are fastest under these conditions. Thus soils in tropical regions develop quickly and are often deep. In contrast, desert soils and those in cold regions are slow to build up and are usually quite thin. The role of water in the formation of soil and in its ongoing development is particularly important. It provides the solution in which chemical reactions take place, initiates weathering through such processes as hydrolysis and supplies the moisture needed by the organisms that participate in weathering. Once the soil has begun to form, the movement of water in the profile brings about the redistribution of materials. In humid areas, where precipitation generally exceeds evapotranspiration, the net movement of water is downwards under the effects of gravity. It takes with it chemicals and fine soil particles in a process called leaching. This is common in mid-latitude maritime areas with a temperate humid climate and in tropical regions with high precipitation. In those areas where precipitation is limited and evapotranspiration high, such as the semi-arid plains of North America, Russia and Argentina, water moves up through the soil, bringing chemicals with it. The movement of water and the material it carries has important implications for soil formation and fertility.

The impact of topography on soil formation is a function of slope and the influence of slope on drainage. When soils are forming the normal agents of landscape erosion, such as water, wind and gravity, continue to remove or rearrange material at the surface. By creating smaller, unconsolidated particles, soil formation actually encourages erosion, but as long as soil is being formed more rapidly at its base by the continued weathering of the regolith than it is being removed by erosion at the surface, the soil profile will deepen. Since erosion is more rapid and effective on slopes, the soils there tend to be thinner and less mature than those on flat land. Topography plays a part in soil drainage also. Soils on slopes tend to be well drained, in contrast to the poorly drained soils common on valley bottoms or other flat land. Human activities that expose soil on slopes to increased erosion or disrupt drainage increase the effects of topography on the characteristics of soils.

Biotic processes involve the activities of plants and animals. Both contribute to soil formation and the effects of their activities are apparent in the soils that develop. This is particularly true of plants, which are incorporated in the soil when they die, in the form of humus. Humus is partially decomposed organic matter, dark brown in colour, and an essential component of fertile soil. It aids the retention of soil moisture, helps to bind soil particles together and bonds with nutrients, helping to keep them in the upper layers of the soil where they are easily accessible to growing plants. Soils that are rich in humus are less likely to dry out and less susceptible to leaching, and as a result are able to retain their fertility. The black-earth soils of mid-western North America and Ukraine, which owe their dark colour to the presence of abundant humus, were among the most fertile in the world when they were first brought under cultivation.

The impact of animals is less obvious, but the nutrients that they contain are also returned to the soil in death, and during life they return organic waste as part of their metabolic cycle. Animal waste is an important constituent of the organic fraction in some soils, such as those of tropical and temperate grasslands, which support large herds of grazing animals. At the lowest end of the biological scale are the micro-organisms. Along with algae and fungi, bacteria make a major contribution to soil development through their ability to bring about the decay of dead plant and animal material, leading to the return of nutrients to the soil in a form usable by plants. Agricultural practices that reduce or prevent the return of organic material to the soil change its character. The reduction in humus levels leads to reduced fertility and retards the ability of the soil to retain moisture during periods of reduced moisture availability. The traditionally fertile black earth soils have suffered this fate, but have survived, whereas the brown earths, characteristic of drier areas and with less humus – hence their brown coloration rather than black – have experienced serious erosion.

SOIL CLASSIFICATION

By examining the characteristics of mature soils, soil scientists have been able to classify them (Strahler and Strahler 1992). A simple classification involved grouping soils according to their relationship with the environmental situation in which they occurred. Climate was seen as central to this type of classification, since it had a major role in influencing weathering, moisture availability and the types of flora and fauna that inhabited a region. Zonal soils, for example, had characteristics that reflected regional climatic conditions; intrazonal soils were not typical of the climate zones in which they occurred, because of the presence of overriding local factors such as geology or drainage; azonal soils were immature and not yet in balance with their environment, although expected to develop into one of the other two types.

The first of the modern soil classifications was developed in the United States in the 1930s. Based on earlier Russian models, it divided zonal soils into pedalfers – soils in which aluminium (Al) and iron (Fe) accumulate – and pedocals – soils in which calcium (Ca) accumulates. Pedalfers occur mainly in humid regions, whereas pedocals are soils of sub-humid, semi-arid and arid climates. Although the system has been superseded, several of the great soil groups identified in the Russian–American classification are well entrenched in the geographical and environmental literature. They include the iron-rich laterites of the tropics, the grey-brown leached podzols of humid temperate latitudes and the chernozems or black earths of the sub-humid grasslands. In 1975 the US Department of Agriculture produced a new, more complex classification, initially referred to as the Seventh Approximation for Soil Classification, because of the number of revisions necessary, but now more commonly designated as the Comprehensive Soil Classification System (CSCS) (Strahler and Strahler 1992). The CSCS is based on ten soil orders (Figure 3.6), which are further divided into sub-orders and great groups to provide a world-scale classification. Sub-groups, families and series, which apply only to the United States, are also included. The names of the orders are intended to provide an indication of the nature of the soil. Aridisols are dry or desert soils, for example, and Oxisols contain high levels of oxides, mainly iron and aluminium. Sub-orders' names reflect specific characteristics of the soil or its environment. For example, soils in the sub-order Humox are oxide-rich soils with humus in the upper horizon.

The CSCS has been widely accepted as a world-scale classification, but, to be useful at the regional level, it requires some modification. The Australian and Canadian soil classifications, for example, are generally compatible with the CSCS, although names and specifications have been modified in many cases to reflect their own regional conditions (see Table 3.1). The Australian classification, based on specific soil data derived from the arrangement of soil horizons in an exposed profile, has fourteen orders (CSIRO 1983). It is based on what is actually there rather than what might be expected from the geographic attributes of a region. As a result, the Australian classification includes an order of Anthroposols – soils resulting from human activities. The Canadian soil classification system has incorporated some elements of the US system, but being concerned with soils in higher latitudes it has no need to include the tropical and subtropical

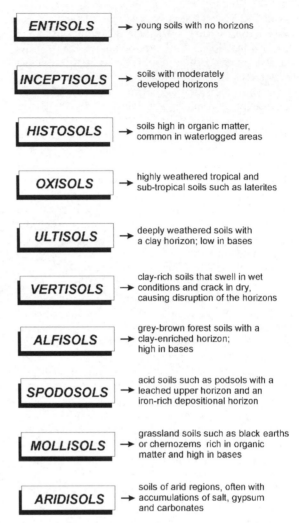

Figure 3.6 **Soil orders of the Comprehensive Soil Classification System (CSCS)**

elements of the CSCS or any other classification. Instead it gives greater attention to cold-region soils such as the cryosols of the far north, which are underlain by permafrost.

Soil classification systems introduce order into an extremely complex and dynamic sector of the biosphere. From a human point of view, an understanding of the characteristics of specific soil groups allows decisions to be made on the type of activity that an area of land can support. This concept of land capability is usually taken as a measure of the potential productivity of the land for agricultural purposes, and includes a range of capabilities, from land with few limitations on

TABLE 3.1 COMPARISON OF SOIL CLASSIFICATIONS IN THE UNITED STATES, CANADA AND AUSTRALIA

United States (CSCS)[a]	Canada	Australia
Entisols	Regosols	Rudosols
Inceptisols		Tenosols
Histosols	Gleysols	Organosols, hydrosols
Oxisols		Ferrosols
Ultisols		Kurosols
Alfisols	Luvisols, brunisols	Ferrosols, chromosols, kandosols, kurosols, sodosols
Spodosols	Podsols	Podosols
Mollisols	Chernozems	Dermosols
Aridisols	Solonets	Calcarosols, sodosols
	Cryosols underlain by permafrost	Anthroposols created by human activities

Source: based on data in A.H. Strahler and A.N. Strahler, *Modern Physical Geography,* New York: Wiley (1992); P. Dearden and B. Mitchell, *Environmental Change and Challenge: A Canadian Perspective,* Toronto: Oxford University Press (1998); CSIRO, *Soils: An Australian Viewpoint,* Melbourne: Commonwealth Scientific and Industrial Research Organization and Academic Press (1983). *Note:* [a] Comprehensive Soil Classification System.

its agricultural use to that which is best left in its natural state. Land capability can be considered in its broadest sense at the soil order level. The highly organic, base-rich Mollisols of the CSCS, for example, are most capable of supporting productive agriculture, whereas the highly weathered tropical and subtropical Oxisols, subject to rapid leaching and oxidation, quickly lose their productive capacity. Within these major orders, however, there are infinite variations, reaching down to the single farm or even field level, that need to be considered if the best use is to be made of the land and the attributes of the soil preserved (Strahler and Strahler 1992). In practical terms, that depends on the farmer who puts his spade or plough into the land, the forester who clears the land of trees or the developer who removes the local ground cover to excavate a mine or enlarge a city. If agricultural activities are carried out with little or no regard for the characteristics of the soil, even the most fertile Mollisols will be degraded and become incapable of providing the returns expected or required by the farmers. Following the clearing of land to produce wood or minerals, the natural character of the soil conditions is also changed, and often the natural character of the soil can no longer be maintained. Thus, in many parts of the world, the balance among such soil-forming factors as geology, climate and biology has been disturbed, and the natural soils represented in soil classifications either no longer exist or are much modified. Classifications

need to be sufficiently flexible to deal with such changes, but the consequences for society are much more serious than amendments to lists or maps, involving as they do questions of soil erosion, food production, water supply and a range of related environmental issues.

PLANTS AND ANIMALS IN THE BIOSPHERE

PLANTS

Plant life varies remarkably in size and complexity, from microscopic phytoplankton floating in the oceans to the drought-resistant cacti of the deserts and the giant redwoods of California. Despite such a range of forms, however, plants have important characteristics in common. Most terrestrial plants, for example, are anchored in the soil by their roots, which provide stability and allow them to absorb the moisture and nutrients that they need for growth. Perhaps their most important characteristic is their ability to promote photosynthesis (Box 3.4). Radiant energy from the sun is absorbed by the leaves of the plant, converted into chemical energy in the form of carbohydrates and then stored in the plant tissue. This ability of plants to produce food energy is essential to animal life on earth. Unable to obtain energy directly from the sun, animals depend upon plants to do it for them, consuming the energy indirectly by eating the plants. Without them animal life would not survive (McKnight and Hess 2000).

As they have evolved, plants have adapted in a variety of ways that allow them to survive and prosper under a range of environmental conditions. Growing in association with each other, groups of plants, with similar physical needs, have created complex communities that are often the most obvious visual feature in the landscape they cover. Although the specific elements involved and the interactions among them are often very complex, the type of natural vegetation that grows in an area reflects the combined effects of climate, water, soil and biotic factors (Figure 3.7).

BOX 3.4 PHOTOSYNTHESIS

Photosynthesis is a biochemical process in which green plants absorb solar radiation and convert it into chemical energy. It is made possible by chlorophyll, the green and yellow pigment that gives plants their colour, contained in specialized structures called chloroplasts located in the cytoplasm of the plant cells. The radiant energy absorbed by the chlorophyll is converted initially into chemical energy in the form of adenosine triphosphate (ATP). That energy is subsequently used in a series of reactions, which bring about the reduction of carbon dioxide to simple sugars. Although complex in detail, photosynthesis can be summarized as a process in which carbon dioxide and water are consumed, carbohydrates are produced and stored, and oxygen is released.

$$6CO_2 + 6H_2O \xrightarrow[\text{chlorophyll}]{\text{light energy}} C_6H_{12}O_6 + 6O_2$$
$$\text{(glucose)}$$

In green plants, the hydrogen for the formation of the carbohydrate is supplied by water, but certain photosynthetic bacteria obtain their hydrogen from hydrogen sulphide and release sulphur as a by-product. Animals depend upon the ability of plants to convert radiant energy into chemical energy in the form of carbohydrates, since they lack the ability to synthesize their food in that way. Thus photosynthesis is the base on which the earth's food supply is built. Photosynthesis also helps to maintain the oxygen/carbon dioxide balance in the atmosphere, but agricultural activities and deforestation have reduced the amount of photosynthesis taking place, allowing levels of atmospheric carbon dioxide to rise and contribute to global warming (see Chapter 11).

For more information see:

Foyer, C.H. (1984) *Photosynthesis*, New York: Wiley.
Schulze, E.D. and Caldwell, M.M. (eds) (1995) *Ecophysiology of Photosynthesis*, Berlin: Springer.

Climate

Climate is the main environmental element responsible for the development and distribution of vegetation (Rumney 1968; although obviously dated in some respects, this volume provides one of the most comprehensive accounts of the relationship between climate and vegetation distribution). The links are so close that in some climate classification systems climate types are described by the vegetation they support. Tropical rain-forest climates or temperate grassland climates are two examples. Although the type of vegetation in any given area will reflect the interplay of the many climatic components of the environment, there are some that make a greater contribution than others. Temperature and precipitation dominate, with secondary elements such as wind, humidity and evapotranspiration resulting from or contributing to the influence of temperature and precipitation. All plant species have optimum heat and moisture requirements, which govern their distribution. Species that have similar requirements grow together in plant associations that have become recognizable as characteristic of certain parts of the earth. The role of temperature is evident in the latitudinal and altitudinal distribution of vegetation groups, for example. Average temperatures decline gradually from the tropics to the poles and as temperatures change so do the vegetation patterns. Boreal forest species that extend across the northern continents in mid to high latitudes tolerate low temperatures that would kill the trees of the tropical forests located adjacent to the equator in South America, Africa and Southeast Asia. Similarly, the cooling associated with increasing altitude ensures that vegetation changes with height in mountainous areas. The role of precipitation is also evident in some areas. Abundant precipitation supports forests whereas lesser amounts support grassland or, if there is little or no precipitation, only the scrubby vegetation of desert areas. Temperature and precipitation do not work independently. It is

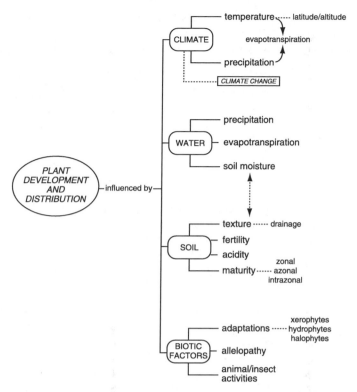

Figure 3.7   Environmental influences on the development and distribution of plants

the combination of high temperature and high precipitation that is responsible for the tropical rain forest and in arid areas vegetation growth is restricted not just by limited precipitation but also by temperatures that cause much of the moisture to be evaporated before it becomes available to the plants. In regions with a distinct cold season low temperatures have an impact on the state and availability of water. When water freezes in the soil it is no longer available to plants and tissue damage from desiccation may occur. The freezing of water in plant tissues also causes plants to die.

In addition to the world-scale climate conditions that influence the distribution of vegetation, there are regional or local conditions that also contribute to variations within plant groupings. These local or regional differences are often associated with topography or hydrology. In mountainous areas, for example, the south-facing side of a valley will support a different type of vegetation from that on a north-facing slope. Differences in the amounts of solar radiation received on the south-facing, sunny slope and the north-facing, shady slope are translated into temperature differences and variations in

moisture regimes, which may be sufficient to influence the vegetation distribution, although the two slopes experience the same general climatic conditions. In northern areas, climatic conditions such as low temperatures, limited precipitation and strong winds are not generally suitable for tree growth, but trees may survive locally north of the tree line in the shelter of gullies or depressions in the land surface where more favourable climatic conditions are present. Variations also occur around large lakes where temperatures are influenced by differences in seasonal energy fluxes between land and water and precipitation totals are higher on the downwind side of the lake, because of the moisture picked up as the winds cross the open water.

The close links between vegetation and climate mean that any change in climate will be followed by a change in the distribution of plant groups. Since plants are generally slow to respond, minor changes that are reversed or ameliorated after only a few years have very little impact. However, during major events, such as the climate changes that produced the Ice Ages, the impact on the distribution of vegetation is significant. During the last Ice Age, for example, vegetation patterns changed significantly. In high latitudes the existing vegetation began to die and ultimately was destroyed by the advancing ice. At the last glacial maximum, some 18,000 years ago, the tundra vegetation that now covers areas north of the Arctic Circle lay along the ice margins in central North America and Europe, some 20° of latitude south of its normal position, replacing the temperate grassland and forests that grew there in non-glacial times. The impact even reached into the tropics, with the rain forests significantly reduced in area and the scrubby desert vegetation much expanded (Gates 1993). As the ice retreated and climatic conditions improved into the Holocene, the vegetation began to migrate polewards again in the northern hemisphere, the deserts contracted and the tropical rain forest spread to cover its former range. The change was not completely smooth, with the overall improvement interspersed with periods of deterioration in which the poleward advance of the plants was halted or reversed, but by some 5000 years ago the current pattern of vegetation distribution had been established. Changes have taken place since then, but not on a scale that compares with those during glacial and postglacial times.

Current concerns about climate change and its impact on vegetation are with anthropogenic climate change. If global warming continues at its predicted rate, it has the potential to cause major change to vegetation patterns. Since the greatest warming is likely to be in high latitudes in the northern hemisphere, the boundaries between the tundra and the boreal forest are likely to feel the greatest effects, with the boreal forest moving northwards at the expense of the tundra. At the same time, the mixed forest and deciduous forest that occupy the areas south of the boreal forest could advance northwards in response to the warming and displace the boreal forest along its southern margins. In terms of vegetation distribution, such changes would be significant, but they also have implications for countries such as Canada and those in Scandinavia that depend upon the products of the boreal forest to contribute to their economic well-being (see Chapter 11).

Water

Water is essential for plant growth (Etherington 1982). Without it, seeds will not germinate; without it, growing plants will wilt and die. The water used by plants is provided initially by precipitation that is held in pore spaces in the soil after percolating down from the surface. It is absorbed from the soil by the plant roots and transported up through the stems, to be transpired from the leaves and returned to the atmosphere. During transpiration, water is lost through the stomata (pores) in the leaves and the cooling that accompanies that process helps to maintain temperatures in and around the plant at tolerable levels. The rate at which transpiration occurs depends upon such factors as air temperature, leaf temperature, relative humidity and wind speed. It is greatest when temperatures are high, the relative humidity is low and winds are strong. As moisture is transpired from the leaves it is replaced by water absorbed through the root system and pushed up through the stems by osmotic pressure (see Box 3.5, Osmosis). Nutrients are also absorbed with the water and carried from the root system through the plant. The continued replacement of water maintains the turgidity of the plant's cells, which allows the plant to maintain its form. If transpiration takes place more rapidly than the moisture can be replaced through the root system, turgidity is lost, wilting takes place, leaves become limp and stems droop. In extreme cases the plants shrivel up and die. This happens most often under drought conditions, when transpiration exceeds precipitation and there is insufficient water stored in the soil to make up the difference. It also happens during the winter months in some areas, when low humidity and strong winds encourage transpiration but the water in the soil is frozen and cannot be absorbed by the roots. Deciduous plants are protected from this through the loss of their leaves during the autumn, but needle-leaf bushes and trees often suffer 'winter kill' – a drying and browning of the leaves – as a result.

Water is also a necessary constituent of photosynthesis, a biochemical process in which green plants absorb solar radiation and convert it into chemical energy (see Box 3.4, Photosynthesis) (Gregory 1989). The conversion is made possible by chlorophyll, the green pigment that gives plants their colour, and involves several stages, but the net result is that water and carbon dioxide are combined to produce carbohydrates and oxygen is released into the atmosphere. Photosynthesis provides an essential link between plants and animals. Not only does it allow plants to create and store food that can be consumed by animals, but it also

BOX 3.5 OSMOSIS

Osmosis is the diffusion of a solvent such as water through a semi-permeable membrane, from a solution of low concentration to one of a higher concentration. The membrane permits the flow of the solvent, but restricts the passage of the solute – chemicals in solution. Restricted by their links to the solute, the solvent molecules in the more concentrated solution are kinetically less active than those in the less con- centrated solution. As a result, they exert less pressure on the membrane than the solvent molecules in the weaker solution, creating a diffusion pressure deficit across the membrane and initiating the flow of the solvent into the more concentrated solution. Ulti- mately, as a result of the solvent flow, the concentrations of the solutions on either side of the membrane will tend to become equal. This can also be expressed as osmotic potential, the flow of solvent through a semi- permeable membrane always being from the solution with the higher osmotic potential (less concentrated solution) to that with the lower osmotic potential (more concentrated solution).

Osmosis, through the semi-permeable membranes that surround individual cells, helps to control the flow of water, the most common solvent in the environment, through living organisms. In plants, for example, the sap that occupies the cells is a solution of salts, sugars and organic acids in water, and a

difference in salinity between adjacent cells initiates osmosis. When transpiration takes place from the plant leaves, the sap in the cells that have lost water becomes more concentrated, creating a pressure deficit that causes water to flow from adjacent cells to replace the amount lost. The exchange of water ultimately reaches the root zone, where the difference in diffusion pressure causes soil water to be absorbed by the root cells. If the water is not replaced, the volume of the sap in the cells remains reduced, they become flaccid and the plant wilts.

Reverse osmosis is used in the desalinization and purification of water. Salt water or contaminated fresh water is pumped through a semi-permeable membrane. The water molecules are allowed to pass through the membrane, whereas the salt and other impurities are retained in the original solution. The process is quite effective, but costly.

For more information see:

Amjad, Z. (ed.) (1993) *Reverse Osmosis: Membrane Technology, Water Chemistry and Industrial Applications*, New York: Van Nostrand Reinhold.
Lachish, U. (1999) 'Osmosis, reverse osmosis and osmotic pressure: what are they?' Viewed at http://urila.tripod.com/ (accessed 16 June 2003).

maintains the appropriate atmospheric environ- ment for animals, by recycling carbon dioxide to release the oxygen they need during respiration.

Plants can be grouped according to their ability to grow under specific moisture regimes. Xero- phytes, for example, are plants adapted to life in arid regions where water is in short supply either permanently or seasonally. Adaptations include long or enlarged roots and fleshy roots and stems that allow as much moisture as possible to be absorbed and stored. Other adaptations are designed to reduce moisture loss by transpiration. These include leaves with a minimum surface area, plus thick skins and sunken stomata. In some cases, xerophytic plants that have grown during a period when moisture was available do not survive the subsequent drought, but leave behind seeds, which remain dormant until moist conditions return and they can germinate. Plants that grow in areas with distinct wet and dry seasons often display xerophytic characteristics – dropping leaves to reduce transpiration, for example – to allow them

to survive the dry season. At the other end of the moisture spectrum are the hydrophytes, which require abundant water to survive. Some like water lilies grow in and on the water, whereas others need soil that contains abundant moisture and an atmos- pheric environment in which the relative humidity is high. Hydrophytes are found along the edges of lakes as well as in swamps, marshes and bogs. In coastal marshes they must also tolerate higher levels of salt in the water. In temperate latitudes the salt marsh hydrophytes are commonly grasses or reeds, but along coasts in the tropics mangroves, a group of salt-resistant hydrophytic trees and shrubs, dom- inate the coastal swamps. Beneath the canopies of the world's rain forests – both tropical and tem- perate – abundant soil moisture and high relative humidity create ideal conditions for a wide range of ferns and mosses.

Comprising a large group of plants between these two extremes are the mesophytes, which can survive neither saturated soils nor very dry conditions. Most of the trees, shrubs and other plants in temperate

Plate 3.1 Water lilies. Growing in or on water, they are perhaps the supreme example of a hydrophytic plant.

regions belong in this group. The range is large, with some plants that can tolerate wetter soils than others or can survive at least a short period of drought. They include trees such as oak, beech and maple, which grow in areas with moderate to high precipitation but where the soil is well drained, or larch and black spruce, which will grow on wetter sites, and species of poplar or aspen with well developed root systems that allow them to survive on drier sites. Much of the natural vegetation in temperate regions has been replaced by cultivated vegetation. The crops chosen must be able to survive and produce in the moisture conditions appropriate for mesophytes. If that is not possible the conditions have to be modified, usually by improving drainage or providing irrigation.

Soil

Soils contribute to the growth of vegetation and its distribution through their physical, chemical and biological properties, which together are called edaphic factors. Their contribution may not always

be obvious, because they often work indirectly through other features. Soil texture makes its influence felt through drainage, for example. A sandy soil with high permeability may support only scrubby, drought-tolerant vegetation even when precipitation totals are moderate to high, because its excellent drainage creates soil moisture deficits. Soils such as clay loams are much more able to retain moisture and support a flourishing vegetation cover. Where the clay content of the soil is particularly high, however, drainage will be poor, causing waterlogging and restricting the plant cover to hydrophytic plants. Chemical characteristics of soils often have a more direct impact on vegetation. Acidity has a particularly strong influence. The short grass cover in areas of chalk bedrock is in large part a reflection of the low acidity of the calcium-rich soils that develop on that rock type. In contrast, the high acidity of soils created from sandstone or granite produces an acid-tolerant plant community that may include coniferous trees or heaths and a variety of mosses in damper sites. Where the soils have a high salt content, from inundation by sea water or the high evaporation rates in arid areas, salt-tolerant or halophytic plants provide the vegetation cover.

Biotic factors

Biotic factors include the influence of the plants themselves and other organisms on the nature of the vegetation patterns that develop in an area. Over millions of years plants have developed mechanisms that allow certain species to survive and prosper in a particular environment. Adaptation to drought, salt-rich or highly acid soils, low temperatures and poor drainage, for example, allows plants with these characteristics to populate areas from which plants that have not adapted are excluded. Some have adjusted to the characteristics of the other plants with which they share an environment. In temperate latitudes, deciduous trees shade the forest floor during the summer when they are in full leaf, thus limiting the types of plants that can grow there to those that require limited sunlight. During the early spring, however, before the trees have leafed out, more light reaches the forest floor, providing a window of opportunity for those plants that can complete their growth cycle quickly. Plants that grow from bulbs, rhizomes or some form of tuber which contain the moisture and nutrients needed for rapid growth have a distinct advantage in such a situation.

Some plants actively discourage harmony, by creating conditions that prevent other plants from competing with them and sharing their space. This is called allelopathy (Brown and Lomolino 1998). Gases given off by one plant may be toxic to others; poisons carried in the leaves and leached into the soil or chemicals exuded from the roots will prevent the growth of competing species. Black walnut and pine trees, sagebrush and sunflowers exhibit allelopathy, allowing them to dominate their communities.

Animals, insects and other small organisms also contribute to patterns of vegetation distribution. By eating tree seedlings, grazing animals are very effective in preventing tree species from regenerating, allowing grass to dominate in areas that would normally support trees. Animals and birds also help in the distribution of plants by carrying seeds with them as they move. Insects contribute to the survival of flora through pollination, but population explosions in certain species can cause major damage to vegetation. Every year in Canada thousands of hectares of boreal forest are destroyed or damaged by spruce budworm infestations (Scott 1995). Insects can also spread disease in combination with other organisms. Dutch elm disease, for example, caused by a fungus that is spread by a beetle, has wiped out the elm in much of Europe and northern North America.

Dominant among the biotic influences on the distribution of vegetation is the human element. Natural plant communities have been replaced completely by cultivated crops, harvested to provide lumber, covered over by concrete and asphalt and decimated by pollution and disease. Thus the present distribution of the earth's flora is not particularly natural. It is important, however, to consider what the natural distributions were or might have been without such interference. The non-human, climatic, hydrologic, edaphic and biotic elements that combine to determine vegetation patterns still exist, although they may no longer create natural patterns of vegetation distribution. They do, however, influence the type of agriculture possible or the rate at which renewable plant resources can be used. Failure to appreciate the role of these elements can lead to the misuse of the land and contribute to serious environmental deterioration.

Plant succession

The grouping of plants at any one site represents a stage in the gradual and sequential development of vegetation patterns associated with the interplay of climate, water, soil and biotic factors in the environment. The process is called succession, and it continues until the various components are in balance with each other (Enger and Smith 2002). The classic approach to succession was developed by Frederic Clements in 1916. He presented succession as a unidirectional process in which one stage inevitably led to another in a relatively predictable sequence. It is perhaps best illustrated by changes in plant communities, but it also applies to other elements such as soil and animals, which are intimately linked with vegetation. In its simplest form, primary succession begins when a community is established on a previously undisturbed site – a lava flow, mud flats exposed by falling water levels or fluvioglacial deposits exposed by a retreating ice sheet, for example (Figure 3.8). When primary succession is allowed to run its course, each invading species is out-competed by its successors, ultimately

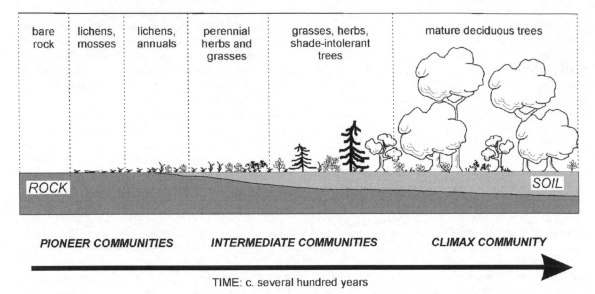

| bare rock | lichens, mosses | lichens, annuals | perennial herbs and grasses | grasses, herbs, shade-intolerant trees | mature deciduous trees |

ROCK   SOIL

PIONEER COMMUNITIES   INTERMEDIATE COMMUNITIES   CLIMAX COMMUNITY

TIME: c. several hundred years

Figure 3.8   An example of primary succession leading to the development of a mature temperate deciduous forest

creating a mature grouping of plants and animals, called a climax community, which owes its composition and structure to prevailing environmental conditions such as soil type and climate. The complete sequence from the initiation of the community, or pioneer stage, to the climax is referred to as a sere, with the various stages in the sequence being seral stages.

The most significant example of primary succession in recent earth history was that following the retreat of the ice sheets some 10,000 years ago (Pielou 1991). Across northern Europe, the northern United States and southern Canada, tundra vegetation colonized the newly exposed land. In time, and with variations caused by local conditions, it was replaced in sequence by increasingly complex communities of shrubs, coniferous forest, mixed forest and eventually a climax community of broad-leaved deciduous forest. In addition to its scale it differed from local or regional plant succession in the extent to which it was driven by long-term climate amelioration. The plant succession that followed the Ice Ages took place over a period of several thousand years. In contrast, primary succession on lava flows in the tropics has been seen to advance from bare rock to tropical forest in as little as half a century.

When viewed at a large scale, the changes in vegetation distribution following the Ice Ages appeared to follow the classic process of plant succession, but regionally the pattern did not always fit, suggesting that the process was more complex than first envisaged (Stiling 1992). Over time ideas on succession were modified. Gleason (1926), for example, introduced a less rigid, less predictable approach in which the individual peculiarities of plants played a greater role in the process, and included the idea of retrogressive succession in which some form of disturbance slows or even reverses progress towards a climax community. Modern approaches to succession include several possible pathways (see, for example, Connell and Slatyer 1977), give greater consideration to disturbances and view the end result as much less predictable. In many semi-arid ecosystems, for example, the vegetation is frequently in a continuing state of flux, driven by disturbances such as drought, fire and insect infestations, and the sequential development towards a climax community cannot occur (Middleton 1999). Despite this, some ecosystems do approach the state in which a dynamic equilibrium exists among the components of the system, creating conditions similar to those expected in a climax community.

Classic climax communities are characterized by their diverse array of species and an ability to use energy and recycle chemicals more efficiently than immature communities. In theory, they are capable of indefinite self-perpetuation, under given climatic and edaphic conditions, but if these conditions change the community will change also. All climax communities can cope with the change characteristic of a dynamic environment – trees in mature forests die and are replaced, for example, and some communities have adapted to annual drought – but most cannot easily deal with major disturbances such as fire, insect infestations, catastrophic weather events or climate change. When the climax community has been disturbed, or the primary succession leading towards that type of community has been interrupted, a new or secondary succession process begins. Climate variations, fire and disease are common natural instigators of secondary succession, but human agricultural and industrial activities can also initiate the process.

Secondary succession will begin on abandoned agricultural land or land cleared by forestry activities, for example. In theory, secondary succession should proceed more rapidly than primary succession, since it does not start from scratch, the soil already being present to receive the pioneering plants and other organisms. A rapid response is not uncommon when the climax community is destroyed by fire. Nutrients from the burned plants are added to the soil and plants which have adapted to deal with fire rapidly colonize the burnt-over area. Grasses, with their growing points beneath the surface, tend to survive more easily than trees and shrubs and are usually the first plants to appear again after the fire. Certain trees, such as jackpine, need high temperatures before they can release their seeds, and although the mature trees are destroyed the seedlings take advantage of the abundant nutrients and light to grow rapidly. Fires are not uncommon in the environment and are now recognized as a natural control on plant communities. Where fires disrupt succession with some regularity, the vegetation is unlikely to form a climax community; rather it remains in a sub-climax state, sometimes called a fire climax (McKnight and Hess 2000).

A different form of secondary succession may be initiated when drainage is disrupted, perhaps by a landslide or through the activities of animals, such as the beaver, which build dams and flood land to create the type of habitat they need. None

Plate 3.2   A beaver dam and pond. Beavers disrupt the existing ecosystem by causing flooding. With time the ponds fill with sediment and secondary succession begins.

of the features formed in that way – swamps or ponds – is permanent. They gradually fill as a result of sediment deposition and plant colonization. With the pre-existing vegetation killed by the flooding or felled by the beavers for food, the secondary succession begins with hydrophytic plants, such as reeds, mosses and sedges, growing around the edges of the water in waterlogged soil. Sediments accumulate around their stems and roots, providing support for more plants. In some cases the mosses and sedges form a floating blanket that gradually covers the surface of the water. Grasses follow to create a wet meadow and, as the sediments gradually replace the water, trees move in. Those, such as willow, that tolerate high levels of moisture are the first colonizers, but gradually the climax community that existed prior to the disruption may be expected to re-establish itself (Enger and Smith 2002).

The changes that take place during the various stages and forms of plant succession include other elements in the environment also. Each climax community has its own population of animals, for example, which has grown along with the plants and responded to the changes associated with succession. Although they are usually less obvious in the landscape, animals are important participants in the various activities, such as energy flow and nutrient recycling, that help to retain balance in the community.

ANIMALS

Like plants, animals exist in a great variety of forms and sizes, from microscopic zooplankton, through more complex insects, worms, fish and birds, to large mammals such as elephants and whales. The largest animals are not always the most important in a community. The human animal, for example, has reached its predominant position among the mammals not because of its size, but because of its adaptability (see Chapter 4). In terms of numbers, the most successful animals are the insects. They reproduce rapidly, but they have also specialized so that they participate at all stages in the development and survival of the community. They help to maintain the community by providing the pollination necessary for plant reproduction, while at the other end of the life cycle they contribute to the processes that lead to the breakdown of dead organic matter, and the return of the nutrients it

contains to the soil to be recycled. Insects belong to a group of animals called invertebrates, or animals lacking backbones. Together with worms, sponges, molluscs, crustaceans and a vast number of microscopic organisms, they account for at least 90 per cent of all animal species and perhaps as much as 97 per cent. The other 3–10 per cent consists of animals with backbones – the vertebrates – which include fish, amphibians, reptiles, birds and mammals (Brown and Lomolino 1998; UNEP 2002).

Because of the great variety of life forms in the animal kingdom, the similarities that might be expected are not necessarily obvious at first sight. A butterfly and an elephant or a sparrow and a whale seem completely unrelated, appear to have little in common, but they do have a number of characteristics which they share and which differentiate them from plants, and other living organisms. They are heterotrophic – dependent on green plants, directly or indirectly, for the food they require to survive. Lacking chlorophyll and therefore incapable of photosynthesis, animals are unable to produce their own food supply from the raw materials – air, water, nutrients and solar radiation – available in the physical environment. Most animals are also motile – capable of spontaneous movement. At a basic level, this allows them the flexibility to find food, water and shelter and to seek out the environmental conditions that are most suited to their needs. Motility is particularly important during times of change in the environment. Seasonal change, for example, leads to the migration of millions of animals, birds and fish every year. In North America alone, as many as 100 million birds migrate annually between their wintering and breeding grounds in spring, and back again in autumn (Figure 3.9). Migration was also characteristic of periods such as the Ice Ages, when the magnitude and time scale of the change were greater. Animals retreated ahead of the advancing ice, but moved back again with the melting of the ice and the return of environmental conditions in which they could survive. Intercontinental migration of people and animals between Asia and North America also occurred at that time.

If global warming continues as forecast, it is likely to be accompanied by animal migration. Already in North America there is some indication that species such as armadillos, opossums and raccoons are extending their ranges northwards, perhaps as a result of the warming that has already

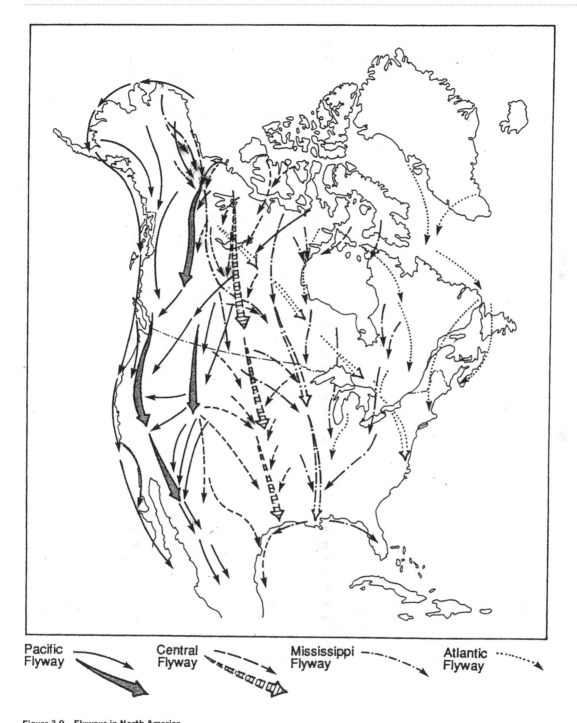

Figure 3.9 Flyways in North America

Source: E.D. Enger and B.F. Smith, *Environmental Science* (8th edn), Dubuque IA: Wm. C. Brown (2002), with permission of the McGraw-Hill Companies.

taken place. Mammals commonly receive most attention when animal migrations are studied, but the migration of some insect species could have even more serious consequences than the spread of the mammals. Insect migration, induced by global warming, has the potential to carry plant and animal diseases into environments where they were previously unknown and where, therefore, the existing organisms have no resistance to them. The environmental and economic consequences for the flora and fauna – natural and managed – could be severe. Vector-borne diseases such as malaria would also spread, with the expansion of the range of the malarial mosquito following global warming, adding to the toll of 1.5 million to 2.7 million deaths caused by the disease annually (Epstein 2000).

Animals respond to physical conditions in the atmosphere such as temperature and precipitation, but the distribution of individual animal species is less closely linked to such factors than the distribution of plants. The boundaries of plant communities can sometimes be tied to a specific isotherm or isohyet, but it is difficult to establish such links for animals, because of their ability to move around in response to short-term changes in local conditions. The winter ranges of animals are often quite different from their distribution at other times of the year, and they may vary from year to year depending upon weather conditions in a particular winter. Harsh winters on the Canadian prairies, for example, commonly caused the herds of bison that were once common to move from the grassland to seek shelter in the adjacent aspen groves. During milder winters, however, they remained out on the open plains. This created food supply problems for the indigenous peoples who inhabited the area and later for the European fur traders who moved in during the eighteenth and nineteenth centuries. The animals were much easier to find in the aspen groves than on the open plains, and as a result maintaining a food supply was often more difficult in mild winters than in the normal cold, snowy winters. Similar variations occur in other areas where the physical conditions differ significantly from season to season. In regions with seasonally variable precipitation, for example, the dry-season and wet-season fauna are often quite distinct.

Through anatomical, physiological and behavioural adaptations, animal species are able to survive remarkably severe environmental conditions

such as those at high latitudes and altitudes or in hot deserts. Mammals have been particularly adept at this. Low temperatures have been dealt with by anatomical adaptations such as the winter coats grown by most high-latitude mammals, by the physiological adaptations associated with hibernation or behavioural adaptations that cause animals to seek shelter or live in colonies to preserve body heat (Stiling 1992). Animals exposed to high temperatures benefit from anatomical characteristics such as large ears that help to dissipate body heat, physiological adaptations that allow increased evaporative cooling and behavioural characteristics that involve sheltering from the sun during the warmest part of the day in a burrow or den and emerging only when temperatures fall during the night.

Other adaptations deal with the moisture element in the environment. The animals that inhabit arid areas could not survive without the adaptations they have evolved to conserve water. Camels, for example, sweat little and excrete only small amounts of water in their urine and faeces. Smaller mammals such as desert rats have similar adaptations. Heat and moisture elements often work together. When temperatures are high and moisture is available, many animals remain comfortable through the cooling that accompanies the evaporation of perspiration from the skin or moisture from the respiratory tract. Evaporation is such an effective cooling mechanism that, if not controlled, in cooler areas it can lead to a loss of body heat sufficiently high to cause hypothermia and threaten life (Oke 1987). To counter that, many animals in high latitudes have developed insulated, waterproof coats that prevent water from saturating their coats or reaching their skin, thus reducing the evaporative heat flow from the body.

The human animal, the most successful of the mammals, has anatomical and physiological adaptations – absence of body hair, efficient perspiration system, upright posture – that are most effective in warm climates, and incorporates a number of self-regulating mechanisms that allow it to maintain the homeostasis, or near steady internal body temperature, required by mammals (Figure 3.10). In addition, through a great range of behavioural and cultural adaptations, from the provision of clothes and shelter to the use of technology to provide direct heating or cooling, it has been able to establish itself almost anywhere on the earth's surface.

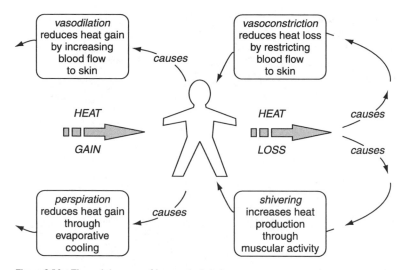

Figure 3.10   The maintenance of homeostasis in humans

In many cases the restrictions that animals face are not the direct result of physical conditions, but rather arise from their dependence upon plants to produce their food for them. Heavy winter snowfall in northern areas, for example, may prevent deer and other grazing animals from reaching the forage they need to survive the winter. Similarly, in semi-arid grassland the effects of drought are usually felt by the animals through the lack of food before they suffer from the lack of water. They have the ability to move to rivers or waterholes, whereas grasses anchored in the soil depend directly on the precipitation falling on them. If it is deficient, they stop growing and die, and the animals have lost their food supply. Animals that do not feed directly on green plants are not immune under such conditions. Foxes and coyotes, which feed on birds, rabbits, mice and other small rodents, fall into this category. Any change in the distribution or productivity of the vegetation in their environment will not impact on them directly, but will ultimately reach them through the small mammals on which they prey. If the grasses, fruits and seeds are not available the rabbits and mice that need them to survive will decline in numbers, and the food supply of the foxes and coyotes will decline also (Figure 3.11). These links between species take the form of food chains or trophic chains. Individual chains are ultimately based on green plants and if the link between plants and animals is broken, the animal community suffers.

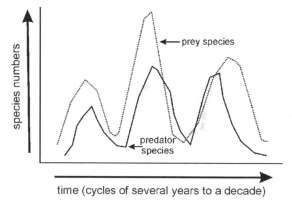

Figure 3.11 A representation of cycles in predator–prey relationships

FOOD CHAINS

A food chain consists of a group of organisms linked to each other through their production and consumption of food. Food is the most obvious element in the chain, its production and consumption easily observed, but the passage of food along the chain also involves the production and consumption of energy. Although not apparent without direct and sometimes complex measurements, the energy aspects of food chains are of major importance in assessing the quality and health of ecosystems. Most food chains are short and linear, with some form of green plant at one end and a carnivore or omnivore at the other. Individual chains are commonly interlocked at various levels to form a web. Different predators, for example, may share the same prey, and consumers may alter their eating habits to become involved in a new food chain if their preferred food source is no longer available. A simple food chain could be made up as follows:

Grass	→	Antelope	→	Lion
(producer)		(herbivore)		(carnivore)
		(primary		(secondary
		consumer)		consumer)

The grass produces the food and energy through photosynthesis. It is then consumed by the antelope, which in turn is consumed by the lion. In this way food and energy are passed along the food chain, with each stage referred to as a trophic level. Ultimately they reach the decomposers, which convert dead organic matter into its constituent parts, releasing nutrients into the soil and energy to be dissipated as heat into the air, water and soil (Enger and Smith 2002).

The conversion process in any food chain is relatively inefficient, with as much as 90 per cent of the useful energy being lost during the conversion from one trophic level to another. The energy assimilated at each level declines from the primary producers at the base of the chain through the herbivores to the carnivores. Some of the primary energy consumed is used to allow the consumers to function and is lost to the environment in the form of heat. In general, about 10 per cent of the energy available at any one level is transferred to the next level up the chain, but the actual value may range from 5 per cent to 20 per cent. As a result, in long food chains the amount of the original solar radiation absorbed by the plants that reaches the ultimate consumer is very low (Figure 3.12). The relationship between trophic levels is usually represented in the form of a pyramid, with the primary producers forming the broad base and a few top carnivores forming the narrow apex.

Linkages among the elements in a food chain are so strong that disruption at one level will be felt along the entire chain. Removal of the producer, for example, will reduce the food supply of primary consumers, perhaps causing a decline in their number through starvation, which in turn reduces the food supply for the predators. At the other end of the chain, a reduction in the number of predators

BOX 3.6 THE COMPONENTS OF FOOD CHAINS

producers Also called autotrophs, these are organisms that produce organic material from inorganic substances. Green plants are primary producers, creating organic products from carbon dioxide and water through the process of photosynthesis.

consumers Also called heterotrophs, these are organisms that feed on organic materials provided by autotrophs or other heterotrophs. Bacteria, fungi and animals belong in this category. *Primary consumers* feed directly on producers (autotrophs) and include a range of organisms from insects and their larvae to fruit and seed-eating birds and large grazing animals. *Secondary consumers* feed on primary consumers (heterotrophs). They include predators, which hunt and kill for food, as well as scavengers that eat organisms that are already dead.

herbivores Animals that feed exclusively on plants. Terrestrial herbivores range in size from caterpillars to elephants.

carnivores Animals that kill and eat other animals. Major carnivores such as the big cats or wolves receive most attention, but fish such as pike and starfish, birds that catch insects, grubs or worms and insects like the dragonfly are also carnivores.

omnivores Animals that consume both plant and animal food. The human animal is the most common omnivore, but others such as rats and bears also fit this category.

scavengers Animals that eat meat from animals killed by a predator or that have died of natural causes or by accident. Scavengers include coyotes, vultures and crows as well as some insects and their larvae. Some animals such as coyotes and crows can be both predator and scavenger, depending upon the type of food available.

predators Animals that hunt and kill other animals.

prey Animals killed and eaten by predators.

For more information see:

Cox, G.W. (1993) *Conservation Ecology: Biosphere and Biosurvival*, Dubuque IA: Brown.

Stiling, P. (1992) *Introductory Ecology*, Englewood Cliffs NJ: Prentice Hall.

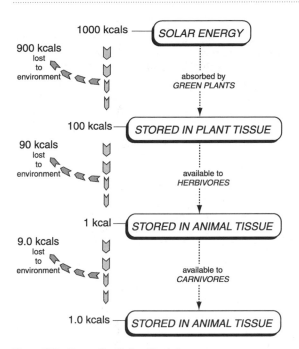

1000 kcals — SOLAR ENERGY

900 kcals lost to environment

absorbed by *GREEN PLANTS*

100 kcals — STORED IN PLANT TISSUE

90 kcals lost to environment

available to *HERBIVORES*

1 kcal — STORED IN ANIMAL TISSUE

9.0 kcals lost to environment

available to *CARNIVORES*

1.0 kcals — STORED IN ANIMAL TISSUE

Figure 3.12   Energy flow in trophic chains

will allow the population of the grazing primary consumers to increase, which places stress on the producers. Such developments are not uncommon in the natural environment, but they are normally resolved with no more than short-term imbalance in the system. In contrast, disruption of the predator–prey relationship by human interference has caused serious environmental problems in some areas. The introduction of rabbits into Australia, for example, created a short food chain with only a limited number of predators. This allowed the rabbit population to increase to such an extent that it competed with other primary consumers for the available food. In some places the abundance of rabbits caused the complete disruption of the vegetation and the initiation of soil erosion (see Chapter 7).

Humans are at the upper end of many food chains, and can adapt to the relative inefficiency of the chains. Farmers in the developed countries grow grain to be fed to pigs or cattle (primary consumers). The pork or beef is then eaten by humans (secondary consumers). Being omnivorous, however, they can take a more energy-efficient approach by shortening the chain and eating the grain before it is processed by the animals. Harvesting the lower levels of a food chain has also been considered in the aquatic environment, where experimental

fishing of large zooplankton such as Antarctic krill has been carried out. Lack of knowledge of the impact of such harvesting on oceanic food chains that depend upon krill, plus negative economic factors, has ensured that it has not gone beyond the experimental stage.

ECOSYSTEMS AND BIOMES

The relationships among the vegetation and animals in a food chain are only part of a much more comprehensive group of interactions that take place among the biotic (living) and abiotic (non-living) elements that make up the environment. The inter-relationships are complex, but they lead to the formation of characteristic communities of organisms that interact with each other and with the abiotic elements of their environment. Such groupings of organisms and the physical environment they inhabit are known as ecosystems. Ecosystems are dynamic entities, driven by the flow of energy within and through them, and being dynamic they can respond to a considerable degree of change while retaining sufficient equilibrium that the basic characteristics of the system are maintained. Despite this, major or prolonged environmental disruption, such as that caused by climate change or fire, may alter a specific ecosystem irreversibly, and bring about its replacement by a system with different characteristics. Human interference has become the most common cause of this type of change.

Ecosystems are commonly divided into terrestrial and aquatic groups (Enger and Smith 2002). Terrestrial ecosystems incorporate continental flora and fauna and the land surface that they occupy, whereas aquatic ecosystems include salt-water and freshwater communities plus those in coastal and interior wetlands. Within any ecosystem, in theory, organisms occupy areas in which the physical conditions best meet their needs, although in practice they are often forced to tolerate conditions that may not be optimal. These areas are called habitats. Although the term may be linked with a particular species – the habitat best suited to elephants, for example – a specific habitat will be shared by a variety of organisms that have requirements in common. The nature of any habitat is determined by a large number of variables, but most can be grouped into climatic, topographic, edaphic and biotic categories. Of these, the climatic factors – light, heat, moisture and wind – are generally considered to be of most importance. The position of an organism within a habitat, defined by its role in the habitat, is its niche. Niche includes not only physical location, but also the functional role of the organism in its community as determined by its needs and its interrelationships with other components of the ecosystem. In occupying a specific niche an organism is making use of the set of conditions that are best suited to its survival. Any change in these conditions, or the conditions in the wider habitat, may threaten the survival of the organism, and through it the integrity of the ecosystem.

TERRESTRIAL BIOMES

The ecosystem concept can be applied at a variety of scales, from the microscopic to the whole earth. Those of intermediate scale – watershed, marshland or local woodland ecosystems, for example – probably receive most attention, but there are also communities of plants and animals occupying major geographical areas, continental or subcontinental in scale, that retain sufficient similarities of form and function that they are recognized as ecosystems. They are called biomes. As with other ecosystems they are classified as aquatic or terrestrial, with the latter group being more complex and diverse, although aquatic biomes include highly diverse and productive ecosystems such as coral reefs, which have been described as the marine equivalent of tropical rain forests (Miller 1994). A biome is usually considered to have the attributes of a climax community – a community that represents the most developed combination of plants and animals possible under the environmental conditions pertaining at a given time in a given area.

Flora and fauna are not absolutely uniform across the entire extent of a biome, particularly where it exists on several continents. Species that inhabit one area, however, can usually survive elsewhere in the biome, as indicated by the ease with which rubber trees from the Amazon were translocated to Liberia and Malaysia, or by the feral camels that originated in the northern hemisphere desert biome, now surviving in Australia. Because of the scale involved it is normal for biomes to include communities that have not yet reached the climax level, but are undergoing the various stages of succession that lead to it. In addition, variations in such factors as altitude, proximity to the ocean

and geology can create individual local communities which share the overall characteristics of a specific biome but differ in detail. These communities are called ecozones.

Terrestrial biomes are normally delineated according to their dominant plant assemblages, but the biome concept also encompasses other organisms plus the interrelationships of the flora and fauna with physical elements such as climate, soil and topography. Although there is no universally accepted classification of biomes, most observers list about ten, including various combinations of tropical and temperate forests and grassland, deserts and tundra. Because there are few sharp boundaries in the environment, adjacent biomes grade into each other in a transition zone called an ecotone, which may be many kilometres wide and include characteristics of the two biomes, or it may be relatively narrow, with the change from one system to the other taking place within as little as a few hundred metres. Where the zone is broad, the ecotone will include plants and animals from both adjacent biomes and perhaps others that are characteristic of the ecotone itself, producing a community that may be more diverse than those adjacent to it.

An important characteristic of biomes is their ability to cope with change. Human interference in the past century or so has introduced so much change into the system, however, that some biomes have not been able to absorb it or recover from it. As a result, maps of biomes do not always reflect the current reality and descriptions of the characteristics of biomes refer to the situation that applied before they were disturbed. If the remaining natural biomes are to be preserved and protected, and the projected changes threatened by global warming are to be managed, it is important to know what currently exists and what might be expected to exist in a particular area. The following descriptions of the earth's major biomes will therefore consider them as they would have existed prior to human interference, but with reference to the nature and impact of that interference as appropriate (Figure 3.13).

Tropical rain forest

Tropical rain forest, also called selva, is the rain forest of low latitudes. It consists of broad-leaved, mainly evergreen trees and is best developed 10°

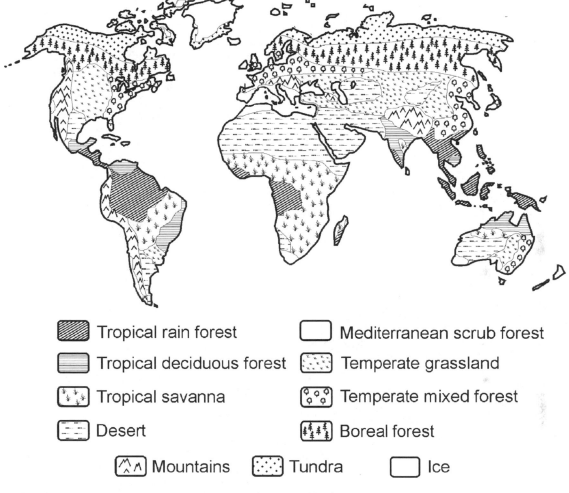

Tropical rain forest   Mediterranean scrub forest

Tropical deciduous forest   Temperate grassland

Tropical savanna   Temperate mixed forest

Desert   Boreal forest

Mountains   Tundra   Ice

Figure 3.13 A simplified representation of the distribution of the earth's biomes

north and south of the equator in the Amazon and Congo basins, West Africa and parts of Southeast Asia, where annual precipitation averages 2500 mm (100 in.) and the mean daily temperature remains at about 27°C (80°F) throughout the year (Figure 3.14). It is arguably the most diverse, most persistent and most stable of the earth's biomes, containing perhaps 65 per cent of the world's plant and animal species (Cuff and Goudie 2002). High temperatures, abundant moisture and readily available nutrients produce rapid growth that continues throughout the year, leading to an abundance of tall trees, often reaching heights of 50–60 m (165–95 ft), and including mahogany, ebony, rosewood, brazil nut, rubber and other important commercial species. The rain forest also supports a large number of plants that have, or are thought to have, medicinal value. Some like the cinchona tree, which provides quinine for use against malaria, have been renowned for many years, whereas the benefits of others, such as the rosy periwinkle, source of an anti-leukemia drug, have been recognized only recently.

The morphology of the tropical rain forest is quite distinctive, consisting of three or more layers stretching from the surface to the main canopy some 50 m (165 ft) above (Figure 3.15). The trunks of the trees tend to be bare beneath the high canopy and some species have buttresses or stilt roots, which are needed to support their great height and

BELEM, BRAZIL

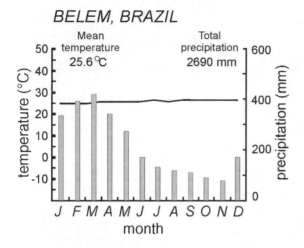

Mean temperature 25.6 °C

Total precipitation 2690 mm

Figure 3.14 Temperature/precipitation graph for a tropical rain-forest biome – Belem, Brazil

Plate 3.3 High cumulo-nimbus clouds, caused by strong convective heating, produce heavy precipitation in tropical and subtropical areas.

relatively shallow root systems. Competition for light plays a major role in determining the structure of the forest. The closely packed crowns of the trees in the main canopy create an effective barrier to the passage of solar radiation and ensure that the forest floor remains in shade. Individual trees or groups of two or three push through that canopy to reach the light they need for growth. Beneath the trees, vegetation is relatively sparse, with only a few shade-tolerant species, such as ferns, and young trees reaching up to become part of the main canopy. Low light levels also ensure that there is little undergrowth on the forest floor. Along rivers, or where falling trees have created gaps in the canopy, sufficient light reaches the ground to encourage rapid growth, producing the jungle-like vegetation popularly associated with the tropics. Also associated with the need for light are the large numbers of epiphytes and climbing plants, such as lianas, which use the trees as a means of support to reach the main canopy.

The abundance of tree species and their rapid growth suggests that tropical rain-forest soils are exceedingly fertile. That is not really the case. An efficient recycling system ensures that nutrients are available to the trees when they are needed, but the nutrients spend little time in the soil. High temperatures and abundant moisture year round support continuous growth in the plants, and nutrients are rapidly moved from the soil to support that growth. Since there are no distinct seasons, leaves and other organic materials are returned to the forest floor throughout the year. The same high temperatures and abundant moisture that support growth also encourage high rates of decay and the nutrients are quickly released to return to the soil, where they are again available to the growing plants. Thus, at any given time, a higher proportion of the nutrients in the system are held in the plant biomass than in the soil. Parasitic plants make use of this by attaching themselves to the trees so that they can receive nourishment without the need to be rooted in the soil.

Although this very efficient recycling system contributes to the position of the tropical rain forest as the most biologically productive of all the terrestrial biomes, it is a delicately balanced system and easily disturbed. The removal of the trees, for example, prevents the return of nutrients to the soil. Heavy rainfall causes the few remaining nutrients in the exposed soil to be leached out and high temperatures promote chemical changes. As a result, if

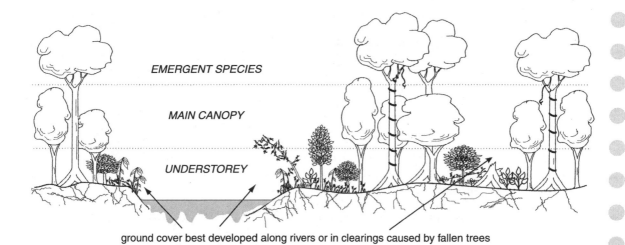

EMERGENT SPECIES

MAIN CANOPY

UNDERSTOREY

ground cover best developed along rivers or in clearings caused by fallen trees
where light can penetrate

Figure 3.15 The morphology of a tropical rain forest

the tropical forest is removed, the soil rapidly loses its fertility and the re-establishment of the forest is particularly difficult (Anderson 1993).

As with the flora, the rain-forest fauna are particularly diverse. Insect species are most numerous by far, ranging from mosquitoes, which spread diseases such as malaria and dengue fever, through a myriad of bees, wasps, moths and butterflies to the various forms of worms and beetles that participate in the decomposition process. Most of the faunal activity takes place in the main canopy, which is populated by a diverse community of small mammals – monkeys, squirrels; birds; reptiles – snakes, lizards; and insects. Larger mammals, such as anteaters, lemurs, pigs, gorillas, jaguars, live at or near the forest floor, but in small numbers compared with those in the canopy. Also common on the forest floor, but much less obvious, are the decomposers, involved in the decay of dead organic matter and as a result contributing to the ongoing flow of nutrients and energy in the biome.

Tropical deciduous forest

Polewards of the tropical rain forest in South America and Asia, but not well developed in Africa, are the tropical deciduous forests (Rumney 1968). Much more patchy in their distribution than the evergreen rain forests, the deciduous forests are found in Central America and the Caribbean, but are perhaps best developed in the monsoon lands

of Southeast Asia, including parts of north-eastern India, Bangladesh, Burma, Indochina, Indonesia, and reaching Queensland on the northern tip of Australia (Brown and Lomolino 1998). Temperatures average close to 25°C (77°F), with a greater annual range than in the evergreen rain forest (Figure 3.16). Annual precipitation totals remain relatively high, ranging around 1500 mm (60 in.), but with considerable regional variation. The most important aspect of the precipitation, however, is its uneven distribution through the year, marked by a distinct period of reduced rainfall. This introduces seasonality into the moisture regime, which

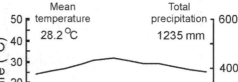

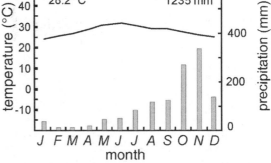

Figure 3.16 Temperature/precipitation graph for a tropical deciduous forest biome – Madras, India

is carried through into the flora and fauna of the region.

Although temperatures remain high enough to support year-round growth, the reduction in the availability of water introduces a dormant period of perhaps one or two months into the growth pattern. The trees respond to the dry spell by dropping their leaves, thus lowering transpiration rates and allowing them to survive the reduced water supply. It means, however, that with the reduced growth the trees in the deciduous forest are not as tall as in the evergreen forests, nor are there as many species represented. One of the characteristic trees of the deciduous forest in Southeast Asia is teak. The additional light that reaches the forest floor allows the development of a dense understorey containing shrubs, grasses such as bamboo and a large number of vines or lianas that climb into the trees. Humidity within the forest and in the canopy tends to be lower than in the evergreen forest and, as a result, plants such as ferns and epiphytes are less common.

Because the change from evergreen to deciduous forest is gradual rather than distinct, and since the animals are mobile, the tropical deciduous forest shares many animal species with the wetter rain forest. The insect, bird and small mammal species are numerous. Many of the mammals are arboreal, spending most of their lives in the canopy, but there are commonly more animals living on the forest floor than in the rain forest. The animal community is less diverse, however, with those animals that are unable to tolerate dry conditions either completely absent or at least absent during the seasonal dry spell.

In many places the tropical forest biomes remain largely natural, but that situation is being threatened increasingly by human activities (see Chapter 7). Most attention has been focused on the destruction of the forests of the Amazon, but areas in Africa, Southeast Asia and parts of Central America and Australia are under threat also (UNEP 2002).

Tropical savanna

In the tropics, temperatures are always high enough to support year-round plant growth. Variations in the distribution of vegetation come about as the result of changes in the availability of moisture. Even in the rain forest, local conditions associated with reduced precipitation – increased altitude or a rain shadow effect, for example – lead to the development of areas of scrub woodland. In general, the level of available moisture declines away from the equator to the north and south, and the ability of the environment to support arboreal species is reduced until the only vegetation that can survive the low moisture levels is grass. These conditions create the tropical savanna biome, best developed in Africa immediately south of the Sahara Desert. In East Africa the reduced precipitation associated with higher altitudes in the tropics allows the savanna to stretch through the entire equatorial zone. Smaller areas of savanna are also present in parts of South America, India and Australia (Mistry 2000).

The climate of the tropical savanna grassland is marked by a pronounced seasonal moisture pattern, created by the movement of the Intertropical Convergence Zone (ITCZ), a thermal low-pressure belt that circles the earth in equatorial latitudes. Its movement into the areas occupied by the biome brings the rainy season, while its retreat marks the beginning of the dry season. In West Africa, for example, the northward movement of the ITCZ in the northern spring brings with it abundant rainfall in the moist, unstable tropical maritime air drawn in over the land from the South Atlantic. As it retreats later in the year, the maritime air is replaced by tropical continental air from the Sahara Desert and the dry season begins. Annual precipitation totals range between 500 mm and 1000 mm (20 in. and 40 in.), with the higher values found closer to the rain-forest margins and the lower closer to the deserts. Characteristically, rainfall in the savanna varies considerably from year to year. Average temperatures range between about 20°C and 29°C (68°F and 84°F), with maxima reaching beyond 30°C (86°F) in the dry season (Figure 3.17).

The typical vegetation of the savanna consists of species of annual grass, which grow tall in the wet season but wither in the dry season. They may provide complete ground cover but towards the drier margins of the biome they may exist as bunches with bare ground between them. The boundary between the tropical forest and the grassland is not sharp. It involves a transition zone, or ecotone, which includes a combination of grasses with a scattering of trees and shrubs in a mixture sometimes referred to as park savanna. The trees and shrubs have developed characteristics such as small leaves and water storage capabilities or thorns instead of leaves to reduce transpiration, which allow them to survive the dry season. When the

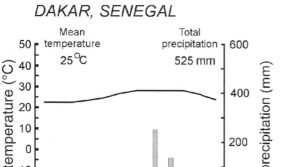

Figure 3.17 Temperature/precipitation graph for a tropical savanna biome – Dakar, Senegal

grasses dry and wither, they collect on the surface, but because of the dry conditions organic processes and bacterial activity are slow to return nutrients to the soil. The accumulation of dry organic material, however, provides abundant fuel for wildfires, which are a common occurrence on the savanna, and the nutrients are returned to the soil in the ashes that remain after the fires. Trees are normally killed by fires, but those that grow in the park savanna have adapted and therefore survive. As well as natural fires, these areas have long been subject to fires set by the indigenous peoples of the area, to drive the game animals they hunt or to provide new grass to encourage the animals to remain in an area. More recently the same approach has been used to provide new grazing for domesticated animals.

The animal population of the savanna varies from continent to continent, but it is dominated by large herbivores and the carnivores that prey on them. The African savanna has the richest and most varied fauna. It supports, or has supported in the past, large herds of wildebeest, antelopes, zebras and elephants plus the lions and hyenas that prey on them. The small mammal population is more limited, but the insect population is large. Many of the insects survive by preying on the grazing mammals, while they themselves are the food of a large variety of bird species. The fauna of the South American and Asian savannas is less rich, and in Australia the tropical grassland biome has its own peculiar population of animals, such as the kangaroo, wallaby and dingo, found nowhere else (Brown and Lomolino 1998).

Desert

Towards the poleward margins of the savanna, precipitation becomes increasingly infrequent or irregular and annual totals of between 100 mm and 400 mm (4 in. and 16 in.) are easily exceeded by high evapotranspiration rates (Figure 3.18). These conditions mark the boundary of the savanna with the desert biome (McKnight and Hess 2000). The permanent aridity responsible for conditions in the desert biome is characteristic of subtropical regions between 25° and 30° north and south of the equator. There, the descending arms of the tropical Hadley cells (see Chapter 2) inhibit precipitation, and the high temperatures produce high evapotranspiration rates that contribute to soil moisture deficits. The Sahara Desert in North Africa is the main product of that circulation pattern, but the deserts of Arabia, India, Australia, South Africa and South America also owe their existence in large part to these conditions. Elsewhere, deserts are created by the rain shadow effects of mountain systems that lie across the prevailing wind directions – as in North America, where the Mojave and Sonoran Deserts owe their aridity to the presence of the Sierra Nevada blocking the moist air masses from the Pacific Ocean. The rain shadow effect also applies to the Gobi and other deserts in Central Asia. The Himalayas prevent the moisture from the Indian monsoon from reaching the area, but the continentality of the region is also an important element. Its distance from the ocean means that the air masses reaching the area have

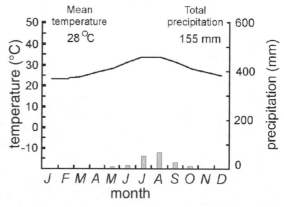

Figure 3.18 Temperature/precipitation graph for a tropical desert biome – Khartoum, Sudan

lost most of their moisture before they arrive. The world's deserts cover some 5 million km², and an additional 40 million km² are arid or semi-arid. In total, almost 33 per cent of the earth's land surface exhibits desert characteristics to some degree or other.

Limited precipitation and high evapotranspiration rates restrict plant growth, and the desert biome is distinguished by sparse vegetation cover, or, in some areas by the complete absence of plant life. Its overall biomass productivity is low. In places the high evaporation rates create very salty soils, which also inhibit plant growth. However, even these surfaces that appear completely devoid of vegetation may bloom immediately following precipitation, as seeds and dormant plants respond rapidly to the availability of moisture. Although the plant community in the desert is sparse it includes a variety of species, including grasses, shrubs such as sage and acacia as well as a mixture of cacti and other succulents that are particularly suited to cope with the lack of moisture. These plants are xerophytes, and are able to survive in the desert because they have adapted to the conditions (Stiling 1992). Some have small leaves with thick, waxy skins and sunken stomata, which together reduce transpiration rates; some have thorns and spikes that help dissipate internal heat; some have long, widespread root systems to capture as much water as possible; some have structures that allow them to store water until it is needed.

Desert animals have also responded to arid conditions by reducing their moisture needs. They are relatively few in number and tend to be small mammals, birds and reptiles. Many have developed nocturnal habits to avoid the heat of the day and its accompanying water stress, while others have physiological adaptations that allow them to reduce daily heat build-up or to survive with minimum water intake. Rats and other rodents, along with snakes and lizards form the bulk of the small animal population, with foxes, coyotes and various species of hawk being the main predators (Louw and Seely 1982).

Mediterranean scrub forest

The Mediterranean scrub forest is a small biome, transitional between the dry deserts and savannas of the subtropics and the more humid conditions of temperate latitudes. It is recognized in half a dozen widely separated mid-latitude locations, but best developed around the Mediterranean Sea. California, central Chile, the Cape region of South Africa, western Australia around Perth and South Australia in the vicinity of Adelaide are the other locations (McKnight and Hess 2000). They

Plate 3.4 The sparse xerophytic vegetation and unique landforms of the desert biome.(Courtesy of Christine Deschamps.)

share the same climatic conditions of hot, dry summers and relatively mild, wet winters. Average annual temperatures range between 18°C and 21°C (60°F to 70°F) with maxima often above 26°C (80°F) during the summer. Precipitation totals range between 400 mm and 900 mm (15 in. and 35 in.), with 65 per cent to 75 per cent falling during the winter months (Figure 3.19). This pattern is produced by the seasonal expansion and contraction of the subtropical high-pressure cells. During the summer months they extend over the regions, creating stable atmospheric conditions with clear skies and little precipitation. In contrast, when they contract during the winter months, travelling low-pressure systems have access to the areas and bring in cyclonic rainfall.

This climate regime promotes a sclerophyllous (drought-resistant) scrubby woodland referred to variously as maquis (Mediterranean), chaparral (California), matorral (Chile), fynbos (South Africa) and mallee (Australia). All share flora that are physically similar, but with local variations in specific plant types (Rumney 1968). Some degree of drought resistance is necessary, for example, to allow the plants to survive the summer drought and most plants have adaptations such as thick, waxy, evergreen leaves or needles to reduce water loss. Dry summers also encourage frequent fires and many plants have adaptations that make them resistant to fire. Around the Mediterranean the most common trees in the biome are scrub oak, olive and pine, with a ground cover of grass, various heaths and other woody shrubs, rosemary, thyme and other

aromatic herbs, and in places a mix of flowering plants such as anemones, lavender and daisies. Oak and pine species are also typical of the biome in California, with shrubs such as buckthorn species of wild lilac and native flowering plants such as lupines and poppies. In South Africa the dominant flora are heaths, while in Australia eucalyptus and acacia species are typical of the biome. Compared with the plant communities, the fauna in the Mediterranean scrub forest biome is relatively inconspicuous. There are no major herbivores. Small rodents that have adapted to the dry summers by living in burrows and seed-eating birds are the most common animals.

Mid-latitude temperate grassland

The temperate grasslands occupy areas in mid-latitudes, where moisture levels are insufficient to support woodland and perennial grass species predominate. Their greatest extent is in the northern hemisphere where the North American prairies stretch from northern Mexico through the United States to southern Canada and east from the Cordillera towards the Mississippi and the Great Lakes (Scott 1995). In Eurasia similar grasslands extend from Hungary in the west through Ukraine and around the northern edge of the Central Asian deserts to Mongolia and north-east China. In the southern hemisphere the pampas of Argentina, Uruguay and southern Brazil belong to the biome, as do the high veldt in South Africa and small patches of grassland in Australia and New Zealand.

The latitudinal and longitudinal extent of the grasslands plus changes in elevation, particularly in the northern hemisphere, ensure regional variations in temperature and precipitation that are translated into differences in soils and vegetation types across the system. In North America, precipitation ranges from about 350 mm to 1200 mm (14 in. to 48 in.), being highest in the south and declining towards the north and west (Figure 3.20). Across Eurasia, precipitation totals are generally lower, ranging from about 600 mm (24 in.) in the west to as little as 200 mm (8 in.) in the interior close to the desert margins. Northwards towards the forest amounts increase to about 400 mm (16 in.). In the southern hemisphere, low precipitation – about 400 mm (16 in.) per annum – occurs in the lee of the Andes in Argentina and the Southern Alps in New Zealand, but in

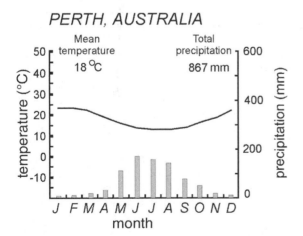

PERTH, AUSTRALIA

Mean temperature 18 °C

Total precipitation 867 mm

Figure 3.19 **Temperature/precipitation graph for a Mediterranean scrub forest biome – Perth, Australia**

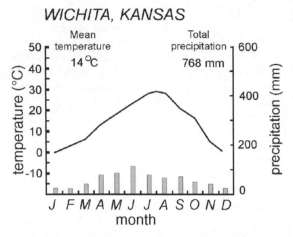

WICHITA, KANSAS

Figure 3.20 Temperature/precipitation graph for a mid-latitude temperate grassland biome – Wichita, Kansas

South Africa averages range around 750 mm (30 in.) and up to 1500 mm (60 in.) in the wetter parts of the South American pampas. Some of the precipitation falls as snow, remaining on the ground for most of the winter, but the bulk of the moisture arrives with the early and midsummer rains. Because summers on the grasslands are hot, evapotranspiration rates are high, and soil moisture deficits are common. Coupled with the general variability of precipitation from year to year, this ensures that the grasslands are prone to drought.

In both Eurasia and North America, average temperature ranges are large. Average winter minima as low as −25°C (−13°F) are offset by summer maxima above 28°C (82°F) in North America, while in the central parts of the biome, in Asia, minimum temperatures may fall as low as −50°C (−58°F). In the southern hemisphere, where the temperate grasslands are closer to the ocean, temperatures are much less extreme and the annual range is not nearly as great as in the north.

The interplay of temperature and moisture creates variations in the nature of the grassland communities within the biome (Archibold 2002). Where precipitation totals are higher and temperatures moderate, long grasses are common. The tall grass prairies in the United States, for example, dominate the southern and eastern parts of the plains close to the Mississippi River, with species that grow to heights of 2 m supported by well developed root systems that bind the soil together into a thick sod. It also includes broad-leaved flowering plants, or forbs, such as black-eyed susan, goldenrod and

sunflowers, which add colour and variety to the community. When the grasses die at the end of summer the organic material and the nutrients it contains are returned to the soil, helping to maintain its fertility and promoting the growth of the following year's crop of grasses. These soils are the chernozems, or black earths, so called because of the dark colour produced by the high levels of humus they contain.

As temperature and moisture levels decline, the tall grass prairie is gradually replaced by a mixed grass community, which averages perhaps a metre in height by the end of the growing season, until eventually that too is replaced by a short grass community that may grow to no more than 20 cm to 50 cm during the season. In the driest locations, bunch grasses are common and bare soil may be exposed. The short grass prairie is also called steppe, the name most often applied to the grasslands of Ukraine and Russia. Low precipitation and high evapotranspiration levels ensure that the soils under the steppe often suffer from a moisture deficit and many of the species that grow there have developed drought-resistant characteristics. As on the tall grass prairies, the steppe communities include a variety of forbs and legumes such as clover and lupines. Because biomass production is much less in the mixed and short grass prairies than in the tall grass community, less organic matter is available to be returned to the soil and that is reflected in its colour. As the grassland changes from the tall grass to short grass, the soils change from black earths to brown earths, the thickness of the sod declines and the general fertility of the soil is reduced. Dead grass lying on the soil surface and grass withering and dying as a result of drought provide abundant fuel for fires. As in the tropical savanna grasslands, fire is very much a characteristic of the mid-latitude grasslands. With their growth points beneath the surface, grasses are unharmed by all but the most intense fires, and using the nutrients available in the ash left on the surface they quickly respond to the first available moisture by sending up new shoots. Fires also discourage the migration of tree and shrub species into the grasslands, thus helping to maintain their character and distribution (Cox 1993).

The largest groups of animals in the temperate grassland biome by far are the herbivores (McKnight and Hess 2000). Most communities include species of deer, but in some regions there are animals found nowhere else. Kangaroos and wallabies are peculiar to Australia, for example, and the guanaco, a species of llama, is native to the South American pampas. In North America the bison dominated the grasslands. Numbering in the millions, they roamed the plains in huge herds until the latter part of the nineteenth century, when they were almost wiped out. Bison were also part of the Eurasian grassland community, along with wild horses, but fewer in number than their North American counterparts. Associated with these herbivores are the animals that prey on them. These include the big cats, such as lions and cougars, dogs, such as dingoes and wolves, and, in North America and Eurasia, bears. Small mammals are represented by rabbits and hares, plus prairie dogs, ground squirrels or marmots, which live in tunnels beneath the surface for safety. Bird populations tend to vary with the seasons, but include seed and insect-eating species and, particularly in North America, migratory waterfowl that move through the region in the spring and autumn. Insects include herbivores such as the grasshopper, along with a variety of biting flies and ticks that target the large grazing mammals.

Little temperate grassland remains, although many places that once supported natural grassland continue to have a natural look to them because the introduced plants species – even the cultivated grains – are usually grasses and the cattle that graze there are not unlike the herbivores that once roamed the grassland. The resemblance is only superficial, however. The habitat is much less diverse than it was and much less resilient to change (see Chapter 7). The native grassland that has survived is highly fragmented and therefore not particularly viable. Its survival will depend upon sustained conservation efforts accompanied by the reintroduction of grassland species where possible.

Mid-latitude temperate mixed forest

The mid-latitude grasslands are mainly a response to inadequate levels of soil moisture during the growing season. When soil moisture availability increases, grasses are replaced by the trees that form the mid-latitude temperate mixed forest. The cool season that is typical of mid-latitude climates stops growth and initiates a period of dormancy during which the deciduous broad-leaved species that make up the bulk of the forest lose their leaves. Needle-leaved trees are also part of the biome, as

are the evergreens of the temperate rain forests (Scott 1995).

The temperate mixed forest biome is most extensive in the northern hemisphere. It reaches its greatest extent in eastern Asia, where it stretches from Manchuria more than 3000 km to southern China and across the Korean peninsula to Japan. In North America, the forest lies east of the Mississippi from the St Lawrence lowlands in the north to Florida in the south. Both these areas are located on the eastern edges of continents, but in Europe the biome is in a western location covering an area from Britain, France and Spain in the west, through central Europe almost as far as the Ural mountains, before being pinched out between the steppe to the south and the boreal forest to the north. In the southern hemisphere, where the land masses are quite narrow in mid-latitudes, the mixed forest is confined to a number of relatively small areas in South America, Australia and New Zealand. The temperate rain forests of western North America, southern Chile and the west coast of New Zealand, with their evergreen forests, can be included in the mixed forest biome (Rumney 1968).

Although the sheer extent of the forests ensures that they will include variations in climate, they have a number of elements in common. All have distinct seasons, but with a relatively limited temperature range. Summers are never exceedingly hot, nor winters exceedingly cold (Figure 3.21). Average summer temperatures range between about 16°C and 21°C (60°F to 70°F) in Europe, 18°C to 27°C (65°F to 80°F) in North America and are highest at

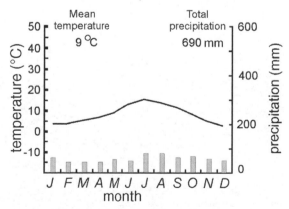

Figure 3.21 Temperature/precipitation graph for a mid-latitude temperate mixed forest biome – Edinburgh, Scotland

about 21°C to 30°C (70°F to 85°F) in Asia. Winter temperatures are highest in Europe at between –6°C and 10°C (21°F to 50°F) and lowest in Asia at between –14°C and 7°C (7°F and 44°F). In North America the equivalent range is about –12°C to 10°C (10°F to 50°F). Precipitation is provided mainly by the low-pressure systems that move through mid-latitudes or by the summer monsoons in Asia. Precipitation is reasonably well distributed through the year, peaking in the summer in Asia and eastern North America, and in the winter elsewhere. There are local variations in precipitation, but overall totals range between 700 mm and 1400 mm (30 in. to 60 in.), declining to about 500 mm (20 in.) towards the continental interiors and reaching 2000 mm (80 in.) in some of the monsoon areas of East Asia. The temperate rain forest in North America and New Zealand receives in excess of 2500 mm (100 in.) in places. In the northern hemisphere some of the precipitation falls as snow, particularly at higher altitudes and in the continental interiors.

The vegetation in the temperate forest varies with the climatic conditions and the soil, but the bulk of the trees are hardwoods such as oak, beech,

ash, hickory, maple, elm and walnut (Brown and Lomolino 1998). Different varieties of these trees grow in Asia along with species – ginkgo, Chinese elm, sugi or Japanese cedar – that are found only there, and in Australia eucalyptus dominates. Towards the southern edges of the biome in North America and China, sub-tropical species such as magnolia and a variety of palms are found among the hardwoods. There also, epiphytes like Spanish Moss appear. Where the soil is particularly well drained needle-leaved species replace the deciduous hardwoods. The southern pines that grow on the sandy coastal plains of Georgia in the US south-east are an example. Other needle-leaved trees, including larch (tamarack) and black spruce, are found in wetter areas. The trees in the temperate mixed forest grow to heights of 25 m to 30 m (75 ft to 100 ft), with interlocking canopies that shade the forest floor. Thus the undergrowth in the mixed forest does not grow well once the trees are in leaf. In the spring, however, there is a rapid development of herbaceous plants that need to complete their growth while the trees are still bare and light levels at the forest floor are higher. With

Plate 3.5   Eucalyptus trees in the Australian temperate forest. (Courtesy of Karen Armstrong.)

abundant precipitation and temperatures that allow growth for most of the year the Douglas fir, western hemlock, sitka spruce and red cedar that are the main species in the rain forest reach heights of 50–60 m (150–80 ft) and sometimes more. The kauri pine, which can grow to a height of 60 m (180 ft), and a variety of large tree ferns are unique to the New Zealand forest (Rumney 1968). Under the rain-forest canopy the forest floor is shaded all year and as a result it tends to be bare of vegetation except along rivers or on the coast.

The mid-latitude temperate forest supports a varied animal community. Acorns, beech nuts and chestnuts provide an abundant food supply for deer, wild pigs, squirrels and a variety of other rodents. Migrant birds and year-round residents feed on the great number of insects that inhabit the forests in the summer and birds of prey feed on these birds and the mice and rabbits that inhabit the forest floor. Larger predators include foxes, weasels and coyotes (McKnight and Hess 2000).

Most of the temperate forest has been cut down over the past several centuries, cleared to provide agricultural land, felled to supply the needs of industry or displaced by urban development. Where it does survive, it is often the focus of controversy and conflict between the forest industry and the conservation community (see Chapter 7).

Boreal forest

As temperatures decline and winters lengthen in higher latitudes the temperate mixed forest is replaced by a forest of needle-leaved conifers. This is the boreal forest, found only in the northern hemisphere, there being no land masses sufficiently large to support it in mid to high latitudes in the southern hemisphere. In North America it stretches in a great curve from Alaska, south-eastwards through the gap between Hudson Bay and the Great Lakes and east to Newfoundland. The Eurasian boreal forest extends eastwards from Scandinavia, widening through Siberia to reach the Pacific Ocean in the Kamchatka peninsula. Together these units almost completely encircle the earth between about 50°N and 65°N. The Siberian section of the boreal forest was traditionally referred to as taiga, and that term has also been applied to the entire biome, but it is currently most commonly used to designate the more open transition between the boreal forest and the tundra (Scott 1995).

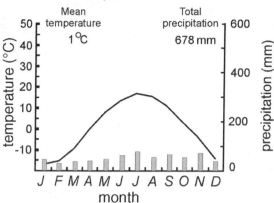

Figure 3.22 Temperature/precipitation graph for a boreal forest biome – Timmins, Ontario

Winter is the dominant feature of the climate in the boreal forest biome. Below freezing temperatures prevail for five to seven months of the year, and although mean temperatures for the coldest month may range around –12°C (–10°F), extreme minimum temperatures fall far below that. Extremes in excess of –40°C (–40°F), for example, are not unusual in the continental interiors of Canada and Siberia (Figure 3.22). Summers are short, but warm, with temperatures averaging between 15°C and 18°C (60°F and 65°F). Occasional hot spells can push extreme temperatures as high as 26°C to 32°C (80°F to 90°F). The strong contrasts between the summer and winter values produce annual temperature ranges greater than anywhere else in the world. The northern limit of the boreal forest – the tree line – generally matches the 10°C (50°F) isotherm for the warmest month.

Precipitation varies considerably across the biome, being highest in coastal locations and declining towards the continental interiors. Totals range between 250 mm and 500 mm (10 in. to 20 in.), with most falling as rain in the summer and the remainder accumulating as snow in the winter. In the interior, the southern boundary of the forest is determined by moisture availability, and there it merges with the temperate grassland, but in the wetter areas it grades into the mid-latitude temperate forest.

The plant community of the boreal forest is characteristically homogeneous, with great stretches consisting of only two or three species. Most trees are needle-leaved, with a pyramidal shape that allows them to shed the winter's snow relatively easily. In North America the dominant species are black and white spruce and larch (tamarack), with jackpine and balsam fir also common. Jackpine is frequently found in drier areas or in areas that have been burned over. The Eurasian species include Scots pine, Norway spruce and Siberian fir, with larch more common than in North America. Both areas include a scattering of deciduous species, including birch, poplar, aspen and willow, in amounts that vary with factors such as local temperature and moisture conditions. Willow species, for example, are common in wetter areas. In the predominantly coniferous areas, the forest canopy shades the forest floor so effectively that there is very little undergrowth. The thin soils carry only a covering of mosses and lichens plus a layer of needles that accumulate faster than they can decompose, their presence contributing to the absence of ground vegetation. Where the canopy opens up, as result of fire or blowdown, grasses, shrubs, such as Labrador tea and blueberries, and young trees quickly colonize the area. The quality of boreal forest is not uniform throughout. The largest trees and densest forest are found along its southern margins. Towards the north, the trees are

smaller and grow farther apart, creating an open lichen woodland community. Eventually, close to the northern boundary, the trees become stunted, before being replaced completely by the flora of the tundra biome (Rumney 1968).

With its short growing season, the productivity of the boreal forest biome is low. Among other things this means that the food supply for animals is relatively limited and the fauna of the forest are neither very numerous nor very diverse. The few large mammals include moose, elk, woodland caribou and other species of deer in numbers that tend to fluctuate quite radically from year to year. A year or two of heavy snowfalls, for example, can decimate the whitetail deer population by limiting their access to food and making them easier prey for predators such as wolves that can more easily deal with deep snow. Some of the smaller fur-bearing mammals have been important commercial species in the past. The beaver and later the muskrat were the mainstay of the fur trade in the eighteenth and nineteenth centuries and continue to be trapped for their skins. Mice and rabbits are also common, kept under control by predators such as weasels, coyotes and wolves. Bears are also predators to some extent, eating meat or fish where it is available, but being omnivorous, they also consume large amounts

Plate 3.6 The boreal forest biome in northern Canada.

of fruit and berries. Because of the harsh winter season only a few birds remain in the forest year-round. These include grouse and owls, but most birds are migratory, passing through on their journeys north and south or spending only the summer in the region, feeding on seeds and fruits or on a seemingly inexhaustible supply of insects. The insect population explodes in the spring, producing millions of bloodsucking mosquitoes and blackfly plus a variety of insects such as spruce bud worms, tent caterpillars and army worms that periodically cause major damage to specific groups of trees.

The boreal forest has been cleared in places for agriculture, but the podzolic soils of the biome, being thin, acidic and of low fertility, are not particularly suited to agriculture. The short growing season is also a limit in most areas beyond the southern margins of the forest. Animals are still harvested for their furs or hunted for food and sport, but, if carried out in a sustainable manner, these activities should have little overall impact on the biome. The major change has been caused by the removal of trees for lumber or to produce pulp and paper. Clear cutting has laid bare thousands of hectares, particularly along the southern margins of the forest, causing serious environmental problems in some areas. However, because of the immensity of the biome, there are still large areas, relatively untouched by human activity, that represent the biome in its natural state.

Tundra

Towards the northern boundary of the boreal forest the trees become smaller and stunted and the forest thins until the few scattered trees that remain eventually disappear. The treeless open landscape that remains is part of the tundra biome. It circles the north pole between the boreal forest and the permanent snow and ice of the polar regions, extending from Alaska across northern Canada to Greenland and again from northern Scandinavia across northern Russian to the Bering Straits. Apart from a few patches on the Antarctic peninsula and adjacent islands, it is absent from south polar regions, excluded by the permanent ice in Antarctica. Elsewhere, it exists at high altitudes as alpine tundra, where environmental conditions resemble those in the Arctic (Bliss *et al.* 1981).

The tundra biome is a product of low temperatures and low precipitation (Figure 3.23). Tem-

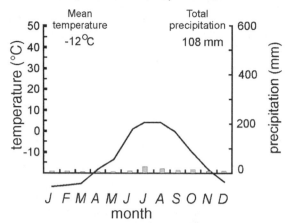

Figure 3.23   Temperature/precipitation graph for a tundra biome – Point Barrow, Alaska

peratures are low throughout the year, but during the winter, when the sun is completely absent or rises above the horizon for only an hour or two each day, extreme minima regularly reach −50°C (−60°F). Average January temperatures often fall below −30°C (−22°F), although in locations close to the ocean, as in Alaska and northern Scandinavia, values as high as −10°C (14°F) are not uncommon. Mean temperatures in the two to four-month summer season range between 0°C (32°F) and 10°C (50°F), with daily maxima only occasionally rising above the latter value in some locations.

Annual precipitation totals are generally less than 200 mm (8 in.), declining northwards to as little as half that total, but higher in locations exposed to the oceans. The limited precipitation is offset to some extent by the low evapotranspiration rates in the summer and by the poor drainage that keeps soils wet. In most places there is a slight summer peak, and because of the low temperatures snow may fall at any time of the year. Wind is an important feature of the climate of the tundra biome. It causes desiccation of the vegetation, particularly in the winter when plants are unable to replace the moisture lost through transpiration.

An important feature in the tundra environment is the presence of permafrost. Beneath the surface to depths of as much as 500 m (1600 ft) in some places, the ground is permanently frozen (French 1996). In most areas, however, there is an active layer at the surface, which is subject to seasonal freezing and thawing, and it is on this active layer that the tundra vegetation grows. During the summer, when the ice in the active layer melts, the water is unable to drain because of the ice remaining below and the soil becomes waterlogged. The tundra vegetation insulates the frozen ground from excess melting, and therefore has an important role in maintaining the stability of the permafrost layer. If the vegetation is removed or destroyed there is an increased inflow of heat into the ground, which causes the active layer to thicken. The additional water, unable to drain downwards into the soil, collects in pools or runs across the surface, causing erosion. The expansion and contraction that accompany the freezing and melting in the active layer also produce a hummocky surface on the tundra, or, where vegetation is absent, patterned ground characterized by stone stripes and polygons.

Vegetation in the tundra biome consists of plants that can cope with the low temperatures of the short growing season and the limited precipitation but high moisture content of the soil. Communities of low-growing mosses, lichens, sedges and saxifrages are most common, but local variations occur as a result of such factors as drainage, exposure or the availability of nutrients. Patches of vegetation separated by areas of bare ground are particularly common in the polar desert of the high Arctic, for example. In contrast, towards the southern margins of the biome the vegetation includes shrubs such as Labrador tea, Arctic willow, various berry bushes and species of birch that have adapted to the conditions by growing close to the ground.

Given the short growing season and the resulting low productivity of the plants, the biome supports a remarkably diverse fauna. The largest mammals are the caribou, reindeer and musk oxen that find enough forage in the tundra's thin vegetation cover. Smaller mammals such as hares, lemmings and birds such the snowy owl and ptarmigan also survive as permanent residents of the biome, but during the summer migratory birds and insects predominate. Predators include polar and grizzly bears, wolves and foxes (McKnight and Hess 2000).

AQUATIC BIOMES

Although the term 'biome' is applied most often to terrestrial communities, there are aquatic, biotic communities that have also been recognized as biomes. They are most commonly divided into marine and freshwater communities, with further subdivision into ecozones according to depth or proximity to shore. All have water in common, but the interplay of a number of other factors determines the nature of the community that will develop. These include water temperature, the extent to which light can penetrate from the surface, the nature of the bed of the water body and the amount of dissolved solids in the water.

Marine biome

The marine biome includes open ocean communities and coastal communities (Figure 3.24). The former are subdivided into benthic (bottom living) and pelagic (open ocean) ecosystems (Enger and Smith 2002).

Benthic ecosystems

Benthic communities include plants and animals that live on or in the sea floor. Where the bottom is solid, barnacles, limpets or corals can attach themselves to the rock and crustaceans such as crabs and lobsters can live on the sea floor. Demersal, or bottom dwelling, fish are an important element in benthic communities. If the bottom is soft, worms, clams and a variety of other bivalves are able to burrow beneath the surface. Plants such as seaweed will grow on the bottom if

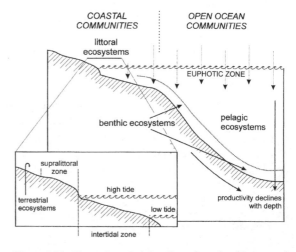

Figure 3.24 The nature and location of marine biomes and ecosystems

the water is shallow enough to allow the penetration of sunlight. Kelp, for example, grows well in relatively shallow water, where there is sufficient light to allow photosynthesis. Little sunlight reaches the deeper parts of the oceans, plants are absent and the animal population is adapted to the low light conditions and high pressure experienced at depth. The concentration of light in the upper levels of the oceans – the euphotic zone – means that temperatures are higher there also, and as a result the benthic community in shallow waters is more complex and productive than in the ocean depths. The effects are best seen on the continental shelves, in areas such as the Grand Banks of Newfoundland, which have long been famous for their high fish populations. Heavy overfishing decimated these populations in the twentieth century, however. In the shallow, warm waters of the tropical oceans, corals are an important component of the benthic community. The extensive reefs that they build are the most productive of any marine ecosystems, supporting a remarkably diverse population of invertebrate and fish species, as well as plants and algae.

Pelagic ecosystems

Pelagic communities include a combination of nekton – organisms that are able to swim – and plankton – organisms that drift with the current. Nekton includes microscopic animals, but also fish and large marine mammals such as seals and whales. Plankton consists mainly of algae and microscopic phytoplankton, which support photosynthesis and as a result form the basis of most marine food webs. To obtain enough light to carry out photosynthesis, they must live in the upper layers of the oceans and as a result most of the other organisms that depend upon them for food live there also. Thus the pelagic ecosystem is most diverse and productive close to the surface. This is particularly true over the continental shelves, where the water is relatively shallow, and run-off from the adjacent land provides a regular supply of nutrients, although the benefits may be offset by increased turbidity where rivers with high sediment loads flow into the sea. Areas where the upwelling of deeper water brings nutrients back to the surface also have a more complex and productive pelagic community (Brown and Lomolino 1998). With time phytoplankton and other pelagic organic materials tend

to sink towards the ocean floor, where they contribute to the food supply of benthic organisms.

Coastal ecosystems

Coastal ecosystems within the marine biome may include attributes of both benthic and pelagic communities as well as elements from terrestrial biomes in some places. They support shellfish and crustaceans, seaweed beds, free-swimming fish, birds, mangrove swamps and tidal salt marshes, for example. They also differ from the open oceans in that they are subject to the daily cycles of wetting and drying associated with tidal movements. Coastal ecosystems are often extremely complex and display considerable local or regional diversity. Rocky coasts, sandy coasts or tidal estuaries support different communities, for example, although they may exist in close proximity to each other along a particular stretch of coastline. They are also more subject to change than open ocean communities. Adjustment to coastal erosion or deposition, changes in sea level, variations in erosion or run-off from the adjacent land is an integral part of ecosystems in coastal areas. Unconsolidated sandy shorelines are particularly susceptible to change, their beaches being built up or combed back depending upon the state of the sea, or moved laterally by longshore drift. This limits the diversity and productivity of the communities they support. In contrast, rocky shores provide a more stable environment.

The communities are complex, but well structured, occupying distinct zones according to their position within the normal tidal range. The supralittoral zone, for example, lies above the normal high-tide limit and is under water only during exceptionally high tides. It often includes species from the adjacent terrestrial biomes. At the other end of the sequence the low-tide zone is normally always under water. In between the two is the intertidal zone, which experiences daily wetting and drying, and includes standing water in rock or tide pools. Each zone has its distinct flora and fauna, which have adapted to the changing conditions created as the tide rises and falls.

Tidal estuaries occupy a distinct position among coastal ecosystems because of the mixing of fresh and salt water that takes place as tides move in and out. Estuarine communities have adapted to these conditions, and are among the most productive in the marine biome. Nutrients are readily available from the streams and rivers that empty into the estuaries and the relatively shallow water allows sunlight to penetrate to the sea floor to warm the water and promote photosynthesis. Overall, coastal communities are more diverse and often more productive than communities in and beneath the open ocean. Because of their location at the land/water interface, however, and their consequent proximity to human activities, they were also among the first to suffer environmental deterioration (Goudie and Viles 1997).

Freshwater biome

The freshwater biome includes the lakes, rivers and streams on the earth's surface. Water in the lakes is relatively stationary or moves only slowly, whereas the water in rivers and streams is obviously in motion. This creates an initial difference between the two environments, but there are also differences created by climatic conditions, sediment levels, sources of water and depth of water. Together these factors create an infinite variety of ecosystems within the freshwater biome (Enger and Smith 2002).

Lakes, particularly if they are large, share some of the environmental characteristics of oceans. They have, for example, a euphotic zone into which light penetrates, to support photosynthesis in the phytoplankton and algae that live there. These organisms in turn form the base of a food chain, which includes zooplankton, insects, shellfish, crustaceans, various fish species and waterfowl. The euphotic zone also contains plants such as bulrushes and water lilies that grow in the littoral zone along the shore. In deeper lakes, where light levels, oxygen levels and temperatures are low, a few freshwater benthic organisms, such as worms and bottom-feeding fish, may be present.

Productivity in lakes varies considerably. Those that have a low concentration of nutrients are referred to as oligotrophic. They are low in organic material and therefore tend to be clear. This is characteristic of young lakes, which have had insufficient time to develop an adequate nutrient supply and recycling system. Recently filled reservoirs, for example, are generally oligotrophic. As lakes age, their organic productivity increases. The breakdown of organic plant and animal material plus the addition of natural wastes by organisms raises the concentration of nutrients in

the system. Water bodies that have a high concentration of nutrients are said to be eutrophic. They tend to be more cloudy or turbid than oligotrophic lakes. This aging process, called eutrophication, is natural, but in many places it has been accelerated by the addition of pollutants such as sewage and agricultural fertilizers.

The ecosystems in rivers and streams are quite variable. In any one river system, for example, the ecosystems in the clear, fast-flowing, low-temperature, highly oxygenated water close to its headwaters will be quite different from those in the more turbid, slower-moving, nutrient-rich water close to its mouth. In theory, it might be possible to recognize a continuum in the communities of micro-organisms, insects, plants, fish, birds and mammals along the length of a particular river, but local variations in width, depth, gradient, turbidity, flow rate and volume make such generalizations difficult. Nutrient levels and productivity in rivers and streams depend very much on adjacent terrestrial ecosystems. Organic material such as dead leaves and animal waste or chemicals leached from the river banks are natural sources of nutrients, but these have been augmented or replaced by agricultural and industrial sources along most of the world's rivers and streams with detrimental effects on the freshwater ecosystems they contain.

Human interference in both the marine and the freshwater sectors of the aquatic biome has involved mainly pollution and resource depletion. The open oceans show only limited evidence of pollution, but it is a serious problem in coastal and the immediate offshore area (Goudie and Viles 1997). Along the coasts pollutants reach the sea from human activities on land. Agricultural wastes, sewage and industrial pollutants are regularly released directly into the coastal zone or arrive with the run-off from rivers. Intermittent events such as oil spills following tanker accidents or the flushing of oil tanks by shipping close to shore can cause such damage to coastal ecosystems that they take decades to recover. Pollution in some form or other is widespread in the freshwater biome. In some cases it is obvious, in the form of organic wastes on, in and beneath the water, but often the pollutants are present in a dissolved form and there is no visible evidence of pollution. Lakes that have succumbed to acid rain, for example, are often clean and clear, but they contain dissolved acids and a variety of other chemicals in solution. There has been some success in dealing with water pollution in both

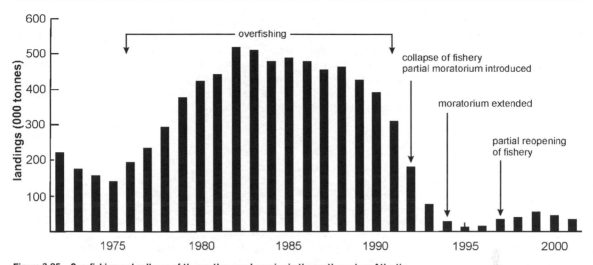

Figure 3.25   Overfishing and collapse of the northern cod species in the north-western Atlantic

Source: based on data in Canada: Department of Fisheries and Oceans, 'Historic Landings'. Viewed at http://www.dfo-mpo.gc.ca/communic/statistics/Historic/landings/HLAND_E.htm (accessed 2 July 2003).

the freshwater and marine sectors through the increased control of effluents. This has produced significant improvements in pollution levels in the lower Great Lakes in North America and species such as salmon have returned to rivers in Western Europe from which they had been driven out by pollution. In many areas, however, pollution control standards have not been established and even where they do exist they are not always enforced. As a result, water pollution remains a worldwide problem (see Chapter 8).

Pollution alters plant and animal communities in the aquatic biome, by killing species directly or by changing their habitats to such an extent that they eventually die or move away. The net result is a change in the number of species and the mix of species in a particular ecosystem. Human activities can produce a similar result by changing the habitat of water bodies and by over-harvesting aquatic species. In the freshwater sector, the demand for water in industry and agriculture or the need to reduce flooding has led to the widespread re-engineering of natural waterways, including the damming and channelling of rivers and the draining of wetlands. Most major river systems have experienced such activities, which change existing habitats in the river basin, forcing plant and animal communities to adapt or be replaced by communities more suited to the new conditions.

An even more direct impact comes about as the result of over-harvesting of aquatic species (Figure 3.25). In theory, it should be possible to harvest these species in such a way that they provide a sustainable yield. In practice, that has not happened. Knowledge of the aquatic environment lags behind that of the terrestrial environment and it is difficult to establish, with any accuracy, what the sustainable yield of a particular species might be. Coupled with major improvements in harvesting techniques this led to over-fishing in many areas. On the Grand Banks of Newfoundland, for example, cod stocks were almost wiped out in the 1980s, whales and sharks are under serious pressure and in coastal areas crabs, lobsters, shrimp and even some shellfish stocks are threatened. Similar situations apply in rivers and lakes, where the over-fishing of preferred species such as salmon and trout has completely changed the species mix. The aquatic biome, particularly the marine sector, is much less productive than was once thought, and although attempts have been made to re-establish sustainability – through harvesting quotas, for example – it is clear that some ecosystems will take a long time to recover.

SUMMARY AND CONCLUSION

The biosphere is the most complex of all the major spheres that make up the environment. It is dynamic by nature, combining elements from the lithosphere, atmosphere and hydrosphere to provide the necessities of life. Part of its dynamism lies in its ability to accommodate and adapt to the change that is an integral part of all environmental systems and it coped with that successfully for hundreds of thousands of years. The evolution of the human species as the dominant animal in the biosphere threatened that situation. It introduced change directly into the biosphere by altering soils, plants and animals, or indirectly by creating disturbances in the lithosphere, atmosphere or hydrosphere that fed back into the biosphere. Human interference takes many forms and has many causes, but in the broadest terms it can be linked with population growth and technology, and these will be considered in the following two chapters.

SUGGESTED READING

Spellerberg, I.F. and Sawyer, J.W.E. (1999) *An Introduction to Applied Biogeography*, Cambridge/New York: Cambridge University Press. A succinct, readable introduction to biogeography with consideration of its applications in resource management, environmental assessment and land use planning.

Paton, T.R., Humphreys, G.S. and Mitchell, P.B. (1996) *Soils: A New Global View*, London: University College of London Press. A non-traditional earth sciences approach to the study of soils, quite technical in places, with examples from Australia and Africa.

McKnight, T.L. and Hess, D. (2002) *Physical Geography: A Landscape Appreciation* (7th edn), Englewood Cliffs NJ: Prentice Hall. An introductory physical geography with a good section on the biosphere.

QUESTIONS FOR REVISION AND FURTHER STUDY

1   Why is the recycling of matter critical to the continuance of life on earth?

2   Debate the proposition that 'Soil is essential to the survival of society'.

3   Human beings are at the top of a great number of food chains. Based on your meals over a week, develop a series of food chains in which you are the final consumer and examine the results. Do you regularly eat from the lower levels of the food pyramid – salads and cereals – or from the upper levels – meat or fish? What are the environmental implications of your food consumption patterns?

4

Demography and World Population Growth

After reading this chapter you should be familiar with the following concepts and terms:

age-specific fertility rate (ASFR)
agricultural revolution
arithmetic progression
baby boom
birth control
carrying capacity
contraceptive pill
crude birth rate (CBR)
crude death rate (CDR)
demographic transition model
doubling time
emigration

exponential growth
family planning
fecundity
fertility
fuelwood
general fertility rate (GFR)
geometric progression
gestation
immigration
industrial revolution
K-strategists
migration
mortality
non-renewable resources

population
population planning
population projections
r-strategists
rate of natural increase (RNI)
renewable resources
replacement fertility rate (RFR)
replacement migration
total fertility rate (TFR)
zero population growth (ZPG)

POPULATION ECOLOGY

A population is a group of individuals, usually of the same species, occupying a specific area. In the case of the human species the area involved is effectively the entire earth, but human population distribution is uneven. Although humans have developed a remarkable ability to adapt to different environmental conditions and to manipulate environments to meet their ends, some regions remain largely unpopulated. Empty spaces such as those in the Arctic and Antarctic, the deserts of Africa and Asia and the mountainous regions of all the continents, for example, are not unaffected by human activities, but in population numbers they contrast sharply with adjacent more populous regions.

CARRYING CAPACITY

In these parts of the earth thinly populated by humans, the population of other species tends to be low also, reflecting the limited resources available to support life. In ecological terms, the carrying capacity of these areas is low. This concept, linking population numbers with the availability of natural resources, originated in ecology, but it is now considered too simplistic for many ecological situations. It continues to be used in conservation ecology and ecological economics, however, and in consideration of relationships between society and the environment (Turner II and Keys 2002). Carrying capacity is a measure of the maximum number of organisms that can be supported in a particular environment, and under natural conditions it represents a theoretical equilibrium state within a dynamic system. If the species in an area are below carrying capacity, for example, populations will tend to increase until some form of balance is reached with the resources available. If the carrying capacity is exceeded, because of the rapid growth in the number of organisms in the ecosystem, for example, there will be insufficient resources to support the excess population, and numbers will decline until equilibrium between the resource base and the population is re-established. Typically, the population fluctuates above and below the level of the carrying capacity before it eventually approaches equilibrium again (Figure 4.1). If provided with an adequate supply of food, animal populations will increase rapidly. If the food supply is then restricted, or if the demand for food exceeds the capacity of the environment to provide it, the population will decline rapidly again. Such cycles are not uncommon in animal communities. Laboratory and field studies have shown that population growth follows a specific pattern in most cases. Growth may start off slowly, but quickly becomes exponential. This period of rapid growth will continue as long as the population is allowed to develop in an optimal environment with unlimited resources. If environmental restrictions are introduced, the exponential growth will be brought to an end. In laboratory experiments, restrictions may include the reduction in the area available for expansion or a reduction in the amount of food available. Such restrictions may also be seen in real life, where, for example, expansion is restricted by natural features in the landscape, or where drought,

floods or other natural disasters diminish the food supply. The resulting increase in the death rate may be augmented by outbreaks of disease caused by malnourishment and overcrowding, while growing dissension within the population, as competition for scarce resources grows, may also lead to increased mortality. As these dynamics are resolved, the population will eventually stabilize, indicating that the carrying capacity of the environment has been reached and population and environmental resources are again in balance.

The carrying capacity of a specific environment is not static. It will vary with the resources available. In areas with marked seasons, for example, the carrying capacity will vary between summer and winter, being much lower in the winter when food resources are not readily available. This does not mean that all individuals in excess of the winter carrying capacity will die off. In most cases they will migrate to areas where there is sufficient food to support them. The mass migration of birds from northern Europe to Africa and subarctic North America to Central and South America is an annual response to reduced carrying capacity during the northern winter. Variations in carrying capacity also occur as the result of longer-term environmental change. In the past this was usually associated with natural physical change in the environment – variations in climate, for example – but more recently, human interference has become a major contributor to change. Natural resource extraction, habitat changes associated with forestry or agriculture and the growth of urbanization all reduce the carrying capacity of the environment for non-human species. In contrast, agricultural activities or waste disposal in a particular environment can increase the food supply and raise the carrying capacity for insect species or small mammals such as mice and rats, causing their populations to increase.

The utility of the carrying capacity concept in ecological studies has been questioned, particularly for environments that are not in equilibrium (Middleton 1999), and it receives less attention in ecology than it once did. In contrast, it has contributed significantly to studies of society–environment relationships at various times in the second half of the twentieth century and even before that in the eighteenth century, through the ideas of Thomas Malthus, which linked food supply and population growth (see Box 4.4, Thomas Malthus). The Club of Rome report, called *Limits to Growth*, published

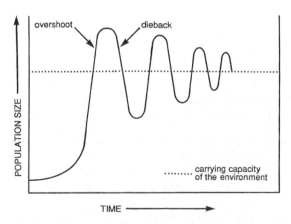

Figure 4.1 The relationship between population growth and the carrying capacity of the environment

in the early 1970s (Meadows *et al.* 1972), was essentially an exercise in relating population growth to carrying capacity, through such elements as resource use, industrial output, food supply and pollution. More recently, the success of sustainable development as a means of providing, among other things, a more equitable distribution of resources requires that economic growth and technological development bring about increases in carrying capacity as they have done in the past.

By transferring resources from one region to another, or by applying technology to increase resource availability, humans have artificially increased the carrying capacity of many areas, to the extent that in some of the more populous parts of the world the human population easily exceeds the natural carrying capacity of the land. Effective as such developments may be, any restriction in the supply of these resources, or a rapid increase in population beyond the availability of additional resources, would set in train the processes that naturally bring population into line with the carrying capacity of the environment. Regular bouts of famine in sub-Saharan Africa associated with drought have indicated the reality of that situation over the past several decades. Although the concept of carrying capacity is usually applied at the ecosystem or regional level, there is also a theoretical carrying capacity for the earth as a whole, applied to the maximum human population that it may support. A realistic value is probably impossible to estimate, since factors other than resource availability – acceptable quality of life, for example – will help to decide what the final carrying capacity will be.

The concept of carrying capacity is applied in a number of socio-economic contexts. In agriculture, for example, the carrying capacity of an area will determine the number of grazing animals it will support, or, when applied to recreational land use, it represents the number of people and types of activity that can be accommodated without environmental disruption. Human interference frequently causes the carrying capacity of an area to be exceeded. Introducing extra grazing animals on rangeland or allowing too many people to use a recreational area will eventually lead to the deterioration of the environment.

The complexity of the relationship between resources and populations, particularly human populations, ensures that issues associated with the concept of carrying capacity will be subject to much debate. On the one hand, the earth is a closed material system, its resources are ultimately finite, and it would seem reasonable to suggest that somewhere in the future there will be an upper limit to the numbers they can support. On the other, human ingenuity has done much to extend and improve resource availability and continues to do so with considerable success (see Chapter 5). The extent to which this ongoing increase in the earth's carrying capacity can continue or whether there is an ultimate limit to population are questions that cannot yet be answered, however, but finding the answers will require further consideration of the nature, causes and effects of population change.

r-STRATEGISTS AND K-STRATEGISTS

Organisms such as insects and small mammals can respond rapidly to conditions that favour them, because of their reproductive strategy. They are called r-strategists, characterized by small size and relatively short life spans during which they mature quickly to produce large numbers of offspring. They do little to support these offspring, but depend upon quantity rather than quality for the survival of the species. Numbers fluctuate wildly. Efficient reproduction allows the population to grow rapidly as long as the conditions are favourable. If these conditions change – food supply is curtailed, for example – the population will crash and remain low until conditions improve again.

In contrast, humans belong to a group of organisms that have a completely different reproductive strategy. These are the so-called K-strategists, organisms that are usually large, have relatively long lives and produce only a limited number of offspring. They invest considerable time and energy providing for the survival of these offspring, so that they in turn can reproduce and ensure the continuation of the species. Along with humans and other primates, larger mammals such as deer, lions, bears and elephants are all K-strategists. They do not respond to change as rapidly as r-strategists, but neither do their numbers rise and fall as wildly. Because of their time commitment to their offspring, they survive best under stable environmental conditions.

Although the human species cannot compete in terms of total numbers and reproductive capacity with organisms such as bacteria or insects, whose body size is small and whose reproductive cycles are short, humans are the most numerous of all the mammals. Only the rat comes close to matching human population totals. Among the higher primates – gibbons, chimpanzees, orangutans and gorillas – the human position is even more supreme, its numbers exceeding the sum total of all other species of primates.

THE EARTH'S HUMAN POPULATION

Spectacular as that situation may appear, the end result would be even more so if human beings reproduced at full capacity. With a gestation period (the length of time between conception and birth) of nine months and a reproductive span of about thirty years, it is technically possible for a woman to give birth to about thirty children. This is a measure of the fecundity, or maximum productive capacity, of a population. Given the time required for physical recovery and for nursing the child, such a high number is unlikely to be achieved, although totals in the high teens and low twenties have been recorded. It has been estimated that, under ideal living conditions, the maximum number of births that might be expected is between fifteen and seventeen (Molnar and Molnar 2000). Fortunately, the number of children actually born to a couple is generally far less than that.

FERTILITY

The actual number of live births is a measure of the fertility of a population. This can be expressed in a number of ways (Table 4.1). The general fertility rate (GFR) is a measure of the total number of births per thousand women within the age range of fifteen to forty-nine. Within that age range, an age-specific fertility rate (ASFR) can be calculated using five-year intervals. Typically, highest fertility occurs in the twenty to twenty-nine age group. One commonly used expression of fertility is the total fertility rate (TFR), which is an estimate of the average number of children that would be born to each woman in a population during her child-bearing years, based on birth rates among all women in the same age group in the population (Hornby and Jones 1993). In the early 1990s, for example, the TFR ranged from 6.0 in Asia to 1.7 in Europe, with a worldwide average of 3.2. By 1998 the world TFR had fallen to 2.9 and according to

TABLE 4.1 REPRESENTATION OF FERTILITY AND MORTALITY IN A POPULATION

Crude birth rate (CBR) Annual number of live births per thousand persons in a population

Crude death rate (CDR) Annual number of deaths per thousand persons in a population

Rate of natural increase (RNI) The difference between the CBR and CDR, usually expressed as a percentage

General fertility rate (GFR) The total number of births per thousand women within an age range of fifteen to forty-nine in a population

Age-specific fertility rate (ASFR) The total number of births per thousand women measured at five-year intervals within the age range of fifteen to forty-nine

Total fertility rate (TFR) An estimate of the average number of children born to a woman during her lifetime

Replacement fertility rate (RFR) A TFR of 2.1

UN Population Division estimates, between 2000 and 2005 the TFR will range from 4.91 in Africa to 1.38 in Europe, with a world average of 2.69 (UNPD 2002). A TFR of 2.1 is called the replacement fertility rate, since it represents the number of children required to replace the parents. The extra 0.1 is required to allow for children who die in infancy and therefore produce no offspring. All these numbers are well below the maximum productive capacity or fecundity of a population, and that was so even when the earth's human population was growing much more rapidly than it is now. Although the earth's human population has grown rapidly over the past two centuries, without the continued difference between fertility and fecundity the rate of change would have been even greater.

MORTALITY

The birth rate of a population is only one element in population change. It might be considered the input element. At the other end of the system is the death rate or mortality rate. Together, birth rate and death rate allow the natural change in the population to be calculated. If the birth rate is greater the change will be positive and the population will grow. In contrast, a death rate higher than the birth rate will produce a declining population.

Death rates tend to fluctuate more than birth rates, with one bad natural disaster often having spectacular effects. For example, in the fourteenth century, the 'Black Death' in Europe left some 20 million dead and another attack in the seventeenth century had a death toll of 10 million. In England, between 35 per cent and 40 per cent of the population died when the Black Death struck in 1348 and 1349, and the population continued to decline for a hundred years after that, before beginning to recover in the latter part of the fifteenth century (Hatcher 1996). The HIV/AIDS epidemic killed 2.4 million people in Africa in 2002 and has the potential to cause millions more deaths in the next decade (UNAIDS/WHO 2002) (see Box 4.1, HIV/AIDS). Smaller-scale disasters involving disease, drought, famine, typhoons and hurricanes may have significant effects on local or regional populations. Famine associated with climate variability in northern Europe in the 1690s brought about the deaths of about 20 per cent of the population of Finland (Jutikkala 1956) and perhaps 10 per cent of the population of Scotland (Anderson 1996a). The infamous Irish potato famine of the mid-nineteenth century brought about the loss of more than 800,000 lives in the five-year period between 1846 and 1851, creating the momentum for an overall reduction in population that has never been replaced (Cousens 1960). More recently, in China between 1958 and 1961 as many as 20 million people died in a famine caused by a combination of drought and government mismanagement of the agricultural system (Jowett 1990), and drought and famine caused the deaths of hundreds of thousands of people in sub-Saharan Africa in the 1960s, 1970s and 1980s.

Although human activity exacerbated some of these natural disasters, humans most commonly contribute to increased death rates through war. The death toll in World War I, for example, was 10 million and in World War II, 50 million (Keegan 2000). Since the latter ended in 1945 there have been some 150 conflicts, in which up to 20 million lives have been lost (Otok 1989). In all wars up to the end of World War I, the main casualties were among the combatants. Since then, civilians have provided a higher proportion of casualties as a result of total war, characteristic of World War II, or the various

guerilla wars and civil wars that flared up in the second half of the twentieth century. Because of the nature of war, most of the dead are in the young adult male group. In World War I, for example, more than half of the 700,000 casualties in the British armed forces were twenty to twenty-nine-year-old males, and within that group a disproportionately high percentage of those killed were officers (Winter 1977). The 300,000 fatalities in the French army between August and November 1914 included deaths among the young men in the twenty to thirty age range that were ten times the normal mortality for that group (Keegan 2000). Although these men were in their prime reproductive years and their loss created socio-cultural changes that might have been expected to influence population dynamics – disrupted marriage patterns, for example – the overall impact on total population was ultimately relatively small. Dips in death rate curves and the uneven distribution of males and females in diagrams of the population structure are common for all the major belligerent nations, but the increase in fertility that immediately followed the war more than offset the decline in numbers caused by the deaths. A similar pattern occurred after World War II, creating the so-called 'post-war baby boom' (Figure 4.3).

NATURAL INCREASE

Birth rates and death rates are commonly expressed as a rate per thousand rather than in total numbers, allowing for ease of comparison between countries and regions. The number of live births per thousand of the population in a given year is the crude birth rate (CBR) while the number of deaths per thousand of the population in a given year is the crude death rate (CDR). A comparison of the CBR and the CDR provides the rate of natural increase (RNI) in the population, usually expressed as a percentage. Nigeria in 1998, for example, with a CBR of 45 and a CDR of 15 had an RNI of 3 per cent. In comparison, the RNI in the United Kingdom was only 0.2 per cent and in both Russia and Ukraine, where the death rate exceeded the birth rate, the RNI was negative (Figure 4.4). The RNI of a nation or region can be used to project the future size of its population in much the same way as an interest rate is used to calculate compound interest for a savings account. Just as the projected interest earnings in a year are added to the

BOX 4.1 HIV/AIDS

Acquired immune deficiency syndrome, or AIDS, is an incurable disease caused by the human immuno-deficiency virus (HIV). The virus suppresses the immune system, leaving the victim open to a variety of deadly respiratory infections and rare forms of cancer. It is transmitted through sexual activity and intravenous drug use, and in the twenty years since 1981, when it was first recognized, some 27 million people have died as a result of the infection. An estimated 42 million people are living with HIV, but the latency period for the virus is between seven and ten years, during which time there are no outward signs of AIDS. At that stage it can be detected only by blood tests, and many poorer nations lack the resources required for large-scale testing. Thus the numbers probably underestimate the reality of the situation. There is currently no cure for AIDS, and although the symptoms can be treated using an array of drugs, the costs of such treatment are high.

The distribution of those infected by HIV is very uneven (Figure 4.2). More than 85 per cent of the victims are in the developing world, with the nations of sub-Saharan Africa, where some 29.4 million are living with HIV, by far the worst affected. In several countries in southern Africa at least one in five adults is HIV-positive. In South Africa the rate rose to almost 25 per cent in 2000, while in Botswana, where the prevalence rate is 36 per cent, the epidemic appears out of control. Only in Uganda has there been any success in turning the epidemic round. There, a concentrated effort has reduced a rate of 14 per cent in the early 1990s to 8 per cent in 2000.

Elsewhere in the world, the numbers infected by HIV are less, ranging from 15,000 in Australia and New Zealand and about 500,000 in North Africa and the Middle East to 1.9 million in Latin America and the Caribbean, 1.0 million in Eastern Europe and Central Asia and 1.5 million in the other industrialized nations. The region causing most concern outside Africa is Asia. This, the most populous region in the world, has more than 6 million people infected with HIV/AIDS, and the potential for a rapid rise in that number is high, particularly in India, where 3.7 million people are already living with the virus, and in China, where existing rates of infection are low, but are rising steeply.

The demographic impact of such conditions is already apparent in some areas. In southern Africa in the late 1980s, for example, the life expectancy was fifty-eight years. That represented some thirty years of improvement, but it is expected that by 2010 life expectancy will have fallen again to about forty-five years, mainly as a result of the AIDS epidemic. Death rates are rising across the region, and in nations with high levels of HIV/AIDS infection, such as Zimbabwe, Botswana and Namibia, increasing mortality will soon equal or surpass the high fertility rates to create zero population growth. Behind such figures are devastating socio-economic impacts. Hidden are the orphaned children, 12.1 million in southern Africa alone, the decline in family income as breadwinners become too ill to work, and the overall physical and mental strain imposed on women.

The effects spread across society to disrupt and even reverse social and economic development. As more personnel are infected, the work force declines or becomes less efficient, costs rise and productivity is driven down. No sector of the population is spared. Businessmen and educators succumb, as do members of the armed forces or workers in mining, transport and manufacturing industries. The economic effects are already being felt in Africa and parts of Asia and will only become worse. Health costs are rising. Although drugs have been developed to mitigate the effects of the opportunistic infections that strike HIV/AIDS victims, they are costly, which puts them out of reach of those in the developing world who need them. There is no cure for the disease, and the numbers requiring some form of medical care will continue to increase until education and prevention become sufficiently effective that the level of infection is reduced. Despite the ominous trend in the number and rate of infections through the 1980s and 1990s, politicians in some countries denied the existence of a problem, delaying an appropriate response and condemning thousands to infection and premature death. In contrast, countries such as Uganda and Brazil have shown that aggressive intervention, involving both education and prevention, can stabilize or reduce the rate of infection, but that too comes at a cost, a cost that cannot be met by many developing nations without outside aid.

The world has faced nothing like the HIV/AIDS epidemic since the influenza epidemic of the early twentieth century or even the Black Death of the fourteenth century. Advances in medical knowledge and expertise should allow this most recent threat to be fought more effectively, but the economic, political and social dynamics of the issue mean that success will be possible only with the increased commitment of money and resources to the problem, put in place

BOX 4.1 – continued

and managed through global co-operation. Without that, the world faces a crisis that will lead to the curbing or even reversal of socio-economic development, a growing gap between rich and poor nations and a serious threat of social and political instability.

For more information see:

Brown, L.R., Gardner, G. and Halweil, B. (1999) *Beyond Malthus*, New York/London: Norton.
UNAIDS (2001) *Together We Can*, Geneva: Joint UN Programme on HIV/AIDS.
WHO (2003) *AIDS Epidemic Update: December 2002*, viewed at http://www.who.int.hiv.pub.epidemiology/epi2002/en/ (accessed 16 June 2003).

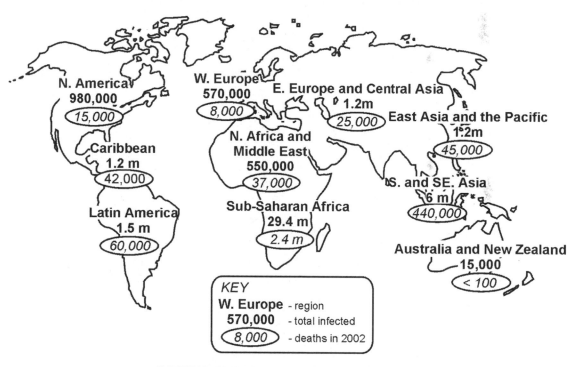

INTERNATIONAL HIV/AIDS SITUATION - 2002

People living with HIV/AIDS - 42 m

New infections 2002 - 5 m

Deaths due to HIV/AIDS 2002 - 3.1 m

Total deaths since outbreak began - 27 m

Figure 4.2   The global distribution and impact of HIV/AIDS
Source: UNAIDS/WHO, *AIDS Epidemic Update*, Geneva: Joint UN Programme on HIV/AIDS and World Health Organization (2002).

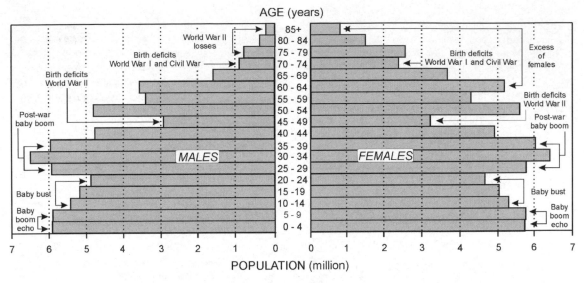

Figure 4.3   The effects of war on population structure. The example of Russia

Source: based on data from the US Census Bureau International Data Base. Viewed at http://www.census.gov/ipc/www/idbpyr.html (accessed 2 July 2003).

principal, to be included in subsequent calculations, the number of births estimated using the RNI for one year is added to the population base to be included in the calculations in succeeding years. Using that approach, population totals can be projected several years into the future, although longer-term projections are likely to be less accurate because of changes in the RNI with time, and need to be revised regularly. Using the same approach, the RNI can be used to calculate the time needed for a population to double in size. For most purposes, doubling time can be estimated by dividing the

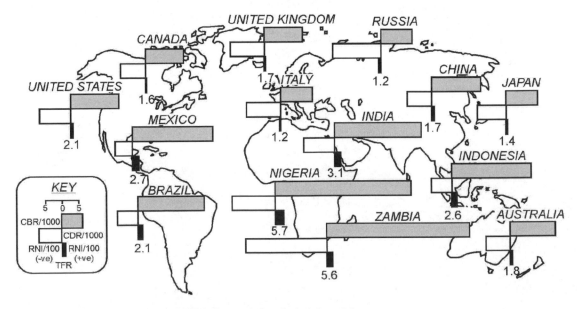

Figure 4.4   The rate of natural increase (RNI) in the population of selected countries

Source: UN Population Division, *World Population Prospects 1950–2050 (The 2000 Revision)*, New York: United Nations (2001).

number 70 by the RNI of the population. In Nigeria, for example, where the RNI is 3.0 per cent, the population will double in only twenty-three years, whereas in the United Kingdom, with an RNI of 0.2 per cent, doubling time is 350 years. For the world as a whole at the current rate of natural increase, the doubling time is close to fifty years. Useful as such figures are for comparative purposes, they have to be treated with caution, since rates of natural increase can change quite rapidly owing to variations in birth and death rates.

MIGRATION

The relatively simple relationship between births and deaths is modified by a third element in the equation. That element is migration. In terms of world population totals, migration may seem comparatively unimportant, since it only brings about a redistribution of population; it does not change it. On a local and regional scale, however, migration will augment or decrease the rate of natural change. Migrants tend to be young people still within their childbearing years. Thus any children they have are

a loss to the population of the country they left and a gain for the country to which they have moved. Traditionally, in modern times, certain regions have been areas of emigration – outflow of population – whereas others have been areas of immigration – inflow of population (Figure 4.5). In the nineteenth and early twentieth centuries, European countries were the main source of emigrants, with 44 million leaving for North America, Australasia and other destinations such as South Africa and South America between 1821 and 1915 (Devine 1999). Although Italy, Germany, Spain and Portugal provided more total immigrants, in terms of emigration *per capita*, Scotland, Ireland and Norway led the way for most of the nineteenth and early twentieth centuries (Devine 1999). Between 1900 and 1910, 15,000 people per year left Finland, mainly for North America, while at the same time emigration from eastern and southern Europe to North America grew and continued strong up to the First World War. Although migration has been a common characteristic of human history, the scale of the migrations of the nineteenth and early twentieth centuries was unparalleled and initiated the worldwide expansion of European culture. Between the

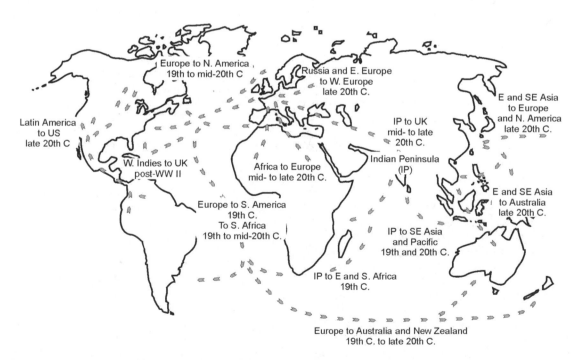

Figure 4.5   Global migration flows in the nineteenth and twentieth centuries

wars and in the 1950s emigration from Europe continued at varying rates, but in more recent years the Indian sub-continent, Southeast Asia and the Caribbean have become areas of major out-migration.

The destinations of these millions of migrants varied with time. In the 1840s and in the early twentieth century Canada was popular with emigrants from Britain. In the 1850s, 1860s and 1870s, Australia and New Zealand were seen as attractive destinations, but the major draw for most emigrants from Europe was the United States. For half the emigrants from Scotland between 1853 and 1914, for example, the United States was the preferred destination, reached either directly or via Canada (Devine 1999). At the end of the twentieth century, all these areas remained popular destinations, but the sources of immigrants had changed dramatically. Between 1991 and 2001, for example, over 1.8 million immigrants entered Canada. Of these, less than 50,000 came from Britain. European countries still provided nearly 200,000, but by far the greatest number of immigrants, slightly over 1 million, came from Asia (Statistics Canada 2003). For Australia, well into the second half of the twentieth century, Europe, particularly Britain and Ireland, remained the main source of immigrants, but by the late 1980s, Asia was providing more immigrants than Europe (Hugo 2001). Similar changes took place in the United States, where, in 1890, 86 per cent of the immigrants were from Europe. Little more than a century later, in 2000, only 14 per cent of the immigrants were from Europe, with 32 per cent from Asia and by far the largest number, 47 per cent, from Latin America, with Mexico at 23 per cent the main source in that region (MPI 2003).

Migration from Europe continues, but some of the countries that were the main sources of emigration in the nineteenth and early twentieth centuries are now experiencing immigration from the developing world. Britain, for example, became the destination of choice for many emigrants from former colonies in the Indian subcontinent, West Indies and Africa, while France experienced a steady flow of immigrants from its former territories in northern, western and central Africa. According to the UN Population Division (2001), that pattern is likely to continue and perhaps accelerate in the near future, as the populations of the nations in Europe, plus Japan and the Republic of Korea, begin to decline. All have below-replacement fertility, which unchanged would bring about significant declines in population over the next fifty years. In Italy, the population of 57 million in 2000 would decline to 41 million by the middle of the century and in Japan it would fall from 127 to 105 million in the same time period. Compounding these declines is an increase in the age of the population. With fewer births and an increase in longevity, the proportion of older people in a population increases; the size of the working-age population declines and the imbalance in the age distribution of the population creates significant socio-economic stress. It is this imbalance, rather than the decline in total numbers, that is problematic. In all these nations, society has developed in such a way that older members of the population are supported by the activities of the working-age group. As the numbers in that group decline and the post-retirement population grows, maintaining an adequate level of support may no longer be possible. In the absence of any major increase in fertility, the UN Population Division has suggested that replacement migration could provide a solution to the problems of declining and aging populations. The numbers involved would be large. For the members of the European Union, for example, migration at a rate of 13 million per year from 2000 to 2050 would be required to maintain the current size of the working-age population. Such numbers are well beyond past experience and would require the reconsideration of existing criteria for international immigration. Even in countries such as the United States, Canada and Australia, which have traditionally been the destinations for most of the world's migrants, the total fertility rates are close to or below replacement values, and the role of replacement migration needs to be considered in setting future population policies in these areas also. Thus, although the difference between birth rates and death rates will continue to determine the total world population, migration will have an increasingly important part to play in establishing the local, regional and national structure and nature of that population.

ZERO POPULATION GROWTH

The change in a population is determined by the interplay of these three elements – birth rate, death rate and migration. The population of a country will grow if, over a period of time, the number of births plus the number of immigrants exceeds the number of deaths plus the number of emigrants. If deaths plus emigrants exceed births plus immigrants, then the population will decline. If both are equal, zero population growth (ZPG) is said to exist. ZPG is sometimes confused with replacement reproduction, in which each family consists of only sufficient children to replace the parents. If adhered to, that situation would eventually produce ZPG, but it would not be immediate because of the time lag involved as age groups move through the system. As long as the numbers entering the reproductive sector of the population exceed those leaving it, the population will continue to grow, even with a birth rate at the replacement level (see Box 4.2, Population structure). To reach ZPG more rapidly, the reproduction rate would have to be reduced significantly below the 2.1 children per family normally considered to be the replacement reproduction rate. Despite this, there are nations in the developed world that have already reached or even surpassed the conditions associated with ZPG.

STAGES OF WORLD POPULATION GROWTH

The pattern of world population growth is complex, but it has not been haphazard. It can be represented by a general model, which compares birth rate, death rate and the resulting population change. The model suggests that world population has gone through a series of sequential stages delineated as follows: (1) high stationary, (2) early expanding, (3) late expanding, (4) low stationary, (5) declining (Figure 4.7). This model of population dynamics, referred to as the 'demographic transition', was based on the European experience of population change during and following the industrial revolution (Molnar and Molnar 2000). Originally derived from the relationship between population and economic change, the different stages have been recognized in other areas where the European pattern of development has been introduced. Differences in timing within and between regions have to be considered, and the factors that have brought about progress through the various stages of the transition also vary with time and place. Although nineteenth-century technology in Europe and twentieth-century technology applied in the developing world were quite different, for example, both brought about the decline in mortality characteristic of the second stage of the transition.

As with all models, the demographic transition model is a simplified representation of a complex phenomenon, but despite its origins it is sufficiently flexible that it can be used to chart stages in worldwide population growth. In addition, passage through each of these stages is associated with socio-economic and (sometimes) cultural change and with the evolution of the relationship between society and environment.

HIGH STATIONARY STAGE

For most of the human existence on earth, the entire world must have been in this phase. High birth rates of as much as 30–40 per thousand per year were countered by similarly high death rates, total population therefore remained low and growth was close to zero. Populations lived at or near subsistence levels, with numbers fluctuating significantly in the short term, much as animal populations fluctuate, and for many of the same reasons. The tendency to population growth in good times was offset by frequent natural events such as famines, epidemics, floods, droughts and climate fluctuations, which, given its low level of technology, society had no means of mitigating. The fossil records from prehistoric times and even early documentary material suggest that, during the high stationary phase, the length of an average generation was very short, with only the rare individual surviving to pass through a full reproductive cycle into old age. Despite this, over the longer term, fertility was sufficiently high to compensate for the high mortality rates and allow the human species to survive.

Few, if any, countries remain in this high stationary phase in the modern world, mainly because of the worldwide spread of technology, which has helped to reduce the death rate and allowed the birth rate to remain high. Countries such as Afghanistan, with a crude birth rate of 49 per thousand and a crude death rate of 22 per thousand between 1990 and 1995 (UNPD 2001), are not long out of this phase, and any increase in the death rate would push them back into it. The rising death rate associated with the HIV/AIDS epidemic in some African countries may have the same effect there (see Box 4.1, HIV/AIDS). However, most developing countries have moved well into the next phase of the model over the past several decades.

BOX 4.2   POPULATION STRUCTURE

Raw population totals tell nothing about the structure of the group of individuals that they represent. They provide no information on the age and gender of the group, or the proportion of males and females, children and adults that it contains, yet these elements have an important influence on the current and future socio-economic impacts of the population. To illustrate and investigate these aspects of population structure, demographers have devised the population pyramid or age–sex diagram. A pyramid represents the male and female members of the population in five-year age groupings or cohorts arranged along either side of a central vertical axis. The horizontal axis indicates the percentage of males or females in each cohort. Since the numbers in any population tend to decline with age, the diagram takes on the appearance of a pyramid (Figure 4.6). The exact shape of the pyramid shows the historical development of the population structure, and indicates the potential for future change.

Those countries currently experiencing rapid population growth, such as the developing nations of Africa, have a population pyramid with a wide base and a narrow apex, indicating a high proportion of individuals in the younger age groups. Perhaps as much as 35 per cent of the population will be under fifteen years old, foreshadowing continued growth as they move into the reproductive sector (fifteen to forty-nine years). If no action is taken, fertility increases and the base of the pyramid widens further. Reduced infant mortality also helps to maintain the broad base. The introduction of family planning programmes in some of these nations has begun to reduce the percentage of children in the youngest cohorts, but as long as the number of people entering the reproductive sector exceeds those leaving it, the population will continue to increase.

The developed nations had similar population structures as recently as the beginning of the twentieth century, but since then significant changes have taken place and their population pyramids differ considerably from those of the developing nations. As a result of the combination of reduced fertility and reduced mortality, the base of the pyramid in a developed nation is narrower – on average, only 19 per cent of the population is less than fifteen years old – and the upper section is wider, giving the pyramid a shape that is columnar rather than triangular. Within this general shape there are variations associated with such events as war, migration, the 'baby boom' and a variety of socio-economic events, which impacted on different nations in different ways (Figure 4.2). Developed nations such the United States, Canada and Australia continue to grow slowly because of immigration, but many of the nations in Europe are experiencing either zero growth or are declining. In their pyramids, this is evident from the very narrow bases, the slowing of the flow through the reproductive sector and the increasing numbers in the post-reproductive sector.

An important element in population growth, reflected in population pyramids, is the time lag involved. Any change at the base of the pyramid will be felt over a considerable period of time as the individuals in that age group progress through the system. The 'baby boom' generation born between 1946 and 1964, for example, forms an obvious bulge in the pyramids of most developed nations. Despite the reduced fertility of that group, it produced a 'baby boom echo' as it moved through the productive sector because of its sheer size. As it continues through into the post-reproductive sector the number of births will decline, narrowing the base of the pyramid, while the increase in the upper age groups will create a pyramid that appears top-heavy.

All these changes have socio-economic impacts and although the time lag involved allows the impact to be predicted, it also means that any significant change at a specific level in the pyramid can ripple through the system for many years.

For more information see:

Miller, G.T. (2000) *Sustaining the Earth* (4th edn), Pacific Grove CA: Brooks Cole.
Molnar, S. and Molnar, I.M. (2000) *Environmental Change and Human Survival*, Upper Saddle River NJ: Prentice Hall.

EARLY EXPANDING STAGE

This period is characterized by a continuation of high fertility (perhaps even increasing fertility), but with a lowering of the previously high mortality rates. Birth rates of 40+ per thousand per year, combined with death rates of 20–5 per thousand per year are quite typical, and the difference between these figures produces large net increases in population. The maximum rate reached when this phase was well established in Britain, for example, was 1.6 per cent per annum (Hornby and Jones 1993). Such a situation may well have existed in the original agricultural hearths such as Egypt and

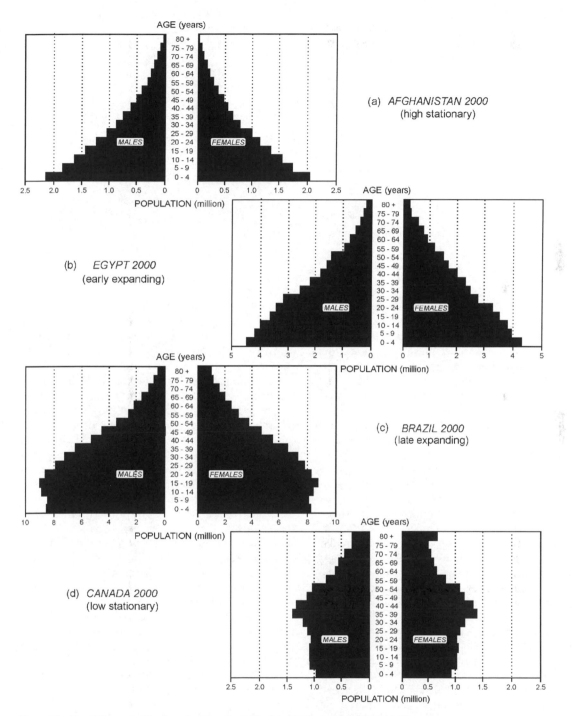

Figure 4.6   Population pyramids characteristics of specific stages in the demographic transition

Source: based on data from the US Census Bureau International Data Base. Viewed at http://www.census.gov/ipcs/www/idbpyr.html (accessed 2 July 2003).

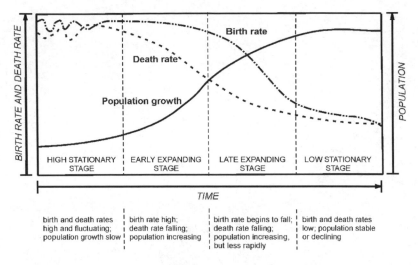

Figure 4.7   The demographic transition model

Mesopotamia, following the discovery of agriculture. These areas were small, however, and their overall effect on world population was limited. The pattern was first identified much later in Europe, where nations gradually moved into the early expanding phase some time in the mid to late eighteenth century, with progress in agriculture providing the initial impetus. The increase in the quantity and quality of the food supply, made possible by the new agricultural techniques introduced at that time, improved nutrition at all levels of society. People were healthier, longevity increased and infant mortality declined significantly. At about the same time the first effects of the industrial revolution were being felt. Although the conditions associated with the establishment of the world's first urban/industrial society were often more likely to increase mortality than to reduce it, ultimately better sanitation, improved housing and an increase in medical knowledge contributed to a decline in death rates, which, coupled with continuing high birth rates, produced the population increase associated with the early expanding phase. Such generalizations hide local and regional variations. In Britain, for example, there was a distinct increase in fertility at the beginning of this stage. This may also have been due to the improvement in the general health of the population, but some researchers have identified more frequent and earlier marriage, associated with economic growth, as responsible for the increase (Hornby and Jones 1993). In France, both birth rate and death rate declined, and in

Germany both remained high. Despite these differences, however, entry into this phase of the demographic transition increased the population of Western Europe from 60–4 million in 1750 to almost 116 million in 1850 (Anderson 1996a).

An examination of crude birth rates and death rates shows that there are a number of nations in Africa and Asia that currently fit the characteristics of the early expanding phase. Ethiopia, for example, has a crude birth rate of 46 and a crude death rate of 21 and in Nigeria the equivalent values are 45 and 15. Pakistan, with birth and death rates of 39 and 11 respectively, and Kenya, with 33 and 13, appear to have recently passed through the early expanding phase. Although the patterns are similar, the causes of expansion in the twentieth century were different from those in Europe in the eighteenth and nineteenth centuries. Improvements in agriculture helped to reduce the death rate but it was little influenced by industrial development, and the major contribution to declining mortality was often from health improvements introduced from the developed nations. Another difference from the European experience was the total population involved. In the eighteenth century European populations were small as the demographic transition began and the rate of natural increase never exceeded 2 per cent. In Asia, Africa and parts of Latin America, many of the nations already had large populations when they entered the early expanding phase. Combined with rates of natural increase that regularly exceeded 3 per cent, this produced significant additions to their populations.

LATE EXPANDING STAGE

Birth rates decline in this phase, but death rates decline even more and a net population increase continues to be registered. Birth rates range typically from 30 to 35 per thousand per year, with death rates less than 20 and often lower than 15 per thousand per year. Britain moved into this phase towards the end of the nineteenth century and remained there until about 1920, during which time the rate of natural increase of the population fell from about 1.4 per cent to about 0.8 per cent (Woods 1996). By that time also, most other Western European nations had moved into the late expanding phase and the pattern was repeated in North America and Australia. The fall in the birth rate seems to have been associated with the continuing industrialization of the economy and the urban patterns of living that accompanied it, but the exact causes remain unclear. The rate fell initially among those in the business or professional classes, perhaps reflecting a reduced need for large families in the new economy, but by the 1920s and 1930s all classes were exhibiting reduced fertility (see, for example, Devine 1999). More attention to, and a wider knowledge of, birth control or family planning may also have contributed to the decline, although only in the later stages of this phase. In reality, there are no clear and definitive reasons for the falling birth rate, but as the late expanding phase progressed family size declined in all the industrialized nations. At the same time, the over-crowding, squalor and disease that had been a feature of many of the new industrial towns and cities of the early and mid-nineteenth century, and had led to increased mortality in some urban centres, were being tempered by modern sanitation and public health services (Woods 1996). The net result was a significant decline in the death rate, particularly among young children. In Scotland, for example, the infant mortality rate declined from 150 per thousand live births in 1850 to 109 per thousand immediately before World War I (Devine 1999). The equivalent figures in England and Wales were 140 per thousand live births in 1880 falling to 95 per thousand in 1916 (Anderson 1996b) with a further decline to 77 per thousand by 1920. In France and Germany the infant mortality rate had

declined to 90 per thousand live births by 1920 and in the United States the rate at that time was 85 (Molnar and Molnar 2000).

Many developing nations, such as Kenya in Africa, India and Indonesia in Asia and Brazil and Mexico in Latin America, now have crude birth and death rates that fit this category. Although the pattern appears similar, their present state is not directly comparable with the situation in Europe in the late nineteenth and early twentieth centuries. Industrialization has taken place in parts of India, Brazil and Mexico, for example, but in other areas the change has come without it. Similarly, urbaniza-tion is proceeding rapidly in developing nations, but without the planning that characterized at least the later stages of European and North American urbanization (see Box 4.3, Urbanization). Much of the change has come about as a result of the intro-duction of improved health care and its impact on the death rate. Sharply reduced mortality and a declining but still relatively high fertility rate have produced increases of between 1.5 per cent and 2.5 per cent. With the higher populations generated during passage through the early expanding phase this has ensured that net growth continues to be high in these nations.

LOW STATIONARY STAGE

The low stationary category is marked by birth and death rates that are both low and roughly equal. By the 1930s, most of the countries in western, northern and central Europe, along with the United States, Canada, Australia and New Zealand, were in this stage. Fertility continued the decline begun in the previous phase until the crude birth rate began to stabilize at or below about 15 per thousand per year. With continued improvements in health services and living conditions, the death rate also declined to similar levels and the net growth in population slowed down significantly, to rates below 1 per cent per annum and sometimes less than 0.5 per cent. While mortality has probably reached its lowest possible level in most of these areas, changing socio-economic conditions have contributed to considerable fluctuations in fertility. Later marriage and the postponement of the birth of the first child contributed to the decline in fertility, as did the introduction of the contraceptive pill in the 1960s. In contrast, the post-war 'baby boom' produced a significant increase in fertility in

BOX 4.3 URBANIZATION

At the beginning of the twenty-first century some 47 per cent of the world's population is living in towns and cities, and the United Nations estimates that by 2030 more than 60 per cent will be urban dwellers. In these three decades, the world's urban population will grow at double the rate of the population as a whole, with the main increase – averaging more than 2 per cent per year – occurring in the cities of the developing nations, where the urban population will grow from 40 per cent of the total in 2000 to 56 per cent by 2030. Combined with very low levels of growth in rural areas – less than 0.1 per cent per annum – this will have a significant impact on the distribution of population in these nations. Urbanization will proceed at a slower pace in the developed world, but the process is already well advanced there, and by 2030 84 per cent of the population will be living in urban areas.

The rate of change and the absolute numbers involved vary from continent to continent, but the overwhelming growth in the developing world is evident. Of the nineteen cities with a population of 10 million or more in 2000, for example, fifteen are in developing nations and, according to United Nations' estimates, by 2015, twenty-three cities will have populations in excess of 10 million, with nineteen being in developing nations (Figure 4.8). Although such cities do illustrate what increased urbanization means in absolute numbers, most of the growth in urban population will take place in communities of less than 1 million. Cities of that size in the developing world will account for 45 per cent of the total urban growth between 2000 and 2015, whereas similar cities in the developed nations will provide only 3 per cent. Compared with the developing nations of Africa and Asia, areas in Europe and North America will continue to have a higher proportion of their population living in urban areas, but it is expected that by 2010 more than half the world's population will be urbanized, with Asia alone having 2.6 billion urban dwellers.

HISTORY OF URBANIZATION

The current level of urbanization and its ongoing spectacular growth is very much a modern phenomenon, but urbanization itself has a long history that can be traced back to the very beginnings of civilization. Although the early civilizations, which grew up in the valleys of the Nile, the Tigris and Euphrates and the Indus, owed their existence to the develop-ment of agriculture, they evolved into societies that built the world's first cities, some 6000 years ago. Cities such as Memphis, Babylon and Mohenjo Daro were the forerunners of Athens and Sparta in Greece and of the city-states of Italy, from ancient Rome to Florence and Venice, which flourished during the Renaissance. Elsewhere, from China to Central and South America, urban settlements developed where local conditions could support a concentrated popu-lation. Although their inhabitants numbered tens of thousands, perhaps hundreds of thousands in a few cases, these cities were the exception until at least the eighteenth century in Europe, and much later elsewhere. Most of the world's population lived a rural existence or inhabited small towns that had grown to have populations of a few thousand because they were administrative centres, market towns or ports.

Modern urbanization grew in step with the industrialization that began in England in the mid-eighteenth century. At that time London was the largest city in the country, with a population of about 800,000, but no other city had more than 100,000. The indus-trial revolution created towns and cities where none had been before and caused existing urban centres to double or triple in size in only a few decades. The com-bination of coal and iron led to the development of heavy industry, producing the raw materials for the manufacture of steam engines, locomotives, ships, bridges and the ancillary items such as rails, wheels, girders and boilers associated with them. The textile industry followed, leaving its cottage industry roots to become a factory industry. It moved in from small towns and rural areas to take advantage of the level of mechanization possible in the cities and attracted by the availability of both a work force and a market in the growing urban population. An expanding network of railways allowed cities to spread and incorporate adjacent settlements, until in some areas individual towns flowed together with no obvious boundary between them. This occurred initially in the English Midlands and central Scotland, but soon spread to areas with similar attributes, such as the Ruhr valley in Germany and the Pittsburgh area in the United States. These were the first urban/industrial conurbations.

The initial growth in the cities came about as the result of the movement of workers from agriculture to industry. Both push and pull mechanisms were involved. The growing number of jobs available in the new cities, and generally higher wages, pulled work-ers from the adjacent rural areas and also the small towns serving these areas. The push from the country

BOX 4.3 – continued

to the town came about as a result of a rural population that had grown with the improvements in agriculture, but was being displaced as mechanization began to reduce the need for human labour. The flow of rural–urban migrants was mainly of young people who through high fertility rates contributed to the growth of the cities by natural increase. The contribution from natural increase would have been greater but for the limits set by poor sanitation, inadequate housing, pollution and malnutrition, which remained part of the urban scene until the twentieth century, and raised urban mortality rates.

When the first census was held in Britain in 1801, the country was still predominantly rural, with only 34 per cent of the population living in urban areas. By the middle of the century, in 1851, the swing to urban living was well established, with 54 per cent of the population living in towns and cities by then. At the beginning of the twentieth century, the urban population made up some 78 per cent of the total, and by its end more than 90 per cent. Similar trends were followed in most of industrialized Europe, where urbanization levels now range from 85 per cent to 95 per cent. In North America slightly more than three-quarters of the population live in towns and cities, while in Australia and New Zealand the level is about 85 per cent.

URBANIZATION IN THE TWENTIETH CENTURY

Hidden within these figures are significant changes in regional and national urban patterns. The second half of the twentieth century saw the geographical spread of many cities through suburban residential development. Started originally with streetcars (trams) and suburban railways, it reached its peak with the widespread adoption of the automobile as the preferred form of commuter transport, particularly in North America. Running parallel to this and contributing to the spread was a decline in population in urban core areas, as those who could, sought the advantages of clean air, open space and the other amenities available in the suburbs. The subsequent decline and decay of the urban core areas has been reversed somewhat by gentrification and the building of high-rise condominiums for those unwilling to face the daily aggravation of traffic jams and ever lengthening commuting times.

As the cities have spread, in some areas they have created conurbations that are immensely large in terms of both population totals and geographical extent (Figure 6.5). The first of these was recognized in 1961 and named Megalopolis. It consisted of a continuous stretch of urban development extending from Boston to Washington DC along the eastern seaboard of the United States. A similar, if slightly less intense, corridor now runs through the American mid-west, from Pittsburgh to Chicago (Chipitts) and, in California, the development between San Francisco and San Diego (Sansan) has many of the attributes of a megalopolis. The pattern is repeated on a smaller scale in Europe, along the lower Rhine and Ruhr in Germany and in England in the London–Liverpool corridor. The largest urban conglomeration in the world is in Japan, where the Tokaido megalopolis, stretching between Tokyo and Osaka along the south coast of Honshu, is home to more than 50 million people.

URBANIZATION IN THE DEVELOPING WORLD

Modern urbanization in the developing world is unlike the earlier growth of towns and cities in that it is driven not by industrialization but by natural growth and migration from rural areas. Industrial activities, such as textile and clothing manufacture or the production and assembly of small appliances for markets in the developed nations, are present in many of the cities, but they are often highly automated or mechanized and do not have a high demand for labour. Rather than a pull from industry, migration to urban areas is driven by a push from the countryside, where the available land is no longer able to support the rapidly growing population. Under these circumstances, there is often no real alternative but to move into the city. Such a move may not bring with it any improvement in opportunities, and many migrants end up living in poverty in the squatter settlements that are part of all major cities in the developing world. Sited on the periphery of the cities, they provide a sharp contrast to the affluent residential suburbs that surround major cities in the developed world. The cities differ in their core areas also. Cities in the developing world lack the transport net that allowed the movement from core to periphery in the industrial cities, and as a result their central cores are more densely populated. Urban infrastructure put in place when the cities were much smaller can no longer cope with modern population pressures. Housing is generally inadequate, the provision of services such as

BOX 4.3 – continued

water supply and sanitation is difficult and organized waste disposal often non-existent. With the potential for disease that all this brings, and with a diet that is usually less nutritious than that in the surrounding rural areas, the urban poor in the developing world face a life of poverty, pollution, malnourishment and ill health that can only get worse as the world's urban population continues to rise.

For more information see:

Brown, L.R., Gardner, G. and Halweil, B. (1999) *Beyond Malthus*, New York/London: Norton.

Hartshorn, T.A. (1992) *Interpreting the City: an Urban Geography* (2nd edn), New York: Wiley.

UN Population Division (2002) *World Urbanization Prospects: The 2001 Revision*, viewed at http://www.un.org/esa/population/publications/wup2001/WUP2001_CH1.pdf (accessed 16 June 2003).

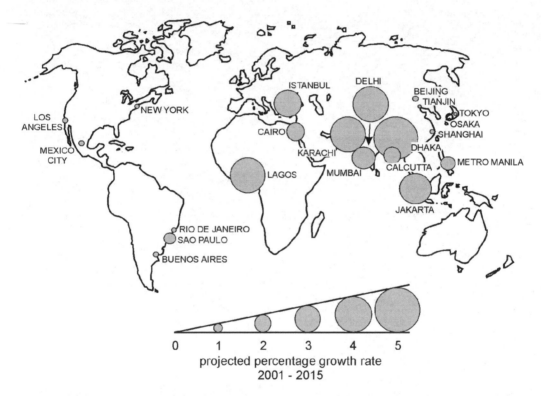

Figure 4.8   Projected annual population growth in the world's megacities (population over 10 million)

Source: based on data in UN Population Division, *World Urbanization Prospects* (2001). Viewed at http://www.un.org/esa/population/pubsarchive/urbanization/urbanization.pdf

the 1940s and 1950s and an echo from that boom has been felt in the 1980s and 1990s.

Significantly, only two or three nations in all of Asia, Africa or Latin America have ever attained the low stationary category. Japan moved into that category in the mid-1950s with the beginning of a rapid decline in fertility, which brought births below replacement level by 1973 (Retherford *et al.* 1996). The decline came about as a result of major social and economic changes following World War II and was encouraged by a government concerned about serious population pressure on the nation's resources. The increased use of contraception and the legalization of abortion had a direct impact on the birth rate, as did changes in Japanese society, which led to major educational and employment gains by women. Together, over less than three decades, these measures cut the crude birth rate by half, which along with continued declines in the death rate brought Japan into line with other industrialized countries in Europe and North America (Molnar and Molnar 2000).

Much smaller than Japan in both area and total population, Singapore in the 1950s and 1960s had a birth rate of 30–45 per thousand per year coupled with a death rate that had declined below 10 per thousand per year by 1960. This gave a natural increase of about 3 per cent per annum and caused severe population pressure on housing, social services and employment. Government-sponsored family planning programmes introduced in the mid-1960s to deal with these concerns were so effective that, by the mid-1970s, the rate of growth was halved. The object of the family planning programmes had been to create a stable population, characteristic of the low stationary phase, but the combination of contraception and abortion, together with a variety of socio-economic improvements and changes in lifestyle happening at the same time, drove birth rates below replacement levels. This raised the fear that the population was drifting into decline and the government was forced to relax some aspects of its family planning programmes (Hornby and Jones 1993).

China, with crude birth and death rates of 17 per thousand and 7 per thousand per year respectively, is technically in the low stationary phase. As with Japan and Singapore this was achieved through the widespread and aggressive use of family planning. So effective were the programmes that in the early 1990s fertility in China fell below replacement levels (Yi 1996). Even with the relative stability of the low stationary phase of the demographic transition, however, the sheer size of the population of China means the addition of some 10 million people to the population every year, and that situation will continue until there is a significant decline in the number within the reproductive sector of the population.

DECLINING STAGE

The traditional demographic transition model had only four phases, but in some nations a fifth phase – one in which the population actually declines – might be considered. In the declining phase, an excess of deaths over births causes the total population of a country to decrease. Sometimes it may be brought on by a high death rate, as in Ireland during the potato famine, or an extremely low birth rate can be a contributing factor, as was the case in Europe in the period between the world wars. In China, between 1958 and 1961, rising death rates and falling birth rates combined to halt a rapidly growing population and send it into decline. Several European nations have moved in and out of this category over the past sixty to seventy years, including Britain, Germany and France. In France deaths exceeded births for ten years between 1936 and 1946, and the situation was a constant source of concern to politicians there and elsewhere. A similar situation prevailed in Eastern Europe in the 1990s, mainly as a result of falling birth rates, and in Russia a falling birth rate plus a rising death rate has caused a serious decline in the population, perhaps by as much as 0.5 per cent per year (Figure 4.2) (Becker and Hemley 1998). In the past, such declines have been turned round with a significant increase in birth rates. In Europe, for example, the dire predictions from the 1930s were confounded by the post-war 'baby boom'. Such a response seems less likely in the twenty-first century, because of the aging of the population in these areas, and it may be that if numbers are to be maintained nations in decline will have to consider replacement migration as a solution to a declining and aging population (UNPD 2001).

THE DEMOGRAPHIC TRANSITION AND THE ENVIRONMENT

As the world's population proceeded through the stages of the demographic transition the relationship between people and their environment changed also. When most of the world was in the high stationary phase, numbers were small and the level of technology was extremely low. As a result, stress on the environment was limited. Demand was for renewable resources, such as water, wood, and plant and animal products. Any damage to the environment, by fire or over-hunting, for example, was quite easily repaired by the environment's built-in recovery processes.

The falling death rate characteristic of the early expanding phase of the demographic transition, probably first experienced some 5000–7000 years ago in the original agricultural hearths (Figure 1.4), owed much to the improved nutrition made possible by developments in agriculture. The available food surpluses also permitted the growth of permanent settlements, with strong central government, a definite division of labour and a need for transport and communication systems. Some of these settlements grew into cities inhabited by tens of thousands of people and in the process the environment was changed. A previously unknown built environment appeared in the landscape; crops replaced natural vegetation; domesticated animals outnumbered the native species; the aquatic environment was altered to allow irrigation. Important as these developments were, the numbers and areas involved were small in world terms and the environmental impact was quite limited. The first signs of future problems were already visible, however. Natural resources such as the soil came under particular stress and in the first agriculturally based societies in Mesopotamia and Central America problems of soil erosion and nutrient depletion were not unknown.

When Britain and Western Europe entered this phase, the impacts of the agricultural revolution were followed by those of the industrial revolution. The demand for renewable resources such as water and wood continued to grow at a rate that exceeded the ability of the environment to recover. In addition, the increasing use of non-renewable resources such as coal and iron created a level of stress in the environment previously unknown. The mining, processing and use of these products changed the landscape, and polluted the air and water. Threats to the environment came not only from the rapidly increasing number of people, but also from the urban/industrial society that they adopted. The atmosphere in most urban areas deteriorated rapidly, becoming filled with smog, while the local rivers, lakes and streams were polluted by the sewage generated by thousands of people living in the growing cities. As such conditions gradually spread around the world along with industrialization and the migration of Europeans, stress on the environment began to be universal.

Environmental disruption accompanied the growth in population in the late expanding phase and even into the low stationary phase when the rate of growth declined. Although the rate of growth in countries now in the low stationary phase is similar to that in the high stationary phase,

the total population numbers are much higher. The relatively simple relationship between society and environment that existed in the earlier phase has also long gone. The needs of an affluent, technologically advanced society place such great demands on the environment that even the stabilization of the population cannot guarantee a decline in environmental deterioration.

DIFFERENCES IN ENVIRONMENTAL IMPACTS IN THE DEVELOPED AND DEVELOPING WORLDS

Additional complications are introduced because not all of the world's regions have passed through the various stages of the demographic transition at the same time and at the same rate. Industrialized nations already in the low stationary or even declining phase, for example, disrupt the environment of nations in earlier phases as they search for and exploit new sources of raw materials not available to them within their own borders (Smith 2000). The developing nations currently in the early or late expanding phases of the transition have not experienced the industrialization, the technological development or the production and consumption patterns that characterized these stages in the developed world. Despite this, population growth rates well in excess of the highest rates ever experienced in Europe in the eighteenth and nineteenth centuries, plus the adoption of industrial and agricultural technology, often under pressure from the developed nations, have ensured that environmental stress is high in the developing world. This is particularly true of renewable resources such as soil, water and forests. Declining soil fertility and soil erosion reduce the ability of the land to produce food, while fragmentation of land holdings and the spread of urbanization take land out of production and increase pressure on the remaining soil resources. New land is then brought into production to offset this, often at the expense of other resources. Harrison (1992) has estimated, for example, that population pressure in the developing countries between 1973 and 1988 led to the clearing of more than 1.1 billion km^2 of forest to meet the demand for additional farmland. The same pressures have forced agricultural settlements to creep up hillsides in countries such as Indonesia, the Philippines and Nepal or to expand into semi-arid lands in the Sudano-Sahelian region of Africa, in China, India and Mexico (United Nations 1994). In both of these circumstances, a common result is environmental degradation associated with the loss of soil fertility and soil erosion.

The environment also comes under threat through the growing demand for the fuelwood that supplies the bulk of the household energy in most of the developing world. Although wood is a renewable resource, it remains so only if the amount harvested is less than that replenished through growth. That point has passed in many areas and the stock of growing wood is depleting rapidly (Myers 1994). The removal of trees from slopes encourages soil erosion, and in areas where fuelwood is no longer available the burning of animal dung or crop waste causes a reduction in organic matter returned to the soil, which lowers its fertility and leaves it open to erosion.

Directly linked with the rapidly growing population in developing nations is urbanization (see Box 4.3, Urbanization). It is not like the urbanization that accompanied the industrialization of Europe in the eighteenth and nineteenth centuries, but it creates similar environmental deterioration. Lack of zoning regulations or lack of enforcement allows the growth of small industry, with its accompanying noise and air pollution, adjacent to residential property. The use of open fires for cooking, burning coal and biomass fuels, contributes particulate matter to the urban atmosphere up to levels that are several times the World Health Organization (WHO) limits. As a result, respiratory disease is common, reaching serious levels in many cities in China, and is particularly high in Calcutta and Mexico City. Noxious exhaust fumes from rising levels of car, bus and truck traffic in the cities of the developing world add to the mix. Even indoors the inhabitants are not safe. Emissions from open-hearth fires or cooking stoves with inadequate ventilation create a major health hazard, especially for women (United Nations 1994).

Urbanization also places great stress on water resources. The water supply is seldom sufficient to meet the needs of the rapidly expanding urban areas and that which is available is often contaminated. The absence of sewage collection and disposal systems in the cities of the developing world contributes to serious deterioration in the aquatic environment, which then feeds back through contaminated drinking water, to expose the population to regular outbreaks of waterborne enteric diseases.

Other waste, such as municipal garbage, is seldom collected, while the indiscriminate dumping of refuse can lead to groundwater contamination and provide a breeding ground for vermin and a variety of pathogens (Falkenmark 1994).

There is abundant evidence to support the claim that population pressures in the developing world make less contribution to global environmental deterioration than increased affluence and technological development in the industrialized nations (see, for example, Ehrlich and Ehrlich 1991; Miller 1994; Smith 2000). However, reality is not as simple as that. Much as technology and economic development have contributed to environmental problems, they have also been used to reduce acid precipitation, ozone depletion and water pollution; the rapid population growth among the developing nations has not been addressed in the same way, if it has been addressed at all, and continues to have an immediate and serious impact on the environment. Where technology or resource planning has been introduced, however, the result has been a much reduced impact. In the Sahel, for example, Mortimore (1989) has suggested that high population densities may not be out of place in areas where proposed soil and water conservation schemes are labour-intensive, and in the Machakos district of Kenya rapid population growth in the second half of the twentieth century was accompanied by environmental improvement as a result of the introduction of erosion control and more appropriate agricultural practices (Tiffen *et al.* 1994). Such studies indicate that population growth does not always lead to environmental deterioration, but these examples tend to be the exception, and the search for a more appropriate balance between population and environment continues.

Stabilization of the population might help, but that has already taken place in the developed world and cannot be accomplished easily in nations continuing to grow rapidly. Even if the population in the developing nations could be moved into the low stationary phase it does not follow that environmental deterioration would cease. Experience in the developed world suggests that stabilization is accompanied by improved economic conditions, which in turn create problems for the environment through the increased demand for resources. According to the Worldwatch Institute this degree of stabilization is unlikely to occur, and if it does it will be as a result of regression into the high stationary phase rather than progression into the low stationary phase. Many developing nations are already suffering from such 'demographic fatigue' in dealing with the effects of a rapidly growing population that their capacity to respond to crises such as drought, famine and disease is very limited. In Africa, for example, the AIDs crisis is spiralling out of control and set to overwhelm more than a dozen nations in the south and central part of the continent alone (see Box 4.1, HIV/AIDS). Under such conditions, population stabilization will be accomplished by a rising death rate rather than a falling birth rate (Brown *et al.* 1999). With the subsequent deterioration in the economic and social infrastructure, the means, and even the will, to deal with environmental issues could be lost, with disastrous consequences.

In short, while there appears to be a simple relationship between environment and population in the early stages of the demographic transition, progression through the transition brings increasing complexity. Direct population pressure on the environment is enhanced by socio-economic and technological change. Stabilization of population alone is insufficient to control environmental deterioration and any progress is likely to require the direct application of appropriate technology.

WORLD POPULATION GROWTH AND TRENDS

Prior to the mid-eighteenth century, the earth's human population remained in the high stationary stage of the demographic transition. As a result, with an average growth rate of less than 0.1 per cent, world population changed little, reaching perhaps 500–800 million immediately prior to the industrial revolution. With the rapid expansion that followed it passed the 2 billion mark sometime in the late 1920s and by the beginning of the twenty-first century stood at about 6 billion (Hornby and Jones 1993; UNPD 2001) (Figure 4.9).

The initial expansion took place in Britain, but quickly spread along with industrialization into Europe and North America. These areas moved through the early expanding phase into the late expanding phase and by the 1920s and 1930s most were in or approaching the low stationary category. Population in the world as a whole continued to grow rapidly, however. The advanced industrial nations, through commerce and colonial expansion, had created conditions in other parts of the world which were to lead to phenomenal growth, largely through the reduction of the death rate. Because of the rapidity of the change, that situation created

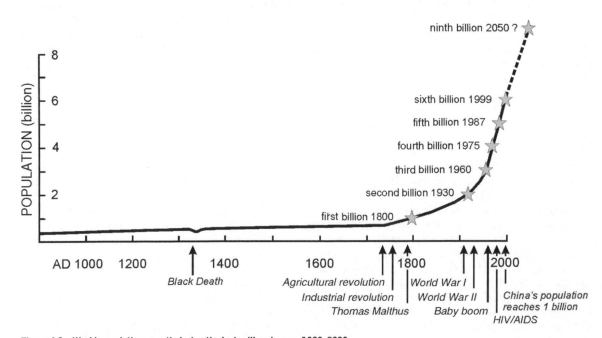

Figure 4.9   World population growth during the last millennium, AD 1000–2000

considerable alarm in the developed world. In the eighteenth and nineteenth centuries the reduction in the death rate came about gradually, and its impact on population growth was buffered somewhat by factors that tended at the same time to reduce the birth rate – a rising standard of living and industrialization, for example – which made children no longer an economic asset. In addition, the rapid growth in Britain and Western Europe was offset by emigration.

The differences are well illustrated by the following examples. In Britain the death rate took seventy years to decline from 22 to 12 per thousand per year. In Japan that same reduction took twenty-seven years, in Mexico about twenty years and in Sri Lanka less than ten years, all following World War II and mainly as a result of better nutrition and health care. Combined with a high birth rate, this produced population growth averaging 3 per cent per annum in all these nations, or about twice the highest rate ever experienced in Britain. With figures such as these, it is scarcely surprising that by the 1960s and 1970s concern about future population growth and its impact on resource use and the environment were high (Hornby and Jones 1993).

The rate of growth of the world's population peaked at 2 per cent per annum in the late 1960s, after which it began to fall, in response to declining birth rates, to reach 1.3 per cent per annum in 1998 (UNPD 2001). Total population continued to rise, however, because the reduced rate of growth was offset by the impact of the large number of young people of reproductive age in the system. This momentum will continue as long as the number of people entering the reproductive sector exceeds the number leaving it (see Box 4.2, Population structure). In world terms, this means that even if reproductive rates are brought down to replacement levels (TFR = 2.1) immediately, population will continue to grow beyond 10 billion before stabilizing, perhaps sometime in the mid-twenty-second century (Shenstone 1997). The annual population increase peaked in 1987 with the addition of 87 million children to the world total. By the first half of the 1990s that number had declined to 81 million annually, and the United Nations has estimated that between 1995 and 2000 the annual increase averaged 79 million. Even with a continuing decline in growth rate, it is projected – based on a TFR of 2.1 – that between 2010 and 2015 the annual increase in population will be 73 million and

by 2050 less than 36 million (UNPD 2001). The net result would be the addition of 2.8 billion people to the world population in the first half of the twenty-first century.

Although it is clear that a significant increase in population should be expected, the totals remain speculative. Past predictions for individual countries and for the world have been consistently wrong. In Britain, for example, it was estimated that, at the rates prevailing in the early 1930s, the population, which stood at 46 million in 1931, would fall below 40 million by 1961 and within a hundred years to 5 million (Branson and Heinemann 1973). The population of Britain in 2000 was actually close to 59 million. More recent predictions have tended to be on the high side and have required regular reassessment. Despite improved reporting and sophisticated analysis, for example, the United Nations reduced its projections of world population totals twice in the 1990s. In an attempt to deal with such variability, the United Nations provides population projections in high, medium and low variants (Figure 4.10). The high variant assumes a TFR of 2.6, which would produce a world population of 10.7 billion by 2050; the medium projection, based on a replacement TFR of 2.1, is of 8.9 billion by 2050; the low variant assumes a TFR of 1.6 and projects a population of 7.3 billion by 2050. The medium variant is the projection most commonly used for planning purposes (UNPD 2001).

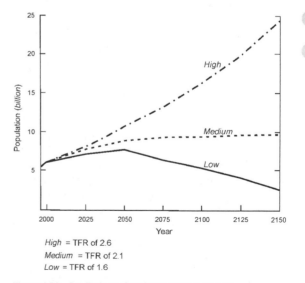

High = TFR of 2.6
Medium = TFR of 2.1
Low = TFR of 1.6

Figure 4.10 Predictions of world population growth
Source: UN Population Division, *World Population Prospects 1950–2050 (The 2000 Revision)*, New York: United Nations (2001).

POPULATION GROWTH AND RESOURCES

Whatever the ultimate rate of growth, and however long it takes for the world's population to stabilize, it is widely agreed that the population increase in the developing world presents the greatest challenge. It is expected that 98 per cent of the growth between 1999 and 2015 will take place in developing nations. These are nations that are already facing serious – perhaps insurmountable – economic, social and environmental problems. Some may already be approaching the carrying capacity of the land they occupy; others face problems of sustainability if nothing is done to slow the rate of growth. A reduction in the population growth should also lead to a reduction in pressure on food supply and other resources, while encouraging the development that would allow the low growth rates to be maintained. However, many of the countries involved are already so fatigued, socially and economically, in dealing with the implications of rapid growth that any recovery will be slow and any development so far into the future that before it is achieved many will have succumbed to famine, resource depletion, uncontrollable political instability and economic decline (Brown *et al.* 1999).

Alarming as such developments may seem, the concepts they represent are not new. As early as 1798 an English clergyman called Thomas Malthus pointed out that population grows not arithmetically but geometrically, like compound interest in a bank account (see Box 4.4, Thomas Malthus). On the other hand, food production can increase only in arithmetical progression, so that population growth would eventually outstrip food production, and famine, disease, pestilence and war would follow.

Shortly after Malthus published his ideas they were apparently confounded. The opening up of new land in the Americas, Australasia and South Africa, coupled with the introduction of new agricultural techniques, allowed food production to keep up with population increases, and in some cases to outdistance it. In places, however, the truth of the Malthusian predictions was all too clear – as in the case of the potato famine in Ireland, where the logistical problems of replacing the lost potato crop, coupled with political unwillingness to try, contributed to the death of several hundred thousand people. Migration from Europe also helped the situation by reducing the number of mouths to be fed at home and providing the labour by which new land around the world could be brought into production. These areas with small populations, but large food-producing capacity, then began to export food to the more populous and developed countries in Europe. All in all, during the nineteenth and early twentieth centuries, food production increased at more than arithmetical progression, contrary to the Malthusian formula. The application of technology to agriculture plus successful research and development improved production from plants and animals, allowing food production to be maintained at levels sufficiently high to meet the needs of the world's current population. Unfortunately, the availability of food varies from one part of the world to another and famines similar to those predicted by Malthus continue to occur. Whatever the historical experience, it remains true that there is a fundamental difference between the maximum rates of population growth and the maximum rates of growth of food production. Despite continuing success in meeting the needs of a growing world population, there are fears that the present level cannot be maintained indefinitely (Molnar and Molnar 2000). It is possible that Malthus was not wrong – only premature.

The human species faces a dilemma in all of this. On one hand, it needs to ensure its survival and on the other it needs to ensure that success in multiplying its numbers does not strain the resources or irreparably damage the environment on which it depends. Successful breeding leads to pressure on resources, and there are neo-Malthusians who see this pressure at a maximum already. If nothing is done to bring down the current rate of increase, society will find itself living in a world exposed to disastrous miseries and Malthusian predictions may yet come to pass (Brown *et al.* 1999). Aware of the serious implications of such a scenario, many nations have introduced policies that mandate some form of population planning and control.

POPULATION PLANNING AND CONTROL

In theory, the reduction in the rate of population growth can be accomplished relatively easily. Since population growth is directly related to the difference between birth and death rates, a combination of reduced birth rate and increased death rate would be most effective. Many primitive societies

BOX 4.4 THOMAS MALTHUS (1766–1834)

Thomas Robert Malthus, an English clergyman and economist, was one of the first to address the problems associated with a rapidly growing population. In 1798 he published a short book entitled *An Essay on the Principles of Population as it Affects the Future Improvement of Society*, in which he set out his concerns and conclusions about the changes he saw taking place at the time. The impact of the original small volume was significant and it grew through a series of revisions between 1803 and 1872 as Malthus developed his ideas further and responded to his critics. Central to the essay was the discrepancy likely to arise between a growing population and its food supply. He recognized that whereas population growth tended to follow a geometric progression (1, 2, 4, 8, 16, 32, 64 . . .), the growth in food production, on which the population depended, was arithmetic (1, 2, 3, 4, 5, 6, 7 . . .) and concluded that population growth would ultimately exceed the available food supply. At that stage, famine, disease and warfare would reduce the population until it was once again within the bounds of available sustenance. As an alternative, Malthus suggested, the practice of 'moral restraint' – abstinence and delayed marriage – could reduce population growth to manageable rates. Although this seemed preferable to poverty, vice and misery, Malthus had little confidence that the population would exercise the necessary restraint and foresaw that the more drastic events would provide the main checks on population growth.

These theories were of interest outside the social, political and economic fields from which they were developed. In reality they dealt with patterns common among animal populations, which tend to expand and contract in concert with their food supply. Malthus saw the human animal as no different from other animals in that respect. This annoyed his fellow clergymen, but it caught the interest of natural scientists, and the most famous of them all, Charles Darwin, acknowledged Malthus's influence on the development of his ideas on natural selection.

On the other side of the world, in China, Huang (or Hung) Liang-chi (1744–1809) had come to the same conclusion at about the same time. Huang was a scholar and intellectual who had been sentenced to death for criticizing the emperor, but had the sentence reduced to exile. He recognized the rapid population growth in China at the time and noted that the means of subsistence could not be expected to match it. Like Malthus he noted the impact of natural checks on the growth of population, which he identified as flood, drought, plague and pestilence, but he also recognized that in the long run they were quite ineffective, since population would begin to increase again once the natural checks had run their course.

In China, the cycle of population increase and decline associated with the inability of the food supply to match the growth of the population continued for some 200 years after Huang's recognition of the problem. In Europe, on the other hand, the Malthusian predictions did not come to pass. The opening up of new lands for European migration and improved agricultural technology kept the natural checks at bay, and the development of birth control and other family planning techniques took the place of 'moral restraint'. Continuing high productivity levels in the twentieth century, coupled with a significant reduction in fertility in the developed nations, seemed to show that Mallthus's fears were unfounded. In the latter half of the century, however, rapidly expanding populations in the developing world, unable to produce enough food for their own survival, are again experiencing starvation, disease and war. For some these represent the main natural checks on population growth foreseen by Malthus, but modern socio-economic and technological conditions are quite different from those in the eighteenth century and the relationship between population and food supply is much more complex.

For more information see:

Dupaquier, J., Fauve-Chamoux, A. and Grebenik, E. (eds) (1983) *Malthus Past and Present*, London/New York: Academic Press.

Ehrlich, P.R. and Ehrlich, A.H. (1990) *The Population Explosion*, New York: Simon and Schuster.

are said to have practised some form of death control such as infanticide or the abandonment of older people when times were hard, but at present any attempt to increase death rates would be looked on with horror. Thus it is birth control, in its broadest sense, that has to bear the burden of reducing population.

Birth control

Effective birth control methods have been available since at least the second half of the nineteenth century, but until then – and even since then in many areas – almost the only birth control methods practised were indirect, usually cultural or social in nature. Often they came about not by deliberate choice, but as a by-product of some physical, cultural or economic activity. Breast-feeding is recognized as providing some protection from conception, for example. Any number of taboos and customs surround sexual behaviour and create a considerable gap between fertility and fecundity. Although these cultural or social methods of population control are less efficient than mechanical or chemical methods, they do work. Even during Victorian and Edwardian times in Britain, when families of fifteen and even twenty were not unknown, the actual number of births was always far below the maximum reproductive capacity of the population.

The range of cultural or social elements that contribute to birth control is wide and includes societal fashions, family decisions, economic position in society, education and employment opportunities (Figure 4.11). In all this, the role of women became increasingly important. Societal fashions, for example, influence marriage customs, which in turn can decide whether a woman's first child arrives early in her childbearing years or is postponed until long after puberty. The latter situation has become more and more common in developed nations, where it is not unusual for women to give birth to their first child after the age of thirty. In some countries – China, for example – the postponement of marriage has become official government policy and, combined with effective birth control, does seem to help to reduce the rate of population growth. Economic conditions that encourage or require women to become part of the labour force also lead to a reduction in the birth rate. In Eastern Europe, for example, following a brief 'baby boom' after World War II, the birth rate declined steadily into the 1980s as the number of women in the work force increased (Hornby and Jones 1993). Elsewhere also, an economic need for women to be in paid employment, combined with other factors such as the growing number of women entering higher education and a subsequent desire to use that education, led couples to postpone starting a family, a decision that made a major contribution to the decline in fertility in the developed nations in the 1970s and 1980s. By providing women with the skills to participate more fully in the workplace, education has had a significant role in population control in the developed world and there is strong evidence that improving access to education for women at all levels – elementary to post-secondary – is one of the best ways to slow population growth. This appears to be particularly true in many parts of the developing world, where women have traditionally received much less formal schooling than their male counterparts. Educated women are more likely to be employed outside the home and to marry later, which leads them to have fewer children. It has been estimated that up to 60 per cent of the decline in fertility in parts of Latin America is due to higher levels of education for young women (ICPQL 1996).

Biological, chemical and mechanical methods of birth control

Modern societal conditions, such as those involving economics and education, which contribute to family planning are indirect and would not work without various direct forms of biological, mechanical and chemical birth control. Biological methods range from total or periodic abstinence to prolonged breast-feeding and include abortion. Although not a method of contraception, since its aim is to put an end to the results of conception rather than prevent it, abortion has been widely used to reduce birth rates, particularly in countries where safer and less invasive methods have not been available. Such was the case in many countries in Eastern Europe, where modern contraceptives were not readily available until the 1980s. It has been estimated, for example, that in Hungary in the 1950s as many as 150,000 illegal abortions occurred

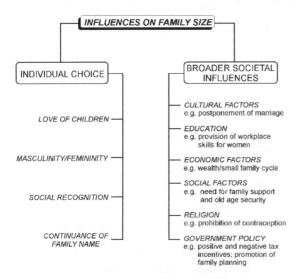

Figure 4.11 Individual and societal influences in family size

annually. With legalization, the number rose to 207,000 in 1969, only falling again in the 1980s – to less than 80,000 – as modern contraceptive methods became available (Hornby and Jones 1993). In China, despite a high level of contraceptive use, almost 54 million abortions were carried out between 1980 and 1984, largely as a result of pressure to have all second and subsequent pregnancies terminated in support of the government's one-child policy (Jowett 1990). Effective as abortion appears to be, it has serious implications for the health of the women involved, particularly when the procedure is carried out illegally. Illegal abortions, which account for some 20 million of the 45 million to 50 million operations performed annually, are often unsafe. Most occur in the developing world, where the death rate is estimated at one for every 250 abortions (ICPQL 1996). Abortion is not an appropriate substitute for modern methods of contraception, but it is likely to continue at relatively high levels in some areas until a combination of economics, education and changing social attitudes allow it to be replaced.

Most of the success in birth control can be attributed to the widespread use of mechanical and chemical methods of contraception. Those providing a physical barrier to conception – condoms and diaphragms, for example – have been available since at least the 1930s, but their overall impact on fertility is difficult to measure and may well have been small (Anderson 1996b). A major advance in contraceptive technology occurred with the introduction of the contraceptive pill in 1963. Its effectiveness and ease of use made it instantly popular, and in developed nations its impact is evident in fertility patterns between about 1964 and 1980 (Murphy 1993). In Britain, for example, the beginning of a twenty-year decline in fertility in the late 1960s is seen as largely attributable to the introduction of the contraceptive pill, with slight reverses in the decline in 1970 and after 1977 being associated with concern over health hazards arising from its long-term use (Murphy 1993; Hornby and Jones 1993). New approaches to contraception continue to be explored. In the past, the emphasis has been on improving effectiveness and ease of use, but more recently, with the growing concern over the spread of the AIDS epidemic (see Box 4.1, HIV/AIDS), health agencies have been promoting the use of condoms, which have the ability to prevent conception as well as slow the spread of the HIV virus.

The most commonly used methods of mechanical and chemical contraception are 94 to 98 per cent effective. They are only effective, however, if they are used. According to UN surveys, 58 per cent of married women of reproductive age (i.e. fifteen to forty-nine) use some form of contraception, 42 per cent do not (UNPD 1999). The level of contraceptive prevalence is spread unevenly across the world (Table 4.2). As might be expected, it is highest among the developed nations. All have prevalence rates in excess of 50 per cent and almost three-quarters have rates of 70 per cent or more. In contrast, values in the developing world range from close to 83 per cent in China – the highest in the world – to as low as 1 per cent in Guinea and Mauritania in Africa. More than half of the nations in Africa have prevalence rates less than 20 per cent; in Latin America and the Caribbean the average is 49 per cent, but in some countries – Haiti and Bolivia, for example – rates continue to be below 20 per cent; in Asia, Thailand and Indonesia, with prevalence rates of 72 per cent and 52 per cent respectively, have successful population stabilization policies, but in India, which may soon surpass China in total population, the rate remains at only 41 per cent. Even in those countries where contraceptive use is low, it does not follow that this reflects the choice of the population. The United Nations recognizes the existence of an 'unmet need' for

TABLE 4.2 LEVELS OF CONTRACEPTIVE USE IN SELECTED AREAS (%)

Country	%
World	58
Less developed nations	55
Africa	20
Asia (except Japan)	60
Latin America and Caribbean	66
Oceania	29
More developed nations	70
Japan	59
North America	71
Europe	72
Australia and New Zealand	76

Source: based on data in UNPD (United Nations Population Division) *World Contraceptive Use 1998*, New York: United Nations (1999).

contraception in many developing nations, particularly those in sub-Saharan Africa (UNPD 1999). Although the necessity to supply that need may be recognized most of the nations involved are unable to set up and finance the appropriate programmes without international aid. Assistance is already provided by the developed nations through a fund created at the 1993 International Conference on Population and Development and was projected to amount to US$5.7 billion by 2000. That figure was not reached. Until it is, countries such as those in sub-Saharan Africa will continue to have an unmet need for contraception and will suffer the consequences of total fertility rates as high as 6.1 and annual growth rates close to 3 per cent (Shenstone 1997; ICPQL 1996).

Direct state participation in population planning

Although decisions on birth control take place ultimately at the family or individual level, the overall management of population growth is a complex and dynamic matter, which is unlikely to be successful without government participation at the national or even international level. The days when a large and growing population was seen as an indication of a nation's success or prestige are long gone. Most governments are now involved at some level in the management of population growth. The approach may be relatively benign, through advertising, propaganda and the provision of family planning materials, for example, or more aggressive, as in China, where the one-child policy placed a direct limit on family size, or in India, where forced female sterilization was common in the 1970s (Jowett 1990; Jacobson 1991). Various forms of negative reinforcement or coercion through social and economic disincentives are also common. These range from penalties for marrying too young to increased hospital costs for second or third children. Families with more than the recommended number of children may also suffer discrimination in the form of reduced tax relief or restricted access to education. Such approaches appear to have been successful in China (Cheng 1991) and in Singapore (Hornby and Jones 1993), for example, but it is not always possible to determine how much of the reduction in fertility is due to direct planning and how much happened as a result of other factors. In Singapore, the family planning initiatives were introduced at a time when social and economic conditions were improving, and that may have contributed to the success of the programmes. A similar combination of family planning and economic change occurred in Japan (Retherford *et al.* 1996) beginning in the 1950s and producing results comparable to those in Singapore.

Family size

It is not surprising that socio-economic change should have such an impact on population growth. Even before direct family planning was available in the late nineteenth and early twentieth centuries, birth rates had begun to fall in the European industrialized nations, but the desire for, or acceptance of, smaller families is a more recent phenomenon. In the past, and in some nations at present, larger families were seen to provide certain benefits. Even children younger than ten years old can contribute to the overall family income, for example, and in pre-industrial Europe young boys and girls routinely worked on farms and in mines. Later they were employed in the textile and other manufacturing industries. In the absence of help from the state, children were insurance against neglect and poverty in old age, and in the days before improved health care a large family was one way of increasing the odds that at least a few of the children would survive to provide some form of parental care. The provision of child labour and family support of the elderly continues to be important in some parts of the developing world. That will tend to slow the trend towards smaller families, but in much of the world, including many developing nations, it has become apparent that social and economic factors combine to generate a situation in which a certain lifestyle can be maintained only if the family remains small. In general, whatever the society, middle and upper-level socio-economic groups tend to have smaller families, while lower socio-economic groups have larger families. With fewer mouths to feed and bodies to clothe, well-to-do families become wealthier. Children receive the benefits of a good education and have a better chance of finding good employment when they grow up. A positive cycle develops. Small families promote wealth and wealth promotes small families. From this it might be argued that an overall improvement in economic conditions would be the best contraceptive. Unfortunately, the world cannot wait that long. Customs and habits do not change overnight. Even with effective, conventional family planning methods change is slow, especially in the developing world, where the present numbers are so high that the time lag in population reduction may be several decades long.

Barriers to population planning

Although the benefits of a rational population policy incorporating family planning are becoming increasingly clear, cultural, economic, religious and political barriers remain in some areas. In rural communities in the developing world, where children are important for family support and old-age security, fertility continues to be high. These are often also the areas in which participation in education is low, particularly for women, and the well established relationship between rising education levels and falling fertility does not come into play (ICPQL 1996). Elsewhere, moral or religious pressure may be brought to bear in such a way that increased births will result. The attitude of the Roman Catholic Church towards population control has aroused considerable controversy. The Church has argued the morality of birth control for many years, and prohibition of most methods continues in place. The doctrines of the Church are being eroded at the local and national level, however, as individuals decide perhaps that it is less morally wrong to practise artificial methods of birth control than it is to bring children into the world to face a life of misery, poverty and disease. Questions of morality also arise in the diversion of funding to armaments in India, Pakistan and some nations in sub-Saharan Africa. Social programmes, such as education and health, suffer and family planning programmes are neglected. To be effective, any population policy must do more than provide the physical necessities of modern family planning. It must also be aware of, and address, the cultural, economic educational and religious issues that have the potential to set obstacles in the way of achieving the stabilization of the population.

APPROACHES TOWARDS A SOLUTION

Although population growth has slowed and steadied in many parts of the world, it will continue to be high in some areas for many years to come. The greatest impacts will be local or regional, but because of the nature of modern economics and

politics the effects will be felt worldwide. Stabilization of the world's population can be successful only through international co-operation, with those nations that have achieved it, and have resources available, helping those that are having difficulty coping with the economic and social consequences of a rapidly growing population. That help may take the form of direct monetary aid to provide access to family planning for a larger proportion of the population, but it must also include consideration of other factors known to be related to population growth. The Brundtland Commission in 1987 recognized, for example, that the sustainable development that it promoted would not be successful without the integration of population policies and socio-economic development programmes (Starke 1990). The International Conference on Population and Development, held in Cairo in 1994, supported that approach, but also identified the improvement of reproductive health and the empowerment of women as important elements in any population policy (Shenstone 1997), while the Independent Commission on Population and the Quality of Life included education and counselling in a list of elements needed to provide an appropriate level and quality of population planning (ICPQL 1996).

SUMMARY AND CONCLUSION

The consideration of population is essential to the understanding of modern environmental issues. In the past, the relationship that human beings had with the environment differed little from that of other animals. Small numbers and primitive technology, which restricted access to resources, limited the progress that society was able to achieve, but at the same time the impact on the environment was minimal. Exponential population growth combined with major advances in technology that increased the demand for resources has radically changed that situation, particularly over the last three centuries, and the environment has suffered serious damage. The relationship between population growth and environmental deterioration is not a simple one – the countries of the developed world continue to exert major pressure on environmental resources, for example, even after their total populations have stabilized or, in some cases, declined. The developing nations, technologically much less advanced, are less demanding of resources, but even a population with limited needs can have a significant impact on the environment if the numbers are sufficiently large, and in the developing world the numbers are large enough to have created demands on land and water resources that cannot easily be met. Serious water and air pollution, which matches or surpasses that common in the industrial nations from the early nineteenth to the late twentieth centuries, is also in large part a function of the rapidly growing population.

Although there is no agreement on what an ideal population for the earth as a whole or a specific country might have to be to achieve a balance between population and environmental resources, by the last three decades of the twentieth century it was widely accepted that the world's population was growing too rapidly and the rate would have to be curtailed. Since then, the United Nations, international aid organizations and environmental groups have encouraged the development of population planning programmes, and government-sponsored or controlled programmes are now common. There is some indication that the rate

of growth has peaked, but given the time-lag associated with population change it will be some time before the world population stabilizes. Even when this happens environmental problems will remain until some balance can be reached between what the population wants from the environment and what the environment is able to provide.

At the regional or national level no one solution to the present problems will apply to all situations and no solution will be immediate. Variations in time and effectiveness have to be expected. Variations take place because population deals with people – not just numbers – and people respond to a variety of stimuli, and not always in logical ways. There may be no one approach that suits all situations, but if no resolution is achieved then the checks and balances which severely impacted the population of Ireland in the 1840s, the Sahel in the 1960s and 1970s, Ethiopia in the 1980s and perhaps Afghanistan and North Korea in the early twenty-first century may well come into play with increasing regularity as the century progresses.

SUGGESTED READING

McCullough, C. (2000) *Morgan's Run*, New York: Simon and Schuster. A well researched novel of forced immigration – the First (convict) Fleet to Australia in the eighteenth century. Provides a particularly interesting account of the response of the convicts to the different environment in Australia, particularly on Norfolk Island.

Ziegler, P. (1998) *The Black Death*, Godalming: Bramley Books (originally published 1969). An account of the devastating impact of the Black Death on the population of Europe and its ongoing social and economic impacts.

Scott, S. and Duncan, C. (2001) *Biology of Plagues: Evidence from Historical Populations*, Cambridge: Cambridge University Press. Challenges the traditional explanation of the Black Death as being the result of bubonic plague – a bacterial disease – and suggests that it was caused by a viral haemorrhagic disease such as ebola. Makes an interesting comparison with Zeigler's account.

Hunter, L.M. (2000) *The Environmental Implications of Population Dynamics*, Santa Monica CA: Rand Corporation. Deals with the complexity of relationship between population growth and environmental change, considering problems such as the ongoing rapid population growth in the developing world and over-consumption in developed nations.

QUESTIONS FOR REVISION AND FURTHER STUDY

1 Debate the proposition that 'The growth in resource consumption, particularly as a result of increased affluence in the developed world, represents a much more serious crisis for the world than the current growth in population.'

2 Using population data available in United Nations publications or on the World Wide Web (see, for example, the US Census Bureau, International Data Base at http://www.census.gov/ipc/www/idbsum.html), draw two population pyramids, similar to those in the 'Population structure' box, one representing the age/gender distribution in a developing country and one showing the age/gender distribution in a developed nation. How do the two pyramids differ? What does their shape tell you about future population growth in these countries?

3   Examine the role of migration in the country in which you live. Does it create net emigration (out-migration) or net immigration (in-migration)? Were the migration patterns the same 100 years ago? If not, what caused them to change between then and now? Do you think the patterns are likely to change again in the future?

5

Society, Resources, Technology and the Environment

After reading this chapter you should be familiar with the following concepts and terms:

Agenda 21
agrarian civilizations
appropriate technology
biomass
Bronze Age
Brundtland Commission
commons
common-pool resources
composting
domestication – animals, plants
electricity

energy – non-renewable,
   renewable
fire
flow resources
fossil fuels
full cost pricing
industrial revolution
Iron Age
Luddites
nuclear energy
OPEC

recycling – glass, metals, paper,
   plastics
reserves
resources – conditional,
   non-renewable, perpetual,
   renewable, undiscovered
stock resources
Stone Age
sustainable development
technology

All organisms, from the simplest plants to the most complex mammals, have certain basic needs that must be met if they are to survive. The extent to which a particular ecosystem can provide these needs is reflected in its carrying capacity (see Chapter 4). In most cases the necessities of life include food, water, various atmospheric gases and appropriate shelter, which are provided in a well balanced ecosystem through the interaction of organisms with the physical components of the environment and with each other. Food required by plants, for example, is obtained direct from the physical environment as solar radiation or in the form of chemical nutrients in the soil, whereas animals obtain their food by consuming organisms, either plants or other animals, with which they share the ecosystem. Human beings, as animals, are involved directly in such processes (see Chapter 3), but, alone among the animals, they have stretched their basic needs into wants, and in so doing have disrupted the ecosystems in which they are involved. Wants take many forms, but include such elements as the demand for the improved quality and

quantity of food or safer and more comfortable shelter, which go beyond the basic needs required for survival, and have evolved to make life easier or in human terms more rewarding. Society has been very successful in ensuring that its wants have been met. With the help of technology, people can live comfortably, almost anywhere on the earth's surface, while the growth in population that has made the human animal the earth's dominant species is also an indication of success in satisfying these wants. It is now clear, however, that in environmental terms the price of that success has been high.

RESOURCES

In the broadest environmental sense, resources are any objects, materials or commodities that can be used by an organism to allow it to survive and maintain its functional role in a community. The concept has become almost entirely anthropocentric, however, being used to refer to almost anything that can be used by society to meet its wants and needs. Since the nature of society differs from place to place and has changed with time, needs and wants have not remained constant, and as a result ideas of exactly what constitutes a resource have varied also. Something that constitutes a resource for one group of people, for example, may be unused by another, perhaps because it is unknown to the second group or it has another commodity that serves the same purpose. Those materials, such as air and water, which are needed and used by all, have always been universal resources. Other necessities of life, such as those that provide food and shelter, have shown considerable variation in both time and place. The potato became a resource in Europe only in the seventeenth century, for example, when it was introduced from the Americas, where in South America it had been a staple food resource for thousands of years. Many foods, including corn (maize) and other grains, cattle and sheep, fish, wild game and waterfowl remained resources of only regional importance until improved transport and the large-scale migration of people allowed them to be incorporated into the food supply of much wider areas. Similarly the resources required to provide shelter often had only limited distribution, depending upon such factors as geology and vegetation patterns. In one area, for example, the main building resource might be brick made from local clay, in another it might be stone, or in a forested area it might be wood. Frequently, some or all of these might be combined to provide shelter, and with time, improvements in transport and construction techniques allowed the best that each could offer to be incorporated into buildings almost anywhere.

For much of human history, the dominant resources included land and water for the production of food, wood to provide construction materials and fuel, and metals, particularly iron, tin and copper, for the manufacture of tools, utensils and jewellery. The major changes in agriculture and industry that began in the eighteenth century ultimately produced a modern technological society with its characteristic reliance on metallic ores and seemingly insatiable demand for fossil fuels. These developments also contributed to the view that resources were primarily agricultural and industrial raw materials. Modern interpretations, however, include a much wider range of elements that are categorized according to the context in which they are being considered (Figure 5.1).

RENEWABLE RESOURCES

Perhaps the most common classification of resources is based on availability, with a simple split into renewable and non-renewable groups. When used in an economic context, renewable and non-renewable resources are commonly referred to as flow and stock resources respectively. Renewable resources are replaced at a rate which is faster than, or at least as fast as, they can be used. The oxygen in the air, the plants and animals in the biosphere, the water in the hydrologic system and energy from the sun are all renewable, for example. They are replaced by efficient recycling systems once they are used, as in the case of oxygen or water, through reproduction, as in the case of plants and animals, or by physical and chemical processes, as in the case of the sun. Modern pressures on such resources have created situations in which they are unable to renew rapidly enough to meet the demands placed on them. Plants and animals are being consumed at rates that exceed their natural rates of reproduction, and in an increasing number of areas around the world water consumption has outstripped the ability of the hydrologic cycle to maintain the natural supply. Other renewable resources are available in a relatively constant supply no matter how they are used. Oxygen is unlikely to be used more rapidly than it can be replaced, for example, and the sun will continue to supply energy for billions of years. Such resources are sometimes referred to as perpetual resources.

Common-pool resources

Some renewable resources are considered as common-pool resources (CPRs). They include the atmosphere and the oceans, some water resources – ground water, for example – and historically pasture and forests. These are available to all and have been used for personal gain with little or no consideration for other users (Boothroyd 2000). The benefits of using a CPR are enjoyed by an individual, at least in the short term, but the detrimental effects of the misuse of the resource are spread among other users, therefore there is much less incentive to use the resources wisely. Air pollution, the depletion of aquifers, the collapse of marine and freshwater fisheries and the onset of soil erosion as a result of over-grazing are all examples of the misuse of CPRs. Garrett Hardin, an ecologist,

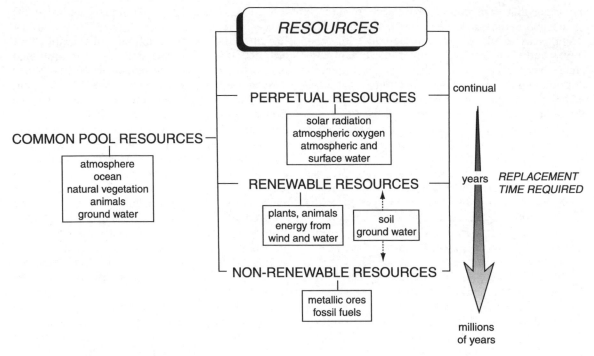

Figure 5.1 The nature of resources and the relationships among them

referred to this as 'the tragedy of the commons' (Hardin 1968), as a result of which CPRs were exhausted, to the detriment of those who depended on them. He considered that the historical concept of the commons was no longer tenable in modern society, mainly as a result of rapid population growth. In this he recognized the Malthusian dilemma of the conflict between population growth and resource availability (see Chapter 4). Hardin considered two possible choices to deal with the problem, both involving the infringement of personal freedoms. Freedom of access to CPRs would have to be restricted, but society would also have to relinquish its freedom to breed. The resulting reduction in the rate of population growth would in theory reduce the threat to resources. In practice, although population growth rates have declined in many areas, there has seldom been an accompanying decline in resource use or misuse (see Chapter 4).

Few, if any, CPRs are completely free of some form of regulation aimed at maintaining their viability. Management of the resources may be private at the individual or group level, or public, involving local, regional or national government. Examples of pri-

vate property systems include irrigation schemes or common pastures, and at the state level, fishing has long been managed by regional and national governments. Global commons need another level of participation, provided through international agreements, fostered by such organizations as the United Nations or the European Union, which involves attempts to control access to and regulate the use of common resources.

The range and scale of CPRs is large and the rules that are created to manage them have considerable diversity, but they do have a number of elements in common, designed to contribute to successful management (Ostrom 1990). These include, for example: clear definition of the extent of the resource and the limits to its use; those who have access to the resource and those who are excluded must be clearly identified; benefits and costs must be considered fair by the participants; users should have the right to derive the rules rather than having them imposed by some other body; rules must be capable of modification if and when conditions change; regulations must be monitored and appropriate sanctions provided and applied for non-compliance; some form of conflict management should be in place; when the resource base is large – at a global scale, for example – units of different sizes will be required and they must have the ability to co-operate across a range of management scales.

Although there will be some variation in implementation in specific cases, common-pool management units that follow these principles are most likely to be successful (Ostrom 2002). Ineffectual control and regulation are not uncommon, however, leading to resource depletion, potential economic or social problems and environmental deterioration.

NON-RENEWABLE RESOURCES

Non-renewable resources are those that cannot be replaced after they are consumed. The term is particularly applicable to fossil fuels, which can be used only once, but it also describes other mineral resources that are present in fixed quantities in the earth's crust. Many metals belong in the non-renewable category, but unlike fossil fuels they can be reused through recycling. Central to the concept of non-renewable resources is the human time frame. Oil, natural gas and coal are being formed beneath the earth's surface at present and new mineral ores are also being created. The processes involved may take millions of years to complete, however, and society can consume these resources much faster than they can be replaced. Thus in human terms they are effectively non-renewable. Because of this, the supply of non-renewable resources is considered to be finite – at some stage they will be completely consumed – which has led to the calculation of estimates of the life span of specific resources. At the current rate of consumption, for example, coal reserves will be worked out in the next 200–300 years, but oil and natural gas supplies may last for only fifty to seventy-five years more and reserves of some essential metals such as gold, silver, tungsten and zinc may be completely depleted within the next two or three decades (Lomborg 2001). Such predictions must be viewed with caution, however. Past forecasts have been confounded by the influence of changing socio-economic conditions on supply and demand and by new technology that led to the discovery of previously unknown deposits of minerals and energy resources or allowed lower-quality ores and smaller accumulations of oil and gas to be exploited. The

percentage of metal required to make an ore viable has fallen steadily over the past century, for almost all important minerals. In 1900, for example, iron ore was considered workable only if it contained 50–60 per cent iron, but by the end of the century that requirement had fallen to less than 25 per cent. Similar values apply to gold and copper, with the average proportion of metal in workable deposits of gold falling from 0.001 per cent to 0.0001 per cent in the twentieth century, and copper concentrations falling from 4 per cent to 0.5 per cent over the same period (Craig 2002).

Consideration of the interplay of economics and technology allows non-renewable resources to be categorized according to their current availability and potential future discovery (Figure 5.2). Reserves are resources that are not currently being exploited, but are present in known quantities that can be extracted or harvested using existing technology under the prevailing economic conditions. Slightly less available are conditional resources, which are known to exist, but are not likely to be used until economic and/or technological conditions change. Undiscovered resources are those as yet unknown but, on the basis of past experience or preliminary exploration, expected to become available. The different categories are not static, but change as the socio-economic, cultural and technological nature of society changes, with the boundary between reserves and conditional resources being particularly elastic. If conditions become sub-economic for a particular ore reserve, for example, it

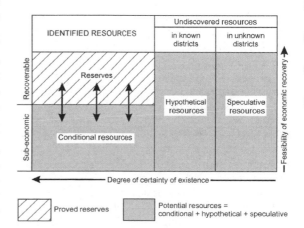

Figure 5.2 Classification of non-renewable resources based on availability and future potential
Source: after Park (1997) with permission.

may revert to being a conditional resource, only to return to reserve status when economic conditions again become favourable for its use.

Given the great range and rapidity of socio-economic change and technological development over the past hundred years or so, it is not surprising that forecasts of the availability and life span of particular resources tend to be inaccurate. There is general consensus, however, that at the beginning of the twenty-first century the rate at which the demand for resources is growing is sufficiently high that it will create shortages of some commodities within decades unless steps are taken to manage their use.

CONSERVATION, PRESERVATION AND SUSTAINABILITY

CONSERVATION AND PRESERVATION

The appreciation of the problems that society faces as a result of the misuse or mismanagement of resources is not new. It was central to the early environmental movement in the nineteenth century, and evolved into two streams (see Chapter 1). One was anthropocentric or technocentric, involving the wise use of resources for the benefit of society. The other was biocentric, aimed at the maintenance of the environmental status quo through the preservation and protection of natural resources such as flora, fauna and physiological features. Both streams have persisted, and are present in the current environmental movement, although neither can claim major success in achieving its aims. Resources have not been conserved or used wisely; few of the world's natural areas have survived, despite the efforts of the preservationists. When it began, the conservation movement was concerned mainly with renewable or biotic resources such as water, animals and vegetation, with some of the earliest attempts at conserving resources being concerned with the appropriate use of forests and water supplies (Ferkiss 1993). The soil conservation movement began in the United States in the mid-1930s as a necessary response to the depredations of soil erosion in southern states such as the Carolinas and Tennessee and on the Great Plains. The United States Soil Conservation Service, established in 1935, was instrumental in developing procedures aimed at the preservation of the quality, quantity and productivity of the soil in an area, using tech-

niques that would maintain soil fertility and slow the rate of soil erosion (see Chapter 6). It also became involved in the rehabilitation of land through the rejuvenation of soil and vegetation. Conservation commonly involves the retention of a vegetation cover, the maintenance of a good soil structure (see Chapter 3) along with a reduction in the speed at which wind and water move over the land, and the adoption of such practices helped the Great Plains to recover remarkably quickly from the ravages of soil erosion in the 1930s. In many areas around the world, however, particularly on land marginal for agriculture, technological and socio-economic factors prevent the implementation of good soil conservation practices, and soil erosion remains a major global problem (Morgan 1995).

The concepts of conservation are now also widely applied to non-renewable resources, particularly those involving energy. Although in the past conservation was seen as having negative connotations, because it disrupted the normal workings of supply and demand and did not fit in with the ideas of growth associated with modern economic development, it is very much a part of the modern approach to resources, which includes not only preservation and protection, but also planning and management to allow both use and continuity of supply of resource materials. It includes both the reduction in unnecessary consumption and increased efficiency in the use of resources so that the production of waste is reduced. All this has the potential to have significant positive impacts on the environment, but there is still a gap between theory and practice.

The world's developed nations attained their status through dedication to economic growth and technological innovation. Growth depended upon the availability of resources plus the expertise to develop them. It provided many socio-economic and cultural benefits, but was also accompanied by consumption patterns that generated large volumes of waste material and led to the destruction of the environment, locally, regionally and eventually globally. The impact remained manageable for the most part as long as the demand for resources was limited by low population numbers and less sophisticated levels of technology, but by the middle of the twentieth century the ramifications of a system that required ever increasing resource consumption began to cause serious concern. It became clear that the availability of some resources was threatened, disproportionately large amounts of these resources were being consumed by a relatively small number of developed nations, the gap between richer and poorer nations was widening and the environmental problems associated with economic growth and development appeared to be out of control. The concept of sustainable development was formulated in an attempt to deal with such issues.

SUSTAINABLE DEVELOPMENT

Sustainable development is development judged to be both economically and environmentally sound, so that the needs of the world's current population can be met without jeopardizing those of future generations. It grew out of the work of the World Commission on Environment and Development (WCED), set up in 1983 by the General Assembly of the United Nations to consider issues involving the relationship between environment and development. The WCED was chaired by Gro Harlem Brundtland, the Norwegian Prime Minister, and as a result became known popularly as the Brundtland Commission. Economy and environment were firmly combined through its promotion of sustainable development. Part of the commission's mandate was to explore new methods of international cooperation that would foster understanding of the concept and allow it to be developed further. To that end, it proposed a major international conference to deal with the issues involved, which led directly to the UN Conference on Environment and Development (UNCED) – the Earth Summit – held in Rio de Janeiro in 1992. The commission's final report *Our Common Future* was made public in 1987 and presented to the General Assembly of the United Nations later that year (WCED 1987; Starke 1990).

Sustainable development was widely embraced as being new and innovative, but sustainability has always been part of all natural systems and reflected in their characteristic equilibrium. It is maintained by the controlled flow of matter or energy through the systems, and disrupted by the restriction of these flows or damage to the components of the system. Natural processes can disturb the equilibrium required for sustainability, but most current threats are the result of population pressure or human activities that result in the excess use of available resources. Non-renewable resources, such as most minerals, cannot normally support sustainable development in the long term, although prudent management combined with recycling can extend their life spans. Renewable resources, on the other hand, should be able to provide a sustainable yield if appropriately managed. However, declining plant and animal stocks, threats to biodiversity, soil depletion and increasingly widespread ecosystem deterioration provide evidence that modern harvesting and resource extraction techniques far surpass the ability of renewable resources to maintain sustainability.

Malthus recognized the dire consequences associated with loss of sustainability, as did early preservationists such as John Muir, although they did not use that term specifically. More recently the concept has expanded beyond its original links with natural systems, and begun more and more to include socio-economic and developmental elements. The reports of the Club of Rome, for example, incorporated the sustainability theme through their consideration of population growth, resource use, food supply and pollution (Meadows *et al.* 1972; Mesarovic and Pestel 1975).

The modern approach to sustainable development

Sustainable development is increasingly multi-faceted, with a need to integrate a wide range of elements. It includes not only the expansion of economic development, but also the resolution of social issues such as poverty and education. It has become people-centred, with local strategies being developed to meet local problems – water supply, for example – while at the same time it retains the scope to foster sustainable patterns of international trade and finance. Sustainable development was central to the UNCED's Agenda 21 (United Nations 1993), and the signatories to that agreement have pledged to work towards the implementation of its recommendations under the auspices of the Sustainable Development Commission. Current sustainability issues cover a very broad spectrum, with the concept of sustainable yield applied to agriculture, energy, fishing, forestry and resource development in such a way that the economic factors are integrated with environmental concerns as far as possible. To make that possible, other UN agencies such as the Food and Agriculture Organization, the UN Environment Program, the UN Development Program and the World Bank have incorporated sustainability into

TABLE 5.1 MAIN RECOMMENDATIONS OF THE WORLD COMMISSION ON ENVIRONMENT AND DEVELOPMENT, THE BRUNDTLAND COMMISSION

Aim	Recommendation
Revive growth	Stimulate growth to combat poverty, particularly in developing countries. Industrialized countries must contribute
Change the quality of growth	Growth to be sustainable and related to social goals such as better income distribution, improved health, preservation of cultural heritage
Conserve and enhance the resource base	Conserve environmental resources – clean air, water, forests and soils. Improve the efficiency of resource use and shift to non-polluting products and techniques
Ensure a sustainable level of population	Population policies to be formulated and integrated with economic and social development programmes
Reorient technology and manage risks	Capacity for technological innovation to be enhanced in developing countries. Environmental factors to receive more attention in technological development. Promotion of public participation in decision-making involving environment and development issues
Integrate environment and economics in decision-making	Responsibility for impacts of policy decisions to be enforced to preserve environmental resource capital and promote sustainability
Reform international economic relations	Basic improvements in market access, technology transfer and international finance to allow developing countries to diversify economic and trade bases
Strengthen international co-operation	Higher priorities to be assigned to co-operation on environmental issues and resource management in international development

Source: based on information in L. Starke, *Signs of Hope: Working Towards our Common Future*, Oxford/New York: Oxford University Press (1990).

their programmes to help them deal with such issues as health, poverty, biodiversity, energy efficiency and land use. Many national, regional and even municipal government organizations have also embraced sustainability as they begin to appreciate its potential to suggest integrated solutions for the environmental and resource issues that they face.

On paper, sustainability can be achieved by following the recommendations of WCED and Agenda 21 (Table 5.1). At the international level, the developed nations must be prepared to alter their patterns of resource use and provide help to the developing nations, but converting theory into practice or ideas into reality is not always easy. The concept of sustainable development is in many ways a guide, which points to a wide range of possible directions that might lead to success. On the way, however, there are likely to be areas of potential tension and conflict. Conflicting views of resource exploitation and conservation held by different sectors of society will have to be reconciled, for example, by conservationists and preservationists accepting planned economic growth using natural resources wisely to alleviate problems of poverty and food supply, and developers appreciating the need to maintain the stability and integrity of natural systems.

The World Bank has already attempted to reduce such problems by attaching environmental and social standards to the development projects that it funds. This top-down approach is a necessary first step, but in itself is probably not enough, and in some cases the short-term impacts of such proposals appear detrimental rather than beneficial. A reduction in the harvesting of tropical hardwoods from the tropical rain forest, for example, is one of the goals of sustainability (see Chapter 7). It would conserve resources, reduce local environmental damage, maintain biodiversity and contribute to the slowing of global warming. For many tropical nations, however, lumbering in the rain forest is an important source of revenue, and sustainability will be achieved only if the issues are defined and presented in such a way that those involved see it as in their own best interests to introduce measures that prevent further damage to the environment and ameliorate existing problems. The approach may be at the national, regional or local level, but for best results it will also have to involve individuals. Experience in dealing with desertification in the semi-arid areas of East Africa, for example, indicates that the best results are achieved when farmers can see clear and immediate rewards in addition to the less obvious longer-term benefits (Pegorie 1990). Success requires education, persuasion (perhaps even coercion) and legislation to provide individuals with the opportunity to participate in the achievement of sustainability. This has worked well in some cases – resource recycling, for example – but in others the appropriate mix has not been achieved: attempts to introduce sustainability into the energy sector through increased efficiency and the greater use of renewable resources have met with limited success. The achievement of even a low level of sustainable development is still some way into the future, but it seems likely that it will be achieved only through a co-ordinated approach involving co-operation among developed and developing nations, international environmental and development agencies, national, regional and municipal governments and including participation by individuals in urban, rural and indigenous communities (see e.g. Chiras 1998; Warren and Pinkson 1998; Woollard and Ostry 2000).

RECYCLING

The need to conserve and enhance the earth's resource base was central to the concept of sustainability in the report of the Brundtland Commission (Starke 1990). Although that will be achieved only through a wide range of possible actions, the reduction and reuse of waste material through recycling has been promoted as an appropriate socio-economic and educational approach to conservation. Its contribution to sustainability through the physical and economic benefits it offers is as yet small, but it can also be seen as having an educational role in that it can be used to make people aware of environmental issues. It provides individuals with the opportunity to participate directly in activities that reduce waste and in many cases allows them to benefit from the results through the use of recycled materials.

Since the earth/atmosphere system is closed in material terms, it includes a number of very efficient sub-systems that recycle elements such as oxygen, carbon, sulphur and nitrogen and compounds such as water, allowing them to be used many times over (see Chapters 2 and 3). When used in the context of conservation and sustainability, however, recycling refers to the recovery of waste material for reprocessing into new products. The reuse of discarded products is often included as part of the recycling process and that will be the case here. Glass milk and beer bottles or compressed gas cylinders are used many times, for example. Eventually the bottles and cylinders will no longer be reusable, but may then be reprocessed to produce new containers or be incorporated in other products.

Although often seen as a modern phenomenon, brought on by the need to conserve resources, recycling has a long history. Farmers who fed food waste to animals were recycling, for example. In the days before large-scale factory farming, the feeding of swill – waste food from homes, restaurants and food processing plants – to pigs was common, and although its use has been banned for health reasons in some countries, elsewhere it continues to be a source of feed for animals on small family farms. Waste pickers, following the traditions of the nineteenth-century 'rag and bone' men, operate on dumps in both developed and developing nations, scavenging materials that can be reclaimed or sold for recycling. The salvaging of recyclable metal products is particularly common, and increasingly, plastics, glass, paper products and organic materials are being recycled rather than being incinerated or dumped into sanitary landfill sites.

Scrap metal

There have been flourishing scrap metal markets for such commodities as iron, steel, aluminium and copper for many years, and a very high proportion of all the gold ever mined probably remains in circulation, because of its high value and resistance to corrosion. The use of iron and steel scrap has been an important component of the ferrous metals industry since the nineteenth century.

Plate 5.1 A recycling depot in central Scotland. As well as plastics and glass for recycling, clothes and shoes are collected for charity organizations.

Manufacturing processes have been designed to use scrap, and in the United States alone some 68 million tonnes of steel were recycled in steel mills and foundries in 2000 (USGS 2002). A major source of steel scrap is the automobile – providing as much as 14 million tonnes per year in the United States – but ship breaking, small appliance recovery and the milling and casting waste produced during manufacturing also provide iron and steel for recycling. The demand for ferrous metal scrap is such that it has become a significant item in international commodity trading (Figure 5.3). Because of the great variety in the form and composition of iron and steel products, most scrap users are both importers and exporters, but there is a net flow of scrap iron and steel from industrial nations in North America and Europe to the Pacific Rim, particularly South Korea, Japan, China and India. The volumes involved vary with the health of the worldwide economy.

After iron and steel, aluminium, copper and lead are the most frequently and effectively recycled metals. The history of aluminium as a recyclable really began only with its widespread use in the aircraft industry in the 1920s and 1930s. During World War II, in Britain and Germany, aluminium pots and pans were collected and melted down to provide raw material for the manufacture of warplanes, and it continues to be recycled in the aircraft and automobile industries, but aluminium's current

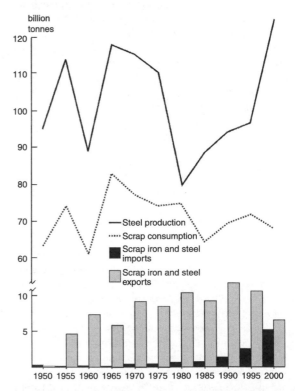

Figure 5.3 Steel production, scrap iron and steel consumption and trade in scrap iron and steel in the United States, 1950–2000

Source: based on data from the US Geological Survey. Viewed at http://minerals.usgs.gov/minerals/pubs/of01-006.

prominent position in recycling is a result of its use as a beverage container. As a result of the relative ease with which they can be collected and reprocessed, plus very effective campaigns launched by the producers and by environmental agencies, an estimated 65 per cent of the millions of beverage cans produced by the industry are recycled every year. The demand for copper in the electrical industry and its use as a constituent of alloys such as bronze and brass have led to the depletion of the best deposits and a decline in the average grade of ore available. This has encouraged recycling, not only of the copper itself, but also of its alloys, and in the United States alone, scrap copper has met as much as 35 per cent of the total demand (USGS 1996). Lead, being soft and easily melted for reuse, has been recycled for hundreds of years. Although its use has declined because of the problems created by lead poisoning, it remains one of the most commonly recycled metals, with as much as 60 per cent of the lead used being from recycled sources, such as old car batteries.

Paper and plastics

The most frequently recycled materials, other than metals, include paper, glass, plastic, organic waste and small amounts of rubber. Some chemicals in liquid wastes, such as those produced by the pulp and paper and photographic industries, are recycled and waste oil from the automobile industry is increasingly being treated and reused. Paper was once a major constituent of landfill sites and was often disposed of by incineration, but paper products are now among the most commonly recycled commodities. Newsprint and office paper are the easiest to recycle, being repulped into a variety of products from new newsprint to cardboard and writing paper. Paper can be repulped only for a limited number of times before the fibres it contains become too short and broken and the quality of the product is compromised. The solution is to mix recycled paper with new pulp. Every tonne of newsprint recycled is one tonne less dumped in a landfill site, and although the recycled paper cannot be used indefinitely, it does reduce the number of trees that have to be harvested, since one tonne of repulped paper is equivalent to the pulp available from seventeen trees (Chiras 1998). As well as the environmental advantages of this approach there are economic advantages, and many

paper manufacturers have begun to include a proportion of recycled material in their product. Major producers have the capacity to use over 1 million tonnes of recycled paper annually, which represents an important reduction in the annual pulpwood cut and a significant contribution to forest sustainability.

Because of such properties as strength, lightness, flexibility, durability, insulating value and resistance to corrosion, plastics are used in a wide range of products, from household utensils to construction materials, and have replaced wood and metal for many purposes. The strength and durability of plastics, while a major advantage when they are in use, create serious disposal problems. Many plastics will last for several hundred years once discarded. Old fishing nets, beer can holders, coffee cups or shopping bags cause aesthetic pollution, but also pose threats to wildlife, which may swallow or get caught in them. For some time now, plastic has accounted for between 7 per cent and 10 per cent of solid waste by weight and between 20 per cent and 25 per cent by volume (Kupchella and Hyland 1993). Until recently, most plastics were disposed of in landfill sites or incinerated, but because of the

Plate 5.2 Collection and recycling of used oil saves resources and prevents pollution.

pressure on land and concern over the release of toxic gases during incineration other methods of disposal have been considered. One of these is recycling and that approach has gained a high level of acceptance among consumers.

Not all plastics are suitable for recycling (Figure 5.4). Thermosetting plastics, which contain resins, cannot be reused because of the permanent chemical changes that take place during their formation. Thermoplastics, however, can be heated or melted and reformed without losing any of their properties. Polythene, polyvinyl chloride (PVC) and polystyrene all belong in this group. In theory all thermoplastics are recyclable, but because of contaminants – for example, colouring pigments – and their use in products containing a combination of plastics or plastics and other materials, it is not possible to return all thermoplastics to their original state. The most commonly recycled plastics are polyethylene trephthalate (PET), used in soft drink bottles, high-density polyethylene (HDPE), used in larger containers, particularly those for corrosive liquids, and polystyrene, used in fast food containers, cups and plates. Plastics are not usually re-formed into their original products. The PET of plastic drink bottles may be reused as plastic tiles, automotive parts, carpet backing or containers for non-food products, while recycled HDPE may appear in garbage cans, flowerpots or so-called 'plastic lumber' used for fencing or house siding.

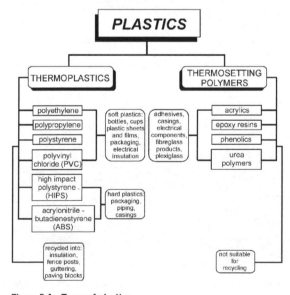

Figure 5.4   Types of plastics

Polystyrene, cleaned and pelletized, is most commonly used for insulation and packaging. Despite the availability of technology to recycle plastics, it is not generally viable without selective collection or some form of automatic sorting. Separation from domestic garbage is not economical and as a result as little as 20 per cent of easily recycled plastics are actually recycled. In 2001, in the United States, for example, only 21 per cent of used plastic bottles were recycled (APC 2002). The remainder continue to be disposed of in landfill sites or incinerated.

Organic waste – composting

A high proportion of the waste produced by human activities is organic in nature. Since it is biodegradable, natural recycling processes gradually integrate it back into environment. In the past when populations were small that did happen, but now, with populations much larger and increasingly concentrated in urban areas, natural processes can no longer cope, and direct waste management is required to deal with the large amounts of organic waste produced. Much of the waste is human sewage or animal waste, and although some recycling does take place through sewage treatment or the use of animal manure on agricultural land, most of the effort in the recycling of organic material is directed at food and garden waste. Much of that is found in municipal refuse, but the food processing industry is also an important source.

The most common method of recycling organic waste is composting. This depends upon the ability of aerobic bacteria and other micro-organisms to decompose leaves, grass, domestic vegetable refuse and other organic waste, including paper. It does not include animal waste, although in some places sewage sludge or animal manure may be mixed with the compost before it is used. Composting may take place in large, unconfined piles or in manufactured wooden or plastic containers. The end product, after six months to several years of decomposition, is a humus-like material that can be used to improve the texture and fertility of soil. It is used mainly by gardeners or horticulturalists, but large-scale composting has been used with some success in tropical countries such as India, where the need to maintain soil fertility is crucial and where climatic conditions promote rapid decomposition. National governments in Europe – for example, in the Netherlands, Germany, Sweden and Italy – have built large-scale

Plate 5.3 Natural composting rates are enhanced by the introduction of temperature and moisture-controlled facilities. The compost in the foreground is produced in days rather than the weeks or months it would take to form naturally.

composting plants, while many North American municipalities have provided composters free or at minimal cost, to encourage what is perceived as environmentally appropriate behaviour and reduce the amount of garbage sent to landfill sites. As with other commodities, the main restriction on recycling organic waste is the collection and sorting of the material to ensure that it contains only waste suitable for composting. It must be done by the consumer, otherwise it is not economically feasible and the waste is sent on to take up space in landfill sites.

Problems and prospects for recycling

Recycling has the potential to provide environmental benefits and support sustainability at all stages of resource development and utilization, from extraction and harvesting to waste disposal (Figure 5.5). The recycling of existing material, for example, reduces the need to exploit new resources. Scrap metal recycling dampens the demand for new ores, which in turn means that landscape, hydrology,

flora and fauna are spared the depredations of mining. Similarly, recycling in the pulp and paper industry benefits the forest environment and makes it easier for companies to work towards a sustainable yield. Indirect rewards include a reduction in energy use for extraction and transport, with less pollution generated at the site and elsewhere. Following extraction or harvesting, energy use and pollution continue to be important elements in the processing of the raw materials. Recyclables also need processing, but in all cases, less energy is needed to convert waste material into a new product than is required to manufacture the product from its raw material. A tonne of steel can be produced from scrap using as little as 25 per cent of the energy required to produce the same amount of steel from iron ore, and for aluminium and copper processing, which is particularly energy-intensive, the savings are even greater, being up to 95 per cent for aluminium (Mannion 1997; Miller 2000). The net result is that less energy is used, helping to reduce demands on the energy resource sector, but also contributing to a reduction in environmental pollution. At the disposal end of the resource use

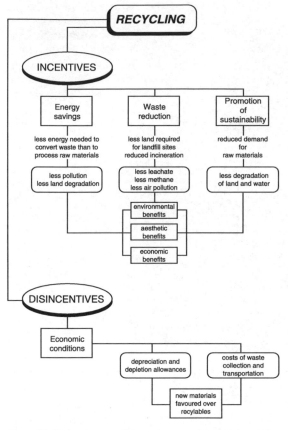

Figure 5.5 Incentives and disincentives associated with recycling

stream, recycling eases pressure on land by reducing the area required for landfill. Since even the best managed sanitary landfill sites disrupt landscape and hydrology and release gases such as methane into the atmosphere, any reduction in waste disposal has clear environmental benefits. Another common form of waste disposal is incineration. Although improved combustion technology and anti-pollution legislation have curtailed their environmental impact, incinerators continue to be significant sources of air pollution. Recycling, particularly of paper and plastic products, contributes to the reduction in the volume of waste that needs to be incinerated, with positive consequences for the environment.

Recycling is generally accepted as being environmentally appropriate and capable of promoting sustainability. In addition, the technology exists to achieve the effective reprocessing of many materials that are suitable for recycling. Why then is it less widely developed than it might be? The problems

faced when introducing or expanding recycling programmes tend to be economic rather than technological or social. The economics of resource use currently favour new materials. Tax incentives such as depreciation and depletion allowances, for example, plus tax deductions for mineral exploration, discriminate against recycling (Smith 2000). Although there are savings in extraction and processing costs when materials are recycled, they are often offset by the costs of collecting them and transporting them to an appropriate processing plant. For many communities, therefore, it is less costly to collect and dispose of the materials as waste than to recycle them. It is only when social and environmental factors are considered that recycling begins to appear worth while, but the costs involved are often difficult to quantify. It is difficult to express aesthetic values or the long-term costs of habitat disruption in monetary terms, for example. As a result, social and environmental concerns usually receive much less consideration than economic factors when recycling becomes an issue.

In attempts to encourage recycling, governments at all levels have used a wide range of incentives and disincentives. These usually take the form of subsidies, taxes or direct legislation controlling the disposal of certain products. Subsidies are used to allow recycling where it would not otherwise be economically feasible but with time is expected to become self-financing. Taxes may be direct, with the revenue being used to subsidize the recycling or disposal process. A number of jurisdictions in North America have taxes in place aimed at dealing with the problem of scrap car tyres, which tend to accumulate in massive dumps, trapping water that provides breeding places for mosquitoes and, in the event of fire, having the potential to cause serious environmental problems through the smoke, gases and oil released. Taxing the use of landfill sites encourages recycling indirectly by making it a less costly alternative to dumping. Legislation directed at encouraging recycling can also include the banning of certain non-recyclable products, such as 'throw-away' bottles and other containers. Individuals are most often involved in recycling through kerbside pick-up – the so-called 'blue box' programmes – or depot recycling, where the waste is dropped off at a central location, both of which have become mandatory in many communities. The main problem with that approach is the sorting of the collected waste. The inclusion of inappropriate materials can make recycling impossible and large quantities of plastics collected for recycling are regularly dumped in landfill sites for that reason.

Although the amount of material being collected for recycling is growing rapidly, it is not being used as rapidly. Changing markets alter the demand for the recyclables and as a result they accumulate in warehouses, and the financial and environmental benefits are not realized. Thus although the potential benefits of recycling are high, and the collection of appropriate materials can be legislated, the process cannot be divorced from the broader considerations of the economic system. At present, the price of an item includes the costs of such elements as raw material extraction, manufacturing, marketing and distribution, but seldom the cost of disposal. One possible method of dealing with that

would be through full-cost pricing, in which the price of a commodity includes the cost of disposal and any other environmental costs that use of the commodity might incur (see Box 5.1, Environmental values). Although the development of full-cost pricing is still very much at a theoretical stage, approaches that include taxes or fees for the use and disposal of specific commodities may represent the first steps in that direction. As natural resources become scarcer and therefore more costly to produce, and waste disposal problems reach crisis levels, economic conditions should become more favourable for recycling and other elements that support sustainability. It remains to be seen, however, if that will occur before the environment sustains irreparable damage from a system that tends

BOX 5.1 ENVIRONMENTAL VALUES

In the modern, market-based financial systems that dominate the global economy, the value of a resource is represented by the price that an individual or a group is willing to pay for it. In large part, prices vary according to the availability of the resource and the demand for it from consumers. Scarcity of supply will cause the price of a resource to rise, for example, as will an increase in demand. This applies to non-renewable resources such as fossil fuels and metals, plus the products manufactured from them, and it is also true of some renewable resources such as trees and agricultural products. Many natural resources, however, such as those that provide environmental goods and services, are not involved in this type of economy. Environmental goods such as fuelwood, meat, fish and fruit collected by individuals and groups for direct consumption do not appear on financial account books, nor does the value that society receives from such environmental services as the absorption of waste products by the atmosphere and the hydrosphere. Since such goods and services are not incorporated in the market system, they are often ignored or undervalued. As a result, the impact of their use may not be immediately evident, hiding potential problems of over-consumption or over-exploitation.

One proposal for dealing with this is full-cost pricing, in which environmental costs – to the extent that they can be quantified – are incorporated in the final cost of the product, causing environmental resources to become subject to the same market influences as other resources. In the case of a metal

product, for example, full-cost pricing might include not only the expenditure involved in mining the metal, processing it and manufacturing the product, but also the costs to the environment of the loss of natural vegetation or the disruption of the groundwater system during extraction, as well as the costs of waste disposal once the product is no longer needed. With food resources, pre-production environmental costs might include soil loss by erosion, harm to wildlife from pesticides and herbicides or the eutrophication of lakes from fertilizer contamination. Delivery of the product to market would incur environmental costs in atmospheric pollution from delivery vehicles, and the disposal of waste products created during processing or following consumption would also generate costs. Introducing such environmental costs would raise prices and lower demand for the product, in turn reducing pressure on resources, while in some cases improvements in efficiency, introduced as a means of combating rising prices, would also lead to less stress on resources. In addition, the revenue provided by this approach could be used to minimize or repair environmental damage. Advantageous as all that might be, lowering demand would have an impact on profits, which may be one of the reasons why few organizations have incorporated the concept into their accounting practices.

Even if full-cost pricing were to be widely embraced, however, calculation of the environmental costs would be complex, because they involve estimating the value of goods and services that are not normally bought and sold. One attempt at calculating

BOX 5.1 – continued

the economic value of the world's environmental goods and services (Costanza *et al.* 1997) has suggested that they are worth on average US$33 trillion per year, or 1.8 times the existing global gross national product. This attempt was acknowledged as a relatively crude initial estimate by its authors, but it does indicate the possibility of evaluating environmental goods and services in monetary terms. Whatever the real value may be, if the costs of providing such goods and services are incorporated into the global economy, the results will be higher commodity prices and the current levels of wages, interest rates, profits and trade would change.

Although there is no indication that this approach will be widely adopted in the near future, there are some indications that the importance of incorporating economic factors into environmental issues is already acknowledged. The 'polluter pays' principle is widely accepted, for example. Simple approaches such as returnable deposits on aluminium beverage cans, lead acid batteries and taxes on tyres to encourage recycling are common. On a larger scale, trading in acid gas and carbon dioxide emissions is already in place and green taxes or carbon taxes continue to emerge from time to time for consideration as a means of controlling resource use.

Some elements in the environment are particularly difficult to value. The aesthetic appeal of landscape, for example, or the psychological benefits of a particular combination of flora and fauna, involve intangible elements that vary from individual to individual depending on such factors as age, gender, state of health and a wide range of cultural factors. Some indication of the value of these elements may be gained by considering the willingness of people to pay to experience them. The number of visitors to a national park, for example, and the fees that they pay give an indication of the value that might be placed on the park, but it is probably not very realistic. The fees may be based on recovering the costs of cutting trails, maintaining camp sites or administering the facilities, for example, and the numbers representative of accessibility to the park rather than any personal assessment of its value. Similarly, the costs of cruises to view the Alaskan glaciers or safaris to East Africa to view animals in the wild are based not on the value of the environmental experience but on transport, accommodation and meals, which reflect market values in the travel and tourism industry rather than the value of the environment to the participants. Such activities do indicate that people appreciate the benefits provided by the natural environment, which may be aesthetic, emotional or intellectual depending upon the individual, and are willing to pay to experience them. The costs involved, however, do not really reflect the inherent value of the environment, or what it is worth to them. The level of satisfaction will vary from person to person, and placing a value on their pleasure or enjoyment, as an indication of the value of the environment, is probably impossible.

Another indication that people value the environment is in their response to conservation groups and wildlife organizations, who regularly raise money to preserve an acre of rain forest, to purchase a wetland, to protect a particular species of bird or mammal or to achieve some environmentally appropriate goal. These groups raise millions of dollars every year by appealing to people who will seldom see the project they are supporting. These donors do, however, appreciate the value of the environment and may receive from their support a degree of satisfaction that they feel exceeds the monetary value of their contribution, although there is no way to quantify that satisfaction.

Assigning economic values to environmental goods and services is complex, but it is an area that is receiving considerable attention from environmental economists. Some environmentalists have philosophical objections to the approach, while others are concerned that the environment may be undervalued because some environmental goods and services are difficult to represent in financial terms. There is a growing feeling, however, that in a culture driven by market forces, the absence of a monetary value for environmental goods and services places the environment at risk and prevents it from being given due consideration in issues such as resource management or development, which in the long run might compromise the sustainability that is considered essential to meet the needs of society now and in the future.

For more information see:

Costanza, R., d'Arge, R., de Groot, R., Farer, S., Grasso, M., Hannon, B., Limburg, K., Naeem, S., O'Neill, R.V., Paruelo, J., Raskin, R.G., Sutton, P. and van den Belt, M. (1997) 'The value of the world's ecosystem services and natural capital', *Nature* 385: 253–60.

Hecht, J.E. (1999) 'Environmental accounting: where we are now, where we are heading', *Resources* 135: 14–7.

Turner, R.K., Perrings, C. and Folke, C. (1997) 'Ecological economics: paradigm or perspective?', in J. van den Bergh and J. van der Straaten (eds) *Economy and Ecosystems in Change*, Cheltenham: Edward Elgar.

to seek low-cost, short-term solutions to resource problems.

THE ROLE OF TECHNOLOGY

In its most basic form technology consists of the combination of tools by which society can accomplish certain tasks. It includes devices that help to make life easier for human beings or allow them to do things that they are not designed to do – such as flying in the air, diving deep beneath the sea or projecting their voice over great distances (Figure 5.6). Although a few animals and birds have

been observed to use natural objects as tools, their impact in ecological terms is minimal (Berthalet and Chavaillon 1993). Early humans differed little from these animals, using stones and sticks to provide food and shelter, but by some 2.5 million years ago they had developed enough to be modifying these natural materials so that they might be used as tools. The descriptive terms applied to early human technological levels indicate the materials from which the tools were made. The earliest was the Stone Age, in which natural rock provided hammers, scrapers, axes and a variety of weapons. Slowly, as some groups developed techniques for fabricating tools from metals, the Stone Age was

Figure 5.6 A timeline for the development of technology, 1750–2000

Source: compiled from various sources.

replaced by the Bronze Age and then the Iron Age. Although each of these represented a major technological step forward for the people involved, the technology was still primitive in modern terms and the impact of the population on the environment therefore very limited.

In some ways, the Iron Age may be considered to have continued up to the present, since iron, in some form or other, remains the most common metal used by society. Current technology, however, is characterized by dependence upon a multitude of resources, and the tools used by society involve complex combinations of metal, plastic, glass, liquid and gas. As a result it is now almost impossible to describe technology in terms of its most common constituent. Modern society has been described variously and popularly as living in a petroleum age, a plastics age, a nuclear age or a computer age, all of which suggest tools or components of tools used by society, but none of which is universally dominant. Only a few countries use tools provided by nuclear technology, for example. In reality, even when stone gave way to metal, many communities remained in the Stone Age – as late as the twentieth century in some cases – and those that

used metals also continued to use stone implements. Today a similar scenario predominates, with different nations using different mixtures of technologies, the choice depending upon a combination of socio-economic, political and cultural factors. The economic differences between developed and developing nations produce different technological mixes, for example. The industrial tools in common use in the economies of the developed nations are missing or available only at a less advanced level in the developing world and nations in the process of increasing their industrial base may use technologies that have been replaced elsewhere. In China and India, for example, development is being driven by a technology based on coal, which in the Western industrialized economies was superseded in the second half of the twentieth century by petroleum-based technology.

Another characteristic of modern technology, which contributes to its complexity, is the speed at which it changes. This is particularly true in the computer field, but it also applies to the technology used in medicine and in communications. In the Stone Age, the form and function of stone axes changed little over tens of thousands of years; in

Plate 5.4 Stonehenge, in Wiltshire, England. A late Stone Age, early Bronze Age monument that indicates what society could accomplish even at relatively low levels of technological development.

BOX 5.2 ISSUES THAT HAVE A NEGATIVE IMPACT ON THE DEVELOPMENT OF NUCLEAR ENERGY

- Increased costs of nuclear fuels as the better-quality ores become depleted
- Capital costs of nuclear plants that are much higher than those of conventional fossil fuel plants
- Citizen opposition to construction
- Questions of safety
- Problems associated with nuclear waste disposal

the twenty-first century computer components can become outdated in a matter of months. Changes in modern technology also tend to result from a planned scientific approach rather than the trial and error of the past. That has allowed society to develop tools to deal with disease, food shortages, lack of shelter and other basic needs very successfully over the past century or so. Given the complexity of modern technology, however, and the speed with which it advances, even the rational, objective scientific approach can lead to adverse effects on society and the environment, as is revealed by the problems associated with nuclear technology (Box 5.2) and the genetic modification of plants and animals (see Chapter 7). There will probably never be agreement on the advantages or disadvantages of science and technology for society, but there can be no doubt that, for better or worse, in the modern world the two are intimately linked.

TRADITIONAL TECHNOLOGY AND APPROPRIATE TECHNOLOGY

Technology involves the practical application of science for the benefit of society. Increasingly, politics, economics, social conditions and other cultural elements influence the nature and direction of technological development, and colour the ways in which it impacts society (Ferkiss 1993). By some, technology is perceived as a positive component of modern society; to others it is at the root of most – if not all – of society's problems. Although such perceptions oversimplify the role of technology, there is some truth in both. The misuse of technol-

ogy, either deliberately or through ignorance, has created the environmental disruption that now threatens society, but technology also has the ability to reverse the disruption and in some cases has already done so. Ironically, many of the problems that society now seeks to manage through a techno-centric approach, involving the application of new and improved technology, were originally created by technology, and past experience suggests that while the continuing search for technocentric solutions is likely to solve some problems, it is also likely to exacerbate others and ultimately create new ones. The impact of technology is now so all-embracing that some observers see society existing in a tech-nological environment rather than in a natural environment (Ferkiss 1993). If that is so, it would appear that the technocentric approach to environ-mental problems is the only way to solve them. Not all groups in society operate at the same level of technology, however, and even in groups with access to identical technologies, it may be the way in which they are used rather than their availability that creates the problem. In such cases, non-technological factors, such as economics and politics, play an important role in determining the impact of technology and it is possible that greater attention to these would allow a less technocentric approach to be successful.

One of the alternatives is the use of what has been called 'appropriate technology'. The idea grew out of the work of E.F. Schumacher. As an economist he was aware that the ever-growing demands being made on resources and the environ-ment could not be sustained, and he put forward his ideas on what might be done to tackle that problem in *Small is Beautiful*, published in 1973. He pro-posed the development of an intermediate tech-nology, which would still allow growth without imposing unmanageable stress on resources and the environment, and out of this grew the concept of appropriate technology. In many ways a precursor of sustainable development, appropriate technology involved a decentralized approach that operated as far as possible with local resources, used less energy and produced less waste. As a result, it was more friendly to the environment than conventional technology. It is difficult to introduce appropriate technology into developed nations, because of their almost absolute reliance on the large-scale, 'hi-tech' use of resources, controlled by governments or multinational organizations. Despite that, tech-nologies that are relatively simple, locally adaptable, resource efficient and environmentally friendly have been introduced successfully in many developed nations. Modern recycling technology and the use of alternative forms of energy would fit that category, for example. Perhaps the most promising future for appropriate technology is in the developing world. Already programmes using environmentally compatible devices – hand tools, methane digesters, wind and water-powered machinery – and locally available resources are operating successfully in Africa and Asia. The demand from the developing world for more advanced technology, plus the willingness of developed nations to service what they see as a growing market, will undoubtedly limit the impact of appropriate technology, and make it necessary in many cases to depend upon the technocentric approach to solve resource and environmental problems.

TECHNOLOGY AND ENERGY RESOURCES

Energy in its various forms has played a major role in the development and application of technology. By developing techniques for concentrating and using massive amounts of energy, human beings have had an impact on the earth that surpasses that of any other species, and throughout history the tools associated with the exploitation of energy resources have been a significant component of the technology available. Because of this, an examination of the human use of energy resources provides useful examples of the changing relationship between society, resources and technology and the resulting impact on the environment.

The sun provides the earth with the continuous supply of energy it needs to run the various physical and biochemical systems which together create the conditions that support life. Some of the solar energy intercepted by the earth is captured by green plants and through the process of photosynthesis is converted and stored in the plants as chemical energy. That energy is then made available to animals when they consume the plants (see Chapter 3). When plants and animals die and decompose, they release their stored energy in the form of heat. This process is responsible for the heat that builds up in a working composter, for example. If, for some reason, decomposition is prevented or is incomplete, the stored energy is retained in its chemical form until conditions are again right for its con-

version. The length of time that such chemical energy can remain in storage is variable. It may last for only a few months – until decomposition halted by low winter temperatures resumes in the spring, for example – or for millions of years, as is the case with the energy stored in fossil fuels. Submerged under water or buried by sediments, the remains of the plants and animals that make up these fuels did not decompose and the energy they contained when they died was preserved. By burning coal and oil, human beings start or restart the process of decomposition and release heat derived from solar energy that entered the earth/atmosphere system millions of years ago back into the active part of the system.

Like all animals, the human animal must consume plants or other animals to obtain the energy it needs for growth, reproduction, locomotion and a variety of bodily functions. For most animals, a high proportion of the energy consumed is expended on the search for the food required to maintain the energy supply. The first human beings were no different. The need for food remains as necessary as it was hundreds of thousands of years ago, although in most cases much less time and effort is expended on obtaining it, and food now provides only a very small proportion of the total energy used by individuals. Nutritionists measure food requirements in kilocalories (kcals), with the daily requirement for individuals being between 2500 kcals and 3000 kcals. These amounts are small in terms of the total energy used by society every day, but they provide a basic unit against which the changing demands for energy can be measured (Table 1.1).

Energy from food and fire

The earliest humans probably consumed some 2000 kcals per person per day, obtained from plants or from animals that had been trapped or hunted. In itself, that is sufficient to ensure survival, but in these hunting and gathering communities the food supply was not guaranteed or regular. Variations in supply occurred as a result of seasonal changes in the weather, natural hazards such as drought, fire and flood, animal population cycles or naturally occurring plant and animal diseases. Most groups ate large amounts when food was plentiful and did without when it was scarce. Although there is no real way of knowing, it seems likely that populations in the very primitive societies of a million years ago fluctuated in concert with food supply, in much the same way as other animal populations (see Chapter 4). Under such circumstances, human impact on the environment must have been minimal, easily accommodated by the built-in flexibility of the system.

The ability of these primitive communities to bring about change was limited by the fact that they had only human muscle power with which to apply the energy they obtained from the food they ate. Later that was augmented by the development of tools and weapons that allowed the muscle power to be used more efficiently (Simmons 1997). Stone axes were used to cut down vegetation; digging sticks allowed easier access to roots; sharper spear and arrow points combined with bows and throwing sticks led to increased hunting success. It was only with the more intensive use of energy in the form of fire, however, that early humans began to have a more significant impact on their environment. In many places, there is evidence of fire in layers of charcoal present in the stratigraphic record, but since fire is common in many biotic communities it is not always clear whether the charcoal represents natural fires or those set by humans. Pollen grains, sediments and archaeological materials found in association with the charcoal sometimes provide clues to its human origins, and the evidence suggests that hunting and gathering groups began using fire deliberately some time between 1 million and 500,000 years ago (Pyne 1997). In addition to basic uses such as lighting, heating and cooking, fire was also used directly in hunting to drive game animals into traps or into areas where they could be more easily slaughtered. Fire removed woody vegetation and encouraged the growth of new grass, which attracted grazing animals and improved prospects for the hunters. Later, when shifting agriculture was developed, the use of fire was the only effective way to clear space in the woods and forests to allow planting. Used in these ways, fire was the first mechanism by which human beings brought about major change to their environment, and they were able to do so by using the energy stored in plants directly, rather than indirectly through food and muscle power. Although fire was an important element in changing the relationship between society and environment, when it was first used communities were small, isolated and nomadic. Natural regeneration allowed vegetation damage to be repaired

and animal populations recovered when hunting pressures were reduced.

The domestication of plants and animals

The environmental impact of the activities of these early human groups remained local and relatively short term. That situation prevailed as long as the earth's population remained small, nomadic and dependent upon hunting and gathering for survival. Some 10,000 to 12,000 years ago, however, a significant change began to take place. A few groups in widely separate locations discovered ways to make their food supply more reliable, reducing the need for their nomadic lifestyle and allowing (perhaps even requiring) a more settled, sedentary existence. It was made possible by the domestication of plants and animals, which began in areas around the western Mediterranean Sea, Asia and parts of the Americas, eventually leading to the development of agriculture and the creation of the first agrarian civilizations in Egypt, Mesopotamia, India and China around 7000 to 5000 years ago (Smith 1995). Similar developments occurred somewhat later in Central America (see Chapter 1; Figure 1.4).

Some of the world's most important crop plants and domestic animals had their origins in the wild species native to these areas (Figure 5.7). In the Middle East, from the western Mediterranean to Mesopotamia, cereals such as wheat, barley, oats and rye were domesticated, as were fibre producers such as flax and hemp (Goudie 1999). Domestic sheep, goats, cattle, pigs and camels originated there also. Farther east, in China and India, rice was first grown as a crop some 8000 years ago, perhaps in several locations, including the lower Yangtze river (Smith 1995). Domesticated species of millet were developed in northern China, along the Huang He, at about the same time. Chickens appear to have been domesticated from jungle fowl in Thailand some 8000 years ago and species of pig and the water buffalo were also derived from native species at several places in east and Southeast Asia. Only a limited number of plants were domesticated in the Americas, but those that were have become important crops. Maize originated in Mexico some 5000 years ago, while the potato was domesticated perhaps as much as 7000 years ago from tubers growing in the northern Andes. Varieties of squashes and beans also had their origins in the Americas. Among the few animals domesticated in

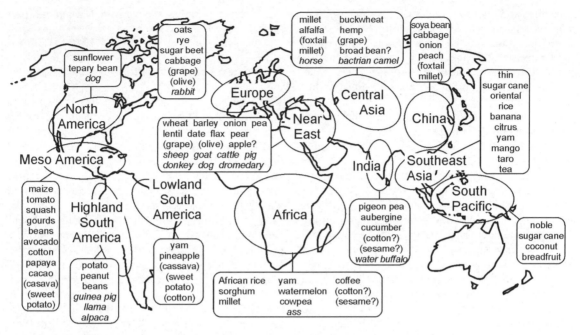

Figure 5.7 Location of major plant and animal domestications. Plants are labelled in plain type, animals in italics; crops thought to have been domesticated independently in more than one area are enclosed within brackets
Source: after Goudie (1989).

the Americas, the llama and the guinea pig in South America and the turkey in Mexico are considered the most important (Mannion 1997).

Some important species were domesticated outside these main centres. The first domestic horses, for example, were bred from herds of wild horses on the steppes of Ukraine, as much as 6000 years ago (Anthony *et al.* 1991). Elsewhere, the sunflower was domesticated in eastern North America and sorghum, yam, coffee and probably cotton were first domesticated in Africa, south of the Sahara (Goudie 1999).

It seems unlikely that it will ever be possible to explain why the domestication of plants and animals and the development of agriculture took place when and where it did. There are some clues, however, that suggest that there were environmental and cultural stimuli, which, acting together, could have created the right conditions (MacNeish 1992). The period immediately following the end of the last Ice Age, during which the domestication of plants and animals took place, was a time of significant environmental change. In the areas that were to support the first agrarian civilizations, the climate grew warmer and drier and, particularly in north-west India, Mesopotamia and Egypt, the river valleys became

the equivalent of oases in an increasingly dry, desert-like landscape. The flourishing vegetation, maintained by a combination of year-round warmth and a regular water supply brought by rivers from adjacent areas with abundant precipitation, supported a large animal population, which in turn attracted nomadic hunting groups from the surrounding drier areas. Sustained by an abundance of resources not available elsewhere, the human population probably increased, and the inclination to settle must have been strong. Even in the lush environment of these riverine plains, permanent settlement would have stressed the native plant and animal communities, and perhaps domestication was a means by which that stress was relieved and a sedentary population could be supported (Mannion 1997).

Elsewhere, in other parts of Asia, Africa and the Americas, the stimuli that brought about domestication were different. In Southeast Asia and Central America there was no threat from desiccation, for example, but the environmental changes associated with the end of the Ice Ages did have an impact on the nature and distribution of flora and fauna there also. The hunting–gathering societies of these areas differed culturally from those in Egypt and

Mesopotamia and were perhaps driven by different cultural stimuli – although the need for food and shelter is universal. There is no individual factor or group of factors, environmental or cultural, that adequately explains the timing and location of the domestication of plants and animals or the development of agriculture. The domestication of so many different species at so many different locations within a relatively short, shared time frame suggests that there could be a common cause, but it remains elusive and may never be discovered. Whatever the cause, this marked the beginning of the manipulation of the environment by human beings. It was the first significant change in the relationship between society and environment, with an impact that has reverberated down through history to the present day.

The birth of agriculture

The development of agriculture is commonly viewed in terms of its socio-economic and cultural effects, with the permanent settlement and local urbanization that it encouraged being the first expressions of civilization. The food surpluses it provided allowed the division of labour that ultimately led to a hierarchical society, with organized religion and a government structure based on a ruling elite. At the peak of their development, much of the activity in these societies took place in urban communities, which served as markets for the produce from the surrounding agricultural land, were centres of government and religion, housed craftspeople and were involved in trade with each other. Although not everyone lived in permanent settlements or cities, there can be no denying that all were influenced by the social, cultural and economic progress associated with these agricultural hearths. Driving much of that progress, and contributing to society's increased impact on the environment, was the change in the amount and form of energy made available by the development of agriculture.

The food surpluses made possible by the adoption of agriculture were also energy surpluses. Bred to produce higher yields, domesticated plants provided more food energy from a given area of land than their natural ancestors, and if the land was dedicated to a specific crop the energy returns were even higher. Similarly, domestic animals confined to an area with a readily available food supply retained more of the energy they consumed than

when roaming free in the wild. In addition to these direct energy benefits, there were also indirect improvements in society's energy budget. With the food production concentrated in the vicinity of permanent settlements, energy did not have to be expended moving around to collect fruits, seeds, nuts and berries or stalking wild game over great distances. In fact it was no longer necessary for most of the community to be involved in the provision of food, as had been the case in most hunting and gathering groups. The energy normally expended on these activities was put to other uses, including clearing land, designing and building permanent structures of brick and stone to house people and animals or to store the surplus food, and, in areas such as Egypt, Mesopotamia and the Indus valley, developing irrigation systems. Most of the work done still involved the use of human muscle augmented by devices such as levers, rollers or wheels, but with the domestication of animals it became possible to use their energy to accomplish tasks that were beyond human muscle power. They were trained for ploughing and a variety of agricultural tasks, for carrying people and moving heavy loads. Probably the most commonly used draught animal was the horse, but communities used what was readily available. Dogs were used in North America, oxen throughout Europe and later North America, donkeys and camels in North Africa and the Middle East, elephants and water buffalo in Southeast Asia. By the time the agrarian civilizations were well established some 5000 years ago, all these benefits translated into a doubling or tripling of society's daily *per capita* energy consumption to 12,000 kcals (Table 1.1). This additional energy increased society's ability to alter the environment. Natural flora and fauna were replaced by domesticated varieties, nutrient balances in the soil were disturbed by agricultural activities, the hydrologic cycle was changed in some areas through the introduction of irrigation, the quarrying of building stone or the digging of clay for bricks altered the landscape and the establishment of towns and cities set the stage for the environmental problems associated with permanent settlement. Formidable and universal as all of these issues were to become, at that time they were still restricted to only a few regions and impacted a relatively small number of people, but they were a portent of what was to come as society found ways of using larger and larger amounts of the energy and other resources available to it.

With time the new agricultural techniques were taken up in other parts of the world. The resulting food surpluses and the energy they provided continued to enhance society's ability to alter its environment. Innovations in metalworking provided bronze and then iron tools that allowed the additional energy to be applied more effectively and permanent settlements, housing increasingly complex societies, became more and more common. In some cases the outcome was monumental, as in the cities of ancient Greece, China and the Roman Empire, but the less opulent results in most other communities were ultimately no less important.

Wood, wind and water

The need for energy to smelt copper and iron was met by using wood directly or by converting it to charcoal. In ironworking the latter was preferred, since it allowed carbon to be incorporated more easily into the metal to improve its properties. Such was the demand for wood and charcoal that deforestation was increasingly common in iron mining areas from the Middle Ages on. Copper and ironworking created local air pollution, and pollutants from the smelting of lead in Roman times have been recognized in the ice of the Greenland Ice Sheet (Hong *et al.* 1994). The use of other sources of energy increased also. The energy in moving air was captured by windmills, which were developed in the eastern Mediterranean, and used for raising water or grinding grain. From there they spread to western and northern Europe, where by the Middle Ages, in areas that depended upon cereal production, the windmill was an important component in most agricultural communities. Where stream flow was suitable, an alternative was the waterwheel, which captured the energy in running water and transferred it by way of shafts, belts and cogged wheels to millstones that ground the grain into flour. Water wheels also drove sawmills, cloth and textile mills and provided the power needed to hammer and shape metal products.

A significant development was the harnessing of wind energy to drive sailing ships. The earliest sailing ships were probably river boats developed on the Nile and in Mesopotamia as much as 4000 years ago, but by about 2500 years ago the ancient Greeks and the Phoenicians in the eastern Mediterranean and the Etruscans in Italy had established a seagoing trading tradition based on sailing ships (Mannion 1997). Somewhat later, shipping was central to the expansion of the Roman Empire. With a simple sail technology that prevented them from sailing close to the wind, or sailing at all if the winds were light or absent, these early sailing ships also depended upon human muscle power. The galleys of Greece and Rome, and later the Viking longships, were combination vessels, using sails where possible, but resorting to oars when winds were contrary or absent. By the middle of the nineteenth century, when sailing ship development was approaching its peak, sail plans and hull design were so advanced that ships like the legendary tea clippers could make very effective use of the energy available in the wind, setting speed and distance records that the early steamships could not match. The environmental impact of the growing use of sailing vessels covered a wide spectrum. Wood was needed to build them. Along with the clearance of woodland for agriculture and the provision of fuel for metalworking, shipbuilding contributed to deforestation originally around the Mediterranean and later in northern Europe. Sailing ships also provided society for the first time with the ability to move larger amounts of material over longer distances. The Greeks and later the Romans were able to bring home grain, wine, animal products and metals from all over their empires in the holds of their ships. The extraction of the metal ores and the spread of cultivated land changed the environment in the colonies, and the wealth that accumulated in the Greek city-states and Rome allowed greater urbanization, which changed the environment there also. Even the efficiently designed and engineered Roman road network could not compete with the empire's ships in the volume of material they could move. Ships moved people, and animals such as rats and cats, over long distances and even caused the spread of disease. The Black Death that ravaged Europe in the Middle Ages was brought to Italy and then to England on sailing ships (see Chapter 4).

By the fifteenth century, sailing ships were carrying European explorers around the world. The colonists who followed brought with them European ideas that were to lead to major environmental change. Although the main impact came through agriculture, which led to the clearing of forests and the introduction of monoculture, often with non-indigenous species, to produce the cash crops needed to make the colonies viable, other environmentally disruptive activities such as the exploitation of minerals – particularly precious metals like gold and silver – and the harvesting of other natural resources such as lumber and furs were also important in some areas. The colonists involved were part of an ongoing assault on the environment that became increasingly worldwide in the centuries that followed, driven by expanding populations and increasingly sophisticated technologies, fuelled by changing energy resources and the way they were used.

Renewable and non-renewable energy use in pre-industrial societies

By the beginning of the fifteenth century the daily *per capita* energy consumption was more than double the level in Greek and Roman times. Human and animal muscles were still providing a high proportion of the energy used by society, but the introduction of mechanical devices to harness the energy in wind and water made a gradual but significant contribution to the total in some areas. Up to that time all of the energy being used was renewable energy – capable of being replaced after use. Food energy was replaced through the annual growth of plants, for example, as was the wood used for fuel. Together, these biotic resources provided biomass energy. The hydrologic cycle ensured a regular supply of flowing water and the circulation of the atmosphere renewed the energy available from the wind. The supply was not completely dependable, however. Wind power was notoriously unreliable in the short term, although that was overcome to some extent by siting windmills in exposed locations. Drought brought reduced stream flow and caused famine, which restricted the availability of food energy. Since their continued supply depends upon seasonal or annual rhythms, which make it difficult to concentrate and use them in large amounts, renewable energy resources are generally considered to be environmentally friendly. They are not impact-free, however, and their use did contribute to a surprising level of environmental change up to the seventeenth and eighteenth centuries, although it pales in comparison with what followed when non-renewable sources of energy were adopted.

Non-renewable energy is the name given to those forms of energy that cannot be regenerated once they have been used. It is commonly applied to fossil fuels – coal, oil and natural gas – which release huge amounts of energy when burned. A rapid increase in the use of fossil fuels from the mid-eighteenth century supported the creation of a technologically advanced society that now dominates its environment. Fossil fuels were used in some areas much earlier than the eighteenth century. There is a long history of coal use in China, for example, and it was used for heating and cooking by the indigenous peoples of the south-western United States as early as the twelfth and thirteenth centuries. At about the same time in central Scotland it was an important source of income for a number of religious communities, which derived revenue from the mining and sale of coal. The production and consumption of coal was well established in Britain by medieval times. Coal mined in the north of England was shipped by sea to London, already a major population centre, which had long since consumed most of the readily accessible fuelwood in its vicinity. The burning of 'sea coal', as it was called, in open fireplaces created an air pollution problem that was to plague the city until the mid-twentieth century (Brimblecombe 1987). The use of oil, the other major fossil fuel, was less common, but there is evidence that the inhabitants of areas as far apart as Alaska and the shores of the Caspian Sea used oil from natural seepages for heating and cooking from a very early date.

Coal

The modern dependence on fossil fuels dates back to the industrial revolution, which began in Britain in the mid-eighteenth century (Hudson 1992). Even before then, however, the use of coal had been increasing as the supply of fuelwood declined. Although, in theory, woodlands were capable of providing an ongoing supply of fuel, clearing for agriculture, poor management and growing demand from the metalworking industries meant that wood was being consumed more rapidly than it could be regenerated. Annual coal production in Britain stood at about 800,000 tons in the early seventeenth century, rose to 1,400,000 tons in the early 1700s, reached a level of 10 million tons by the end of the century and continued to grow (Mannion 1997). Sought initially for the thermal energy it provided, coal did not reach its full potential until it was combined with the steam engine, which allowed the thermal energy it released when burned to be converted into kinetic energy to drive machines. Used first of all in the 1770s to pump water from coal mines that were being dug deeper and deeper as the demand for the product continued to grow, steam engines were also introduced into other areas, particularly the textile industry, by the end of the century. Using cables and pulleys, stationary engines hoisted ore from mines or hauled wagons up inclines, while in the manufacturing industries they were designed to drive hammers and cutters. A major development occurred when steam engines became mobile, powering locomotives and ships. The first steam locomotives, and the railways over which they travelled, were developed in Britain in the early nineteenth century, followed by rapid expansion in the 1830s and 1840s. Planned originally to carry coal from the mines to the markets, it was soon realized that, at a time when the road network was still limited, railways had the potential to carry large volumes of raw materials and manufactured goods quickly and economically from one part of the country to another. Before long they had become important people movers also. The continued availability of coal as a low-cost source of energy and the heavy industrial base of the economy ensured that railways continued to be the main form of land transport in Britain and Europe well into the twentieth century.

The first steamships could not compete in terms of speed and capacity with the existing sailing ships, but ultimately the new technology prevailed. By the middle of the nineteenth century steamships were leading a major expansion in global trade. Benefiting from its early industrialization, Britain led this expansion, shipping manufactured goods from its growing port cities and importing the raw materials needed to maintain industrial progress in the returning ships. When Britain could no longer feed its growing population from its own agricultural production in the middle of the nineteenth century the shortfall was easily made up by shiploads of grain, beef and other more exotic foods imported from around the growing empire. Steamships also contributed to the spread of European culture around the world, by transporting millions of European emigrants to North and South America, South Africa and Australasia up to the outbreak of World War I (see Chapter 4).

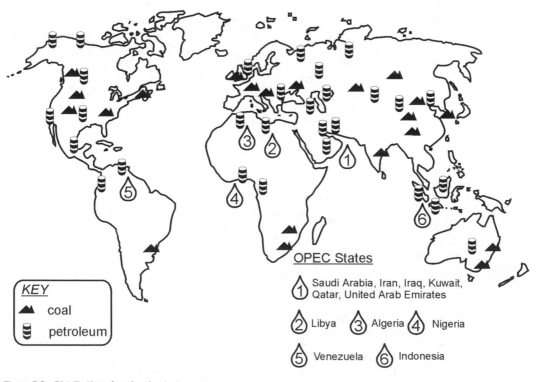

KEY
▲▲ coal
🛢 petroleum

OPEC States

① Saudi Arabia, Iran, Iraq, Kuwait, Qatar, United Arab Emirates

② Libya ③ Algeria ④ Nigeria

⑤ Venezuela ⑥ Indonesia

Figure 5.8 Distribution of coal and petroleum resources

Not everyone appreciated the advantages of the new devices. In the early nineteenth century in England members of the Luddite movement, who were opposed to the mechanization of trade, destroyed some of the new machinery, and groups in France and Germany made similar attempts to slow modernization. Any success was short-lived, however, and the changes that had begun in Britain spread to the coalfields of western Europe, east to Russia and across the Atlantic to North America as the nineteenth century advanced (Figure 5.8). The adoption of coal as the main source of energy was not the only factor in bringing about the industrial revolution. Population growth, the introduction of new techniques in agriculture, innovations in metalworking and manufacturing all contributed to the creation of the urban industrial societies that moved forward into the twentieth century, but, without the extra energy made available by the use of coal, many of the innovations would not have been possible, or in some cases not even needed. The average *per capita* energy use of close to 100,000 kcals available by the beginning of the century made possible the maintenance of a new industrial society, but it also set in train a series of changes that within a few decades had created a group of serious environmental problems.

Electricity

Coal continued to be the main source of energy into the twentieth century, but its dominance was being increasingly challenged by electricity and petroleum. Electricity is a secondary form of energy, generated in a number of ways, one of which is thermal generation – the conversion of thermal energy released by burning a fuel into electrical energy (Figure 5.9). The most common fuel used is coal and in some ways thermal electric generation is an extension of coal technology. The steam produced by heating water in coal-fired boilers was used to turn generators rather than to power steam engines. That could be done in a central location and the electricity then distributed along power lines. With time steam engines were replaced by electric motors, which provided advantages such as greater efficiency and cleanliness. Without electricity, technological developments that have had significant socio-economic and cultural impacts would have been severely restricted or would not have occurred at all. Trams or streetcars and, in larger cities, underground railways, powered by electricity, became a feature of early twentieth-century urbanization. Certain industries became more footloose when, relieved of reliance on steam power, they no longer needed to remain on or close to the coalfield industrial areas. In Britain, for example, growing markets drew industry to London and the southeast, away from the heavy industrial areas of the Midlands and the north, and industries dependent on imported raw materials were attracted to port locations. In the United States the aircraft industry, with no ties to traditional coal and steam technology, was dispersed across the country. Industries that have had a significant social and

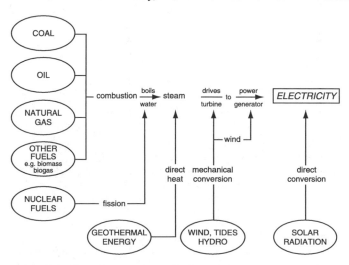

Figure 5.9   Various methods of electricity production

cultural impact in the world were also made possible by electricity. The current extensive use of telecommunications and computer technology, for example, would not have been possible without the reliable supply of relatively cheap electrical energy. Labour-saving devices, from elevators to a myriad of small appliances, have enhanced the quality of life for many, particularly in the developed world, and the provision of better lighting, readily available hot water, washing machines and refrigerators has contributed to improvements in health and nutrition. Availability and consumption of electricity, and the accompanying benefits, have lagged in the developing world, where an estimated 2 billion people do not have access to commercial electricity. Demand in developing countries grew at 8.3 per cent in the 1980s, however, at a time when the increase in worldwide demand averaged 3.5 per cent, and it is projected that by the mid-twenty-first century 60 per cent of global electricity consumption will be in these countries (Preston 1995).

The demand for electricity varies with the health of the world economy, but since at least the 1950s there has been an overall increase and this has led to the regular addition of new generating plants. Most have been fossil fuel plants, but where the physical requirements can be met hydro-electric generating plants have been built. Using the renewable energy available in falling water, they are usually seen as more friendly to the environment than coal-burning

plants, but they are not completely free of problems. A major development between the mid-1950s and mid-1970s was the construction of nuclear power plants to help meet the growing demand for electricity. By the early 1990s, however, the nuclear energy industry was more or less stagnant, in part because of costs, in part because of reduced demand and in part because of safety concerns. It recovered somewhat in the years that followed, and by 2000 more than thirty new plants were under construction (see Box 5.3, Nuclear energy; Figures 5.10 and 5.11; Table 5.2).

For some purposes – in the electronics field, for example – electricity is the only suitable form of energy; for others it is the energy of choice – for space and water heating, for example. It allows the use of modern technology that contributes to the saving of time, energy and space; it is efficient and clean at the point of use; it permits flexibility through the separation of the primary energy source and the point of consumption. Although in many ways it appears the ideal form of energy, there are hidden costs. The production of electricity is remarkably inefficient. Only about 30 per cent of the chemical energy in coal is converted into electrical energy in fossil fuel plants and in nuclear plants the efficiency is not much better (Figure 5.12). The conversion process in coal-burning plants causes air pollution (see Chapter 10) and contributes to global warming through the

BOX 5.3 NUCLEAR ENERGY

FISSION OR FUSION

The commercial production of nuclear energy depends upon nuclear fission, a reaction in which the nucleus of an atom of a heavy metal such as uranium splits into two relatively equal parts, emitting neutrons as it does so and releasing large amounts of energy in the form of heat and radiation. It is the heat released in the reaction that is used to produce electricity. If not effectively contained, the radiation is a major environmental hazard. An alternative source of nuclear energy is nuclear fusion, the process responsible for supplying the energy produced by the sun and other stars. During fusion, the nuclei of light atoms fuse to form a heavier nucleus and as a result of that reaction large amounts of energy are released. In the sun, the fusion of the

nuclei of hydrogen atoms creates helium atoms, producing in the process enough energy to maintain the surface temperature of the sun at about 6000°C. Some scientists see nuclear fusion as the solution to all society's energy problems, but as yet the technology to make it commercially possible is not available.

In nature, the fission process may take place spontaneously, but in a planned reaction it is initiated by bombarding the nucleus of a fissionable atom such as uranium with a neutron. This causes the release of additional neutrons and the subsequent chain reaction is accompanied by the continued generation of thermal energy (Figure 5.10). The development of this process allowed the creation of the atomic bomb, in which an uncontrolled chain reaction releases energy so rapidly that it causes an explosion

BOX 5.3 – continued

equivalent to thousands of tonnes of TNT. When controlled within a nuclear reactor, the fission process provides the basis for the commercial production of nuclear energy.

TYPES OF NUCLEAR REACTOR

Nuclear reactors differ in their detailed engineering, but all designs have essential elements in common (Figure 5.11). The nuclear fuel is the core of the reactor system. It is usually the ^{235}U isotope of uranium in the form of natural or enriched uranium oxide, housed in a shielded reactor vessel. A moderator such as light water, heavy water or graphite is used to slow the neutrons, thus increasing the efficiency of the fission process, and control rods are used to manage the rate of the reaction. Some form of coolant is required to prevent the system from overheating. In light or heavy water reactors the moderator may also act as the coolant, but carbon dioxide, helium and liquid sodium are also used. The heat released during the reaction is transported via a heat exchanger to a boiler where steam is produced. From that point in the process, the nuclear system is no different from that in a conventional thermal power plant, where high-pressure steam-powered turbines are linked to generators that produce electricity. All reactors incorporate systems for refuelling the core and containing or disposing of nuclear waste products created in the fission process.

Most reactors are burner reactors that consume fuel, but some are breeder reactors that produce additional fissionable products during the fission process. Loaded with ^{235}uranium or ^{238}uranium, an unmoderated reactor will ultimately produce ^{239}plutonium. Plutonium is the main ingredient in nuclear weapons, but in terms of energy output it is similar to uranium and can be used as a fuel. Former warhead plutonium is being burned in some reactors in the United States, for example. In the breeder reactors more fuel is produced than is consumed, which suggests that they could be an important source of fuels for burner reactors. Costs are high in time and money, however, and there are safety concerns with breeder technology. Only Britain, France, Russia and Japan have operated breeder reactors, but, other than in Russia, interest in that approach to energy production is declining rapidly.

According to the International Atomic Energy Agency (IAEA), the main international regulatory body for the nuclear industry, in 2000, 483 nuclear power plants, with an installed capacity of 351 GW(e) (1 GW = 1×10^9 watts), were in operation. An additional thirty-three nuclear plants were under construction at that time. Although the energy available in one tonne of ^{235}uranium is equivalent to that in about 3500 tonnes of coal, not all nations are interested in nuclear energy. Those that have nuclear programmes are grouped in four main areas. In North America, with 118 plants operating in Canada and the United States; in western Europe, where France meets 76 per cent of its electricity requirements from nuclear energy; in the former Soviet Union, where Russia and Ukraine continue to construct plants; in Asia, where Japan, with limited access to fossil fuels, has fifty-four plants in operation and continues to add capacity.

Some 17 per cent of the world's electricity is generated using nuclear fuels, but the number is likely to decline and the IAEA estimates that by 2015 it will be only 12 per cent. Plants are being closed down faster than they are being constructed in all areas except Asia. Reactors built in the 1960s and 1970s have reached the end of their operational lives, and the rising costs of construction, the cost of enhanced safety measures and higher fuel costs compared with fossil fuels mean that their replacement is not economically feasible. Disposal of nuclear waste products is a major environmental concern (see Chapter 6) and events such as the Chernobyl disaster in 1986 have had a major negative impact on public confidence in the safety of the nuclear industry. At the beginning of the twenty-first century the nuclear industry is declining. Its contribution to global energy may well have peaked in the second half of the twentieth century, but its impact will continue to be felt through a legacy of waste material that will remain hazardous for many centuries to come.

For more information see:

Ahearne, J.F. (1993) 'The future of nuclear power', *American Scientist* 81: 24–35.

Hodgson, P.E. (1999) *Nuclear Power, Energy and the Environment*, London: Imperial College Press.

International Atomic Energy Agency (2000) *Power Reactor Information System*, viewed at http://www.iaea.or.at/programmes/a2/ (accessed 18 June 2003).

Pryde, P.R. (2002) 'Chernobyl', in A.S. Goudie (ed.) *The Encyclopedia of Global Change*, New York: Oxford University Press.

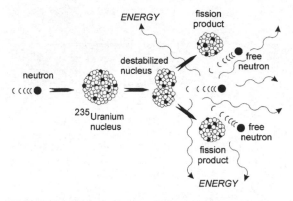

Figure 5.10   Schematic diagram of the nuclear fission process

release of carbon dioxide during combustion (see Chapter 11). Nuclear plants produce neither of these, but they have their own problems. The radioactivity present in nuclear waste creates special disposal requirements (see Chapter 6). Tonne for tonne, nuclear fuels contain much more energy than conventional fuels, but when the construction, operating and decommissioning costs of nuclear plants are taken into account, nuclear energy is arguably the least cost-effective method of producing electricity. Together these have ensured the stagnation, if not the ultimate demise, of the nuclear power industry. Compared with fossil fuel and nuclear power generation, hydro-electric generation is often seen as the bright spark in the

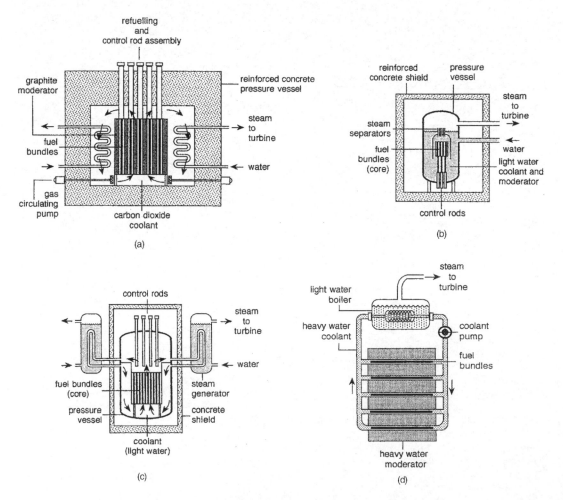

Figure 5.11   Types of nuclear reactors: (a) a gas-cooled reactor, (b) a boiling water reactor, (c) a light water reactor, (d) a heavy water reactor

TABLE 5.2 DISTRIBUTION OF NUCLEAR PLANTS AND THEIR CONTRIBUTION TO THE GENERATION OF ELECTRICITY

Country	No. of units	Total MW(e)	Share of generation (%)
France	59	63073	76.4
Lithuania	2	2370	73.7
Belgium	7	5712	56.8
Slovak Republic	6	2408	53.4
Ukraine	13	11207	47.3
Bulgaria	6	3538	45.0
Republic of Korea	16	12990	40.7
Hungary	4	1755	40.6
Sweden	11	9432	39.0
Switzerland	5	3200	38.2
Slovenia	1	676	37.4
Japan	54	44289	33.8
Armenia	1	376	33.0
Finland	4	2656	32.1
Germany	19	21283	30.6
Spain	9	7524	27.6
United Kingdom	33	12498	21.9
Czech Republic	5	2560	20.1
United States	104	98071	19.8
Russia	30	20793	14.9
Canada	14	9998	11.8
Romania	1	655	10.9
Argentina	2	935	7.3
South Africa	2	1800	6.6
Netherlands	1	450	4.0
Mexico	2	1360	3.9
India	14	2503	3.1
Brazil	2	1901	1.9
Pakistan	2	425	1.7
China	3	2167	1.2

Total operational units: 438
Total capacity: 353489 MW(e)

Source: International Atomic Energy Agency, *Power Reactor Information System.* Viewed at http://www.iaea.or.at/programmes/a2/ (2000) (accessed 18 June 2003).

economically and environmentally challenged industry. Hydro-electric generators can be close to 90 per cent efficient and do not heat or pollute the water they use. Once the initial construction costs have been covered, therefore, hydro plants can be profitable. The construction of dams, the diversion of rivers and streams and the filling of reservoirs do have a significant impact on the environment, however, and environmentalists do not look favourably on major hydro-electric projects.

Electricity has made an immense contribution to the socio-economic development of a modern society based on high technology. The rapid and continuing technological advances it has supported in computer science, electronics and information technology have become essential to the world economy and have contributed, or have the potential to contribute, to a better quality of life for the world's population. These developments have come with an environmental price tag, however, which is often ignored, but which will ultimately have to be paid, perhaps through an increase in the production of electricity from renewable resources.

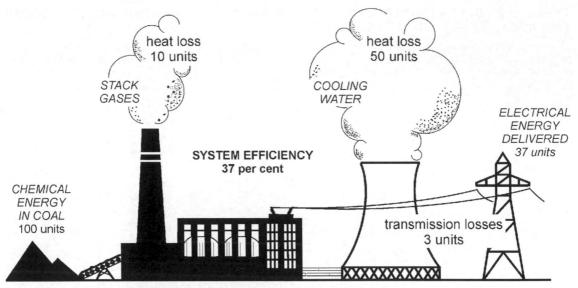

Figure 5.12   Conversion efficiency in a coal-fired thermal electric generating station

Plate 5.5   A fish 'ladder' built to allow migrating fish such as salmon to bypass a hydro-electric dam. The disruption of fish migration is one of the environmental problems associated with dams.

Petroleum

The main challenge to coal as a direct source of energy came from petroleum, including liquid oil and natural gas (Figure 5.8). The first commercial wells in North America were drilled in the late 1850s in Pennsylvania and south-western Ontario, mainly to provide a source of kerosene (paraffin) for lighting and to replace increasingly scarce whale oil. At that time, wood was still the main source of energy used in the United States, but by the end of the nineteenth century the combination of gasoline with the internal combustion engine was well established and the stage set for petroleum to become the dominant source of energy for transport. Oil consumption received a boost during World War I, when it was used to power ships, submarines, aircraft, tanks, trucks and cars and following the conflict its popularity continued to rise. It gradually replaced coal as the fuel of choice in shipping and in railway locomotives, but the main boost came from the rapid growth of the aircraft and automobile industries, which would not have been possible without gasoline-powered engines. Petroleum has several advantages over coal as a primary energy source. It provides more energy per unit volume, is cleaner to use and is more easily stored and transported than coal. It tends to produce less particulate matter and less carbon dioxide than coal when it is burned, although the photochemical smog associated with petroleum

Plate 5.6   A coal-fired thermal electric power station. The waterside location, the fuel storage space requirements and the release of pollutants set such power stations in conflict with the environment, in this case a lake-shore marsh ecosystem.

combustion is every bit as dangerous as the sulphurous smog produced by the burning of coal.

Through the second half of the twentieth century, the socio-economic and cultural impact of petroleum use intensified. The automobile easily surpassed the flexibility that had originally been provided in transport by the railway and the built environment began to reflect that, particularly in the developed world. Existing road networks were upgraded and new systems built. Suburban expansion took place in the vicinity of most major cities and within the cities problems of traffic flow and air pollution became serious. Raw materials and finished products were increasingly transported by road, and airlines took over the long-distance movement of people from railways and ships. Urban and even rural landscapes are now dominated by the infrastructure needed to support the continuing use of petroleum. Although the main use of petroleum continues to be in transport, it is also used in the production of electricity, and particularly in the form of natural gas is a major source of energy for heating water and for space heating. So complete is the dependence on oil that even when opportunities arise to reduce it they are

seldom followed up. In the mid-1970s, for example, the developed world experienced an oil crisis brought on by an oil embargo imposed by the Organization of Petroleum Exporting Countries (see Box 5.4, OPEC) (Figure 5.8). It produced serious shortages for the Western industrialized nations and caused a major upheaval in the global economy. To combat this, energy conservation was encouraged and attempts were made to substitute other energy sources for oil. Much more time and effort was spent searching for new sources of oil outside the jurisdiction of OPEC, however, and with considerable success. As a result the embargo, and the response of the non-OPEC nations to it, brought about radical changes in the geography of oil production and consumption, but did little to reduce dependence on petroleum.

Along with its many economic and technological advantages, the use of petroleum has the potential to cause serious environmental disruption. As with any other hydrocarbon, combustion produces air pollution, but often more serious, because they are unexpected, are the oil spills that take place on land and in the sea (see Chapter 8). Common sources of oil pollution on land include motor vehicle

BOX 5.4 THE ORGANIZATION OF PETROLEUM EXPORTING COUNTRIES

The Organization of Petroleum Exporting Countries (OPEC) is a group of the world's major petroleum producers and exporters (Figure 5.8). Recognizing the importance of oil as a source of future development funding, a number of Middle Eastern states and Venezuela came together in 1960 with the intention of using their petroleum resources to advance their economic interests. The addition of other members, with similar interests, from Asia and Africa created the current group of eleven members – Algeria, Indonesia, Iran, Iraq, Kuwait, Libya, Nigeria, Qatar, Saudi Arabia, United Arab Emirates and Venezuela. Ecuador and Gabon are former members of the organization. OPEC had little success in the 1960s, in large part because there was an abundance of crude oil available, and it was not taken seriously by the Western industrial economies at first. That changed in October 1973, when the Arab-dominated OPEC supported Egypt and Syria in the Yom Kippur War against Israel. Its support took the form of an oil embargo against Western nations seen to favour Israel. Oil exports were reduced to some countries, and banned completely to the United States and the Netherlands. The embargo, which lasted until March 1974, was accompanied by a series of rapid unilateral price increases which doubled the price of crude oil almost overnight and led to 1973 being referred to as 'the last year of cheap oil'.

By 1973 the economies of the world's industrial nations had become seriously dependent upon cheap imported oil, mainly from the OPEC nations who at that time supplied more than 80 per cent of all imported oil. As a result, the combined effects of the embargo and the price increases were immediate and almost universally devastating. The global economy rapidly went into recession, characterized by double-digit inflation, rising interest rates and spiralling debts incurred by the oil-importing nations. To combat the situation, energy conservation was encouraged and an effort was made to substitute other energy sources for oil. The main response, however, was increased exploration for non-OPEC sources of oil, encouraged by the higher price of oil, and in the second half of the 1970s production increased in Canada, Mexico, Alaska and the North Sea. The improved supply plus reduced demand for oil brought about a stabilization of the price by 1978, but imported oil was still an integral part of the industrial economies and, partly as a result of OPEC's manipulation of the supply and partly as a result of the shutdown of Iranian oil production during that country's Islamic revolution in 1979,

prices spiked again in the early 1980s, helping to produce a repeat of the recessionary conditions of the 1970s. Oil prices declined after that, and by the mid-1990s, allowing for inflation, the price of oil was about the same as it was in the mid-1970s.

The OPEC oil embargo was the major political and economic event that showed the power of energy-rich nations in a world dependent upon petroleum products. The embargo, and the response of the non-OPEC nations to it, brought about radical changes in the geography of oil production and consumption. It caused a massive flow of wealth into the oil-producing nations, much of it spent on the creation of modern urban infrastructure, particularly in Saudi Arabia and the Gulf states. While the price increases caused recession in the industrial economies of the developed world, they had a devastating effect on developing nations that did not have easy access to oil and helped to widen the gap between the developed and developing world. OPEC established a Fund for International Development in 1976 to provide grants and loans aimed at alleviating the problems of the developing world.

OPEC operated very successfully in the 1970s and 1980s, but since then internal disputes and significant changes in the world's petroleum economy have reduced its importance. However, the member countries produce 40 per cent of the world's crude oil output and control perhaps as much as 75 per cent of the world's petroleum reserves, and as a result their potential to influence international economics and politics remains high. OPEC also has the potential to have an impact on current environmental concerns such as global warming. It is generally unwilling, for example, to support the environmental initiatives associated with the Kyoto Protocol since they would have a detrimental impact on the revenues of its member states. In this, OPEC is no different from some other non-OPEC oil producers, but it has the economic clout to use its concerns to retard even more the already slow progress towards implementation of the Protocol.

For more information see:

Danielsen, A.L. (1982) *The Evolution of OPEC*, New York: Harcourt Brace Jovanovich.
Organization of Petroleum Exporting Countries (2001) *OPEC: General Information*, Vienna: OPEC Secretariat.
Shihata, I.F.I. (1982) *The Other Face of OPEC: Financial Assistance to the Third World*, London/New York: Longman.

Plate 5.7   A chemical plant that produces feedstock for the plastics industry from natural gas.

operations and maintenance, refineries, pipelines, petrochemical plants and other industrial operations. In the oceans, pollutants are provided by shipping activities – both tanker and non-tanker – and by offshore petroleum exploration and production. The classification into land-based and ocean-based spills is not perfect, since petroleum products released on land can be carried into the oceans in run-off, and the oil slicks created by spills at sea can be washed up on shore. Oil pollution receives most attention following major spills, both on land and in the oceans, but the regular ongoing small-scale contamination, from leaking underground storage tanks or from the bilge water discharges of ocean shipping, for example, commonly exceeds the irregular contributions of major pipeline ruptures or oil tanker accidents. Since petroleum acts as a feedstock for other industries – plastics and pesticides, for example – it also contributes to the environmental problems associated with them.

Patterns of modern energy consumption

Although petroleum is the dominant source of energy in modern society, most energy use involves a mixture of sources. Coal, electricity and even human or animal muscle power are also used to meet specific needs. The fossil fuels have many advantages in terms of availability, efficiency, cost and ease of use, but they also contribute to most of the earth's major environmental problems. In an attempt to reduce the latter, consideration has been given to redeveloping renewable resources, which were used prior to the industrial revolution and are generally seen as less threatening to the environment. Additional benefits include the ability of renewable resources to contribute to economic growth in the developing world through sustainable development and the diversification and security of supply needed to meet real or politically contrived shortages in conventional fuels.

Renewable energy

Current renewable energy use is much more technologically advanced than in the pre-industrial era, but the sources are the same – flowing water, wind, biomass and the sun – plus a number of more exotic forms such as geothermal energy, provided by the earth's internal heat, and energy from the oceans,

Plate 5.8 A solar collector plate, used increasingly in remote areas to meet basic electrical needs.

in the form of tidal power or wave power. The sun is the ultimate source of renewable energy (Figure 5.13). It is a very effective nuclear furnace, able to provide the earth/atmosphere system with a relatively steady flow of energy as a result of ongoing nuclear fusion. Solar energy can be used directly through solar panels or photovoltaic cells, and once in the earth/atmosphere system it drives the hydrologic cycle, which ensures the renewal of water power, it helps create the pressure differences that cause the wind to blow and brings about biomass renewal through photosynthesis. Following the industrial revolution, renewable energy continued to be used in many areas – wood was the main fuel used in the United States as late as 1850 – but the more concentrated, more reliable and more efficient energy available from fossil fuels caused the renewable resources to decline in importance. Only water, with its ability to provide electrical energy, retained and even increased its importance. Interest in renewable energy increased in the 1960s, when it appeared that oil and gas sources were facing depletion, and in the 1970s, following the petroleum price increases initiated by the OPEC oil embargo of 1973. Growing environmentalism also favoured the re-adoption of renewable energy resources, since they were perceived as causing less environmental damage than fossil fuels. They are also capable of sustainable development, but they are not completely environmentally friendly. Biomass burning can create air pollution and in some developing

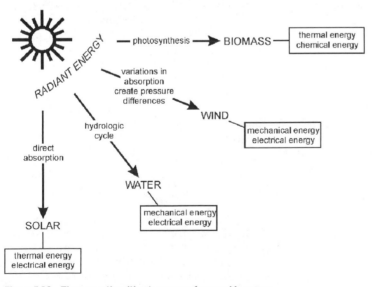

Figure 5.13 The sun as the ultimate source of renewable energy

countries, such as Nepal and parts of the Sahel, where renewable resources did not completely succumb to the dominance of fossil fuels, serious problems of soil erosion and desertification have followed the removal of forest, bush and scrub to meet the fuel needs of rapidly growing populations. Although the generation of electricity using falling water is less damaging to the environment than its production in thermally powered stations, hydro-electric schemes have a significant effect on the hydrologic cycle through the creation of large reservoirs and changes in stream flow patterns. The development of wind energy systems is often accompanied by noise and aesthetic pollution.

Renewable versus non-renewable energy

About 20 per cent of the world's current energy consumption is supplied by renewable energy, including hydro-electricity and biomass (Figure 5.14). The United Nations estimates that by 2050 renewables could supply 60 per cent of the electricity required and 40 per cent of the energy provided directly by fuels (Johansson *et al.* 1993). Fossil fuels would con-

tinue to supply the bulk of the energy consumed, and it is widely considered that despite their many desirable traits renewables are unlikely to replace or even seriously challenge the dominance of non-renewable energy resources, since they cannot supply energy in the quantity and with the efficiency demanded by modern society. That may well be so, although the technology exists to allow renewable sources to provide all of the world's energy (Hill *et al.* 1995). The infrastructure needed to make that possible would take forty to fifty years to construct and socio-economic conditions would also require major reconstruction. Because renewable energy production often involves new, small-scale and experimental technology and lacks the decades or even centuries of development plus the economies of scale enjoyed by conventional energy sources, it can be costly. Subsidies can be used to help offset these costs, but in economic terms that is taken as an indication that renewables are not a viable option to replace fossil fuels and other energy resources. The real problem, however, may be in the calculation and comparison of unlike costs. Fossil fuel production is also subsidized, directly or indirectly, in many places through exploration incentives,

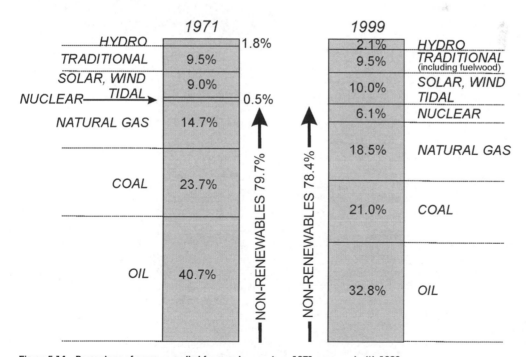

Figure 5.14 Percentage of energy supplied from various sectors: 1971 compared with 1999
Source: based on data in World Resources Institute, *Earthtrends*. Viewed at http://www.earthtrends.wri.org/ (accessed 2 March 2003).

depreciation allowances on equipment and other beneficial tax policies. In addition, external costs associated with the production of conventional fuels, including site rehabilitation, pollution remediation and waste disposal are not commonly included in the final cost of the energy they provide. If these factors were to be included, then the gap between the viability of renewable and non-renewable sources of energy would be much smaller (Hill *et al.* 1995).

Already, solar energy, wind power and biomass are making important local contributions to energy supply in the developing world, from the relatively simple production of biogas from animal waste to the technically advanced production of electricity from photovoltaic collectors to pump water and provide lighting and refrigeration in areas remote from a central source. Not only does this improve the quality of life, but it also contributes to sustainable development in these areas. Industrial, economic, cultural and political inertia tends to militate against the increased use of renewable energy in developed nations, however. There, almost all aspects of life are built around the intensive use of conventional energy, either directly as fossil fuels or indirectly through the use of thermally generated electricity. Industry depends upon a ready supply of relatively cheap energy and the population is largely urban, living in cities that have gradually evolved to meet the requirements of the gasoline-powered automobile, and with a lifestyle characterized by the use of a multitude of labour-saving appliances and entertainment equipment powered by electricity. Massive investment over the past century has created a capital-intensive energy infrastructure in the developed world and has ensured that economic and political advantages lie with those companies and nations that control or have easy access to conventional energy supplies. Thus the amount of energy supplied from renewable resources remains low (Figure 5.15) and that situation is unlikely to change for several decades at least. Improvements in technology and efficiency at all levels, from exploration through extraction and refining to combustion, have consolidated the hold that conventional fuels have on the energy market. Ironically, rising efficiency has tended to increase rather than decrease their use, since greater efficiency tends to reduce price, which in turn encourages consumption and discourages conservation (Herring 2000).

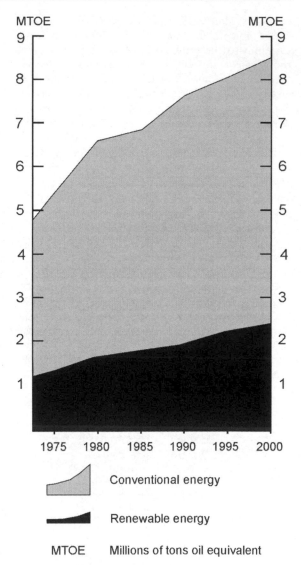

Figure 5.15   Consumption of conventional and renewable energy
Source: based on data in World Resources Institute, *Earthtrends.* Viewed at http://www.earthtrends.wri.org/ (accessed 2 March 2003).

Fossil fuel supplies are not limitless, however, and current levels of consumption suggest that although there is enough coal available to meet demands for at least the next century and a half or even longer (Hill *et al.* 1995), oil and gas reserves will be exhausted before the end of the twenty-first century (Masters 1995). Renewable energy resources must be integrated into the mainstream long before then to prevent serious economic and political disruption similar to that which accompanied the oil crisis of the mid-1970s but on an even grander scale. Amelioration of the increasingly serious environmental problems associated with the continuing use of fossil fuels also requires an increase in the use of renewables. As well as improving the quality of life locally, this also has implications for global issues. Renewable energy either produces no new carbon dioxide or simply recycles the gas already in the system. The adoption of renewables on a much larger scale would therefore help to mitigate the problems of global warming and the climate change associated with it. Environmental concerns such as those may well initiate the increase in the use of renewable energy, if governments honour the commitments they made in the last decade to sustainable development and the reduction in greenhouse gases. The factors that drive current energy consumption patterns do not favour renewables, however, and until that changes successful implementation of these pro- grammes will be difficult. Government legislation will undoubtedly be necessary, but even that needs public support, and as yet society appears to be willing to accept the status quo and the con- sequences of continued reliance on depleting and polluting energy resources.

Society's use of energy illustrates well the ways in which the environmental impact of resource use depends upon the complex interplay of a variety of elements. It extends beyond simple physical relationships to encompass social, cultural, eco- nomic and political issues initiated by human needs and wants that vary from time to time and from place to place. If they are to be understood, environmental issues must be viewed in this multifaceted context, and any attempts at finding solutions to environmental problems must also take that into consideration.

SUMMARY AND CONCLUSION

Society's impact on the environment is strongly related to its need for resources and that in turn is influenced by population and technology. In the past, when populations were small and the level of technology was low, the demand for resources was limited and the human impact on the environ- ment was minor. The need was for food, water and shelter, which could be met from resources eas- ily renewed or replaced by the environment after use. As technology advanced – and population grew along with it – the demand for resources increased and the nature of the resources changed. By the mid-eighteenth century a new technology based on non-renewable resources was emerging. Driven by coal and iron, it brought economic, social and political advances to society, but through increased resource extraction, industrialization and urbanization it also caused increased pressure on the environment. The landscape was changed and the waste produced during the use of the resources fouled the air and waterways. As economies grew, both fiscally and geographically, and technological advances, particularly in the energy field, encouraged the use of more resources, the impact on the environment increased also until it became global. At the same time, the internationalization of economies and politics and the spread of a culture driven by resource consumption also contributed to environmental deterioration.

Attempts at reducing that impact have involved consideration of sustainable development and resource recycling, as well as a return to renewable resources, which are seen to be less harmful to the environment. While the technology exists to allow the introduction and development of these approaches, where it has been done it has met with only limited and mixed success, in part perhaps because of the complexity of the relation- ship between society and the resources it uses. That relationship extends beyond simple physical and technological links to encompass social, cultural, economic and political issues initiated by human needs and wants that vary from time to time and from place to place. If the interplay of issues involving society, resources, technology and the environment is to be understood, it must be viewed in this multifaceted context, and any attempts at finding solutions to environmental problems must also take that into consideration.

SUGGESTED READING

Freese, B. (2002) *Coal: A Human History*, Cambridge MA: Perseus. A fascinating account of the socio-economic, cultural and political aspects of the use of coal and its impact on global society over the centuries.

Gruebler, A. (1998) *Technology and Global Change*, Cambridge: Cambridge University Press. The causes of technical change and its impact on the global environment. Full of interesting material, some for the specialist but, overall, very readable.

Ristinen, R.A. and Kraushaar, J.J. (1999) *Energy and the Environment*, New York: Wiley. An introduction to energy concepts, resources and applications that leads to consideration of current environmental problems associated with energy use.

QUESTIONS FOR REVISION AND FURTHER STUDY

1   What attempts are being made to reduce energy consumption in your school or university? How are they linked with the broader aspects of energy conservation – what might be the ultimate impact of switching off all the lights in the library overnight, for example? Do you see additional ways in which energy consumption might be reduced? If so, why do you think they have not already been adopted?

2   How valid is the assertion that modern technology is at the root of existing environmental problems, but only technology can provide solutions?

3   Prepare a list of materials that are commonly recycled. Which of them do you as an individual already recycle and which do you dispose of as garbage? What might be the advantages to you and to the environment as a whole if you recycled more?

6

Human Use of the Land and its Environmental Consequences

After reading this chapter you should be familiar with the following concepts and terms:

assessment	land – as commodity, as place, as resource	secure landfill
beneficiation		sheet wash
concrete recycling	leachate	site rehabilitation
construction aggregates	metallic ores	soil erosion
contour ploughing	NIMBY	strip mining
decommissioning wastes	nuclear waste	subsidence
environmental impact	open-pit (open-cast) mining	suburbanization
gangue	overburden	surface mining
green belt	pits and quarries	underground mining
gully erosion	radioactivity	urban sprawl
hazardous waste	resource extraction	waste disposal
infrastructure	ribbon development	wind breaks
	sanitary landfill	

The world's land area covers some 150 million km^2, or 29 per cent of the earth's surface, and at the beginning of the twenty-first century that land supports 6.3 billion people. These figures are deceptive, however, for not all of the land is suitable for human habitation. Antarctica's 14 million km^2 have no permanent inhabitants, the 9 million km^2 of the Sahara Desert and the 2 million km^2 of Greenland are mainly empty, with only a few isolated settlements, and the world's mountain areas are sparsely populated. In sharp contrast are locations such as Hong Kong where some 6 million people live together on only 1100 km^2 of land. It might appear that the greatest human impact on the land will be in those areas where population density is greatest, but society's influence is now ubiquitous and even the most isolated land can experience the environmental effects of human activities. Society expects the land to provide mineral resources, sustain agriculture, support urban and industrial development, absorb waste and provide space for recreation. Poor planning or the failure to recognize the limits of the ability of the land to handle any of these expectations can lead to serious environmental problems.

Land can be viewed in a number of ways (Box 6.1). It is most obviously a physical entity, comprising the rock and soil that form the surface morphology of the lithosphere. The land surface supports the plants and animals of the biosphere, and is the natural foundation on which society has developed agriculture and industry and built the complex structures that comprise the modern urban/industrial environment. Land also includes the upper part of the earth's crust from which minerals and fossil fuels have been extracted to help society reach its present stage of development. More than a combination of these physical components, however, it involves a complex mix of cultural, socio-economic and political elements that contribute to a variety of environmental problems initiated by society's use or misuse of the land.

BOX 6.1 CONTRASTING VIEWS OF LAND

What is land? Land can be viewed as:

- **A physical entity** including rock, soil, morphology and minerals
- **Place** providing a location, country or point of origin
- **Space** somewhere to live or pursue a variety of activities
- **Resource** with the ability to meet a great number of human needs
- **Commodity** with a value as real estate to be bought and sold

Source: based on information in Environment Canada, *Stress on Land*, Canadian Government Publishing Centre (1983).

In human terms, land is both place and space. Most individuals identify with a particular place, although not necessarily or only their current home. Many immigrants, for example, identify with the country they left and that identity can continue down through several generations. The concept of land as space is quite variable, depending upon the activities for which the land is required as well as personal and group expectations. At an individual level, the inhabitants of a densely populated urban area such as Tokyo or Hong Kong will view space differently from a rancher on the North American prairies or in the Australian outback. The availability of space in the past, and the willingness of people to make use of it, allowed the redistribution of population on a scale not now possible. During the rapid growth of industry, and the urbanization that accompanied it, in the eighteenth and nineteenth centuries, pressure on the land was relieved by the movement of millions of people from the crowded landscapes of Europe to the open spaces of North America, South America and Australasia. Similar pressures and the attractions of readily available, cheap land contributed to the rapid movement of population west of the Mississippi following the American Civil War in the late 1860s.

In some cases the attraction lay not in the land itself, but in the resources it contained. Gold rushes in California, Alaska, the Yukon and Australia in the nineteenth century drew thousands to land previously only sparsely populated. Trouble occurred when the existing population objected to newcomers usurping their land and resources. In South Africa, for example, the discovery of gold in the Transvaal in the late nineteenth century was a harbinger of political conflict between the established Boer population and the British that would lead to the Boer War. Throughout history, the acquisition or control of land for the space it provides or the resources it contains has been an important aspect of state or national policy, achieved by political or military means. In modern times, the political approach is more common, but military force may sometimes be required. During World War II, for example, the German push eastwards into Romania, Ukraine and southern Russia was in part to ensure access to the oil and coal available in these areas and the invasion of Norway was carried out to prevent the Allies from disrupting supplies of Swedish iron ore. Similarly, there are those who would argue that the underlying reason for the Gulf War in 1991 and the Iraq crisis of 2002–03 was the need for the Western industrial nations to maintain access to Middle Eastern oil and the economic and political power associated with it.

In the past, land was held in common, with no individual ownership, and, in theory, all members of a community had equal access to the benefits the land had to offer (see Chapter 5). Although common land still exists, in modern society land has become a commodity, owned by individuals or groups, to be bought and sold like any other resource. Its value varies depending upon such factors as its physical attributes, availability and location, which together determine the land use that will provide the greatest economic return. With land considered as a commodity and its use determined by its economic value, the stage is set for conflict. The use that will provide the greatest monetary return from a piece of land is not necessarily the one that is environmentally most appropriate. The Arctic National Wildlife Refuge (ANWR) in Alaska, for example, which occupies 7.7 million ha of land in the Arctic tundra, is, according to petroleum geologists, underlain by commercially viable reserves of oil. Exploitation of the oil beneath the Refuge would provide a much greater economic return from the land than its present use, but at a major environmental cost to the physical landscape and the flora and fauna of the region. Similar disputes regularly flare up around the logging of old growth forest and the development of hydro-electric power. In such cases, if the value of land as a commodity is the main determinant of its use, the environment tends to lose, in large part because many of the benefits retained when the environment is preserved are difficult to quantify (see Box 5.1, Environmental values). Profit from a barrel of oil or a truckload of lumber is easy to predict, but the aesthetic value of a landscape or the physical and mental benefits of a wilderness experience cannot always be assessed in real monetary terms. Thus the environment is at a disadvantage when the traditional cost–benefit analysis approach is used to determine the use of a piece of land. In theory, environmental impact assessment and land use planning should ensure that minimal environmental disruption is caused by the use of the land, but in some cases, socio-economic considerations are allowed to override strict environmental factors.

THREATS TO THE LAND RESOURCE

Whether land is viewed as a physical entity, as place, space or commodity, in its broadest sense it is a resource. For thousands of years it has met society's needs and wants, either directly through its rocks, minerals and soils or indirectly through the plants and animals it supports. As with all resources, demands on the land have changed with time and place. In the past it could accommodate these changes because populations were small and the level of technology was low, but with the burgeoning populations and increasingly advanced technology of modern times, it cannot always meet the demands placed on it. Human use of the land has introduced a level of change that often exceeds that of natural processes, such as the weathering and erosion with which the land can normally cope. Mining and quarrying operations frequently move more rock at a greater rate than would be possible in nature, except perhaps under the effects of catastrophic events such as earthquakes and volcanic eruptions. In some areas, human-induced soil erosion exceeds natural sediment loss from the land. Because of the integrated nature of the components of the earth/atmosphere system, the impact of such change is carried through into the hydrosphere, atmosphere and biosphere.

The human activities that create pressure on the land take many forms, but they can be fitted into

Figure 6.1   Threats to the land from human activities

three broad groups (Figure 6.1). Some involve the removal of materials from the land, either deliberately in the case of mining or quarrying, for example, or inadvertently as in the case of soil erosion ('Resource extraction and depletion'). Some of the materials extracted will be relocated either in their original form or fabricated to some degree or other to become part of the built environment ('Infrastructure and waste disposal'). Included in this category are the waste materials generated during extraction and processing or created through activities in the built environment. The third group involves activities that interfere with the flora and fauna supported by the land surface and these will be examined in detail in Chapter 7.

PRESSURE ON LAND: RESOURCE EXTRACTION AND DEPLETION

RESOURCE EXTRACTION

Society has a long history of extracting minerals from the land, to be used in construction, to make tools or in many cases for decoration. They are essential components of modern industrial economies, and since individual countries seldom have all of the mineral resources they need, they have always been important elements in international trade. With the earth/atmosphere being a closed system in material terms, the survival of the developed nations at their current level and the economic advancement of the developing nations requires that these resources continue to be extracted from the land. The mining and mineral extraction industries of the future will continue to share some of the characteristics of the past, but they are also compelled to consider the broader environmental impact of their activities. The search for high-grade raw materials has always been part of the mining industry and is unlikely to change. For many minerals, if not most, the better-quality resources have been found, however, and are already being worked or in some cases have been exhausted. As a result, the industry is working with lower-quality deposits, which are more costly to extract, often have marginal profitability and require the extraction of greater volumes of unwanted crustal material. The resulting waste is a source of a number of environmental problems and the modern mining industry has evolved to deal with such problems, often under the pressure of environmental legislation. Environmental impact assessments are required before major mineral extraction projects are allowed and the results of the assessment may lead to the abandonment of the project, or its modification to reduce the perceived impact. In theory, a project should be allowed to proceed only if the impact statement indicates minimal environmental disruption, but in some cases, often for political or socio-economic reasons, a potentially disruptive development is permitted. In developing countries, for example, it is not uncommon for a higher level of environmental pollution than would be allowed by an environmental impact assessment to be traded off against reduced unemployment and poverty.

The imposition of environmental regulations on the extraction industries is often considered to add to production costs, and that appears to be true with established operations that may be working with obsolete technology or limited capital. There is growing evidence, however, that in new projects the use of innovative technology and appropriate managerial strategies to deal with environmental issues can reduce both production and environmental costs (Warhurst 1994). Promising as that may be, the reality is that it is impossible to extract minerals from the land without some degree of environmental disruption. It occurs at all stages in the development of a resource, from the initial preparation of the site, through the extraction, transport and processing of the material, to the final abandonment of the site (Figure 6.2). A sig-

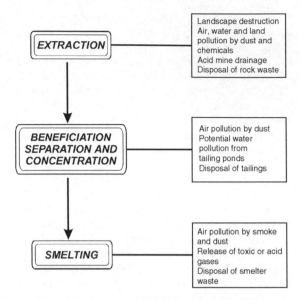

Figure 6.2   Processing of metallic ores and related environmental problems

nificant problem in areas that have a long history of resource extraction is the presence of old mines, quarries and waste dumps left by companies that ceased operation before the passage of current environmental regulations. They continue to contribute to environmental degradation and represent a major regulatory and economic challenge. Most minerals are extracted from underground mines or surface mines, with the latter including quarries and pits (Figure 6.3). The method used at any given location is determined by a variety of factors, including the nature and location of the resource, the technology available and a number of socio-economic considerations that range from the market value of the commodity to the land use in the area adjacent to the operation and the environmental conditions associated with the site.

Underground mining

When a mineral deposit is not readily accessible from the surface it is extracted by underground mining (Figure 6.3). A vertical shaft is sunk from the surface into or adjacent to the deposit and horizontal galleries are driven to remove the mineral, which is then raised to the surface. In hilly or mountainous areas, where the strata are dipping, the minerals may be reached by a sloping shaft,

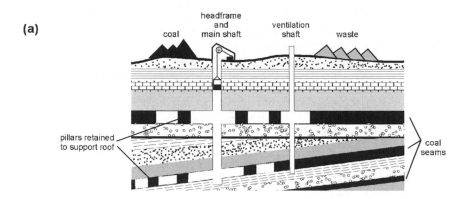

(a)

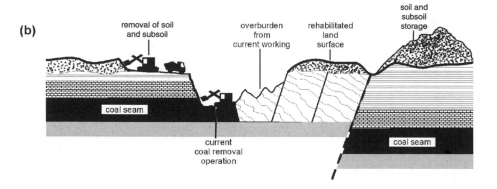

(b)

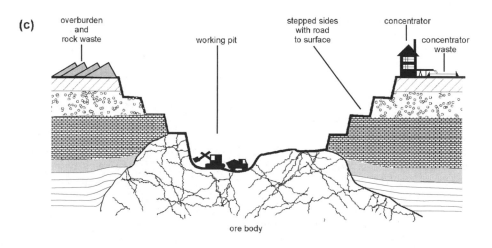

(c)

Figure 6.3   Methods of mineral extraction

or adit. Coal has long been one of the most common minerals obtained by underground mining, although the proportion extracted by surface mining is growing. Globally, about two-thirds of hard coal production comes from underground mines, but the amount varies regionally. Traditionally underground output in Britain was high, being more than 70 per cent as recently as 1990, but it has declined to about 54 per cent in 2001 (UK DTI 2002), which compares with less than 40 per cent in the United States (Buchanan and Brenkley 1994). It is not possible to remove all of the deposit in underground mining. Depending upon geological conditions, as much as 50 per cent has to be left to support the roof, and in a working mine ongoing maintenance is required to prevent rock bursts or cave-ins. In coal mines where the geology is suitable – seams thick and horizontal, for example – longwall mining allows as much as 75 per cent of the coal to be removed. Self-advancing hydraulic supports temporarily hold up the roof until the coal is removed by a mechanized coal cutter/conveyor belt system. As the cutter moves forward into the seam, the hydraulic supports are advanced and the roof is allowed to collapse behind them. In some mines, where the metal content of the deposit is particularly high, the pillars of ore left to support the roof are removed immediately before the mine is finally abandoned and the roof allowed to fall in. In such a case, as much as 90 per cent of the ore can be recovered. Once the mine is no longer operative, the rocks move and adjust naturally to fill the voids left by the extraction of the minerals. As a result of this, in many mining areas there are problems with subsidence, which creates fissures or hollows at the surface, changing the shape and drainage of the land and creating problems of shifting foundations, broken gas and water lines or cracking walls in residential areas.

Surface mining

Minerals that are present close to the earth's surface can be extracted from open pits or by way of strip mining (Figure 6.3). Whatever form it takes, surface mining involves the removal of the overlying soil and rock – the overburden – to expose the ore beneath and allow its extraction by a combination of blasting, dragline or shovel excavation and trucking. Techniques vary according to the nature of the terrain and the minerals being extracted. Open pits are most common with ore bodies that have a restricted horizontal dimension, but extend to some depth beneath the surface, whereas strip mining is usually most common for minerals such as coal that are present as horizontal or near horizontal strata, perhaps only a few metres thick but extending over a considerable area. In hilly terrain contour stripping is practised, with the overburden being removed along the contour of the slope. In the past, the excavated material was simply dumped on the downslope side of the cutting, destroying vegetation and filling streams with sediments and chemicals eroded from the unconsolidated slope. Overburden is now removed and stored so that it is available for the rehabilitation of the site when extraction is complete, but the coal mining areas of the eastern United States in Kentucky and West Virginia, worked in the days before the introduction of environmental legislation, still suffer the depredations of unrestricted surface mining. Where the terrain is relatively flat, area strip mining is practised. A strip of overburden is removed and stored and the exposed mineral excavated. When the extraction is complete, a new strip is cleared and the overburden dumped in the previously excavated area. This continues until the deposit has been worked out, and the original overburden is returned to complete the process (Chiras 1998).

Open pits vary in size, depending upon the extent of the ore body or the equipment available to excavate it. The numerous small gold mines that exploit exposed veins or alluvial gold deposits in the Amazon are dug by hand, whereas the mammoth Bingham Canyon copper mine at Bingham, Utah, which is 4 km in diameter and 0.8 km deep, reached that depth as a large, rich deposit was followed down from the surface using massive earth-moving equipment. Once the overburden has been removed the exposed ore body is drilled and blasted to loosen the rock, which is then carried up ramps to the surface by huge dump trucks capable of carrying more than 100 tonnes in every load. The ore body is followed down by way of a series of benches or terraces until it dies out or the pit becomes too costly to operate. In some cases, roads are driven into the side of the pit to allow the deposit to be extracted by underground mining. In open pit mining, the pit is only part of the operation. When metallic ores are mined the metal is incorporated in a mix of unwanted minerals called gangue, and before it can be used it must be separated out from the mixture. Since the proportion of metal in the ore

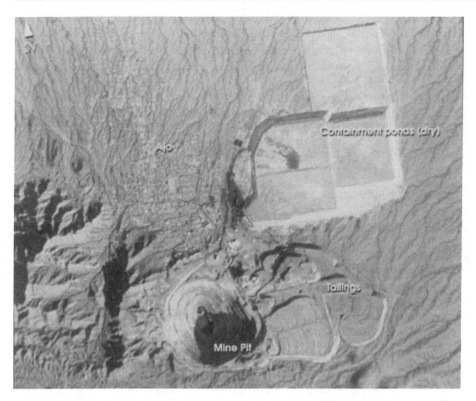

Plate 6.1 The New Cornelia mine near Ajo, Arizona. Copper is extracted from a pit that is one mile wide and more than 1000 ft deep. The containment ponds and tailings dumps contribute to environmental problems in all pit mining activities. (Courtesy of NASA: Visible Earth.)

is commonly quite low and transport of unwanted gangue would increase the cost of the metal, the initial separation is normally carried out at the mine site. The process is called beneficiation and involves the separation of the ore from the country rock by crushing, magnetic separation and flotation to produce concentrated ore and leave behind large amounts of rock waste (Wills 1992).

A combination of high productivity and relatively low production costs means that surface mining is generally less costly than underground mining, and as a result it can be used to extract lower-grade ores and smaller deposits profitably. It can also be technically and economically feasible where excavation sites are restricted by adjacent land uses involving agriculture, residential and industrial infrastructure, but in these cases environmental considerations may prevent development. Overall, surface mining has a greater environmental impact than underground mining, but whatever precautions are taken, neither type of operation can

be accomplished without disrupting the environment, and that disruption often continues long after mining has ceased.

Waste disposal and site rehabilitation

The problems associated with the disposal of huge volumes of waste are shared by both underground and surface mining. Even in coal mines that are working thick seams of coal there is other rock that is removed in the mining process. It is brought to the surface to be dumped, where, as well as being aesthetically unpleasant, it creates physical and chemical problems for the environment. At least with coal the waste material is usually easy to separate from the mineral product. In Britain, for example, about 80 per cent of the colliery waste is dry and solid, excavated when shafts are being sunk or underground roadways are being driven, or included with the coal from adjacent strata

(Buchanan and Brenkley 1994). With metallic ores, the proportion of metal in the matrix of country rock is frequently less than 1 per cent, and the volume of waste that remains is consequently high. Mining waste can create local environmental problems such as air pollution, in the form of dust from the finely crushed tailings produced in the beneficiation process, and water pollution from the tailings pond in which the fine sediments are allowed to settle (Figure 6.2). Returning the waste underground has been considered and has been accomplished in some cases. It can be done during the operation of the mine or following its closure. The technology is available, but the cost tends to be high and if an environmentally appropriate means of surface disposal is available it is usually pre-ferred. Underground disposal would help to reduce subsidence, but in cases where the waste includes active chemicals its return to the mine has implica-tions for groundwater contamination.

Modern mining operations tend to use the best available technology and are subject to environ-mental legislation from site preparation, through the operation of the mine to waste disposal and site rehabilitation when the operation ceases. Even when there is strict adherence to that legislation, it is impossible to prevent environmental change during mining. Vegetation is removed, animals desert the area, the local hydrology is disrupted and the operation generates air and water pollution. During rehabilitation, overburden may be replaced in the order in which it was removed, but it is never exactly the same. The original solid rock, blasted and broken up during its removal, is returned as unconsolidated rock and the soil stripped and stored is returned as a layer lacking its former profile and structure. The very shape of the land is changed as thousands of tonnes of material are moved from one part of the land surface to another. With time, natural processes take over the rehabili-tation and eventually the site will be incorporated once more into the local landscape.

Legislation such as the US Surface Mining Control and Reclamation Act of 1977 has had mixed success, but it has the potential to reduce the impact of current mining activities (Smith 2000). Problems arise, however, with operations that were abandoned before the environmental legislation was put in place. In the past, when a mine shut down, the costs of waste disposal, site rehabilitation and environmental protection were effectively trans-ferred to the public sector and in reality little was

done to deal with the problems created. Cleaning up the waste from these abandoned sites now would be extremely costly, and it is only in the United States that a serious attempt has been made to address the problem. The Superfund was established in 1980 as a multibillion-dollar fund to be used to clean up abandoned hazardous waste dump sites that threaten the environment and public health (Smith 2000). Although mining waste is not the major source of these wastes and not necessarily the most hazardous, there are hundreds of thousands of locations around the world, many in the United States, where mining waste is contributing to air, water and soil pollution. Under the Superfund program in the United States, fines to pay for the clean-up can be levied on the companies which dumped the waste, but the owners cannot always be identified and lengthy litigation can stall the recovery of the costs. Modern environmental legis-lation should prevent an increase in the abandon-ment of waste sites in the developed world, but the legislation is often less stringent in developing countries where the need for foreign investment, improved employment opportunities and increased export revenue influences the regulatory regime (Warhurst 1994).

Pits and quarries

Operations that extract minerals from the land through pits and quarries tend to be smaller than those involved in underground and surface mining, with almost all of the material extracted being used and relatively little waste being produced. Quarries often take advantage of natural rock outcrops, working existing cliff faces by drilling and blasting or using natural jointing and cleavage in the rock, the actual technique varying with the rock type and the intended product. Sand and gravel are com-monly extracted from fluvial deposits such as those in river terraces or from a variety of fluvioglacial land forms deposited as the ice of the last Ice Age melted. In coastal locations, they are frequently extracted from beaches or dredged up from the sea bed immediately offshore. Clay is dug from pits in recent lacustrine or marine deposits or from geological formations that may be several million years old.

The products of these pits and quarries have a wide range of industrial uses. Sand has long been a raw material for glass making and for moulds in the metallurgical industries, for example. Limestone is used as a flux in iron smelting and has many uses in the chemical and agricultural industries and some types of clay – kaolin, for example – are important resources in the china and pharmaceutical industries. The bulk of the products are used in the construction industry, however, as construction aggregates.

Rock quarries have produced building stone for centuries and much of the character of the older sections of cities, particularly in Europe, is a result of the nature of the local building stone. Granite, sandstone and certain types of limestone were commonly used in buildings in the past, with more exotic forms such as marble being used for institutional or ceremonial structures. Slates, which split easily into relatively thin sections, were a common roofing material and flagstones – fine sandstones which split like slates – were used for flooring and paving. In areas where hard rock was unavailable an alternative for building was the clay brick. Formed from soft pliable clay and then fired in a kiln, the bricks produced were a durable, easily worked alternative to stone, and bricks continue to be one of the most common building materials. Both stone and brick have to be held together using some form of mortar, which is pliable when being worked, but dries into a hard, waterproof and durable product. In the past, most mortars were a mixture of sand, lime and water, sometimes with straw included, but the most common modern form is Portland cement, which also includes clay in the form of silica and aluminium oxide. Along with sand, gravel and water Portland cement is also a main ingredient of concrete, which has been used in some form or other since Roman times and has become perhaps the most common building material now in use. In Britain in 1997, for example, more than 70 per cent of sand and gravel production was used in concrete (UK DETR 2000) and in the United States in 2000 concrete production consumed close to 45 per cent of the sand and gravel extracted (Bolen 2000) although in neither case was the use confined to building construction. Most modern buildings include a mix of different construction materials. Stone has become too expensive for most purposes and where it is used it is often only a decorative facing. Similarly, brick is often used as a facing for concrete, steel or even wooden framed buildings and the slate so common on roofs in the past has been replaced by wood or asphalt shingles, although neither has the durability of the slate.

Large volumes of sand, gravel and crushed rock are used in the transport industry, as a foundation for roads and airport runways, as the roadbed ballast on railways and as a constituent of the concrete used for such structures as bridges and harbour facilities. In 2000, more than half of the sand and gravel extracted in the United States was used in road-related construction (Bolen 2000). Since most of the products of pits and quarries are high bulk/low value commodities, they are costly to transport and when a project is being planned the availability of easily accessible rock or sand and gravel is an important consideration. In many cases it is economic to open short-term quarry operations to provide the appropriate quality of crushed rock or to dig gravel pits with the necessary screening facilities to extract the necessary size and quality of fill for road foundations. This is particularly so where long distances are involved. Along some sections of the Trans-Canada Highway, for example, which was built across Canada in the 1950s, there are small abandoned gravel pits no more than several kilometres apart, from which the fill for the highway was excavated. Some have become impromptu, illegal garbage dumps, some have been converted into highway rest stops, but most remain much as they were when they were abandoned, with the natural vegetation gradually taking over the site again. As well as providing the base for the roads, gravel and crushed rock are mixed with asphalt to build up a durable road surface.

The number of working sand and gravel pits in a given region varies considerably from year to year, depending in part on the changing number and location of construction and highway projects. Since they can be worked with a relatively small investment in basic earth-moving equipment and mobile screening and sorting facilities, small pits can be brought on stream or shut down relatively easily. Despite this, there is increased interest in recycling these commodities, mainly for economic and environmental reasons. Demolition and construction waste containing concrete and asphalt is increasingly recycled in the United States, particularly in areas where raw materials are scarce and transport costs high (Bolen 2000). The original aggregates can be used again after crushing and screening. Broken concrete and bricks can be used as hard-fill and slates and bricks can be salvaged for reuse in their original form, although the cost of removing mortar from the bricks may be prohibitive

(Lowton 1997). In Australia the recycling of concrete is being examined as a means of including the construction industry in a broader attempt to achieve ecologically sustainable development (Mak 1999). Financial savings accrue from the reduction in energy use in raw material extraction and transport, while the reduction in primary extraction means less environmental damage to the land and recycling reduces the amount of construction waste disposed of in landfill sites.

Although pits and quarries tend to have less of an impact on land than surface mines or open pit operations, they do cause significant local environmental disruption. Site preparation changes the environment immediately through the removal of vegetation and soil. When a quarry or pit is being worked it is visually intrusive and a source of dust and noise pollution. Since most of the products extracted from pits and quarries are transported to their destination by road, there is an increase in heavy truck traffic in and around the extraction site. This spreads pollution to areas along the local roadways, creating at best a nuisance and at worst a health hazard for those living in adjacent communities or using the roads.

When pits and quarries are excavated to depths below the local water table, they fill with water. This is particularly so along river valleys where river terraces can be an attractive source of water-washed gravel. Since the water table is close to the surface in such locations, the pits fill rapidly to become ponds or small lakes. These water-filled hollows are difficult to rehabilitate and usually remain as part of the new, post-extraction landscape. Even when the workings are dry, pits and quarries leave obvious holes in the land, which become a permanent part of the local morphology. There is very little waste generated on site with which the excavations can be filled, and even when the topsoil is returned and the site is landscaped it is difficult to disguise the old working face of a quarry or the hollows formed by the removal of several thousand tonnes of gravel. Sand and gravel removal in coastal areas can create serious environmental problems. The removal of beach sand, for example, can disrupt the dynamics of erosion and deposition, causing coastal retreat or shoaling, and in some cases initiating wind erosion of coastal dunes. In a number of places along the south coast of England beaches have been lost and coastal retreat has been aggravated as a result of the removal of sand and gravel (Lowton 1997).

Land that is capable of being developed as a source of sand and gravel is often suitable for other uses also, and as a result conflicts arise over the most appropriate land use for an area. The well drained, level terraces along river valleys may be seen not only as a source of sand and gravel but also as ideal land for agricultural or urban development. In southern Ontario, Canada, large areas of agricultural land are underlain by high-quality sand and gravel (McLellan 1983). The demand for these deposits has caused conflict between pit operators and farmers. Much of the demand comes about as a result of the growing population in the area and the need to provide residential development and the infrastructure that accompanies it. That too creates problems for the aggregate producers, for the very development that they are supporting is covering land and denying them future access to sources of sand and gravel.

The establishment and operation of pits and quarries, particularly in the developed world, is now surrounded by a mass of legislation covering all aspects of the activity, from the initial environmental assessment to the rehabilitation of the site (Figure 6.4). Appropriate planning, suitably enforced, should allow the extraction industries to coexist with agricultural, residential, recreational and other land uses. Major improvements have occurred in the last twenty to thirty years or so, but conflicts still arise, often as a result of economic or political considerations. The development of improved rehabilitation techniques has allowed agricultural land to be brought back into production relatively quickly, but raw material extraction is commonly viewed as a socially unacceptable land use adjacent to urban areas – although it may well have pre-dated urban development in a specific area – and this is a frequent source of conflict, often associated with the NIMBY (Not In My Back Yard) syndrome. NIMBY refers to the situation in which homeowners are unwilling to have any activity that might reduce amenity or property values in the vicinity of their homes. The term also implies that they would be unwilling to forgo the benefits ultimately associated with the activity and would not complain if it happened to be located in a backyard belonging to someone else.

Even with the best possible rehabilitation techniques, it is impossible to extract minerals from pits and quarries without altering the environment, and as a result the extraction industry often faces antagonism from environmental groups. Since the demand for the products of mines, pits and

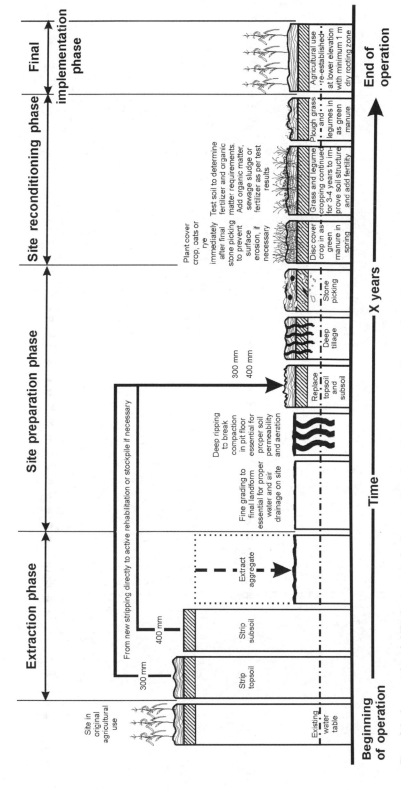

Figure 6.4   A progressive rehabilitation sequence for the restoration of a sand and gravel extraction operation to an agricultural after-use

Source: Figure 3, 'Stress on land in Canada', from W. Simpson-Lewis, R. McKechnie and V. Neimanis (EDS), Environment Canada 1983, reproduced with permission of the Minister of Public Works and Government Services, 2003.

(a)

(b)

Plate 6.2 Contrasting results of quarrying. (a) No rehabilitation of any kind. (b) A former limestone quarry has been converted into gardens which have become a major tourist attraction. The chimney of the limekilns can be seen in the background.

quarries continues to grow, such conflicts will remain. Legislation has improved the situation over the last several decades, however, and if it is developed further, progressive improvements will help to alleviate some of the pressure on the land created by the extractive industries.

SOIL EROSION

Mining and quarrying disturb the environment through the deliberate removal and redistribution of materials from the land. Materials can also be removed inadvertently, however, as a result of poor planning or unsuitable land use. Soil erosion is a prime example of the environmental consequences of a range of inappropriate human activities which unintentionally lead to the removal of soil from the landscape. Unintentional as the consequences may be, they are no less serious.

Soil erosion is a natural part of landscape formation and change. In many places – temperate, humid areas, for example – the rate of erosion is sufficiently slow that it is matched by the formation of new soil and the quality of the environment is retained. Elsewhere, climatic or morphological

conditions can increase the rate of erosion to create peculiar local landscapes such as those of the Badlands in the western United States and Canada. Natural soil erosion is greatest in areas such as these, where unconsolidated sediments are directly exposed to the elements. Winds easily erode the exposed sands of the desert, for example, and in the sparsely vegetated landscapes of the semi-arid regions of the world, bare soil is easily eroded by infrequent, but often intense precipitation. Erosion rates are also high in areas with well marked seasonal wet and dry periods, such as those that experience monsoon and Mediterranean climates (Figure 2.21).

In modern usage, soil erosion usually refers to accelerated erosion associated with human activities that have caused the topsoil to be eroded at a rate greater than it can be formed. Global rates of soil erosion are difficult to measure, but perhaps as much as one-third of the world's agricultural topsoil is being eroded faster than it is being regenerated (Chiras 1998). The human contribution to soil erosion comes about most often as a result of activities that lead to the removal of vegetation and the direct exposure of soil to the elements (Table 6.1). Most arable agriculture involves

Plate 6.3 Serious gullying in unconsolidated sediments in the semi-arid south-western United States. (Courtesy of Christine Deschamps.)

TABLE 6.1 PRECURSORS OF SOIL EROSION

Natural		
Unconsolidated sediments Sparsely vegetated soils Climate change		
Anthropogenic		
Agriculture:	■	Reduced soil fertility
	■	Overgrazing
	■	Inappropriate land use
Forestry:	■	Clear cutting on slopes
	■	Soil compaction
	■	Road building
Mining:	■	Removal of vegetation
	■	Disruption of surface morphology
Recreation:	■	Destruction of vegetation
	■	Soil compaction

activities that leave the soil exposed for extended periods of time, and therefore vulnerable to erosion. In addition, ploughing, harrowing and rolling contribute to the break-up of soil aggregates, producing smaller particles that are more easily eroded. Compaction by machinery during cultivation and harvesting damages the soil structure and reduces the amount of pore space in the soil, which in turn reduces infiltration capacity and increases run-off. Any reduction in soil fertility also encourages soil erosion, through the loss of the humus and nutrients that help to bind the soil particles together. In pastoral agriculture, over-grazing may cause sufficient damage to the vegetation cover to initiate erosion.

Beyond agriculture, forestry, mining and recreational activities also encourage increased soil erosion. Clear cutting of forests, particularly on steep slopes, increases the volume and rate of run-off and the unconsolidated cuttings and banks of gravel roads built to extract the timber are susceptible to erosion. Modern logging methods involving mechanized tree harvesting using skidders and trucks destroy the surface vegetation and compact the forest soil, both of which increase run-off and can contribute to erosion. Mining activities that involve the removal of vegetation or disruption of the land surface also help to initiate soil erosion. Although mining normally causes greater dis-

ruption of the surface than even accelerated soil erosion, the run-off from spoil tips or soil stored for future rehabilitation can still contribute to environmental deterioration in adjacent areas. Even apparently benign land uses such as recreation can initiate soil erosion. In mountainous areas such as the Alps, for example, the sheer volume of walkers in some areas has destroyed vegetation and compacted the soil along frequently used paths. These become preferred routes for flowing water, which erodes the paths and creates deepening gullies on the hillsides (Price et al. 1997). The increased use of all-terrain vehicles (ATVs) or off-road vehicles (ORVs) for recreation also promotes soil erosion, mainly as a result of their disturbance of soils and damage to vegetation in fragile ecosystems. Frequent travel by wheeled vehicles causes soil compaction, reducing the infiltration capacity of the soil, increasing surface flow and the potential for soil erosion. Similar consequences ensue when the soil is opened up to direct attack by heavy rainfall, following the destruction of vegetation by ORVs. Studies in the desert and semi-desert areas of the south-western United States have shown that in areas of heavy ORV activity, erosion is ten to twenty times greater than in undisturbed areas (Cox 1993).

The nature and rate of soil erosion depend upon a number of factors, including rainfall intensity and run-off, slope, soil erodibility and vegetation cover. The direct impact of raindrops on bare soil can be sufficient to break down soil aggregates and disperse the finer constituents such as silt, clay and organic matter. That material may block up pore spaces in the soil, reducing infiltration capacity and encouraging increased run-off. The impact of rain on the soil is most noticeable during thunderstorms or other heavy rainfall events, but even less intense precipitation over a longer time period can cause significant erosion. If the soil has been compacted, or if it is saturated so that no more water can be absorbed, surface run-off is greater and the potential for soil erosion increases. Steeper slopes add to the problems, by promoting faster run-off. Not all soils exposed to these conditions will suffer serious erosion. Some soils are more easily eroded than others. Soils with a good structure have a mixture of particle sizes, held together by organic material and nutrients in strong aggregates that resist erosion. They also have abundant pore spaces, which facilitate infiltration and reduce run-off. Loams (see Chapter 3) have these characteristics and as a result tend to be less easily harmed by erosion than

Plate 6.4   The building of gravel access roads and the removal of trees by the forest industry provide conditions for the onset of soil erosion.

soils containing a preponderance of silt or very fine sand. Cultivation practices that break down the soil aggregates, reduce the organic content of the soil or fail to replace soil nutrients tend to increase the erodibility of even the most resistant soils. The potential for soil erosion depends very much on the nature of the vegetation cover. Vegetation protects the soil from the direct impact of heavy precipitation and slows down the rate of surface run-off. The roots of the plants also help to bind the soil particles together, helping them to resist erosion. When vegetation is deliberately removed, during logging, for example, or destroyed by human activities, the potential for soil erosion is increased. As a result of agricultural activities, crops have replaced natural vegetation in many parts of the world. Crops, however, do not provide such complete or continuous cover as natural vegetation. For periods during cultivation the soil is exposed to the elements, and even when they have begun to grow, crops such as maize and potatoes do not provide complete ground cover. The bare soil that remains between the rows provides a natural pathway along which running water can initiate erosion. Even when the land cover is not removed – where natural pasture is used for grazing, for example –

poor land management and overstocking can lead to the destruction of the vegetation and the exposure of the soil to erosion. Although serious erosion is often associated with agricultural activities that lead to overstocking, excessive mono-cropping and the ploughing of marginal lands, the potential for erosion exists wherever agriculture is practised, because it cannot be carried out without disrupting the natural balance in the biosphere.

The physical processes that cause soil erosion involve both water and wind. On moderate slopes, erosion by water is often initiated by the impact of large raindrops causing smaller particles to move. If the precipitation continues, sheet wash occurs in the form of a relatively shallow flow covering the whole slope, normally because the infiltration capacity of the soil cannot cope with the intensity of the precipitation. With time the water flow may begin to develop a pattern of small channels or rills in which the erosion becomes concentrated. Ultimately continuing erosion will cause some of the rills to develop into gullies. Gully erosion is common on steeper slopes, where the velocity of the flowing water is greater, or on slopes where the vegetation has not been removed completely, but where paths, roadways or even the bare soil between row crops

Plate 6.5 The perils of ploughing up and down even a relatively shallow slope. Sheet wash and gullying have carried topsoil from up-slope to be deposited in the foreground.

TABLE 6.2 SOIL EROSION: PREVENTION AND REVERSAL

Prevention

Agriculture:
■ Appropriate land use (e.g. pastoral versus arable farming)
■ maintenance of soil fertility
■ contour ploughing
■ terracing
■ planting of windbreaks
■ improved water use (e.g. well managed irrigation and run-off)

Forestry:
■ controlled harvesting (e.g. reduced clear-cutting on slopes)
■ less intrusive use of mechanization

Mining:
■ management of topsoil removal and storage
■ attention to local run-off patterns

Recreation:
■ planned and appropriate land use
■ controls on intensity of use

Reversal

Agriculture:
■ gully infilling
■ restoration of soil fertility
■ stabilization of eroded areas

Forestry:
■ gully infilling
■ slope stabilization
■ reforestation

Mining:
■ land rehabilitation
■ return of topsoil
■ re-establishment of plant cover

Recreation:
■ slope stabilization
■ re-establishment of plant cover

allows the flow to be concentrated along a specific pathway. Wind erosion is most common in drier, open areas and mainly involves finer soil particles that are small enough to be carried in suspension or drifted along the ground.

Soil erosion causes the productivity of the affected area to be impaired, but its environmental impact can extend into adjacent areas. Soil removed from one area by wind or water can be deposited in sufficient quantities in nearby areas to cover crops or disrupt existing soil processes. Eroded soil can fill drainage ditches, and sediments carried into streams and lakes can cause shoaling, disrupting transport or damaging fish habitat. In some cases the problem is compounded by the presence of pesticides, herbicides and fertilizers in the eroded material.

Attempts at reducing or reversing soil erosion commonly involve the retention of vegetation cover on the land, the maintenance of a good soil structure and a reduction in water and wind speeds (Table 6.2). Vegetation cover can be retained by employing forest practices that do not include clear cutting or logging steep slopes. The prevention of overgrazing or the adoption of agricultural practices – minimum tillage techniques or the planting of cover crops such as clover or alfalfa, for example – that do not allow the soil to remain exposed for long periods of time also helps. A good soil structure requires the organic content and nutrient levels in the soil to be maintained. Both help to bind individual particles together into larger aggregates that are less easily eroded. Since organic materials and nutrients are essential components of a fertile soil, maintaining soil fertility is good soil conservation practice. Contour ploughing – ploughing across the slope rather than up and down – of relatively shallow slopes and terracing of steeper slopes reduces the volume and speed of water flow downslope, and therefore reduces erosion. Windbreaks, which slow the wind, and cultivation practices such as strip cropping help to reduce soil erosion by wind. The adoption of such conservation practices helped the Great Plains of North America to recover from the ravages of soil erosion in the 1930s, but in many areas, particularly on land marginal for agriculture, technological and socio-economic factors allow poor soil conservation practices, such as over-grazing and inappropriate cropping, to continue, and soil erosion remains a major global problem.

PRESSURE ON LAND: INFRASTRUCTURE AND WASTE DISPOSAL

INFRASTRUCTURE

At the beginning of the twenty-first century, some 47 per cent of the world's population is living in towns and cities, and in the developed world the number is close to 80 per cent (see Box 4.3, Urbanization). The physical infrastructure that has grown up to support these urban dwellers has had a significant impact on the land. Buildings are the most obvious elements in the infrastructure, but the transport links that join the towns and cities, the systems that deliver energy to them and the structures that help dispose of the waste they create all contribute to pressure on the land. Major physical features in the landscape, such as steep slopes or water bodies, will influence the nature of urban development, but once a town or city is established, its impact on the land is pervasive. The existing soil and vegetation are stripped away to be replaced by bricks, concrete or asphalt; buildings increase the irregularity or roughness of the surface; drainage is disrupted by the large areas of impermeable surfaces that are created; stream channels are plumbed into pipes or culverts; wetlands are drained; the aesthetics of the landscape are changed; green space is lost (Stutz 1995). In addition to the visible changes there are others that are hidden below the surface. Unseen beneath the streets are webs of power, telephone, television and data communications cables, plus gas, water and sewage pipes, without which modern cities are unable to function.

Pressure on the land is increasing as urbanization proceeds and brings change to larger and larger areas of the landscape. In large urban conglomerations such as those in the eastern United States, in the lower Rhine valley in Europe and the Tokaido area in Japan, the natural environment is entirely lost and urban sprawl is characteristic of most centres in the developed world (Figure 6.5). Among the developing nations, urbanization is being driven by rapid population growth (see Chapter 4) but, because of economic conditions, the development of the urban infrastructure cannot keep pace with the needs of the growing population and the poverty-stricken squatter settlements that have grown up around major cities in the developing world contrast sharply with the affluent residential suburbs in the developed nations. Urban sprawl in the developed world overwhelms the environment

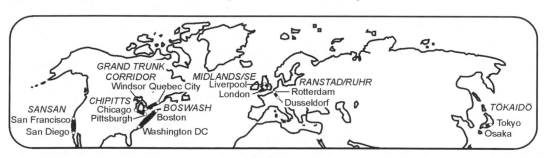

Figure 6.5   Major urban conglomerations in the developed world

Plate 6.6   The result of urban sprawl: houses spreading out over agricultural land, a problem common in most developed countries.

with the physical presence of the urban infra-structure. In the squatter communities around cities in the developing countries it is the sheer weight of numbers coupled with, at best, a rudimentary infra-structure that creates the problem. Conflicts arise when urbanization replaces the existing land use. Highly productive soils may be covered by the con-crete and asphalt associated with residential and industrial development or, as is often the case in the developed world, taken out of production to pro-vide recreational facilities such as golf courses. The net result of urbanization is a permanent change in the land, accompanied by a decline in local food production (although, if appropriate land is avail-able, agricultural production may increase around some towns and cities as a result of the growing demand for food from the urban community) or, where the natural environment is being displaced, a reduction in wildlife habitat and potential increases in run-off and flooding.

In most developed countries, land use planning is in place to determine the best possible use of land in an area. Administered by regional or local governments, these plans incorporate legislation such as zoning by-laws to control development. With land treated as a commodity to be bought and sold, however, decisions are often influenced

by economic factors and the actual use of a piece of land is commonly the result of a compromise among environmental and socio-economic factors. A frequent source of conflict arising from such decisions is the conversion of good-quality agri-cultural land to residential or industrial use. One planning approach that received widespread con-sideration in the second half of the twentieth century was the concept of the green belt, an area around large cities in which development was pro-hibited, allowing it to retain its natural or rural landscape (Figure 6.6). It was an important element in planning for the growth of London, England, and although it has received less attention in recent years, it continues to be part of the planning process in countries such as the Netherlands (Huisman 1997) and in parts of the United States as part of land use planning for sustainable development (Chiras 1998).

Where land use planning is absent or ineffectively administered, the results of development can create serious, even life-threatening, problems for the inhabitants of an area. Failure to appreciate the hazardous nature of floodplain development, or willingness to ignore it, for example, regularly leads to major property damage and loss of life in river basins in both the developed and the developing

(a) *COMPACT CITY*

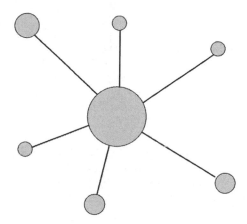

- surrounding towns and villages of various sizes
- limited road network of variable quality
- city self-contained

(b) *COMPACT CITY WITH SATELLITES AND GREEN BELT*

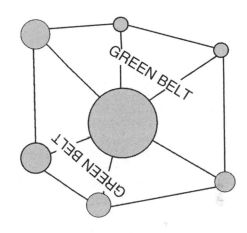

- satellite towns act as dormitory communities
- road network improved and extended
- countryside protected by green belt

declining
sustainability

(a) ———→ (b)

(c) ———→ (d)

(c) *RIBBON OR CORRIDOR DEVELOPMENT*

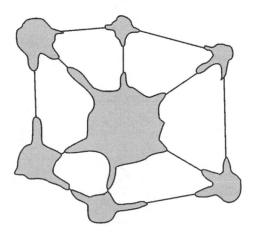

- growth along main roads between city and satellite communities
- quality of road network improved and density increased
- countryside threatened by infilling between main roads

(d) *URBAN SPRAWL*

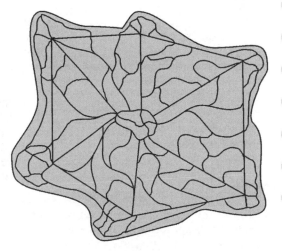

- high-density road network
- residential development and service functions completely replace natural environment
- main city and satellites merge to form single metropolitan area

Figure 6.6 Forms of urban development and their level of sustainability

world. Flooding is a natural process, but whether it be in Bangladesh or along the Mississippi, its impact on humans is a result of inappropriate land use. The rapid population growth in Bangladesh means that there is little or no alternative to occupying hazard land, but in the developing world such situations are not uncommon. The squatter settlements that climb the hills behind Sao Paulo or Lima are there because the rates of population growth and urbanization have far exceeded the planning capacity of the local governments. As a result, built without controls on location, design safety or structural integrity, the settlements and those who live in them are frequently destroyed by landslides and slope failure. Even in wealthier cities, such as Hong Kong, however, where government housing projects have almost eliminated the problems associated with squatter communities, the existence of steep slopes used inappropriately leads to landslides, with accompanying property damage and loss of life. The nature of urbanization in the developing world, driven as it is by natural growth and migration from rural areas, coupled with a general absence of planning, allows the mixing of residential and industrial activities in urban areas, exposing the population to a variety of health problems associated with the emission of toxic materials or the release of hazardous waste into the community.

In addition to the ever-increasing footprint of urbanization, pressure on the land is created by the infrastructure associated with the transport links that join the towns and cities. The main links are provided by land and air transport, each of which has its peculiar infrastructure. Water transport also has an impact on the land, but it is more indirect and usually incorporated in the main urban infrastructure, as is the case with dock facilities in port cities, for example.

Roads

Although even single-lane gravel roadways can have a significant impact if built without due consideration of the existing environmental conditions, the greatest pressure on the land comes from the great web of major and superhighways that has grown up in, around and between the world's major cities. Constructed from asphalt and concrete and perhaps sixteen lanes wide in places, they have been responsible for modern suburbanization by promoting the daily flow of automobile traffic in and out of the cities. The first of these modern highways were the Autobahns built in Germany in the 1930s, followed in the 1950s and 1960s by the motorways in Britain and the interstate highway system in the United States, and at the beginning of the twenty-first century these road systems, with their flyovers, underpasses and clover-leaf junctions dominate the landscape of the developed world, indicating the reliance of modern society on the automobile. They do not exist in isolation, however, and where such superhighway construction has taken place the pre-existing road network has had to be upgraded or expanded to cope with the additional traffic generated by the new system. These roads and highways not only cover the existing surface with an impermeable layer, but also incorporate additional land in medians, noise barriers and rights of way reserved for future expansion. Perhaps more important, they cut through ecosystems, creating barriers that prevent the normal movement of animals, and disturb the balance of the system (Alberti 2000). Breeding populations change and predator/prey relationships are disrupted. Animals trying to cross these barriers present a serious traffic hazard in some areas and in most cases face injury or become road-kill as a result of their attempts. In places that have been identified as major crossing points for animals underpasses have been constructed to reduce the danger for animals and for drivers, with mixed success.

The relationship between urban growth and the development of a highway infrastructure is complex, involving positive feedbacks in which expansion in one element tends to require expansion in the other. Development beyond an urban boundary, for example, which increases the traffic on the existing road network, often leads to public pressure to increase the capacity of the net by building new roads or widening those already there. If the pressure leads to investment in additional capacity, travel beyond the city boundary is made easier, people take the opportunity to live in a lower-density urban environment and housing developments, shopping malls and a variety of service industries follow, all of them contributing to urban sprawl. Before long, the road system is as congested as it originally was and public pressure to add capacity increases again. Further success leads to more urban sprawl, and so on. This phenomenon has been referred to as the 'black hole' theory of highway investment, characterized by continuing investment in the system until all of the money is gone, but with no long-term reduction in traffic congestion (Plane 1995). Ultimately a point is reached at which the costs of continued investment outweigh any potential benefits, but by that time the additional infrastructure has completely destroyed the pre-existing environment.

The initial development may be completely dispersed around the perimeter of the existing urban area, or it may take the form of ribbon or corridor development along major transport routes. In some cases smaller communities at a distance from the main centre become growth points linked to the city by road or rail (Chiras 1998). Following initial ribbon development it is common for the surviving open space to be filled in, creating the same overall impact as the urban sprawl associated with uncontrolled expansion around the perimeter of the city. With appropriate planning, however, including the provision of green space to separate them from the main centre, satellite communities can provide a reasonable balance between improved quality of life and environmental disruption. If planning provisions are absent, or inadequately enforced, it is not uncommon for urban sprawl to envelop these outlying centres, with a consequent loss of amenity and the imposition on the land of an urban infrastructure with all its associated environmental problems (Figure 6.7).

Airports

Air traffic increased rapidly in the second half of the twentieth century, and to keep up with the demand airports were expanded and new ones were built. As with road building, airports completely change the existing natural environment. Multiple asphalted concrete runways several kilometres in length are supported by taxiways, parking aprons and terminal buildings, which provide large areas of impervious surfaces. At New York's John F. Kennedy Airport, for example, these structures cover 5–6 km², but the impact extends beyond that. The entire airport covers an area of nearly 20 km², the additional space being taken up by open areas adjacent to the runways and taxiways or by storage and maintenance facilities. Most modern international airports cover 10–20 km² – London Heathrow has an area of 12 km² and Toronto Pearson 18 km², for example – which results in the complete disruption of local ecosystems. The

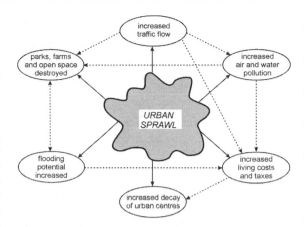

Figure 6.7 Environmental implications of urban sprawl
Source: based on information in Sierra Club, *Sprawl Factsheet.* Viewed at http://www.sierraclub.org/sprawl/factsheet.asp.

original vegetation is destroyed during the construction phase and animals that cannot cope with the changed habitat are forced to move. Native vegetation has been replaced by short grasses in most cases and in time it provides a habitat for a new combination of animals. That, however, can also raise problems. The insects and small rodents that commonly colonize the new vegetation attract birds, which represent a danger to the aircraft through potential bird strikes, and therefore must be controlled. Thus despite the large areas of greenery that are characteristic of airports, their environment is almost completely artificial, bearing little or no relationship to that which previously existed.

Modern airports are a frequent source of conflict over land use. Because of their need for space and accessibility, airports were constructed originally outside the urban areas they served. Once built, however, they attracted ancillary activities such as hotels, warehousing facilities and a variety of aviation-related activities, all of which required the expansion of the transport network. The outward expansion of urban areas also brought suburban development into the vicinity of the airport. In theory, appropriate land use planning would have prevented this, but it seldom did, and despite being first on the scene, airport facilities and operations came under attack from those occupying the new developments. Noise, air pollution and traffic congestion were the main concerns. Although improvements were made in all these areas, through improved flight scheduling, quieter, more energy-efficient aircraft and better planning of road networks or the introduction of rapid rail transport, conflicts resurfaced as soon as attempts were made to increase airport capacity. Other alternatives such as building an entirely new airport beyond the present urban area have also met opposition, usually from those who own the land that would become part of the new airport and from environmentalists who wish to prevent the ecological destruction that would accompany the building of the new facility. The combination of public opposition and new environmental impact legislation has complicated the nature and rate of airport expansion. Public opposition helped to prevent the development of a new airport to serve Toronto in Canada (Beattie 1983) and arguments over the provision of an additional airport for London, England, have gone on for decades. Occasionally violence erupts, as in the case of Narita Airport, built to serve Tokyo in 1978, but opposed by local farmers and students. Twenty-five years later Narita is already at capacity and requires an additional runway. Attempts to replace the existing Kingsford-Smith International Airport in Sydney, Australia, with a new facility in the Sydney basin appear to have been stalled, in part because of public opposition. In Sydney, as a means of increasing the airport's capacity where there was no land available, the existing runways were extended out into Botany Bay. Lack of suitable land in Osaka, Japan, was dealt with by a similar but larger-scale approach. Kansai International Airport was built entirely on an artificial island created specially for the purpose in Osaka Bay. Similarly, in Hong Kong a new airport was constructed on land in part reclaimed from the sea. Ingenious as these solutions may be as exercises in engineering, from an environmental point of view reduced stress on the land is offset by the increase in pressure on the ecology of the offshore zone.

Energy infrastructure

The seemingly insatiable demand that modern society has for energy (see Chapter 5) creates serious environmental damage to the land during resource extraction, but, through the facilities and structures

Plate 6.7 Kingsford-Smith Airport, Sydney, Australia. In the absence of appropriate space or land, the runways have been extended out into Botany Bay. (Courtesy of Andreas Zeitler.)

developed for its transmission and use, energy also contributes to the disruption caused by the urban infrastructure. Facilities such as thermal generating stations, oil refineries and coal handling facilities are included in the urban imprint on the land, but the impact of energy use is also seen in the transport sector. The existing networks of roads and airports, for example, with all of the environmental disruption they cause, have become an integral part of the infrastructure upon which society depends, sustained by dependence on petroleum-powered cars, trucks and planes. Networks of electricity transmission lines and oil or gas pipelines are also part of the infrastructure, spreading out across the landscape from production facilities to consumers. Neither transmission lines nor pipelines appear likely to have as much impact on the land as a major highway or a large international airport. Pipelines are buried underground and transmission cables are carried high above the surface. They do not segregate the landscape in the way that road systems do, nor do they produce the noise and air pollution associated with highways and airports. They are not free of problems, however. The density of transmission towers and power lines can create aesthetic pollution in places and the flow of electricity creates electrical fields that may be responsible for health problems among those living in close proximity to the lines (Randerson 2002). To allow easy access to the transmission lines for maintenance or repair, vegetation along the right of way is usually trimmed regularly or sprayed with selective herbicides, which disrupt the normal development of the environment. Pipeline rights of way are similarly treated and, although the pipelines are buried, in areas of natural vegetation the location of the pipeline is evident from the difference in vegetation cover along its path.

To help meet the needs of a modern industrial society dependent upon petroleum as its main source of energy, pipelines transfer millions of litres of petroleum and millions of cubic metres of gas across great distances every day with few problems (see Box 6.2, Pipelines). Along with motor vehicle operations and maintenance, refineries, petrochemical plants and other industrial operations, however, pipelines are one of the main sources of oil pollution on land. They do leak and human error during operation can allow oil to escape. Most oil spills on land are small – generally less than 4000–5000 l – but occasionally they reach disastrous proportions. In late 1994, for example, a major leak

in a pipeline carrying oil from the Arctic to central Russia allowed some 200,000 tonnes of crude to spill on to the Siberian tundra (Pain and Kleiner 1994). Such a massive spill compares with the largest oil tanker spills (see Chapter 8). However, nothing as yet is comparable to the volumes of oil released into the environment as a result of the Gulf War in 1991, when the defeated Iraqi forces sabotaged some 800 oil wells in Kuwait. Some they set alight, others they allowed to spew oil over the surrounding desert. The Kuwaiti oil fires burned for several months; at their peak, the spills amounted to more than 7 million tonnes of oil, and formed lakes that covered some 49 km^2 of Kuwait. As late as 1995, despite enormous amounts of time and money spent on the clean-up, an estimated 0.5 million tonnes remained (Pearce 1995b).

Although the pollutants released in an oil spill are mainly hydrocarbons, and therefore subject to biodegradation by bacteria and other organisms in the natural environment, the process is slow and the impact of the pollutants may be felt for many years after the initial contamination. As a result, the area damaged by an oil spill requires some form of direct clean-up or rehabilitation. Free oil in pools can be pumped into tanks, but the soil remains contaminated. In Kuwait, after the liquid oil had been removed from the surface of the desert, the soil and sand beneath were still polluted to a depth of 1.5 m (Pearce 1995b). The longer the oil remains in the soil, the greater the possibility that it will be carried into the groundwater system and, to prevent the spread of the pollution, contaminated soil is normally removed if the spill is relatively small. Even if it is possible to remove the soil, however, the local environment is destroyed and the problem of disposing of the excavated material remains. Despite ongoing attempts to improve clean-up technology, complete rehabilitation of a site is seldom possible, and, as with all pollution problems, prevention rather than response after the event is the most appropriate response.

INFRASTRUCTURE AND DEVELOPMENT

The creation of an appropriate infrastructure was an integral part of growth among the developed nations. It contributed to economic expansion and in turn benefited from the results of that

BOX 6.2 PIPELINES

An essential element in the infrastructure of the modern world is the pipeline network. It is particularly important in the developed nations, where it is used to distribute the oil and natural gas upon which all forms of activity depend either directly or indirectly. Pipelines also carry water and a range of chemicals, but by far the largest number are used to deliver energy products. In the United States, for example, 65 per cent of the petroleum and almost all the natural gas is transported across the country by way of pipelines. The pipes range in size from more than a metre in diameter, along the main or trunk lines, to only a few centimetres in diameter for distribution to individual users. In addition to the pipe itself, pipelines include a combination of pumping or compressor stations and control devices to manage the flow of oil or gas through the pipe. In any network, there are pipelines that bring crude petroleum or natural gas from the production wells to the refineries or gas processing plants. The refined or processed product is then pumped along the main lines, usually over long distances, to the markets, where it is distributed through a dense network of smaller-diameter pipes to the consumer. Together these various pipes have a combined length of more than 2 billion km in the United States alone, well beyond the next longest systems in Canada and Russia, which have about a quarter of a billion kilometres each.

Once a pipeline has been built, it can operate continuously with minimum human input. Computer monitoring allows any anomalies that might lead to leaks or blow-outs to be detected at an early stage and dealt with before they become serious. The efficiency with which they can deliver petroleum and gas products has made a major contribution to development. A Very Large Crude Carrier (VLCC) carries on average about 1.5 billion barrels of oil, taking perhaps two or three weeks to travel between the Middle East and North America. Added to that is the time for loading and unloading plus the return journey in ballast. The largest, most efficient pipelines can deliver a similar amount in about ten to fifteen days with no turn-round time, no break in the flow and no concerns over storms at sea, collisions or delays in loading or unloading. It is not possible to build pipelines from petroleum sources such as the Middle East to the main consumers in North America, Europe and Japan, but pipelines are becoming longer and longer. The lines that bring oil and gas from the Alberta in western Canada to the main markets in eastern North America are between

4000 km and 5000 km long and a new pipeline more than 3000 km long designed to bring oil from the Russian Urals to the Adriatic is almost complete. Other lines more than 2500 km in length are being built or are planned to deliver Russian oil and gas to China. One of the advantages of the North Sea as a source of petroleum products is its proximity to its major markets. As a result the pipelines there tend to be shorter, but there are more than 7000 km of pipe in the network that lies beneath the sea, carrying oil and gas from the fields in the centre of the basin to Britain, Norway, Germany, Denmark, Belgium, France and the Netherlands. Once ashore, the gas is piped into existing grids so that natural gas arriving in northern France, for example, can be sent on to Italy and Spain, and gas from the North Sea can pass through Britain for delivery to Ireland.

ENVIRONMENTAL IMPACTS OF PIPELINES

The greatest environmental impact associated with most pipelines is during the construction phase, when vegetation is removed, a trench several metres deep is dug, the pipe is laid and the trench backfilled. Although the trench itself is relatively narrow, the impact of the excavation and fabrication equipment extends outwards on either side, and in places storage yards, living quarters and maintenance facilities may cover several hectares. Once the construction is complete it is often possible to rehabilitate the land to such an extent that there is no indication of the presence of a pipeline. In agricultural areas, for example, the land in most cases can revert to its original use and in urban areas the smaller distribution lines are completely hidden, so much so that care must be taken to ensure that the pipes are not damaged during road repairs or other excavations. Elsewhere, in forested areas, for example, the pipeline right of way is routinely kept free of trees to allow easy access for maintenance or repair and pipeline routes can be followed for hundreds of kilometres across country under these circumstances. Pumping or compressor stations are also part of the system, but they may be as much as 100 km apart and have a relatively minor impact on the environment.

Overall, pipelines have a good environmental safety record, although it is not perfect. Leaks and spills do occur. Most pipes are welded sections of steel and are wrapped and treated to prevent corrosion. In addition, they are regularly inspected by introducing

BOX 6.2 – continued

sensors into the flow to detect corrosion or weaknesses in the welds that could promote leaks. When leaks do occur, they tend to be relatively small, because of early detection through sensors that recognize minor changes in pressure within the pipe and close valves or shut down pumps to stop the flow. Spills in excess of 10,000 barrels of oil are uncommon, for example, whereas even relatively minor tanker accidents can release twenty times that much oil. When leaks occur in natural gas pipelines the gas commonly ignites, and while that creates air pollution the impact on the land is limited. Small as they may appear in comparison with other sources of pollution, however, pipeline leaks can cause serious environmental problems when they pollute soil, waterways or wetlands, for example. Soil pollution can often be dealt with relatively easily, by the removal of the contaminated material and the remediation of the site, but once into the water system, oil can have serious, long-term effects on vegetation, fish, wildlife and drinking water supplies, not only at the location of the original spill, but also downstream and in the groundwater sector.

PIPELINE EXPANSION IN ENVIRONMENTALLY SENSITIVE AREAS

As the demand for petroleum continues to grow the pipeline network will expand and it is inevitable that its environmental impact will increase. It will be greatest in areas that are particularly environmentally sensitive. There is concern among environmentalists, for example, that an 800 km pipeline under construction in West Africa to transport natural gas from the Niger delta to the neighbouring states of Benin, Togo and Ghana will cause irreparable damage to the mangrove swamps and fragile wetland ecosystems of the coastal regions through which it will pass. The mangrove swamps are already under threat from the changing sea levels expected to follow global warming and there is concern that once they are damaged in this area they will not regenerate. Being laid in both onshore and offshore locations, the pipeline will not only cause damage to the land and coastal vegetation, but will also disrupt fishing activities. Similar concerns have been expressed over a 1500 km long pipeline to bring oil from Chad to the Atlantic coast in Cameroon, where the fragile coastal ecosystems would be under threat from infrastructure developments and potential spills. Despite the existence of environmental

legislation to deal with such problems, in place in all the countries involved, the environmental impact of the project has received much less attention than it would have received in the developed world.

ARCTIC PIPELINES

Another environmentally sensitive area that has received increased attention as a source of petroleum in the past thirty years or so is the Arctic. In North America and in Russia, the area north of the Arctic Circle has the potential to produce massive volumes of oil and natural gas. Since the Arctic Ocean is blocked by ice for most of the year, the delivery of the petroleum to southern markets is most likely to involve the building of pipelines. The prototype of Arctic pipelines was the Trans-Alaska pipeline built between 1974 and 1977 to transport oil over 1200 km from the North Slope of Alaska to Valdez on the south shore, crossing two rugged mountain ranges and thirty-four major rivers on the way (Figure 6.8). During construction, and in its operation, the pipeline faced major physical and environmental restraints, which required special treatment to ensure efficient and safe operation. To reduce damage to the fragile ecological balance of the tundra and the underlying permafrost, for example, most of the construction was confined to the winter months. The presence of permafrost also necessitated construction techniques not normally required in temperate regions. The buried pipe has to be insulated to prevent the heat from the flowing oil melting the surrounding ice, and in places, where the permafrost was particularly unstable, the pipe required additional refrigeration or had to be raised above ground on trestles. Near its southern end the pipeline crosses several earthquake-prone areas. There the pipe had to be placed on a series of elevated, movable saddles that would allow it to flex without breaking in the event of an earthquake.

During the planning stage, environmentalists expressed serious concern about the impact of the construction and operation of the pipeline on the wildlife population of the area. The pipeline crosses a number of caribou migration routes, and it was felt that the activity and noise associated with the operation would cause the animals to change their migration patterns, affecting not only the viability of the herds but also the livelihood of the indigenous peoples of the region who depended upon them as a food source. Similarly, the upheaval caused at pipeline river crossings was seen as likely to damage

BOX 6.2 – continued

fish habitat such as spawning beds. A major environmental concern was the possible rupture of the pipe and the escape of thousands of barrels of oil into and on to the tundra. The inclusion of constant monitoring equipment, automatic valves to stop the flow in the event of a break, plus regular inspections and maintenance, has so far prevented any serious spills.

Although it is impossible to undertake a project of the magnitude of the Trans-Alaska pipeline without altering the environment to some extent, after nearly thirty years of operation it has been remarkably problem-free. Much of the credit for that is due to the intensive environmental planning and environmental impact assessment that preceded the construction of the line and the lessons learned will be applied to the new pipelines that will inevitably be built from the Canadian and Russian Arctic oil and gas fields.

Pipeline networks are efficient systems for moving high volumes of petroleum products over long distances, and as long as the demand for petroleum remains high the pipeline network will continue to grow. Although pipelines cannot be described as environmentally friendly, their overall environmental impact has been limited by a combination of technological progress in construction, operation and monitoring and legislation that requires appropriate environmental planning and environmental impact assessment.

For more information see:

Africa News (2001) 'Concern over West African gas pipeline', viewed at http://www.waado.org/Environment/OilPipelines/ConcernOnGasPipeLines.html (accessed 13 May 2002).

Coates, P.A. (1991) *The Trans-Alaska Pipeline Controversy: Technology, Conservation and the Frontier*, Bethlehem PA/London/Cranberry NJ: Lehigh University Press/Associated University Presses.

Pain, S. and Kleiner, K. (1994) 'Frustrated West watches as Arctic oil spill grows', *New Scientist* 114 (1950): 8–9.

Pearce, F. (2000) 'Trouble in the pipeline', *New Scientist* 166 (2243): 20.

Figure 6.8 The route of the Trans-Alaska pipeline

expansion. Despite the environmental stress that it has created, the development of a modern urban infrastructure contributed to an enhancement of the quality of life for most people in the developed world, by improving living and working conditions, increasing mobility, supporting better transport and supplying their growing needs for food, water and waste disposal. Economic conditions in the developing world have not allowed the creation of an efficient infrastructure and as a result they have not been able to take advantage of the benefits that it can bring. Rapid population growth has overwhelmed the existing infrastructure, the quality of life has declined and continuing environmental deterioration has become the norm. There is a contradiction in this. In the developed world, infrastructure has contributed to stress on the land through the environmental problems it creates, whereas among the developing nations environmental problems often arise as a result of the absence of an appropriate infrastructure. Somewhere between these two extremes there should be a level of infrastructure development that allows the socio-economic needs of the population to be met with minimal damage to the environment.

In the developing world, as new infrastructure is introduced, a combination of sustainable development, land use planning and appropriate technology (see Chapter 5) should allow that level to be attained. Investment in housing and other property, highways and energy distribution systems is required to allow the developing nations to compete in the world economy and improve the quality of life for their citizens. Economic development and the provision of a suitable infrastructure cannot be separated. The challenge will be to develop the necessary infrastructure with a minimal impact on the land and other elements in the environment. Among the developed nations the existing infrastructure has already had a significant impact on the land, and that cannot easily be changed. Much of that infrastructure is old, however, and consideration of sustainability along with land use planning, during the repair, upgrading or replacement of that infrastructure could help reduce its impact. The infrastructure problems associated with continuing suburban sprawl, the new roadway construction it brings and the consequent increase in energy consumption, have yet to be adequately addressed (see Box 6.3, Urban development and

Plate 6.8 An attempt to improve sustainability in an urban area by building compact row housing. The land has been recovered from industrial and commercial activities.

BOX 6.3 URBAN DEVELOPMENT AND SUSTAINABILITY

Urban development represents the most concentrated human impact on the environment. In most cases the natural landscape and hydrology are completely changed, and the urban atmosphere is subject to periodic pollution episodes. In addition, the environmental footprint of most cities extends well beyond their urban boundaries. No city can feed its population or meet its energy needs from within its boundaries or from its immediate environs, for example. Decisions made or materials consumed in London, New York, Sydney or Toronto can have environmental impacts half a world away, in Africa or South America, perhaps. Under these circumstances it might appear that sustainability has no place in the consideration of urban development. Sustainability requires development that is both economically and environmentally sound so that the world's current population can meet its needs without jeopardizing those of future generations. It involves the efficient use of resources and consideration for the environment, but in addition it includes a number of human factors such as the improvement of quality of life and implies the acceptance of social equity or justice for present and future populations. Although the concept was originally developed for consideration on a world scale, it is adaptable and can be applied in an urban setting. In its simplest form, a sustainable city is one that allows the efficient use of resources – energy and water, for example – attempts to minimize its environmental footprint and provides quality living conditions for its inhabitants.

THE NATURE AND NEEDS OF URBAN PLACES

Urban places include a complex web of interrelated systems, designed to process raw materials, provide goods and services, move people and products and dispose of the waste products that accumulate when large numbers of people congregate in relatively small areas. The ability of a particular city to achieve or maintain sustainability depends very much on the efficiency with which these systems function. A city that allows its water supply system to leak, for example, or discharges raw sewage into streams, lakes or the sea cannot be considered to be following sustainable principles. A city that fails to dispose of its garbage in an environmentally sound manner or makes no attempt to recycle its waste products

is unlikely to achieve sustainability either. Energy is also an important element in attaining and maintaining sustainability, because it is needed to make the systems work. Electricity is required to provide light and power machinery or appliances; petroleum is used for transport; natural gas is increasingly the main source of energy for home and office heating. The innovative use of energy has improved the quality of life for urban dwellers, but in most communities inefficient heating and lighting systems waste energy, heat is lost through insufficient insulation in buildings and inefficient road networks or transport systems lead to excessive fuel use. Energy is a major contributor to the environmental footprint of a city, creating environmental disruption both where it is produced and where it is used. Furthermore, most of the energy currently used is obtained from non-renewable resources and any that is wasted is energy that will not be available for future generations. Thus a community that is using its energy inefficiently cannot be considered to be following the path towards sustainability.

URBAN FORM AND SUSTAINABILITY

In the study of sustainable development in urban areas particular attention has been paid to the form of cities. Since urban areas have developed for a variety of reasons, individual cities vary in the form they take. A city dependent upon heavy industry will have a form that differs significantly from one that is predominantly a service provider, for example. Modern cities do have a number of characteristics in common, however. Areas within the urban area tend to be segregated according to function – industrial, business and retail, residential – mainly as a result of the zoning approach to land use planning that most communities follow. In the past these functions were much more mixed, with workers tending to live close to their places of employment and retail facilities existing within residential neighbourhoods, rather than being concentrated in the downtown area or, as is the case now, located in suburban shopping centres. Towards the middle of the twentieth century, when conditions within most of the cities in the developed world were becoming increasingly congested and urban air pollution was recognized as a serious problem, many people began to move out towards the city edges, where more space was available and the air was clean, leaving the city core to deteriorate.

BOX 6.3 — continued

This form of urban development increased the environmental footprint of most cities, bringing with it a decline in sustainability. It was made possible by the rapid increase in the use of cars as the preferred means of transport and the impact of the automobile is imprinted on all urban areas in the developed world in the form of a road network designed to accommodate the movement of large numbers of people in and out of the city using their own private transport. This has contributed enormously to the lack of sustainability associated with the modern city, not only through the infrastructure it requires, but also through the increased and often inefficient use of energy and the consequent rise in air pollution levels.

It is widely accepted that current forms of urban development, in which cities tend to spread relentlessly across the land that surrounds them, are not sustainable. There does not appear to be any one form that provides all the requirements of sustainability, but there are certain urban forms that encourage more efficient use of land and resources (Figure 6.6). For hundreds of years the compact city was the norm. Often delimited by walls to provide defence, these cities contained high densities of buildings and of people and supplied all the services required. Although they had to import goods over longer distances, they met most of their needs from the land immediately surrounding the city. As a result, their environmental footprint remained small. A return to such compact cities would improve sustainability. The reduction of urban sprawl would allow the adjacent land to remain in its natural state or as farmland, diminish the infrastructure required for transport, water and power supply, reduce the amount of energy used for transport and generally reduce the physical impact of the city on its environment. Although it would be impossible to reduce the size of existing cities, the rate at which they are expanding could be reduced by applying some of the characteristics of the compact city to their core areas. Particularly in North America, the central parts of many large cities deteriorated when suburbanization accelerated in the middle of the twentieth century. Urban renewal in these areas could provide an overall improvement in the urban environment, an increase in residential property in the downtown area, the revitalization of the retail and service sectors and an expansion of the city's tax base, which would help the financing of further developments. The urban renewal approach can also be applied to industrial areas, to replace the facilities used by heavy industries with new units which support modern 'high-tech' companies involved in the electronics industry or other light manufacturing enterprises. Urban renewal is already a feature of many big cities in the developed world, and as it continues it will increase their sustainability. The compact form advocated as particularly suitable for sustainable development is not without its drawbacks, however. For social and psychological reasons, for example, an appropriate balance has to be struck between structural and population density on the one hand and open space on the other. Early attempts at dealing with urban decay by replacing slum properties with high-density tower blocks improved physical living conditions, but were accompanied by the significant increase in social problems. The redevelopment of traditional neighbourhoods, containing a mix of housing types, plus retail and service facilities and employment opportunities served by good public transport, would help to alleviate some of these problems. With the greater concentration of people, heating units and cars, combined with the changes in urban morphology and climatology that encourage the accumulation of particulate matter and gases (see Chapter 10), compact cities tend to be more polluted than suburban areas. That too can be dealt with through such devices as smokeless fuels, high-efficiency furnaces, reductions in energy waste — using residual heat from steam power plants to heat houses, for example – the introduction of renewable energy and the development of efficient public transport systems.

URBAN TRANSPORT SYSTEMS AND SUSTAINABILITY

Transport is a key element in urban development. The current dependence on the automobile, however, is a major deterrent as cities try to improve or retain sustainability and most plans that attempt to achieve a sustainable urban form pay considerable attention to the impact of transport on sustainability. Every day there is a concentrated flow of cars into and out of the cities in the developed world, carrying workers to the workplace in the morning and back home in the evening. Millions of people are involved, but the car is, in fact, a very inefficient people mover compared with most forms of public transport, particularly during the daily commute when cars frequently have only one or two occupants. In addition, cars require major investment in

Plate 6.9   Streetcars (trams) and light rapid transit vehicles – in Melbourne, Australia, in this case – are increasingly being used in an attempt to reduce the use of cars and the problems they bring to urban areas. (Courtesy of Karen Armstrong.)

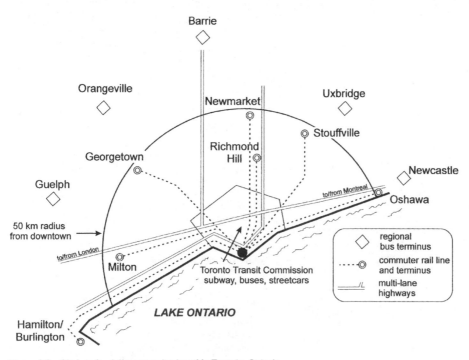

Figure 6.9   Choices for daily commuter travel in Toronto, Ontario

BOX 6.3 – continued

infrastructure, from highways leading into and out of the city to parking spaces on high-value land in the downtown area, and make a significant contribution to pollution. Despite the apparent advantages in increasing the use of public transit, however, it has proved very difficult to move commuters from their cars to buses, subways and light, rapid transit rail systems. It is not necessarily a question of economic cost, for public transport is usually cheaper when compared with the overall costs of maintaining, fuelling, insuring and parking a car. The main advantages of the automobile lie in time saving, convenience and a flexibility that is generally missing from mass transit. Increasing transit ridership and reducing automobile use are central to many plans for improving urban sustainability, but attempts to take them beyond the planning stage have met with mixed success. In continental Europe public transport has retained some of its former importance and is being expanded, and in North America there have been some successes. In Toronto, Canada, for example, effective transport planning has enabled 75–89 per cent of the workers employed in the downtown area to travel to work by bus, subway or rail (Figure 6.9), but, because the numbers involved are so large, highway congestion remains a problem. In southern California, the epitome of a car-dependent society, the city of San Diego has had considerable success in moving large numbers of people into and out of the city using a light rail system. Elsewhere, attempts to increase the use of public transit have not worked as well and in the United States transit use has declined rather than risen, with a 35 per cent reduction in ridership by workers between 1970 and 1990. Despite such setbacks, the introduction of a well planned, efficiently run public transport system would do much to improve the sustainability of most large urban areas.

SUSTAINABILITY IN URBAN AREAS IN THE DEVELOPING WORLD

Most of the studies of sustainable urban development have investigated the situation in developed nations. Any possibility of planned urbanization in the developing world appears hopeless, yet it is there that the most rapid urbanization is taking place. Rapidly growing populations have overwhelmed the existing infrastructure and such features of the cities as lack of clean, running water, the absence of sanitation, lack of basic public services and the socioeconomic inequality represented by the presence of luxury development and squatter settlements, often in close proximity to each other, are visible evidence of their unsustainability. Any attempts to deal with these issues through urban land use planning, however, are unlikely to be successful until the problems of rapid population growth have been dealt with. Even then, success in achieving sustainable urban development in the developing world is likely to be a long-term project. The world community has embraced the concept of sustainable development and as more and more of the world's population becomes urbanized it is inevitable that much of the effort in achieving sustainability will be aimed at urban areas.

For more information see:

Chiras, D.D. (1998) *Environmental Science: A Systems Approach to Sustainable Development*, Belmont CA: Wadsworth.

Hanson, S. (ed.) (1995) *The Geography of Urban Transport* (2nd edn), New York: Guilford Press.

Hartshorn, T.A. (1992) *Interpreting the City: An Urban Geography* (2nd edn), New York: Wiley.

Williams, K., Burton, E. and Jenks, M. (eds) (2000) *Achieving Sustainable Urban Form*, London: Spon.

sustainability). Attention to these and similar issues would reduce or modify the impact of the new infrastructure required as development continues. During the next thirty years, the need for new infrastructure will surpass all that currently exists in the developed world (Hjerppe and Berghall 1996). If it is put in place with insufficient planning, and the inadequacies of the existing infrastructure are carried through, the cumulative degradation of the land will continue.

WASTE DISPOSAL

Waste is any material, solid, liquid or gas, that is no longer required by the organism or system that has been using it or producing it. Waste is a fundamental part of any ecosystem, and the natural environment includes a series of very efficient waste disposal systems, which involve the recycling of the waste products. Organic waste, such as that produced by animals, is reduced by insects, microbes and fungi into its constituent chemicals, which are reabsorbed into the environment. Dead plants and the leaves discarded by deciduous trees in the autumn are treated in a similar fashion. Problems arise when waste is produced in such quantity that the normal disposal systems cannot cope or when it takes such a form that existing systems can dispose of it only slowly or in some cases not at all. Such situations are most often associated with human activities. Population growth, new lifestyles and rapidly changing technology have contributed to an increase in waste and created serious waste disposal problems many of which contribute to pressure on the land, and diminish its role in the environment.

Since almost any substance can become waste, there are many ways of classifying it. There are some common groupings, however. Wastes can be classified according to their origin (for example, clinical waste, domestic refuse, agricultural waste, industrial waste, nuclear waste), form (solid, liquid, gas) or properties (inert, toxic, carcinogenic). Most wastes fit into a number of such groupings. Government organizations also develop classifications for special purposes such as waste management, pollution control, safety or taxation. One class of waste that receives much attention is hazardous waste, defined as waste particularly harmful to the environment or to society. Hazardous wastes may be dangerous because they are toxic, biologically active, flammable, corrosive, radioactive or a combination of these factors. The extent of the hazard

BOX 6.4 LOVE CANAL

In North America the name Love Canal is synonymous with the dangers of uncontrolled dumping of hazardous waste. In the 1940s and early 1950s, near Niagara Falls, New York, waste materials from a pesticides and plastics operation were dumped in an area excavated for a canal that was never completed. Some 20,000 tonnes of hazardous chemicals, sealed in drums, were dumped in the canal and when full the dump was backfilled with clay and covered with topsoil. Some time later a school was built on the property and the adjacent area was developed for residential use. By the 1970s it became clear to the residents that there were problems with the site. The area smelled of chemicals, toxic waste began to migrate into basements, storm sewers, gardens and the school playground. Children playing in the area suffered chemical burns, and adverse health effects such as eye and throat irritation, rashes and headaches became common. There was also some indication of an increase in more serious disorders, such as various forms of cancer, nerve and kidney disease and birth defects.

After twenty years underground, the barrels containing the waste chemicals had corroded, allowing the contents to escape into the local environment. There was little immediate response to the problems, but persistent lobbying by the residents plus intense and unfavourable coverage by the media eventually forced the state government of New York to take action. Environmental tests indicated the presence of more than eighty different chemicals, and health surveys revealed that the incidence of birth defects, miscarriages, respiratory disease, nasal and sinus infections and urinary tract disorders was well above the national average. Some doubts have been raised about the statistical significance of these tests, but as a result of such evidence the site was declared a disaster area by the federal government in 1980, the school was closed and the residents were evacuated. After twelve years and some $250 million spent on the clean-up and rehabilitation of the site, about two-thirds of the area originally contaminated was declared fit for habitation and by 1993 houses were again being built and sold in the area. Following the Love Canal incident, it became clear that the dump was only one of many. At a minimum, tens of thousands of similar sites exist in the United States alone and a comparable number is likely for Europe, where the situation in the former Eastern Bloc nations and the republics that were once part of the Soviet Union is particularly serious. To escape the constraints of modern environmental controls, some companies have contracted to dump waste products in the developing world, creating the potential for future problems in those areas.

For more information see:

Chiras, D.D. (ed.) (1998) *Environmental Science: A Systems Approach to Sustainable Development* (5th edn), Belmont CA: Wadsworth.
Gibbs, L. (1982) *The Love Canal: My Story*, Albany NY: SUNY Press.
Kolata, G.B. (1980) 'Love Canal: false alarm caused by botched study', *Science* 208: 1239–42.

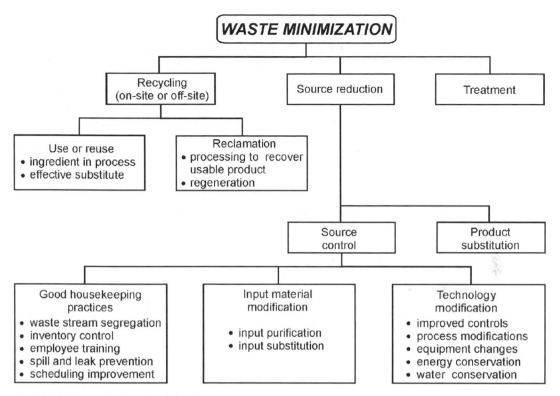

Figure 6.10   Approaches to waste minimization

posed by the waste will depend on the amount involved, its durability and particularly on the methods used to store or dispose of it. Most problems caused by hazardous waste can be traced to ignorance of or disregard for these factors (see Box 6.4, Love Canal).

Waste materials should be disposed of in such a way that the impact on the environment and society is minimal. Dumping and incineration are well established methods of waste disposal that in the past did not meet these criteria, but which have been modified to satisfy modern standards. In the past, domestic sewage was dumped directly into lakes, rivers and the sea, and natural processes were allowed to integrate it back into the environment. Today, the volumes of sewage produced are so great that natural disposal is no longer an option, and it has to be treated mechanically and chemically before it is released (see Chapter 8). Similarly, the uncontrolled dumping of solid waste in gravel pits, in abandoned quarries, on waste land and at sea has been replaced by disposal in sanitary landfill sites. Modern incinerators can attain a higher combustion efficiency and can be fitted with scrubbers or filters to eliminate hazardous emissions and reduce the air pollution that was once characteristic of waste incinerators (see Chapter 10). Certain hazardous wastes require special disposal techniques. Hospital waste and PCBs, for example, are burned in specially developed incinerators, while nuclear waste requires its own unique systems to prevent the escape of radioactivity. Waste disposal is frequently combined with recycling (see Chapter 5). Glass, metal and plastic can often be reused, and are collected separately or removed from the waste prior to final disposal, which eases pressure on sanitary landfill sites and reduces the amount of incinerator ash. Another approach designed to reduce the need for waste disposal is waste minimization, in which an attempt is made to reduce the amount of waste produced in the first place (Figure 6.10). It may involve recycling or waste-to-energy incineration, but re-engineering the process that generates the waste or redesigning the product can also achieve minimization.

Sanitary landfill

Sanitary landfill sites have been developed for the disposal of domestic refuse or garbage by controlled tipping or dumping. They are operated in such a way that they reduce or remove the problems associated with uncontrolled garbage disposal in rubbish dumps or garbage tips. Sanitary landfill sites are designed to minimize the problems of litter and odours that constitute a public nuisance at uncontrolled sites, and to deal with the potential public health threats from rats, birds and flies, which are attracted to dumps. In North America, visits by larger animals such as bears are also reduced by controlled waste disposal. If properly managed, sanitary landfill sites also reduce the indiscriminate dumping of hazardous waste.

The form of individual sanitary landfill sites varies according to such factors as topography, geology, groundwater hydrology, land availability and the volume of waste requiring disposal, but all are operated using similar techniques that work towards minimum exposure of the waste (Figure 6.11). Incoming garbage is spread and compacted in a selected working area and at the end of the day's operation the waste is covered with a layer of compacted soil some 15 cm thick. When the available space at a site has been completely filled, a layer of soil between 50 cm and 60 cm thick is spread over the surface, vegetation is reintroduced and landscaping is carried out to complete the rehabilitation. The rehabilitated land is most commonly used as open-space parkland or for a variety of recreational uses, but in some places agricultural uses such as grazing are allowed.

Despite the many advantages of the sanitary landfill approach to waste disposal, it is not without its problems. Burying the garbage creates anaerobic conditions that lead to the formation of methane gas during the decomposition process. In the past, seepage of methane has caused explosions in buildings sited on former landfill sites. As a result, building in such locations is seldom permitted, and some form of ventilation system is required at most sites to allow the gas to dissipate. In large landfill sites, the amount of methane produced may be sufficient to make its collection and use as a fuel worthwhile. The drainage of liquids – leachate – from landfill sites has the potential to contaminate the local surface water and groundwater supply. This is normally dealt with by appropriate planning that takes into account local drainage patterns, but in some cases it may be necessary to seal the base of the landfill site, using clay, for example, to prevent the drainage of leachate into the groundwater system. The leachate may be pumped to storage for subsequent safe disposal.

Concern about gas and leachate production during the disposal of hazardous waste has led to the development of secure landfill sites (Figure 6.12). To prevent the waste from escaping into the environment through leakage, secure landfill sites are usually sealed with a heavy plastic or rubberized liner and may be compartmentalized, with the products stored in containers to prevent the mixing of wastes. Where the materials are not

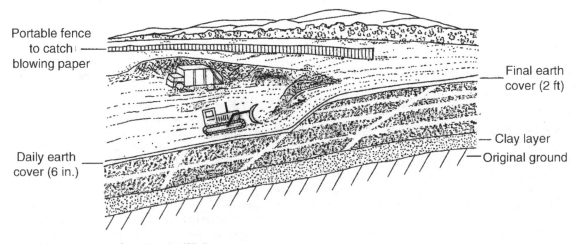

Figure 6.11   An example of a sanitary landfill site

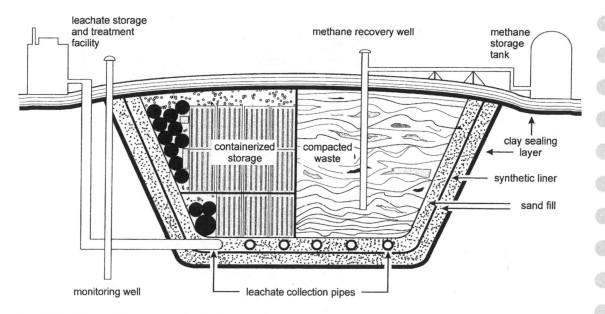

Figure 6.12   Characteristics of a secure landfill site

containerized, anaerobic decay can lead to the formation of methane as in regular landfill sites. When
full, the site is sealed with a plastic and clay cap
to prevent the percolation of precipitation. The
contents of secure landfill sites must be recorded
and the sites monitored for leachate and gas production for an extended period of time to ensure
that the integrity of the system is maintained.

Despite their many advantages over traditional
waste disposal techniques, sanitary landfill sites
are not completely environmentally friendly. They
destroy the existing environment and take agricultural land out of production for an extended
period. They cause noise, and increased traffic
flow puts pressure on the local road network.
Perceived as reducing adjacent land values, and
often mistakenly identified with old-style open tips,
sanitary landfill sites are frequent targets of public
opposition, particularly in the planning stage,
when they are subject to the NIMBY syndrome.
As a result of strong political and environmental
opposition to the creation of new landfill facilities,
many jurisdictions are beginning to reconsider their
waste disposal priorities. Recycling, for example,
can remove large amounts of paper and plastic
from domestic and industrial garbage, reducing
the volume of waste being placed in landfill and
allowing existing sites to remain in operation for a
longer period of time.

Nuclear waste

The disposal of waste products from the nuclear
industry creates special problems because of the
radioactivity they contain. Radioactivity is radiation that is released as a result of the decay of
the atomic nuclei of certain elements. Uranium
and thorium are naturally unstable, for example,
and their atomic nuclei disintegrate spontaneously,
releasing radioactivity. The process is referred to as
nuclear fission and elements such as uranium and
thorium are said to be fissionable. Nuclear fission
can also be initiated by bombarding the atomic
nucleus with a neutron, and this is the method used
in the production of nuclear energy (see Chapter 5).
Adjacent materials absorb the radioactivity released
during fission, and exposure to it can have harmful
consequences for humans, although, under controlled conditions, nuclear radiation is also used
in medicine to treat certain cancers. Uranium is
the principal element used in the production of
nuclear energy, and radioactive waste is created at
all stages of its mining, refining and use as a fuel.
The products of the mining and milling processes
needed to produce the fuel for the nuclear reactors
are usually considered to be low-level wastes, but
rock waste and the liquids used in processing the
uranium must be stored until their radioactivity is
reduced to acceptable levels. Other low-level waste

includes contaminated clothing, filters used to remove radioactive particles from the air and from liquids, piping and contaminated fluids. In the past, these low-level wastes were routinely disposed of in landfill sites or in the sea, but in most cases they are now stored in sealed drums until they lose their radioactivity. The problem wastes are those that emit high levels of radioactivity. When the nuclear fuel cycle is complete, the spent fuel is removed from the reactor. This so-called spent fuel still contains fissionable uranium, but it also contains other radioactive products such as strontium, caesium and iodine, as well as new, highly radioactive elements called actinides such as plutonium, neptunium and curium, which are fissionable. The spent fuel may be reprocessed to obtain the remaining uranium and other fissionable products, but even after that radioactive waste will remain.

Once the used fuel is removed from the reactor, it must be cooled, contained and shielded. Preliminary storage in water bays can deal with all of these. The heat is dissipated in the water and a 1 m thick concrete bay containing 3 m of water will provide both shielding and containment. All nuclear plants have these bays, and once the wastes have cooled and lost an appropriate amount of their radioactivity, they can be transferred to above-ground sealed concrete storage bins. Some of the waste products retain their radioactivity for long periods – millions of years in some cases – and require permanent storage. As yet there has been little progress on this aspect of nuclear waste disposal, in part because the industry has not been growing in recent years and has therefore produced less waste than expected. Thus the urgency to find suitable disposal systems has declined. The amount of high-level radioactive waste continues to grow, however – almost 50,000 tonnes in the United States alone at the beginning of the twenty-first century – and decisions cannot be delayed much longer. The most promising approach appears to be deep burial in stable geological formations (Figure 6.13). Atomic Energy of Canada has been exploring the possibility of burying nuclear waste in the ancient rocks of the Canadian Shield and the US government has identified a site at Yucca Mountain in Nevada as suitable, and sufficiently stable for long-term storage (see Box 6.5, Nuclear waste disposal).

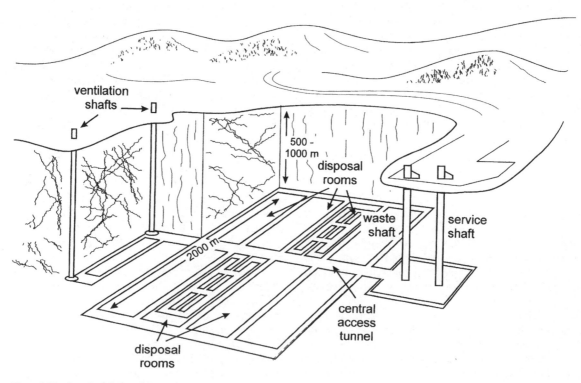

Figure 6.13   Deep burial of nuclear waste

BOX 6.5   NUCLEAR WASTE DISPOSAL. YUCCA MOUNTAIN

The nations that use nuclear energy are facing a growing problem of what to do with the waste material that is generated in their power plants. Since the waste is commonly highly radioactive, it must be disposed of in such a way that it poses no hazard to society, and disposal methods must take into account the fact that much of the waste will remain radioactive for thousands of years. The greatest problems are with high-level waste and spent reactor fuels, which are currently stored on site in reactor pools or in dry storage in concrete structures also on the reactor site. Although these systems are considered safe, there is general agreement that as the amount of high-level waste increases some form of permanent safe storage must be found.

The sites must be in locations that provide shielding from radioactivity, are stable and not subject to earth movements (which could breach the containment and allow radioactivity to escape) and are free of water (which could carry radioactivity in the event of a leak in the containment). Given the longevity of the radioactivity in nuclear waste, the site must also be able to meet these requirements for thousands of years. Among those nations actively involved in considering such sites, deep geological disposal is the most common approach. Canada and Finland, for example, are proposing the burial of waste in deep mines in the ancient rocks of the Canadian and Baltic Shields respectively and China is examining a site in a granite massif in the Gobi Desert. These sites provide the shielding and stability required and if excavated to an appropriate depth the water table is not a problem. In deserts such as the Gobi, the precipitation is so limited that concerns over water in the site are non-existent.

A similar approach has been followed in the United States, where the Department of Energy has been studying the problem for at least two decades and has identified Yucca Mountain in Nevada as the most suitable site for deep disposal (Figure 6.14). Yucca Mountain is located about 100 miles (160 km) north-west of Las Vegas on the federally owned land of the Nevada Nuclear Test Site. It is in an area where the precipitation is less than 7.5 in. (approximately 200 mm) per year, most of which is lost through evaporation and does not become involved in the groundwater system, and the water table is about 1000 ft (300 m) below the surface. The most recent volcanic activity near the site occurred some 12 million years ago and although fault lines are present they have not been active for 5000 years.

Experiments carried out at the nearby Nevada Test Site have shown that earth movements usually cause little damage to underground facilities and the proposed waste repository could be made earthquake-proof. The disposal of radioactive waste in double-layered, corrosion-resistant containers within natural rock chambers would ensure that no radioactivity would reach the surface. As a result of its investigations, the Department of Energy in 2002 recommended the Yucca Mountain as a site suitable for the construction of a radioactive waste repository and the Congress supported that recommendation. As yet no licence has been granted and the choice of site remains controversial.

There is continuing public opposition to the construction of a site at Yucca Mountain from individuals and environmental groups across the country and, in Nevada, state politicians have attempted to have the project stopped, but without success. Despite the results of the various tests carried out and the consideration given to safety issues, the perception remains among most of the citizens of Nevada that the repository would pose public health and safety risks. It would also be expected to impact on the tourism industry, which is a major contributor to the economy of the state. Legislators have argued that since Nevada has no nuclear plants and does not benefit from plants elsewhere it should not be expected to store nuclear waste from those plants. Having had to live with the radioactivity created by nearly 1000 nuclear weapons tests in the Nevada Nuclear Test Site, many citizens in the state feel that they have done their share already for the nuclear policy of the United States and should not be expected to do more. In an attempt to prevent further development, the state has launched a constitutional challenge to the disposal of nuclear waste at the Yucca Mountain site.

Other concerns include the transportation of the waste material to the site. Thousands of tonnes of radioactive waste would be moved to the repository every year once it was in operation, much of it over great distances from the industrial north-east of the country. Although radioactive materials are regularly transported by the nuclear industry without accidents and with no release of radioactivity, the level of activity would rise with the opening of the facility and with it the potential for accidents. Setting up routes from the nuclear plants would also be controversial, since many communities would not wish to have radioactive material passing through regularly by

BOX 6.5 – continued

road or rail, and the political cost of dealing with such problems would be high.

The growth of the nuclear industry and the resultant production of waste slowed in the 1980s and 1990s, which means that the need for such a central disposal facility is perhaps less urgent than it was. With the increase in terrorism, however, there is an argument for consolidating the waste, since it is easier to secure one location than eighty to 100 separate sites. Those who support the development of the repository claim that the technology exists to provide effective, safe storage of nuclear waste and since the waste will ultimately have to be consolidated, it would be appropriate to begin as soon as possible. Those who oppose the development are not convinced about the claims of safety and see the societal and environmental costs as too high. With the two groups polarized, the final decision on the development of Yucca Mountain is likely to be a political one which in the end may please neither group.

For more information see:

Murkovski, F.H. (1998) 'Yucca Mountain: the scientists' choice', in D.D. Chiras (ed.) *Environmental Science: A Systems Approach to Sustainable Development* (5th edn), Belmont CA: Wadsworth.

Nebel, B.J. and Wright, R.T. (2000) *Environmental Science* (7th edn), Upper Saddle River NJ: Prentice Hall.

Titus, D. (1998) 'Yucca Mountain repository: no place in Nevada', in D.D. Chiras (ed.) *Environmental Science: A Systems Approach to Sustainable Development* (5th edn), Belmont CA: Wadsworth.

The following web pages also contain information on the different views on the development at Yucca Mountain (all were accessed 16 June 2003):

http://www.state.nv.us/nucwaste/ The position of the State of Nevada.

http://www.ocrwm.doe.gov/ymp/index.shtml The official position from the Office of Civilian Radioactive Waste Management: Yucca Mountain Project.

http://yuccamountainfacts.org/ News, opinions, local events and information about the proposed waste dump at Yucca Mountain, Nevada – representing a citizens' point of view.

The British government has considered nuclear waste burial in the rocks beneath the Irish Sea. Such disposal methods are not universally accepted, however, and have been the source of much conflict and controversy (Chiras 1998). Secure as they may appear, with burial taking place at depths in excess of a kilometre beneath the surface, they still involve some potential for the escape of radioactivity into the environment.

Since the sources of waste and the preferred disposal sites are often well separated, the possibility of contamination during the transportation of the waste must also be considered, although work in North America and Europe suggests that safe transportation is possible using specially designed containers. Nuclear plants that began operating in the 1960s and 1970s are now reaching the end of their working life and as they begin to close they will not be demolished like normal industrial structures. Because of the radioactivity they contain, the buildings will remain isolated on site or perhaps dismantled and placed in more secure and permanent storage. These decommissioning wastes will join the growing volumes of operating waste as a legacy of the nuclear industry. The environmental lobby against the use of nuclear energy is strong and vocal, in part because of the damage caused by nuclear reactor accidents and because of the radio-

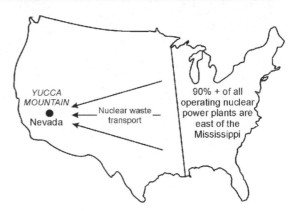

Figure 6.14 Location of the Yucca Mountain Nuclear Waste Disposal Site

active waste that is produced. Even if all existing nuclear plants were to be closed immediately, the waste that they have already produced would remain around for thousands, perhaps millions, of years as a permanent threat to the environment.

SUMMARY AND CONCLUSION

Human impact on the land takes many forms, ranging from the complex infrastructure of urban areas in the developed world to the garbage left behind by visitors to such isolated places as the Arctic or the Himalayas. In these cases the impact is direct and visible, but in others it may be indirect and not immediately apparent, as in the physical and chemical changes to the soil that precede the onset of soil erosion. Whatever form it takes, the net result is an increase in the human element in the landscape at the expense of the natural components, and, because of the linkages in the environmental system, changes in the land ultimately contribute to changes in the biosphere, hydrosphere and atmosphere. It is impossible to support more than 6 million people on the land without some change in the environment, and perhaps because the land has received less attention in the past than some of the other components of the environ-

ment, much of that change has had serious consequences. Despite the seeming permanence of its rocks and land forms, the land is a dynamic medium, but human indifference to its limitations and ignorance of the processes involved have reduced that dynamism and diminished the capacity of the land to maintain its contribution to the healthy environment upon which society ultimately depends. There is little chance that past indignities to the land can be rectified to any great extent, but a growing attention to land use planning with the environment in mind, backed by enforceable legislation, is intended to slow the current deterioration, remedy at least some of the past problems and allow the land to contribute to the sustainable development needed to balance future societal and environmental needs.

SUGGESTED READINGS

Morgan R.P.C. (1995) *Soil Erosion and Conservation* (2nd edn), Harlow: Longman. Comprehensive introductory account of soil erosion with interesting and useful sections on the conservation of soil and erosion repair.

Mannion, A.M. (2002) *Dynamic World: Land Cover and Land Use Change*, London: Edward Arnold. Focuses on the relationship between society and environment as revealed by changing patterns of land cover and land use.

McDougall, F.R., White P.R., Franke, M. and Hindle, P.

(2001) *Integrated Solid Waste Management* (3rd edn), Oxford: Blackwell. Comprehensive and detailed account of all aspects of waste, from its generation and disposal to recycling. Includes cases studies of the economic and environmental assessment of waste management systems.

Hardoy, J.E., Mitlin, D. and Satterthwaite, D. (2001) *Environmental Problems in an Urbanizing World*, London: Earthscan. An authoritative account of the environmental problems facing growing cities in the developing world, including possible solutions.

QUESTIONS FOR REVISION AND FURTHER STUDY

1   Agriculture appears to emulate nature in its use of plants and animals that have attributes and needs in common with their natural counterparts, yet agricultural activity is the source of many serious environmental problems. Why?

2   Compare the form, function and growth of cities in the developing world and in the developed world. Are the main threats to the environment that accompany urbanization the same in each of these areas or do they differ?

3   The government has decided that a waste disposal site to accommodate nuclear waste is to be built near to a small town. If as an environmental scientist you are hired to represent the citizens opposed to the development at a public hearing, what arguments will you use to persuade the government representatives that such a disposal site is not appropriate in that community? If a decision were made to go ahead despite your objections, what safeguards would you expect to be included in the development to minimize its potentially hazardous impact on the environment and the community? (This can be set up in the form of a mock hearing, with students playing the parts of government or industry representatives, environmental consultants and citizen opposition, presenting their arguments in front of the class.)

7

Threats to Wildlife and Plants

After reading this chapter you should be familiar with the following concepts and terms:

annual allowable cut	fire suppression	reforestation
ANWR	fuelwood	reserves
biological diversity	global species totals	selective cutting
bushmeat	habitat	shifting agriculture
certified forest products	herbivores	slash-and-burn
CITES	increasers	softwood lumber
clear cutting	ITTO	species introductions
debt-for-nature	mass extinctions	Statement of Forest Principles
decreasers	NFAP	TFAP
deforestation	nomadic herding	tropical hardwoods
Dustbowl	old-growth forest	tropical ranching
endangered species	over-grazing	Virgin Lands Scheme
extinction	plantation forests	wilderness
extinction rates	protected areas	
extirpation	Red List	

WILDERNESS

Central to the concerns of the early environmentalists, particularly in nineteenth-century North America, was the preservation of the fast disappearing wilderness – the combination of plants and animals they saw as representing the natural condition of the biosphere in a particular area. The concept of wilderness was as old as civilization itself, but their view of it, as something beneficial to society rather than a threat to progress, was new and radical. Hunting and gathering communities, if they thought about it at all, probably saw themselves as part of the natural environment, part of the combination of plants and animals that made up their habitat (see Chapter 1). The development of herding and then arable agriculture brought some human control to the landscape, and with it society began to see itself as separate from the other elements of the environment – including the

'uncivilized' humans – in those areas that remained beyond their control (Nash 2002). These wild lands became the wilderness, representing a threat to humankind, and, in the thousands of years since the emergence of the first agrarian civilizations, much of society's energy has been spent in pushing back the wilderness and replacing it with a more human or civilized landscape. That negative perception of the wilderness still prevailed in the nineteenth century and, laudable as the efforts of the early environmentalists were, they represented only a minority of the population at that time. As a result, what success they had was limited at the time to the creation of a few parks and even there the landscape was not always free of some form of human intrusion. Through the nineteenth and twentieth centuries, as the world's population expanded and technology advanced, natural landscapes were replaced at an unprecedented rate by the seminatural landscapes of agricultural development or

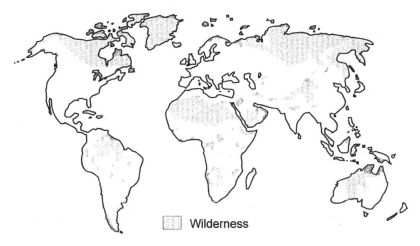

Figure 7.1   Distribution of wilderness in the late twentieth century
Source: McCloskey and Spalding (1989) with permission.

the un-natural conditions of industrial and urban environment. As a result, the natural habitat of as much as two-thirds of the earth's habitable surface has been altered by human activity, and even where nature still prevails, there is often a human imprint of some form or other (Walker 2001). Wilderness does survive in areas where efforts have been made to preserve it (Figure 7.1), or where inaccessibility provides some protection, but even these areas are threatened and wilderness protection has again become central to the environmental movement, as the seeds sown by the early environmentalists a hundred years earlier found more favourable conditions under which they could germinate.

Areas currently considered as wilderness are the remnants of natural biomes that were once distributed across the earth's surface in accordance with the environmental conditions prevailing at a specific time in a specific place. These were complex, dynamic communities of plants and animals that interacted with each other and with the abiotic elements – soil, water and climate – of their environment (see Chapter 3). Being dynamic they were able to respond to natural change without losing the basic characteristics of the system. Their ability to adapt was not infinite, however. Given time they would recover from disturbances such as fire and volcanic activity but major climate change caused large-scale changes in the distribution and composition of plant and animal communities in the past. Modern threats to the environment are associated with a wide range of human activities

which alter the ecosystems to a greater or lesser extent. They may be replaced completely, when forests are cut down to allow agricultural development or when urban sprawl covers them with concrete and asphalt; they may continue to exist in small, scattered fragments where farmers have preserved patches of trees as wood lots; they may be retained for recreational purposes to provide a 'wild nature' experience for visitors. Over the past several centuries such activities have become so disruptive or intrusive that few truly natural ecosystems remain. Plants and animals characteristic of a particular ecosystem may survive these disruptions, but they lose their ability to act as an integrated, dynamic environmental unit and become even more vulnerable to threats from ongoing human activities. These threats originate in agriculture, forestry, industry and urbanization as well as in the unsustainable use of plant and animal products, which combine to destroy habitat and put pressure on the earth's biological diversity (Figure 7.2).

HABITAT DESTRUCTION

A habitat is a specific environment that provides a population or a community of organisms with the conditions under which they are able to survive. Habitats owe their characteristics to a combination of abiotic and biotic factors (see Chapter 3) and range in scale from worldwide through continental to regional, local and even microscopic. Normal

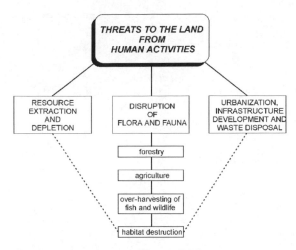

Figure 7.2 Threats to vegetation and wildlife

THREATS TO BIOMES FROM HUMAN ACTIVITY

All biomes suffered as populations grew and advances in technology allowed society to take a more aggressive approach to resource exploitation. Inhospitable areas, consisting of bare rock and ice or suffering from too little water or too much, and inaccessible areas such as high mountains or isolated islands received less attention, but even they were not free from human interference if they were able to provide for society's growing demand for resources. The earth's least productive habitats – for example, in the tundra and desert biomes – seemed to have little to offer and development passed them by.

Tundra

In the Arctic tundra of Eurasia and North America, the indigenous population lived within the limits of the resources available in the biome, either by hunting and fishing or by herding domesticated reindeer – or a combination of all three. The human impact on the environment was relatively limited and almost all of the tundra remained in a natural state for hundreds, perhaps thousands, of years. That situation still applies over much of the biome, but major environmental damage has occurred locally and regionally as a result of mineral exploration and exploitation. Inhospitable and inaccessible as it is, once the tundra was found to be an untapped storehouse of petroleum and natural gas as well as highly valued minerals such as gold, diamonds and copper, it was rapidly threatened by the same problems of industrial resource development that were already common elsewhere. Vegetation was destroyed, the drainage disrupted, the land surface eroded and the wildlife driven from its habitat. Oil spills caused immediate damage, but, even with direct clean-up, the slow recovery rates in the Arctic environment ensure that their impact continues to be felt for many years (Pain and Kleiner 1994).

The fragile nature of the tundra biome and the need to preserve it has been recognized in the formation of 405 protected areas across Eurasia and North America (Figure 7.3), but even these areas can be threatened in the search for resources. That has led to conflict between environmentalists and indigenous communities on one side and

environmental dynamics ensure that natural change is an integral part of all habitats. The changes may be small-scale or temporary, such as those caused by a forest fire or hurricane, for example, or much more dramatic and long-lasting, such as those that follow major climate change. The impact on the organisms that occupy the habitat affected also varies, from minor to catastrophic. They may be temporarily excluded from a particular area, but survive in an adjacent habitat with similar characteristics to their own, returning only when the original habitat is re-established. This temporary exclusion of organisms is called extirpation. At the catastrophic end of the scale, the complete loss of habitat may bring about the extinction of a particular species. Even in the wake of major natural disasters, habitats will tend to become re-established after adjustments that take into account the new conditions created by the change. Following volcanic eruptions that destroy vegetation and change the shape of the land through explosions and lava flows, for example, habitats that have characteristics similar to those that existed prior to the eruption can re-form in as little as a few decades. Increasingly, however, habitat change or loss is being caused by human intervention in the environment, which tends to be ongoing and persistent, giving natural habitats little opportunity to recover. New habitats suitable for human beings and the organisms that associate with them are created, but they lack the vitality of natural habitats and suffer a number of serious environmental problems.

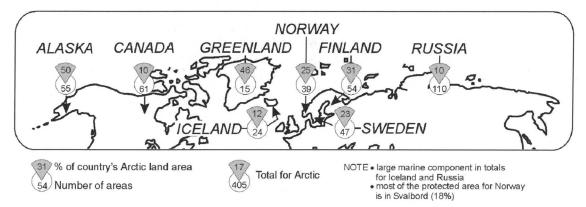

Figure 7.3 Protected areas in the Arctic
Source: based on data in UNEP (2002).

resource developers on the other. The demands of the oil industry for drilling rights in the Arctic National Wildlife Refuge (ANWR) in Alaska, for example, have been opposed by those who recognize the potential for serious environmental damage if development is allowed in the Refuge (Figure 7.4). Although regulations exist to limit such damage, it is impossible to establish and operate a drilling

site without disturbing local ecosystems and, since the tundra environment is notoriously slow to recover, the impact of any development is likely to be long-lasting. Much of the concern over the opening of the ANWR for petroleum exploration has centred on the wildlife community in the area. The Refuge is home to at least 160 animal species, including foxes, grizzly bears, caribou, musk oxen

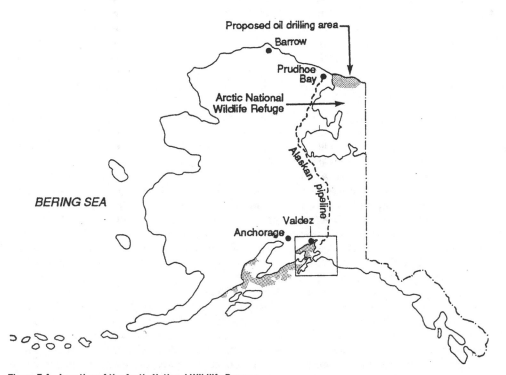

Figure 7.4 Location of the Arctic National Wildlife Reserve

and migratory birds and was set up to preserve the characteristic mix of fauna in the Arctic tundra along the north slope of Alaska. Petroleum exploration and development would interfere with the situation by destroying habitat and disrupting migration patterns. The Refuge is the main calving ground of the Porcupine caribou herd, which winters in northern Canada and migrates to Alaska in the spring. Birth rates and calf survival rates have declined in other oil development areas in the Arctic, and some observers suggest that the herd could decline by 20–40 per cent if development is allowed, changing not only the natural balance in the ecosystem, but also causing hardship to some 7000 indigenous people in both Canada and Alaska who, despite the adoption of some aspects of a modern lifestyle, still remain very much a part of the environment and depend on the caribou to ensure the survival of their way of life (Draper 2002). Although the air and water pollution caused by the oil developments at Prudhoe Bay would indicate otherwise, petroleum companies claim that the loss of habitat and environmental damage to the Refuge would be minimal (Miller 1994). This appears to have been accepted by the US federal government, which has indicated that exploration may be allowed to go ahead in the future, although doing so would contravene several international wildlife preservation agreements signed by the United States (Draper 2002).

Even without the direct habitat disruption that accompanies mining and oil production, the habitat of the Arctic tundra has been subject to change. Pollutants carried into the area in Arctic Haze have accumulated in plants and animals to levels that match, and sometimes exceed, those in the industrialized south. A much greater threat, however, will come from global warming, which is expected to have a particularly strong impact in high latitudes (see Chapter 11). The trees of the boreal forest will colonize the southern edges of the tundra, for example. Higher temperatures will change the amount of precipitation that will fall as snow and, perhaps more important, cause the permafrost to melt. The resulting changes will impact on energy and moisture budgets, by altering the surface albedo, creating more open water and increasing soil moisture levels. This in turn will cause changes in the nature and distribution of vegetation and in the animals that depend upon it for food and shelter. In coastal areas, the formation of sea ice in the autumn is taking place later and later every year

and already causing problems for the polar bear population, which spends winter hunting on the ice. Changes in the timing and pathways of the annual caribou and wildfowl migrations are likely to follow. Although all habitats will be affected in some way by global warming, the tundra biome appears particularly vulnerable and is likely to be in an ongoing state of flux for many decades until greenhouse gas emissions are stabilized and global warming is slowed.

Desert

At the other end of the temperature spectrum lies the hot desert biome and, despite very obvious differences, environmental problems with similar roots to those of the tundra can be recognized there also. Although both have habitats that are vulnerable to change, both are lightly populated with indigenous groups that in the past lived in reasonable harmony with their environment. Like the tundra, the desert biome was spared much human interference in the past because of its inhospitability. The population lived a nomadic lifestyle for the most part, with permanent settlement restricted to areas where water was readily available, either at oases or along river systems, as in Egypt and Mesopotamia. Although the early agrarian civilizations in Mesopotamia and elsewhere contributed to habitat degradation through soil erosion and salinization, the extent and nature of such environmental problems grew considerably when modern, technologically sophisticated society began to take an interest in the areas. Gold and precious stones lured miners to the deserts of Australia and the south-western United States, where the impacts were at most local. In contrast, the oil developments in the deserts of Saudi Arabia and around the Persian Gulf have largely destroyed the desert biome in that area, replacing the limited flora and fauna with the industrial infrastructure of the petroleum industry and the built environment of growing urban centres. In 1991, during the Gulf War, the sabotage of some 800 oil wells by the defeated Iraqi forces allowed 7 million tonnes of oil to spew out over the desert in Kuwait. Despite a massive clean-up effort, the plants, animals, insects and microscopic organisms of the desert habitat were completely destroyed in these areas, and if recovery from such conditions is possible it could take hundreds of years. Fortunately such events are uncommon, but the destruction and contamination of habitat by the everyday activities of the oil industry is ongoing.

The disruption of traditional herding practices has caused over-grazing of already sparse vegetation and the demand for food from the growing population has increased the area under irrigation, replacing the plants of the naturally arid habitat with cultivated crops. The large mammals that inhabit the desert, such as wild goats, gazelles and the cheetah, have been displaced from their natural habitat and have come under increasing threat from hunters armed with automatic weapons and riding in four-wheel-drive vehicles. The cheetah may already have been extirpated, and the Arabian bustard is under threat, but the Arabian oryx, once extinct in the wild, has been reintroduced using captive stock (UNEP 2002).

Elsewhere, the human impact on the desert has been less traumatic, but no less damaging to the flora and fauna. The desert has bloomed in some areas, where the provision of irrigation water has allowed the cultivation of high-value flower, soft fruit and vegetable crops. In the United States, the demand for vacation and retirement facilities in California, New Mexico and Arizona has made it economically feasible to bring water from the neighbouring mountains or to pump it up from beneath the ground to meet the needs of new residential areas with their swimming pools and golf courses. To those living or vacationing there, or those planning and constructing the hotels, condominiums and golf courses, the creation of this human imprint in the desert is seen as a great improvement. In reality, as a form of habitat destruction, it compares with uncontrolled deforestation, but has received much less attention from environmental groups. Residential expansion into the desert on such a scale is the exception, however, and most of the world's deserts are sufficiently inhospitable or remote that the potential for further development remains minimal.

Grassland

Of all of the biomes, the world's grasslands and forests have experienced the greatest habitat changes. The distribution of grasslands, both subtropical and temperate, tends to reflect precipitation availability. They develop in areas where there is generally insufficient moisture to support trees, but within the grassland biomes there are variations from the lush, tall grasses of the areas with higher precipitation totals to the stunted clumps of short grass in the drier areas (see Chapter 3). Where precipitation is unevenly distributed through the year, grassland habitats display seasonal variations, with grasses growing rapidly during the rainy season and dying off in the dry season. In the past the large numbers of herbivores that populated the grasslands, from the wildebeest of East Africa to the bison of North America, responded to these changes by migrating with the seasons to ensure an adequate food supply. They in turn were followed by predators, including the nomadic hunters who depended upon them. Even after the animals had been domesticated and the hunters evolved into herders, the patterns did not change all that much. The animals still followed the rain. In the Sahel region of West Africa, for example, the nomadic herders moved north and south with the seasons in step with the movement of the Intertropical Convergence Zone that brought the rain (see Chapter 9). Tropical savanna grasslands are particularly suited to nomadic herding, which uses the resources of the existing habitat without altering them to any great extent. The herders may use fire to burn off old grass and encourage new growth, but fire is a natural component of these grasslands and the habitat recovers whether the fires are natural or deliberately set.

Although cattle are preferred as grazing animals, differences in the type and availability of forage mean that sheep, goats and camels are more common in some areas. Sheep, like cattle, need grass and tend to be raised in the wetter areas, whereas goats and camels can subsist on poorer-quality forage such as short grass or thorny shrubs. In the Sahel, the composition of the herds has changed, with an increase in the numbers of sheep and goats at the expense of cattle and camels, in part because of the reduced availability of forage during the drought years of the last several decades, but also as a result of socio-economic factors (Turner 1999). The impact of this form of pastoral agriculture on habitat is relatively minor, but it does bring about change. Domesticated grazing animals are in competition with the indigenous ruminants and must be protected from natural predators such as lions and wild dogs, for example. In East Africa, the establishment of wildlife reserves to maintain the natural animal population of the grasslands and encourage tourism has created conflict by reducing the area directly available to the herders. There also, population pressure has brought the introduction of arable farming to the wetter parts of the grasslands, which also provide the best grazing, creating conflict between the herders and farmers (Mannion 2002).

The grassland habitats are under particular stress in Africa, immediately south of the Sahara Desert, as a result of overpopulation, farming practices that encourage over-grazing, land use conflicts and the ever-recurring drought in that area. The most vulnerable areas are along the desert margins, where the flora and fauna are in precarious state of balance with the variable precipitation. Human interference in these transitional areas disturbed that balance and contributed to large-scale habitat destruction. Particularly in the latter part of the twentieth century, large areas of the grassland biome were lost to the encroaching desert, with serious and sometimes fatal consequences for the population (see Chapter 9).

The other major grassland biome, located in temperate mid-latitudes, has suffered much greater change than its tropical equivalent. Being extensive,

with a subdued, or at most a gently rolling, landscape, fertile soils and a climate eminently suitable for both pastoral and arable farming, the temperate grasslands, particularly in North America, attracted thousands of agricultural colonists and soon became the world's major supplier of grain and meat products. This intensive exploitation as croplands and pasture rapidly changed the original habitat and very little natural grassland now remains (Figure 7.5).

Before the arrival of the Europeans, the North American grasslands covered approximately 1 million km^2; now only 10 per cent of that remains, and then only in preserves or in areas that have reverted to their natural state after being abandoned by farmers (Brown and Lomolino 1998). The grasses disappeared completely where they were ploughed up and replaced by grain, but even where the land was used for grazing the habitat was altered. Domesticated livestock tended to be selective in its feeding, grazing preferred species, which consequently declined in number (so-called decreasers) while the non-preferred species became more numerous (so-called increasers) and the

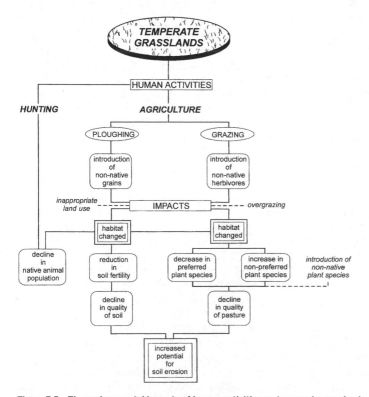

Figure 7.5 The environmental impacts of human activities on temperate grasslands

overall quality of the pasture declined. Non-native grasses were introduced to counter that, and they gradually displaced the natural flora. Changes that were initially less obvious occurred in the soils. The rich black earths, deprived of the annual input of organic material, gradually lost their fertility and became degraded. The cultivation of land unsuitable for cropping – in the drier, short-grass prairie or steppe, for example – led to serious problems of land deterioration in North America and Eurasia, culminating in the former with the Dustbowl conditions of the 1930s (see Chapter 9). Several decades later, in the 1950s and 1960s, the Virgin Lands Scheme, in what was then the Soviet Union, had a similar effect when the steppes of northern Kazakhstan and western Siberia were ploughed up to produce grain. In addition to destroying the grassland habitat, this led to a twofold to fivefold increase in the number of dust storms in the area (Middleton 1999). Comparable changes were taking place at about the same time in China, where an estimated 51.3 million ha of grassland has been lost

since 1949, mainly in the semi-arid short-grass steppe of the northern plains in the provinces of Inner Mongolia and Gansu. Railways built into the plains in the 1950s encouraged the migration of farmers who ploughed up grassland that was not really suitable for cultivation, leaving it vulnerable to desertification (Edmonds 1994).

The loss of habitat that accompanied all of these changes, plus the hunting of some species almost to the point of extermination, seriously threatened the viability of the grassland animal community, and in some places altered it completely. As a result of its wholesale slaughter for food, hides and sport, the bison population of the Great Plains plummeted from 30–60 million head at the beginning of the nineteenth century to a few hundred animals by the end (Cox 1993). By the 1870s, numbers had declined so rapidly that many of the indigenous people, who had depended upon the bison for generations to support their way of life, were threatened with starvation and all that remained of millions of animals were piles of bones, which were eventually shipped

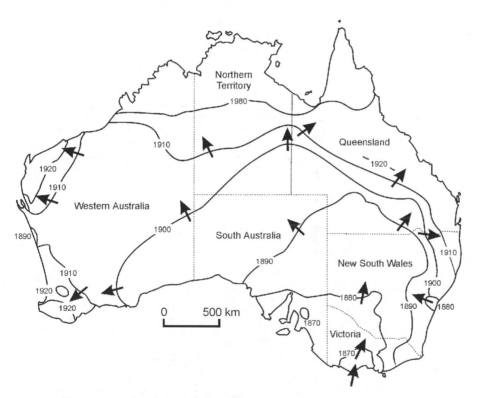

Figure 7.6   The spread of rabbits following their introduction into Australia
Source: after Brown and Lomolino (1998) with permission.

east to be made into fertilizer (Bryan 1973). Elk and pronghorn antelope, the other large herbivores on the plains, faced similar threats, which reduced their numbers significantly in some areas. All of this caused a major disruption of the food chain, and as their food supply dwindled the predator population of wolves and cougars declined also. At a smaller scale, the widespread control of the prairie dog population threatened its main predator, the black-footed ferret, with extinction (Cox 1993).

In the Siberian grasslands, the saiga antelope suffered the same fate as the North American bison. The mainstay of the nomadic population of the area, the saiga was pushed to the brink of extinction in the late nineteenth century by over-hunting and the loss of habitat as the cultivation of the plains

increased. The development of herding techniques similar to those used for reindeer in the north and the introduction of conservation schemes brought the numbers back to about 700,000 in the 1970s. By the end of the century, however, the population had declined by more than 70 per cent as a result of competition from domesticated grazing animals and the inefficient enforcement of hunting and poaching regulations in the disorder that accompanied the dissolution of the Soviet Union (Middleton 1999).

In the Australian grasslands the threat to habitat came with the introduction of the European wild rabbit into an area where it had no serious predators (see Box 7.1, Species introductions). Brought to Australia in the 1850s, within a hundred years the rabbit population had exploded and spread to

BOX 7.1 SPECIES INTRODUCTIONS

Introduced species, exotic species, alien species or invaders are all names used for species that did not evolve in the habitats in which they are currently found, but rather moved into them naturally or were moved into them deliberately. The migration of species is a normal form of environmental activity in the biosphere. When climate changes, for example, some plants and animals may be forced to move in an attempt to regain the environmental conditions most favourable for their survival. Others may find the conditions that suit them are becoming more extensive and they migrate along with those suitable conditions. If the food supply of a predator in a particular area declines, it may move into an adjacent area, where its introduction will change the existing ecological mix. In such cases the species are being introduced naturally into new areas. Their survival will depend upon their ability to deal with competition from existing species, but the local environment ultimately adjusts to deal with the new situation, sometimes preventing the introduced species from surviving, at other times incorporating it in the system. Problems arise when the introduction is so rapid, or takes place in such large numbers, that species are not easily integrated into the existing environment. In some cases the new species is so alien that the existing environment has no natural mechanisms with which it can cope and it suffers considerable disruption. In theory, given sufficient time, the ecology of these areas will reflect the new reality. In the meantime, however, habitat may be lost and the biodiversity of the area diminished.

Human activities can create both of these situations – a relatively benign introduction or one that is more disruptive – and has done so for hundreds of years, sometimes deliberately or sometimes accidentally. All species introductions have an environmental impact, but commonly they are considered in terms of their socio-economic consequences – whether they enhance or disrupt economic activities or cause lifestyle changes, for example. The number of introduced species is very large, covering all life forms, from bacteria and viruses through plants and insects to fish and mammals, and the following list includes examples that can be considered in terms of the nature of the introduction – deliberate or accidental – and the consequences – benign or disruptive.

DELIBERATE AND BENIGN INTRODUCTIONS

- The introduction of *domesticated animals* such as sheep and cattle is now generally considered benign although in the past these animals had a significant impact on the ecology of the temperate grassland in North America and Australia. Now the environment in these areas has been modified to suit them and they fit in.
- *Potatoes* from the Americas to Europe.
- *Rubber plants* from South America to Southeast Asia and Africa.
- *Sitka spruce* from North America to Europe for reforestation.

BOX 7.1 – continued

DELIBERATE BUT DISRUPTIVE INTRODUCTIONS

■ *Rabbits* from Europe to Australia. Rabbits had no major predators in Australia and easily out-competed existing small grazers to cause serious damage to vegetation.

■ *Africanized honey bee (killer bee)*, introduced to Brazil to improve the honey crop, escaped and has spread through Central America and the southern parts of the United States, where it out-competes the established European honey bees.

■ *Kudza*, a Japanese vine, was introduced into the south-eastern United States to provide ground cover that would combat erosion but now chokes out native vegetation and severely disrupts biological diversity.

■ *Coypu*, a South American rodent introduced into Britain to promote the fur industry, escaped and its numbers grew sufficiently that it contributed to erosion and flooding – by burrowing into river banks – and caused crop damage. The original escapes took place in 1937, but the coypu was not eradicated in Britain until 1989.

■ *Gypsy moth* was introduced into Massachusetts in the late nineteenth century to encourage an American silk industry. When it escaped it became a serious threat to forests in the eastern United States.

ACCIDENTAL AND BENIGN INTRODUCTIONS

Most attention given to the accidental introduction of species involves those that have caused environmental disruption. Many organisms are introduced accidentally into new environments and leave no negative evidence of their presence, either because they do not survive or they find an environment into which they can fit comfortably. If environmental conditions change, however, they may create problems.

ACCIDENTAL AND DISRUPTIVE INTRODUCTIONS

■ *Rats* and *cats* escaping from visiting ships have decimated existing populations of small mammals and birds since the time of the early European explorations. In New Zealand, for example, feral cats have caused the extinction of at least five species of native birds since the arrival of the Europeans.

■ *Rinderpest*, a viral cattle disease endemic to Asia, was introduced accidentally to Africa in the 1800s, where it devastated the local cattle population as well as wild ungulates.

■ *Smallpox* was brought from Europe to North and South America, with catastrophic results to the indigenous population, who had no resistance to the disease.

■ *Zebra mussels* and *sea lamprey* were introduced into the Great Lakes when the St Lawrence Seaway was opened, creating serious problems for local species and costing industry more than a million dollars annually for control measures.

The accidental introduction of parasites, bacteria, viruses and other microscopic pests is considered a serious threat to the environment and to human welfare in many areas. Such introductions are made easier by the globalization of trade and the speed with which the organisms can be unknowingly spread – being carried on jet aircraft, for example. Although most countries have inspection processes aimed at dealing with the importation of unwanted organisms, none of them is perfect and accidental introductions will continue.

For more information see:

Drake, A., Mooney, H.A., di Castri, F., Groves, R.H., Kruger, F.J., Rejmanek, M. and Williamson, M. (eds) (1989) *Biological Invasions: A Global Perspective*, Chichester: Wiley.

Stiling, P. (1992) *Introductory Ecology*, Englewood Cliffs NJ: Prentice Hall.

Talbot, L.M. (2002) 'Exotic species', in A.S. Goudie (ed.) *The Encyclopedia of Global Change*, New York: Oxford University Press.

almost all parts of the continent except the far north (Figure 7.6). In the grasslands, rabbit grazing has caused irreparable environmental damage through the removal of vegetation and the initiation of soil erosion. The loss of vegetation has also threatened the survival of the native birds, mammals and insects that depend on the plants for food and shelter, and the rabbit has been implicated in the decline of many native animals, particularly small mammals such as the bilby and various species of bandicoots and rats. With ten rabbits eating as much as one sheep the rabbit causes economic problems also, the estimated cost to the agricultural industry being about US$600 million per year (CSIRO 1997). Despite decades of attempted control through shooting, poisoning and the deliberate introduction of diseases such as myxomatosis and rabbit calcivirus, the population remains at an estimated 200 million and rabbits are likely to remain Australia's most serious animal pest for the foreseeable future. In addition to the rabbit, other introduced species such as feral goats, European foxes and feral cats also contribute to the disruption of the continent's grassland habitat (Szabo 1995).

The mid-latitude temperate grassland biome has changed so much that it is a community that really now exists only in the history books. Attempts at conservation and reclamation have met with some success, but the areas involved are minimal compared with the former extent of the biome. Future changes such as those expected to accompany global warming are also likely to have a significant impact on the grasslands (see Chapter 11). With increasing aridity in North America and Eurasia, some areas would become deserts, but along its northern edge the grassland would gain at the expense of the forest, creating, initially at least, an ecosystem that might have many of the attributes of the natural grasslands that once occupied the area.

Forest

Forests are the most complex and most diverse of the earth's terrestrial ecosystems, covering some 3800 ha or about one-third of the earth's land area. Tropical and subtropical forests account for about 56 per cent of the total, with the remaining 44 per cent being temperate and boreal (UNEP 2002). Human activities, both deliberate and inadvertent, contribute to a continuing decline of the forested areas, and deforestation is high on the list of serious environmental problems. Technically, deforestation is the permanent removal of trees from the environment in areas capable of supporting natural forest cover. This occurs when the land is given over to permanent agriculture or settlement, for example. The reality is less simple, however. Land

intended to be cultivated permanently can revert to forest if cultivation ceases, and in areas once considered permanently deforested trees have been reintroduced through reforestation schemes. Much of the land in the Scottish Highlands, for example, was deforested for 100–200 years before new woodland began to be planted in the twentieth century. On a more popular level, deforestation is commonly equated with the large-scale clearing of forest as part of a commercial forestry enterprise or for some other economic purpose, such as the expansion of settlement or the development of agriculture. It is in this context that the term will be used here.

Concern over deforestation first received international attention at the UN Conference on the Human Environment (UNCHE) held in Stockholm in 1972. Following a series of initiatives to assess the extent and nature of the problem, the need to develop sound policies for forest use and management planning was recognized in the Statement of Forest Principles that emerged from the Rio Earth Summit in 1992 (Box 7.2). Since then, forest decline has stabilized and even reversed in the developed nations, but continues to be a problem in developing countries, where, for example, the deforestation of tropical forests averaged 1 per cent per year through the 1990s (UNEP 2002).

Large-scale deforestation has occurred in the past as a result of climate change. During the Pleistocene ice ages, for example, large areas of forest were destroyed in northern Europe, Asia and North America, and climatic fluctuations since then, such as the Little Ice Age, have also contributed to changes in the distribution of forest. In all of these cases, when conditions again became suitable for their growth and survival, the forests recolonized the areas from which they had been excluded. Current concern over deforestation and the general destruction of forest habitats centres on the ability of people to cause such environmental change that the forests are unlikely to recover and their decline will be long-term or even permanent. Direct threats to forest habitat come from the spread of agriculture, commercial logging and the harvesting of fuelwood.

Although logging often receives most of the attention when deforestation is being discussed, in the 1990s agricultural expansion was responsible for about 70 per cent of reduction in forest area and half the wood harvested around the world is consumed as fuelwood, mainly in the developing

BOX 7.2 MAIN COMPONENTS OF THE STATEMENT OF FOREST PRINCIPLES SIGNED AT THE EARTH SUMMIT IN RIO DE JANEIRO, 1992

■ The vital role of forests in maintaining the ecological balance at local, national, regional and global levels should be recognized and activities harmful to the health and survival of forest ecosystems should be controlled

■ Countries have the right to use their forests for socio-economic development, but should ensure that activities involved cause no environmental damage beyond the extent of their national jurisdiction

■ Development should be consistent with the requirements of forest conservation and sustainable development

■ National policies and strategies should be integrated into a framework that can meet the needs of local communities, indigenous peoples, non-governmental organizations, women and other individuals as well as the broader international community

■ Efforts to create a supportive international economic climate conducive to the sustained and environmentally sound development of forest resources in all countries should be promoted. The special financial and developmental needs of developing nations should be recognized

■ The profits from forest products – particularly those arising from biotechnology products and genetic materials – should be shared with the countries in which the forests that provided them are located

■ Because of their potential to provide fuel and industrial wood resources, particularly in the developing world, plantations should be recognized and promoted as sustainably and environmentally sound forms of forest use

■ Efforts to introduce and increase reforestation, afforestation and forest conservation must be promoted, to maintain and improve the ecological balance in the forest and combined with sustainable forest management to allow continued productivity

■ Management and sustainable development of forests must include protection of primary/old-growth forests as well as cultural, historical, religious and other unique and valued forests of national importance

■ Transfer of environmentally sound technologies must be promoted, facilitated and financed to allow developing nations to manage, conserve and develop their forest resources better

■ Trade in forest products should be based on non-discriminatory, multilaterally agreed rules that allow open and free international trade in forest products; unilateral measures should not be used to restrict trade in timber and other forest products

■ Local processing should be encouraged and tariffs or other impediments to marketing finished goods reduced or removed; forest conservation and sustainable development should be integrated with economic and trade policies, while environmental costs and benefits should be incorporated into market mechanisms

Source: based on information in UN Report of the United Nations Conference on Environment and Development (1992)
Viewed at http://www.un.org/documents/ga/conf151/aconf15126-3annex 3.htm

countries (UNEP 2002). In North America and Europe, in the 1960s, 1970s and 1980s, significant problems arose in forested areas as a result of industrial pollution, mainly in the form of acid rain (see Chapter 10), but with occasional catastrophic events such as the explosion of the Chernobyl reactor in 1986 also contributing to significant regional habitat change (Mnatsikanian 1992). In many areas, particularly in the developing world, the conversion of forests is motivated by a group of underlying socio-economic causes, including poverty, population growth and expanding markets for forest products, which together create conflicts between the need to conserve forest resources

and the need to exploit them for economic development.

Whatever the causes, changes in forest areas extend beyond the destruction or removal of the trees. They destroy the habitat for other plants and forest-dwelling animals and reduce its overall biodiversity. By opening up the environment, and altering such elements as radiation, temperature, wind and moisture, deforestation also produces significant micro-climatological changes. The higher levels of solar radiation reaching the surface cause temperatures to rise, accelerating the decomposition of organic matter. Temperature ranges and wind speeds increase following the removal of the trees and the intensity of the rainfall reaching the surface is higher. The latter causes more rapid leaching of the nutrients already in the soil plus those released from the decomposing organic matter. Run-off also increases, creating a greater potential for soil erosion, especially where slopes are steep, and that may be translated into flooding or more rapid siltation in the local stream systems (Figure 7.7).

Temperate forest

Deforestation is a long-standing phenomenon, which was not always perceived as a problem. In the temperate forests of the northern hemisphere, for example, the destruction of the forest was equated with the advance of civilization, allowing as it did the spread of agriculture and a consequent increase in population. As early as 6000 years ago, in Mesolithic and Neolithic times, the Eurasian temperate forest was already being opened up by prehistoric agriculturalists and about 3500 years ago the improved technology made possible by the spread of metalworking during the Bronze Age allowed a major increase in the rate of forest clearance (Mannion 2002). There is some speculation that these original forests were less dense and less inhospitable than was once thought, more akin to the parkland that has developed in the New Forest of southern England, containing a mix of

mature trees, copses and grass-covered glades kept open by grazing animals (Vera 2002). This more open landscape would have been more easily colonized by the early agriculturalists.

The temperate mixed forest was the forest of Grimm's fairy tales, the Sherwood Forest of the Robin Hood legends and the backdrop for James Fenimore Cooper's tales of the eighteenth- and nineteenth-century American frontier. The forests described in these stories are long gone. The soils underlying the forest were quite fertile, and from medieval times in Europe and after the arrival of the Europeans in North America, the forest was pushed back increasingly rapidly, with crops replacing the trees. Patches survived as royal hunting preserves, but the forest disappeared over most of lowland Europe. In less accessible areas, such as the hills and mountains, it remained longer, but even there it was threatened. The hardwoods were ideal for construction and shipbuilding and as the wood on the plains was used up the woodcutters moved into the hills. These woods were also the main source of the charcoal required for the smelting of iron ore and the making of steel before the advent of coal in the industrial revolution.

A similar pattern emerged in North America, where the forests of the coastal plain gradually disappeared from the eighteenth century on, and as the pioneers pushed westwards in the nineteenth century the trees were cleared from the St Lawrence/ Lower Great Lakes, the Ohio valley and most of

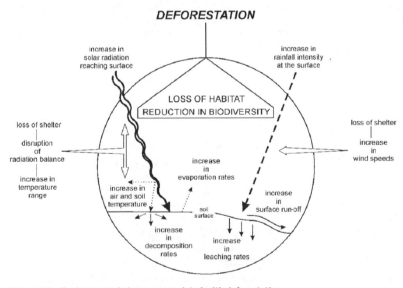

Figure 7.7 Environmental changes associated with deforestation

the area east of the Mississippi. This was the largest temperate mixed forest on earth, but in the 200 years following European colonization the area lost more woodland than had been lost in Europe in the previous 2000 years (Goudie and Viles 1997). It survived only in parts of the Appalachians, where narrow valleys, steep slopes unsuitable for farming and the region's general inaccessibility discouraged settlement. Although not unaffected by human activities, such as logging and recreation, the forests that cover the ridges of the Appalachians from Maine in the north to Georgia in the south, although no longer virgin forest, provide the best examples of what the primeval temperate mixed forest might have been like. The relatively small areas of temperate forest in the southern hemisphere have also largely disappeared. When the Europeans arrived in New Zealand in the early nineteenth century, 70 per cent of the land was forested. By the end of the twentieth century only 16 per cent was forested (UNEP 2002), and the clearing of the temperate forests of the South Island to make way for agriculture made a major contribution to that decline.

A locally significant eco-zone within the temperate forest biome is the temperate rain forest found in west-coast locations in North America, southern Chile and the west coast of New Zealand. These forests contain the oldest and largest trees in the temperate forest and are much sought after by lumber companies. As a result, they have been heavily logged, especially in the Pacific North West in the United States, where only about 17 per cent of the original rain forest remains and that only surviving because it is growing in National Forests (Cox 1993). The trees that have survived are centuries old and from an economic point of view their value is not increasing, because they have reached maturity and are not significantly adding to their biomass. These old-growth forests are regenerating naturally, however, with young trees growing in the gaps where old trees have fallen and the nutrients from the old trees being returned to the soil as they rot. They provide some of the last remaining natural habitat for a diverse community of insects, birds and mammals, some of which are threatened species, and aesthetically they have few peers among the temperate forest ecosystems. As a result, the remaining old-growth forests are seen by many as prime candidates for conservation. In British Columbia, Canada, where the bulk of the surviving natural temperate rain forest is located, the opposing

Plate 7.1   The natural environment in a temperate old-growth rain forest. It includes a complete mix of dead trees, mature trees and young trees. Where trees have fallen, the increased light levels allow a lush undergrowth of ferns.

viewpoints of the logging community and the conservationists led to ongoing conflict in the 1990s. At Clayoquot Sound on Vancouver Island, for example, a coalition of environmentalists and First Nation peoples challenged the right of forests products companies to continue clear cutting of the old-growth forests around the Sound. After a decade of direct protests – such as blocking logging roads – accompanied by a very successful media campaign, by 1995, the protesters were able to get the provincial government and the forest companies to back away from clear cutting (Fleming 1997) and establish a new strategy focusing on old-growth conservation. That success was capped in 2000 with the establishment of Clayoquot Sound as a 1000 km^2 UNESCO Biosphere Reserve (UNEP 2002), which should ensure its survival as a natural wilderness.

At the opposite end of the moisture scale from the temperate rain forest is the seasonally dry environment of the Mediterranean scrub forest (see Chapter 3). Its current landscape of olives, vines, shrubs and herbs is almost completely the result of deforestation, particularly in Europe, but all areas where it is found have suffered change. Because of its equable climate, this biome has always attracted people and as a result the natural vegetation has been altered considerably. Around the Mediterranean Sea, in Europe, North Africa and the Middle East, the change was already under way in Greek and Roman times, with the natural mixed evergreen and deciduous forest – including the famous cedars of Lebanon – being removed to provide land for agriculture, to provide fuel and to provide construction materials (Ponting 1991). Grazing animals, particularly sheep and goats, introduced into areas too dry for arable agriculture, gradually destroyed the shrubby vegetation, allowing grasses to predominate. Over-grazing and poor cultivation techniques have led to the threat of desertification in parts of North Africa, Spain and Greece and the clearing of land for residential purposes has exposed land to soil erosion. Land laid bare by frequent fires fed by dry scrub and grasses also increases the potential for erosion. Although fire was always an integral part of the biome, the disruption of the habitat and the replacement of fire-resistant tree and shrub species have increased the frequency and intensity of both natural fires and those started by human activities. Human interference is likely to continue and, coupled with the potential changes in temperature, moisture availability and fire frequency expected to accompany global warming, even those parts of the biome that have survived in a relatively natural state may well be lost in the not too distant future.

Boreal forest

The greatest of the forest biomes outside the tropics is the boreal forest, accounting for 33 per cent of the world's total woodland cover (see Chapter 3). It is confined mainly to higher latitudes in North America and Eurasia (Figure 3.13), although a few patches are found in mountain locations such as the Western Cordillera in the United States or the highlands of Mexico. Having a short growing season, it is the least productive of the forest biomes, with a mean biomass per unit area about two-thirds that of the temperate mixed forest and less than half that of the tropical rain forest (Brown and Lomolino 1998). The areas in which the boreal forests grow have almost all been recently glaciated, and as a result the soils are generally thin and of limited fertility. In addition, the drainage system has still not recovered from the disruption caused by glaciation, which, coupled with moist climates and low evaporation rates, ensures an abundance of wetlands in the boreal forest biome. Thin, wet soils with low nutrient levels do not attract agriculture, and although the southern edge of the forest was cleared in places for farming in the nineteenth century in North America and earlier in Eurasia, most operations were small-scale and unsuccessful. Once abandoned, the fields were colonized again by natural species, creating a secondary succession that eventually led to a return of the forest. Thus agricultural settlement has never been a major threat to the boreal forest. Animals, such as beaver and muskrat, are still harvested for their fur or, like moose, deer and bear, hunted for food or sport, but if carried out in a sustainable manner, these activities should have little overall impact on the biome.

Local disruption of the environment is associated with mining activity, hydro-electric development and recreation and tourism, but the major change has been caused by the removal of trees to provide construction lumber or to produce pulp and paper. Clear cutting has laid bare thousands of hectares, especially along the more accessible southern margins of the forest, causing serious environmental problems in some regions. Despite this, because of the immensity of the biome, there are still large areas, relatively untouched by human activity, that represent the boreal forest in its natural state, and some of these have been targeted for protection and preservation by environmental groups and government organizations.

Canada and Russia are the main producers of wood from the boreal forest and, as the demand for softwood lumber continues to increase, it is in these countries that the pressure on the forest environment will be greatest. The former harvests 175 million m³ and the latter 181 million m³, and both countries export between 10 per cent and 20 per cent of their production (UNEP 2002; FAO 1997). Until the mid-1980s clear cutting was the most common method of harvesting, and in Canada approximately 90 per cent of the timber harvested continues to come from clear cutting. In Russia the equivalent figure may be around 73 per cent, but statistics are not considered reliable because of the illegal logging that is widespread in some areas (UNEP 2002). Clear cutting is now much maligned, because of the social and ecological costs it incurs, but it is considered to be the most economical

method to log in areas with the limited species mix of the boreal forest. To some extent it is also considered to mimic natural disturbances such as wind storms or fire, which create large openings in the forest, and it allows easy access for replanting (Dearden and Mitchell 1998). In the case of natural disturbances, however, the nutrients remain in the area and contribute to the recovery or the forest. Large clear cuts are aesthetically unpleasant, disrupt energy and moisture budgets, unbalance nutrient cycles, disturb the local wildlife and diminish biodiversity. Although most clear cuts are now routinely replanted, these disruptions during and following harvesting reduce the survival rate of the seedlings or require that they receive ongoing attention – fertilizer and herbicide application, for example – until the new trees are well established. Modern forest management techniques attempt to reduce the environmental disruption, by limiting the size of the cuts, providing buffers along waterways and corridors to allow wildlife to move between patches of unaltered habitat. Where these approaches are recommended rather than

Plate 7.2   The edge of a clear cut in the boreal forest. The surface debris or 'slash' is typical.

mandatory, the temptation is always there to ignore them when they may mean the difference between profit and loss. Techniques such as patch cutting (in reality small clear cuts) or strip cutting, which leave wooded areas to provide a source of seeds for the regeneration of the forests, have been introduced, and although they appear more environmentally appropriate, the areas involved are minor compared to those that are clear cut.

Unlike Canada and Russia, Scandinavia does not have the advantage of large areas of relatively undisturbed natural forest. Only about 5 per cent of the boreal forest in Norway, Sweden and Finland is old growth. The majority is second growth, restored either naturally or through plantation, which has been managed for the past century or so (Taiga Rescue Network 1998). While there are advantages in working a plantation forest, with its limited number of species, all of the same age and planted for ease of management and extraction, plantations have reduced biodiversity, lacking as they do the mix of birds, animals and other organisms of the natural forest. Despite this, there is growing attention to regeneration and plantation forestry as a management tool and its adoption allowed North America and Europe to record a net increase in forest cover in the last decade of the twentieth century (UNEP 2002).

Forest management in the boreal forest

Concern that current harvesting rates in the boreal forest are unsustainable has led to increased attention being given to all aspects of forest management, from seeding and planting to cutting. In its simplest terms, maintaining a sustainable yield requires every tree cut down to be replaced by a seedling, but that is only the first step. Since the boreal forest trees can take fifty to 100 years to reach maturity, some allowance must be made for that time lag and the harvest rate adjusted accordingly. This is commonly achieved through the calculation of an annual allowable cut (AAC), which reflects the estimated annual increment in harvestable timber in an area. The AAC is based on such factors as the age, mix and growth rates of species in the forest and makes allowance for protected areas from which no timber can be cut. If the AAC provides an accurate representation of the biological productivity of the forest, and the actual harvest does not exceed the AAC, the sustainability of the forest can be preserved (Dearden and Mitchell 1998) (Figure 7.8). Natural events, such as forest fires, major insect infestations and climate change, or human activities, such as failure to maintain appropriate levels of regeneration, can reduce the productivity of the forest, however, and if no

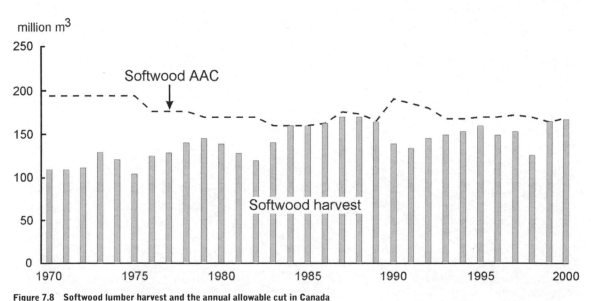

Figure 7.8 Softwood lumber harvest and the annual allowable cut in Canada

Source: based on data from the National Forestry Database Program, Canadian Council of Forest Ministers. Viewed at http://nfdp.ccfm.org/cp95/image_e/fig4e.gif (accessed 10 January 2003).

adjustment is made to the AAC, sustainability will be threatened.

Modern forest management in the boreal forest extends beyond consideration of appropriate harvesting practices. It also involves the maintenance of the ecological functions and overall integrity of the forest through attention to biodiversity, water resources, soils and landscapes, and in the planning and implementation processes includes not only representatives of the forest industry, but also environmentalists, naturalists, hunters and a variety of other user groups. An appropriate plan will be one that allows multiple use of the forest and is acceptable to all user groups (Scrase and Lindhe 2001). In some cases, the preservation of ecologically significant or sensitive areas requires the exclusion of harvesting and the restriction of other uses, such as hunting. The area involved is relatively small considering the total area of the boreal forest. In Canada about 13 per cent of the forest is protected, but a study by a Canadian Senate committee (1999) recommended that 20 per cent of the boreal forest be set aside to preserve and protect ecologically and culturally significant areas, and some progress has already been made. Although the old-growth forest in Scandinavia has declined to only 5 per cent of the total, very little of it is protected and its continuing viability is threatened by continued fragmentation (Taiga Rescue Network 1998). In Russia about 21 per cent of all forests – including boreal and temperate – are under some form of protection, but controlled logging is allowed in some of these areas and the struggling economy has led to an increase in illegal harvesting. The designation of areas in the Komi Republic and around Lake Baykal as UNESCO World Heritage Sites has protected vast expanses of forest from logging (UNEP 2002).

Current and future threats to the boreal forest

The forests currently under greatest threat are those along the southern borders of the biome. They are closest to the areas of industrial development where the demand for timber continues to grow. The more northerly areas are under less pressure. The forest there is thinner and, given its relative inaccessibility, its development is not economically viable. It is unlikely to escape the changes in the biome expected to be caused by global warming, however (see Chapter 11). If computer predictions are to be believed, future global warming will be greatest in mid to high latitudes in the northern hemisphere, exactly the location occupied by the boreal forest. Warming might well increase the productivity of the forest, and it would extend northwards at the expense of the tundra, as it has done at various times in the past (Ball 1986). At the same time, it would come under pressure from the temperate mixed forest to the south, causing its southern boundary to retreat northwards. The net effect would be the eventual relocation of the biome to more northerly latitudes, with the overall environmental impact depending upon such factors as the speed and uniformity at which the change took place. Under several possible scenarios, the biome might become wider, narrower or even be wiped out completely in some places (Figure 11.18).

Tropical forest

Although deforestation continues in the boreal forest, the bulk of the clearing of forests outside the tropics had taken place by the beginning of the twentieth century, and through reforestation the area of forest in the temperate and boreal zones has actually expanded. In contrast, deforestation in the tropics is in its early stages, but is proceeding at a rapid pace. While concern about the destruction of the tropical forest is almost universal, the rate at which it is occurring and the areas that are most affected remain uncertain, because of different definitions of deforestation, different approaches to measuring change in the forest, and different methods of data collection and compilation (Middleton 1999). Most recent estimates of the rate at which tropical forests (both rain forest and seasonal monsoon forest) are being destroyed fall between about 10 million ha per annum (Aiken and Leigh 1995) and 15 million ha per annum, which includes land converted into plantations (FAO 2001). The Food and Agriculture Organization of the United Nations has estimated a loss of 15.2 million ha per annum during the 1990s (FAO 2001), up from 11.3 million ha per annum in 1980 (FAO/UNEP 1982). In percentage terms these values represent a loss to tropical forests of between 0.6 per cent and 0.8 per cent per annum. The accuracy of such data is very much a matter of debate and this allows environmental groups to continue to quote higher, sometimes exaggerated values, whereas groups interested in development are drawn to the lower rates of depletion (Lomborg 2001).

Data from the FAO (1997) indicate that between 1990 and 1995 forest cover was lost at a rate of 4.6 million ha per annum in South America, 3.6 million ha per annum in Africa and 3.0 million ha per annum in Asia. There is considerable variation within these regions, however (Table 7.1). Brazil dominates the deforestation statistics for South America, for example, with a rate more than eight times that of any other nation on the continent, while the Indonesian rate is more than double that of second-ranked Myanmar and exceeds that in all tropical nations other than Brazil. Totals can also be deceptive. The more than 2 million ha lost every year in Brazil represents only 0.4 per cent of its total forest area, whereas only 15,000 lost in Burundi in tropical Africa accounts for 9 per cent of the forest area in that country. The numbers also vary widely from year to year. In the Brazilian Amazon alone, for example, 2.9 million ha were lost in 1994 and 1.6 million ha in 1999 (Fearnside 2002), while in 1998 some 4 million ha were destroyed by fire (FAO 2001). The most rapid deforestation in Africa is in Côte d'Ivoire, where for the past thirty-five years forest has been cleared at a rate of 300,000–400,000 ha per annum and there is now virtually no natural, unmodified forest remaining in the country (Middleton 1999). In Asia, in the 1990s, fires increased the area of forest destroyed. More than

TABLE 7.1 DEFORESTATION IN SELECTED TROPICAL COUNTRIES, 1990–2000

Country	Annual change (000 ha)	Annual rate of change (%)
Africa		
Top three by area lost		
Democratic Republic of Congo	532	0.4
Nigeria	398	2.6
Côte d'Ivoire	265	3.1
Top three by percentage lost		
Burundi	15	9.0
Rwanda	15	3.9
Togo	21	3.4
South America		
Top three by area lost		
Brazil	2309	0.4
Peru	269	0.4
Venezuela	218	0.4
Top four by percentage lost		
Ecuador	137	1.2
Brazil	2309	0.4
Peru	269	0.4
Venezuela	218	0.4
Asia		
Top three by area lost		
Indonesia	1312	1.2
Myanmar	517	1.4
Malaysia	237	1.2
Top four by percentage lost		
Pakistan	35	1.9
Sri Lanka	35	1.6
Myanmar	517	1.4
Philippines	89	1.4

Source: based on data in FAO, *State of the World's Forests 2001*, Rome: Food and Agriculture Organization (2001).

9.7 million ha were lost to fire in Indonesia in 1997 and 1998 (FAO 2001), increasing a rate of change that was already high because of the demand for agricultural land from a rapidly growing population. Throughout Southeast Asia the amount of cleared land grew rapidly in the second half of the twentieth century, so much so that in Vietnam, where forests covered 45 per cent of the land in the 1940s, only 17 per cent is now wooded and in the Philippines, which was 50 per cent forested in 1950, only 25 per cent of the land was forest-covered in the early 1990s (Middleton 1999). Between 1990 and 2000, however, Vietnam added an average of 52,000 ha to its forest area every year (FAO 2001).

Logging

The main causes of tropical deforestation are commercial logging, the clearing of land for agriculture and the cutting of fuelwood (Figure 7.9). Tropical hardwoods such as mahogany, ebony and teak have been in high demand in the developed world for a long time, and developing nations see them as an important source of revenue. Logging increased rapidly in the second half of the twentieth century as demand for wood grew in the developed world. In Malaysia the production of logs was six times higher in 1990 than in 1960 and in Sarawak there was a fifteen-fold increase over the same time period (Aiken and Leigh 1995), with most of the production being exported to Japan. Similar developments occurred in West Africa, where they were encouraged by tax incentives and the sale of logs contributed to the economy. Commercial logging is the most immediate source of income for many developing nations in the tropics, and returns can be high, but unless it is controlled it is not sustainable over the long term. The rapid pace of logging has left little remaining high-quality lumber in countries such as Nigeria and Côte d'Ivoire, for example, leading to a drop in output and revenue. The demand for foreign currency in the developing nations (often to cover debts incurred in their attempts to achieve some degree of development) combined with the demand for forest products in the developed world has led multinational companies to become involved in working the tropical forests, with major concessions granted to them in countries such as the Philippines and Brazil. With their traditional focus on short-term gain, it is unlikely that these multinationals will contribute to the development of sustainability in the tropical forests.

Given the intricate nature of tropical biomes, damage to the environment by logging extends beyond the direct removal of the trees. The network of roads needed before logging can begin contributes to the fragmentation of the forest ecosystem and encourages erosion, while improving access for hunting, farming, mining and other activities that can alter the ecosystem. Modern mechanized logging using bulldozers and skidders causes serious damage to the soil through compaction and the disturbance of the ground vegetation and the upper horizons of the soil, which together increase the potential for erosion, particularly on steeper slopes. Even if measures are taken to reduce erosion, the effects on the soil are generally detrimental. Exposure to higher levels of solar radiation and

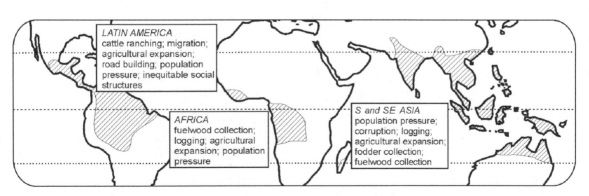

Tropical forest (includes evergreen and deciduous forest)

Figure 7.9   Causes of deforestation in the tropics

moisture encourages the more rapid release of nutrients in the surface layers. Percolating water rapidly leaches them out, however, and since little organic matter is returned to the soil following the removal of the trees, the soil loses its fertility. The nature of the tropical forest and the method used to log them can lead to a considerable amount of collateral damage. Unlike the boreal forest with its dense stands of two or three marketable species, commercially viable trees in the tropical forest are usually widely scattered among non-commercial species. In most cases it is easier to clear-cut an area to reach the desired trees than to select them individually. The net result is damage to a much larger area of forest than necessary and the destruction of trees that are often left to rot after they are felled. With this approach, as little as 5 per cent of the cut timber may be actually used (Park 1992). The net result extends beyond the obvious environmental damage to the forest to the disruption of the lives of the birds and animals that live in it and produces an overall decline in biodiversity.

Selective logging, which targets only the species being harvested, is preferable since it more closely represents the natural fall of trees in the forest. It is not entirely problem-free, however. Falling trees may bring down neighbouring trees or smash their branches and foliage, while the skidding of the felled trees using tractors rips up the undergrowth and the underlying soil. Such problems can be addressed by techniques such as directional felling, which minimizes damage to the adjacent canopy, and the removal of felled trees by cable or draught animals or even by helicopter, which reduces damage to the forest floor. Selective harvesting is not widely used, however, and damage to the forest may be more extensive than is indicated by the total area logged. It has been estimated, for example, that in the Amazon, between 1.0 million ha and 1.5 million ha of land may be damaged every year during logging operations and yet not appear in the annual deforestation inventories (Nepstad *et al.* 1999).

Agricultural expansion

Although logging is commonly highlighted when environmental groups and the media draw attention to the problems of the disappearing tropical forest, in many areas the greater threat is the removal of trees to allow the expansion of agriculture, in much the same way as the temperate forests were cleared centuries ago to provide land that could be farmed to meet the demands of a growing population. The shifting agriculture, long practised by the indigenous peoples in the tropics, was generally compatible with the environmental conditions in the forests. Small patches of forest were cleared and cultivated, almost as gardens, until the declining soil nutrients and competition from weeds reduced yields to levels that could not support the community. At that stage, the area was abandoned and the farmers moved to a new location, allowing the original clearings to be colonized again by the forest.

Modern techniques involve the clearing of land on a much larger scale, often through a slash-and-burn approach, which may provide good returns initially, as the nutrients from the burned vegetation are recycled, but which is ultimately unsustainable (Middleton 1999). Almost 70 per cent of the tropical forest cleared in the 1990s became agricultural land, worked under permanent rather than shifting cultivation (UNEP 2002). Even where shifting agriculture continues, as in Indonesia, for example, population growth and the demand for land have been so great that the land abandoned during the cultivation cycle is unable to recover completely before it is needed again, causing a cumulative deterioration in the environment (Mannion 2002). The large-scale slash-and-burn approach to clearing that is the initial step in the process led to the major fires in Indonesia that caused persistent pollution in Borneo and over the adjacent Malaysian peninsula in 1996 and 1997. While the initial clearing of the forest for agricultural land was for local food crops, increasingly the focus is on cash crops, such as soya beans, cassava, pepper and even wheat, which are commonly grown for export. The cultivation of the coca plant – very much a cash crop in modern society – to provide cocaine for the drug market in the developed world has been responsible for about 10 per cent of deforestation in the Peruvian Amazon (Park 1992).

Plantation agriculture using trees native to the tropical forest has a long history. The British colonial powers introduced rubber to the Malaya in the 1870s, for example, and the French later did the same in Indochina (Aiken and Leigh 1995). Oil palm, coconut palm and banana plantations followed. In West Africa cocoa and oil palm plantations were common, and in Central America companies from the United States set up banana plantations. Coffee is also a major plantation crop in Southeast Asia, East Africa and Central America, but Brazil remains the largest producer. Since most of these crops require the planting of trees, they appear to offset deforestation by replacing the lost trees. However, although the net area forested may be retained, where plantations are created on land cleared from the primary forest, the result is a change in the flora and fauna of the area and an overall decline in biodiversity.

A major component of the expansion of agriculture in tropical areas in Central and South America is livestock farming. In Brazil and Costa Rica, for example, large areas of forest have been cleared to establish North American-style ranching (Mannion 2002). One of the incentives was to provide cheap beef for the US market and the link was dubbed the 'hamburger connection' because the beef ended up being sold in fast-food outlets. Much of it could not meet US health standards, however, and as a result of successful intervention from environmental groups, the trade is much reduced (Park 1992). The greatest expansion of ranching has taken place in Brazil, mainly as a result of government attempts to develop Amazonia (Anderson 1993). Subsidies in the form of tax relief and tax holidays were offered to developers, including multinational companies, willing to open up the area for cattle farming, and the programme was so effective that an estimated 50 per cent of the deforestation in Brazil is attributed to the expansion of ranching in the Amazon basin (Fearnside 1993; Mannion 2002). Livestock farming is not sustainable in the Amazon, however, and the result has been major environmental deterioration. Once the trees are removed, the tropical soils lose their fertility quite quickly, allowing grazing land to survive for only five to eight years before the pasture begins to degrade. Since it is cheaper to open up more forest than to buy fertilizer for the land or to adopt soil conservation techniques, the pastureland is abandoned, sometimes in such a degraded state that the original forest vegetation may never properly recolonize the area (Park 1992).

Fuelwood

In the developing world as much as 80 per cent of the wood production is burned to provide energy for heating and cooking (Park 1992). Fuelwood is a particularly important source of energy in Africa, where, in Cameroon, for example, 78 per cent of the timber cut is used as fuelwood (Goudie and Viles 1997). As with logging and agricultural development, the removal of plant material changes the forest habitat and disrupts the nutrient flow in the ecosystem, reducing soil fertility and increasing the potential for additional environmental deterioration. The actual impact varies according to the types of wood used and the methods by which it is extracted, however. In some areas – parts of India and Pakistan, for example – the supply cannot keep up with the growing demand and forest land is being lost, but elsewhere – in parts of Africa, for example – fuelwood cutting is sustainable and it is not a major factor in deforestation (Leach and Mearns 1993). In the tropical forests, the wood used for fuel takes various forms, including branches and twigs gathered from the growing forest, debris remaining after sites have been cut over by loggers or cleared for agriculture and charcoal made from trees cut down in the forest. In Zambia, some 43,000 ha are cleared annually to provide the 100,000 tonnes of charcoal used there and in Madagascar the collection of wood for fuel – amounting to 1–2 kg per person per day – is the main cause of forest clearing (UNEP 2002). The edge of the forest is gradually pushed back under pressure from the fuelwood cutters and collectors at a rate that varies with the demand. Where population pressure is low the use of fuelwood is sustainable, but where population pressure is high consumption regularly exceeds the annual replacement growth and the net result is a loss of woodland. The situation is aggravated where the forest is burned to clear land for agriculture and potential fuelwood is lost, but in many areas where land is being cleared, co-operation between farmers and charcoal burners allows the cleared wood to be saved for conversion into charcoal. The overall impact of fuelwood cutting and collecting on deforestation in the tropical forest is not clear. In some areas it contributes directly to the disappearance of the forest, but it is often so closely linked with other activities such as the clearing of land for agriculture that it is difficult to identify its specific contribution. Attempts to reduce fuelwood use have included the introduction of more efficient cooking stoves and charcoal kilns, which is important in areas where demand exceeds supply. Whatever its cause, the consequences of deforestation in the tropical rain forest are extensive. Local habitat is destroyed, directly threatening the survival of plant and animal species and contributing to changes in local and micro-climates, hydrology and nutrient cycles. On a global scale, deforestation has the potential to contribute to climate change through the disruption of the carbon cycle (see Chapter 11).

The human impact of deforestation in the tropics

The indigenous peoples who originally inhabited the world's forests have also suffered in a variety of other ways as result of deforestation. This was particularly true of the inhabitants of the tropical rain forests. Over thousands of years they had adapted to life in the forest and developed a life-style that allowed them to make the best use of the resources it had to offer. The forest environment was not always benign, but that lifestyle included coping mechanisms by which the inhabitants could deal with the changes to be expected in a dynamic system. These might include seasonal migration to maintain their food supply, or to avoid seasonal flooding, for example, or longer-term migration associated with shifting agriculture. The forest dwellers had a sustainable relationship with the forests, using as many of its resources as they needed with no serious disruption of the environment. The loss of the forest or its fragmentation made their lifestyle difficult to maintain and the loss of habitat was every bit as serious for them as it was for the other inhabitants of the rain-forest biome. Deprived of their traditional hunting and growing areas, their culture eroded, exposed to diseases such as measles and smallpox, to which they had no resistance, and sometimes deliberately slaughtered by the incomers, their numbers declined dramatically. Before the Europeans arrived in the Amazon basin the indigenous population may have numbered as much as 10 million; less than 400 years later at the end of the twentieth century the population had fallen below 250,000 (Layrisse 1992). Indigenous populations in Africa and Southeast Asia suffered a similar fate. Although media attention has focused on the fate of some of the tribal groups in South America, such as the Yanomami and Kayapo tribes in Amazonia, it has also drawn wider attention to the problems that indigenous people face, and led to consideration of priorities and possible solutions. Attempts at resisting pressure from modern economically and technologically powerful government or private organizations have been generally unsuccessful (Park 1992). There is,

however, a growing realization that the impact of deforestation must be mitigated before it is too late, and indigenous people who have had a sustainable relationship with the forest for thousands of years may well have an important contribution to make to that process.

Solutions to tropical deforestation

As with many modern environmental issues, solutions to the problems of deforestation in the tropical rain forest can be sought and sometimes found in the application of technology. Selective logging, erosion control, agroforestry, plantation fuelwood production and reforestation, for example, all involve techniques that have the potential to limit the rate of deforestation and rehabilitate areas that have already been damaged. These approaches have been successful in some areas, but their impact has done little to slow the overall rate of deforestation. In part, this is because deforestation is more than a technical problem. It involves socio-economic, political and cultural factors also, and if these are ignored the application of technology is unlikely to be successful.

In some ways the issues are not unlike those that faced the early environmentalists in North America in the nineteenth century, with the conservationists willing to allow the extraction and use of the forest resources, while the preservationists preferred it to remain as untouched wilderness. The preservationist ethic continues relatively unchanged, and is reflected in the creation of areas of protected forest. The conservationist approach has evolved into sustainable development, in which the forest can be worked for its resources as long as its productivity can be managed to provide returns in perpetuity, with little or no damage to the characteristics and integrity of the forest. The situation has become polarized to some extent, with calls for preservation coming from governments and environmental groups in the developed world, whereas the developing nations see the forest as an important resource base, which needs to be exploited if they are ever to improve their economic situation. While that right cannot and should not be denied them, it might be possible to adopt a minimum interventionist approach in which preservation and development can be balanced, and attempts at reconciliation of the two viewpoints tend towards that type of solution.

By the early 1980s solutions were being sought at the international level. The International Tropical Timber Organization (ITTO), formed in 1983, was concerned with the management, harvesting and marketing of tropical rain-forest products, to the benefit of producing and consuming countries. Although not in its original mandate, sustainable development became an important part of its platform (UNEP 2002). It was followed in 1985 by the Tropical Forestry Action Plan (TFAP), which paid attention to conservation, agroforestry, fuel-wood management and reforestation, plus the establishment of an institutional framework for training and research. Using the TFAP framework, National Forestry Action Plans (NFAP) were pro-posed in some sixty countries, and the future of the tropical forest appeared to have brightened. The TFAP was successful in raising public aware-ness of the issues and in bringing together non-governmental organizations, such as environmental groups, and aid agencies, such as the FAO and the World Bank, to participate jointly in policy and decision making. It tended to follow a top-down approach, however, with little or no input from those living in the forest, and was criticized for being underfunded and badly directed. Concrete successes were few – only six of the proposed NFAPs were initiated, for example – and in 1991 it was revamped and renamed the Tropical Forestry Action Program, with a revision of its objectives to include greater attention to conservation and sustainable development (Park 1992). Some of the policies formulated in the TFAP were carried through to the Rio Earth Summit in 1992 and included in the Statement of Forest Principles and Agenda 21.

It is one thing to sit in a conference or on a committee and develop policy; it is entirely another to convert that policy into action. Action can be initiated at different levels. International aid agencies such as the World Bank have become increasingly environmentally conscious and now take environmental conditions in the rain forest into account when making loans or grants. As early as 1988, for example, the World Bank refused to fund dams planned for the Altamira hydro-electric power project in Amazonia, in consideration of the damage that would be caused to the forest and the people living there by the subsequent flooding (Park 1992). Aid agencies have also been pro-active in dealing with the high levels of debt that face many of the developing nations and prevent them from coping with their rain-forest problems. One moderately successful scheme is the 'debt for nature' approach. In this, an environmental group or government agency purchases part of the debt of a developing nation – usually at a discounted price – and offers to forgive the debt if the debtor agrees to set aside and manage a section of rain forest. Debt-for-nature has been applied successfully in Bolivia and Costa Rica, having the effect of reducing foreign debt in those countries while helping to reduce deforestation (Middleton 1999). Some see it, however, as an infringement of national sovereignty and an example of foreign interference in natural resource development, despite the apparently positive consequences for all involved.

Multinational companies participate in many rain-forest activities that directly or indirectly cause deforestation, from logging and mining to plantation agriculture and ranching, usually with the connivance or co-operation of national govern-ments. The products from these activities commonly end up in the developed world, where they are pro-cessed and sold. In an attempt to reduce or control the market for these rain-forest resources, and through that the conditions under which they are produced, non-profit international organizations, including the Forest Stewardship Council and the Rainforest Alliance, have developed the concept of certified forest products. Certified wood comes from forests in which management standards encourage regeneration, the conservation of biological diver-sity and the maintenance of soil quality to protect the long-term health and viability of the forest. Some 15 million ha of tropical forest are currently certified and if manufacturers in the developed world can be persuaded to use only certified wood then that area would grow, helping to reduce rain-forest deterioration (Cuff 2002b). Similar schemes promoting the labelling of sustainable forest products or the outright banning of hardwood imports have met with some success in Britain, but are opposed by hardwood producers such as Malaysia and Indonesia, which see them as trade barriers and a deliberate attempt to restrict devel-opment (Park 1992).

In the final analysis, policies to reduce deforest-ation will be successful only if they are implemented on the ground, and it is not always clear that that is happening. There is no shortage of legislation, but it becomes meaningless if enforcement is lax. Ambitious management schemes die because funding is inadequate or has been diverted to other uses. Fraud and corruption are not unknown, either. Other underlying causes such as high debt levels, poverty and population pressure, which force developing nations to make immediate gains from the forest with little consideration of the future, will also have to be addressed. Pressure for the preservation of the rain forest comes mainly from the nations of the developed world, and if they are serious about the issue they must be prepared to compensate the developing nations for the loss of short-term income that preservation entails. Improved aid packages that give more attention to sovereignty and national aspirations, marketing schemes that support responsible development and technological transfer to improve efficiency and reduce waste would all help. Although there are some signs that improvements are being made, and it seems unlikely that the dire predictions common in the late twentieth century, that saw the complete disappearance of the tropical rain forest within decades, will come to pass, the overall prognosis is still not good.

LOSS OF BIODIVERSITY

One of the main environmental impacts of the changing habitat created by human activities such as agriculture and deforestation is a reduction in biological diversity, or biodiversity, as it is more often called. Biodiversity refers to the variety of life forms that inhabit the earth. It involves habitat diversity, plant and animal species diversity within various habitats and the genetic diversity of individual species (see Box 7.3, Genetic engineering). The large-scale slaughter of wild animals, the over-harvesting of plants, the introduction of exotic species and the destruction of habitat worldwide by agriculture, industry and urbanization are threatening biodiversity at a time when its importance is becoming increasingly apparent. Particular concerns are expressed over the habitat destruction in areas such as the tropical rain forest, the threats to large mammals such as tigers and elephants or the near extinction of such marine species as the northern cod or the blue whale, but these are perhaps only the more extreme examples of a ubiquitous problem.

There is no general agreement on the total number of species in the biosphere, but 1.75 million have been named and described, representing only a small proportion of an overall number estimated at between 5 million and 50 million species (Brown and Lomolino 1998). Within that wide range, a total of 14 million species is perhaps the most commonly accepted (UNEP–WCMC 2000). Whatever the real numbers may be, the species are not evenly distributed throughout the biosphere. Tropical rain forests are richest in species, containing more than 50 per cent of the total (UNEP 2002). Species diversity tends to be greater in all tropical biomes, declining in less complex habitats such as grassland, or in biomes experiencing more extreme environmental conditions as in the desert or Arctic tundra. The number of species is only one measure of biodiversity, however. The overall distribution or relative mix of species is important also (Maguire 2002). Where, for example, specific flora and fauna are restricted to small geographic areas – endemic species – they are more vulnerable to environmental change, and reduction in their numbers can produce a rapid decline in biodiversity. Taking into account the combination of richness and balance of species, an area with ideal biodiversity would include many species of reasonably equal abundance, with appropriate allowance for the normal numerical relationships that exist between species – predator and prey, for example. Areas with greater biodiversity appear to be more resilient to disruption and recover more rapidly, but as change becomes more intense and more frequent it is likely that even the most diverse systems will suffer, as is already clear from the local, regional and even world-scale disruption caused by human activities.

Much of the popular attention displayed towards biodiversity has its origins in the emotions, fuelled by concern over individual species, usually photogenic mammals such as tigers or pandas or those such as gorillas that have anthropomorphic characteristics. It is not always clear that biodiversity and the reasons for valuing it are well understood (Wood 2000). Modern societies value most things on a monetary scale and their approach to biodiversity is no exception. Many of the goods and services obtained from the environment are associated with biodiversity and have a price tag or revenue attached to them. Continued returns from the forest industry, for example, which generates some US$50 billion per annum in North America alone, depend upon the maintenance of biodiversity (Mannion 2002), as does the estimated US$10 billion annual trade in wildlife products worldwide (UNEP 2002). Various attempts at calculating the overall worth of biodiversity have been based on

BOX 7.3 GENETIC ENGINEERING

Since plants and animals were first domesticated some 10,000–12,000 years ago (see Chapter 5), humans have been using selective breeding to produce organisms with desirable characteristics. Individual sheep with particularly well developed woolly fleece, for example, were bred with each other to produce more of the same, and over a number of generations, with controlled breeding, that trait became dominant in more and more individuals and became recognized as characteristic of a particular breed of sheep. That approach produced most of the domesticated plants and animals important to agriculture until well into the twentieth century. It was also central to the sport of horse racing, which used the stud approach along with accurately recorded bloodlines to ensure that the genes from winning thoroughbreds were passed on to future generations. Prize-winning show dogs and cats are in great demand as breed stock for much the same reason. Cross-breeding of dissimilar varieties within a species produced hybrid organisms that were often stronger than the parents or possessed characteristics that included the best attributes of both. Hybrid varieties of food crops, bred to cope with short growing seasons, or moisture deficiencies, for example, contributed to the major increases in crop yields that took place in the mid-twentieth century.

GENETIC MODIFICATION THROUGH MOLECULAR BIOTECHNOLOGY

For thousands of years this form of classical biotechnology, based on selective breeding and hybridization, allowed society to develop plants and animals that possessed the characteristics that they desired. By the middle of the twentieth century, however, advances in molecular biotechnology allowed the characteristics of organisms to be altered at the genetic level and that paved the way for genetic engineering. Genetic engineering, or genetic modification, is the manipulation of the genetic make-up of an organism for some specific purpose. Genes contain deoxyribonucleic acid (DNA), a complex organic polymer, which controls the characteristics of specific organisms or parts of organisms, and by transferring genes or segments of genes between organisms these characteristics are also transferred. The results are usually intended to benefit society, but genetically modified (GM) organisms have also been developed for use in biological warfare. Even

apparently beneficial modifications may become inadvertently harmful, however. A series of scares involving genetically engineered crops in the food industry has raised public concern over the potential impact of uncontrolled gene flows from genetically modified organisms into the natural environment.

APPLICATIONS OF GENETIC ENGINEERING

The modification of the genetic structure of organisms has applications in a number of fields. In health it has been seen as having great potential in disease control, through the development of gene-based medicines, for example, or the manipulation of gene sequences associated with a particular medical condition. The engineering of organisms to dispose of waste products, cleanly and cheaply, has significant implications for the environment. The most extensive application of genetic engineering at the beginning of the twenty-first century, however, has been in agriculture, particularly in the production of food crops. Genes have been transplanted between species and even between plants and animals. Plants have been modified to improve flavour, to increase resistance to disease and pests, to mature more rapidly or more slowly, to increase drought-resistance, to increase tolerance of salt and to survive low temperatures. These developments have met with considerable success. Maize, soya beans and oilseed rape (canola) have received most attention, particularly in North America, but genetically modified forms of other food crops such as wheat, rice, cassava, sorghum and millet have also been developed or are in the process of being developed. Non-food crops such as cotton have been modified to improve their resistance to pests.

The apparent advantages of these genetically modified plants led to their rapid adoption, particularly in the developed world. In the last five years of the twentieth century the acreage planted in transgenic crops increased by almost fifty times, and they were seen as having great potential to increase the world's food supply. The process was not without controversy, however.

PROBLEMS ASSOCIATED WITH GENETIC ENGINEERING

The control of the seed pool by a small number of companies in developed countries was seen as

BOX 7.3 – continued

detrimental to the transfer of the biotechnology to the developing world, for example, and the cost recovery and profits expected by these companies were seen as having the potential to push the developing nations even deeper into debt. The greatest concern arose out of the perceived environmental consequences of the widespread adoption of genetically modified crops. Pollination is central to fertilization and reproduction in plants, and, like natural varieties, genetically modified plants continue to produce pollen, which is distributed on the wind or by insects. This opens the door for the genetic contamination of plants in the areas where modified crops are being grown. Adjacent non-modified cultivated crops may receive pollen from the altered variety and related natural plants in the vicinity may receive foreign genes from the modified species. As a result, they acquire some of the characteristics of the genetically engineered variety and the overall nature of the natural gene pool is altered. The resulting loss of genetic diversity would have serious consequences. The modern reliance on a small number of food crops of limited genetic diversity exposes producers to potentially major crop failures. In the past, this was dealt with by reintroducing genetic material from the natural reservoir, but the contamination of this source would mean that new genetic material would no longer be available and revival of the crop species would not be possible. Other concerns include the development of super-weeds or super-bugs that have become resistant to the characteristics built into the new varieties. In China, for example, after only five years of production of modified cotton, the cotton bollworm is already becoming resistant to the toxin

engineered into the cotton seed to kill it. As a result, farmers once again have to resort to pesticide use to deal with the problem.

Although genetic engineering of organisms appears to have much to offer as a means of increasing food production or improving agricultural efficiency, many of the early promises are now being questioned, and opposition is widespread at both the popular level and in the scientific community. Public pressure in some areas, for example, has caused food retailers to refuse to stock products containing genetically modified materials. Environmental organizations such as Greenpeace and Friends of the Earth view genetic modification as having the potential to cause irreversible damage to the environment and to biodiversity, and express concern that the genetic characteristics of modified organisms may already be well established in some areas. At the international level, in 2000, the Cartagena Protocol, a supplementary agreement of the Convention on Biological Diversity (1993), included clauses to address the risks posed by the accidental release of living genetically modified organisms.

For more information see:

Greenpeace (2002) *Genetic Pollution: a Multiplying Nightmare*, viewed at http://archive.greenpeace.org/~geneng/ (accessed 27 March 2003).
Russo, V.E.A. and Cove, D. (1998) *Genetic Engineering: Dreams and Nightmares*, Oxford: Oxford University Press.
Willett, J.D. (2002) 'Biotechnology', in A.S. Goudie (ed.) *Encyclopedia of Global Change*, New York: Oxford University Press.

the human use of the environment in the form of such diverse activities as the harvesting of flora and fauna, waste disposal and ecotourism. They give values of between US$3 million and US$33 trillion per annum (Pimental *et al.* 1997), with the higher estimates exceeding the value of the current world economy (see Box 5.1, Environmental values).

In reality, not all of the various components of the environment participate in the world economy, and assigning values to many of them is complex, often arbitrary and essentially meaningless. Many minor organisms provide essential services such as the decomposition of organic material and the pollination of plants, or participate in biochemical cycles that control the gaseous content of the atmosphere and nutrient levels in the soil. Their

contribution is almost impossible to cost in economic terms, and as a result is often unappreciated. There are also environmental resources as yet undiscovered that would be lost if biodiversity is allowed to decline. Much has been made, for example, of the presence of plants in the tropical rain forest that have the potential to provide a multitude of pharmaceuticals some time in the future (Park 1992). With continued deforestation and the subsequent loss of biodiversity, a multi-million-dollar contribution to public health may be lost. Similarly, biodiversity can provide individuals and groups with the aesthetic pleasure associated with landscape and its associated flora and fauna. While some indication of the value of this can be obtained from tourism or recreation revenues, much of the benefit that people derive is mental or spiritual and as such not easily measured by contemporary monetary yardsticks. Thus assigning a monetary value to biodiversity and its components is not easy, but in a world dominated by considerations of profit and loss it is perhaps inevitable that, like any other commodity, biodiversity will be subject to economic analysis.

Tropical forests are renowned for their biodiversity, supporting half of the earth's species on only 6 per cent of the world's land area, but, as a result of human activities, it is the rain-forest biome that is suffering the most rapid extinction of plant and animal species, and its diversity is under serious threat (Miller 1994). Much attention has been focused on large mammals such as tigers and gorillas, but small mammals are threatened also, and insects, which tend to be more specialized in their requirements, are particularly vulnerable (Middleton 1999). In addition to the destruction of their habitat, some animals are threatened by hunting. Road networks set up to facilitate logging also allowed hunters into previously inaccessible areas. In Central Africa this has led to an expanding trade in bushmeat, initially to feed workers in logging camps, but also to supply the growing demand from urban areas. Some 33 million people live in the Congo basin and to meet their protein needs hunters take 5 million tonnes of bushmeat out of the forest every year (Pearce 2002a). Such a rate is unsustainable and is setting many species, including gorillas, chimpanzees, monkeys and the duiker, a small antelope, on the road to extinction. Bushmeat harvesting is also a problem in Southeast Asia, in Sarawak, for example, but less so in South America, in part because of the lower population in that area (Spinney 1998). Ironically, many of the species involved are on endangered lists and supposedly protected.

Many plant species have already been lost, and given the diversity of the forest it is possible that some species were wiped out even before they were identified. Ten species of palms have become extinct in Malaysia and about half the species identified on the Malaysian peninsula are endangered. More exotic plants, such as orchids, are being plundered by plant hunters and collectors (Aiken and Leigh 1995). Tropical forests have provided a range of important food crops such as bananas, coffee, sugar, rice, cocoa, pineapples and a variety of nuts. As a result of selective breeding and hybridization to improve such characteristics as growth rates and yield these plants have lost some of their genetic variation and are more vulnerable to damage by pests and disease than their wild counterparts. Should that damage be a major threat to a specific crop it might be necessary to go back to the tropical forest gene pool to find plants with disease or pest-resistant characteristics that could be introduced into the domesticated varieties. If that gene pool is depleted by the large-scale destruction of species the chances of success will be much reduced.

Another concern is the loss of medicinal plants. Antibiotics, heart drugs, hormones, tranquillizers, anticoagulant and cancer-fighting drugs have been derived from rain-forest plants, the active ingredients being first isolated, then produced synthetically. Perhaps as much as a quarter of the prescription drugs used in the United States have their origins in the rain forest, and the potential for the future discovery of drugs to fight cancer and AIDS would be significantly reduced by the continued destruction of the forest (Park 1992). Commercial sales of these drugs reach hundreds of millions of dollars every year, but it is often forgotten that in many cases they have been used in their natural form for years by indigenous peoples. Some 75 per cent of the world's population, including a high proportion in the rain forest, continue to rely on traditional medicine for health care (UNEP 2002). They would suffer directly from the loss of medicinal plants.

A more ecocentric approach to the consideration of biodiversity would pay less attention to socio-economic values and more to environmental values. If it is accepted that all the components of an ecosystem interact with each other to the mutual benefit of the entire system, then the components must share some common values. In making decisions about biodiversity it would seem reasonable to take these intrinsic values into account. Unfortunately, the nature and extent of the relationships between individual components of a specific ecosystem are not always well known and therefore any value they may have to each other, and to the community as a whole, is difficult to establish. The situation might be resolved by attempting to preserve all biodiversity without regard for socio-economic consequences, and that is the approach taken in nature reserves in many parts of the world. It is not a realistic option on a universal scale, however, and some reduction in biodiversity is commonly accepted as part of the price that has to be paid for progress in human development.

That pragmatic approach can be used to justify

considerable reduction in biodiversity, and while the growing acceptance of sustainable development has allowed greater accommodation of biodiversity, it is not always clear where the line should be drawn. From the point of view of humans and other large mammals, for example, insect pests and disease vectors appear to have no intrinsic value. What effect would their removal have on biodiversity? The smallpox virus was eradicated with no apparent negative effects (although its survival as laboratory samples has given rise to concern over its potential use as a weapon of bioterrorism) and the goal of public health authorities worldwide is to repeat the process with other insect, bacterial and viral pests, all of which contribute to biodiversity. The impact of species loss is far from clear, and although there are undoubtedly some key species that need to be preserved at all costs (UNEP 2002), there may be others that can be lost with little or no impact on the environment. Molecular genetic surveys of some sub-species of sparrows in Florida, for example, have shown that they are not evolutionary distinct and their loss would have little effect on biodiversity (Stiling 1992). Past experience with evolutionary change over millions of years suggests that, following the extinction of plants and animals, the biosphere adjusts to the new reality, and biodiversity recovers from the loss. Whether that would continue to apply under the stresses imposed on the environment by modern society remains to be seen, however, and allowing humans to choose the species that would not be missed might be a dangerous path to follow.

The conventional environmental wisdom is that any activity or event that causes the removal of a species from an ecosystem, or its replacement by some other species, poses a threat to biodiversity. Removal or replacement may be temporary or permanent, and although both have become increasingly common in recent decades, they are a fundamental part of any evolving ecosystem. The elimination of a species from an area as the result of such factors as environmental change, over-predation or disease is called extirpation. It is less serious than extinction – the elimination of all individuals in a species – since members of the species survive in other areas and may return if the conditions that led to their displacement change. Although this is a natural process, human intervention is often required for the successful reintroduction of an extirpated species. For example, wolves that had been wiped out in Yellowstone National Park in the western United States were reintroduced in the form of breeding groups captured in a similar habitat in Canada.

In some cases extirpation may well be a precursor of extinction, if it causes the numbers in a species to fall to a critical level where reproduction rates cannot maintain the population. The situation is generally more precarious for large mammals – K-strategists (see Chapter 3) – than smaller organisms – r-strategists – but when numbers reach that critical level the species becomes endangered (Figure 7.10). The total number of endangered species is not easily estimated, but the Convention on International Trade in Endangered Species (CITES) (see Box 7.4, CITES) lists some 3000 animals and about 24,000 species of plants sufficiently endangered that trade in them is either completely prohibited or strictly regulated (UNEP 2002). The World Conservation Union (IUCN 2003) lists 3500 vertebrates, 2000 invertebrates and 5700 plants in its Red List of threatened species. Large mammals such as elephants, tigers, whales and gorillas have received most attention, but smaller creatures from migratory birds to minnows and spiders have also been identified as endangered, and some species are at greater risk than others (Figure 7.11). Exotic plants, such as tropical orchids or the cacti of the south-western United States, also make headlines from time to time, when dealers, collectors or poachers are apprehended, but thousands of other plant species are endangered by everyday human activities such as agriculture and forestry. Some endangered species have been brought back from the verge of extinction – the whooping crane, the American bison, the peregrine falcon, for example; others are being preserved in zoos or wildlife refuges. The number of species under threat of extinction appears to be growing rapidly, but the methodologies used in sampling and analysis are not always scientifically rigorous – sometimes depending upon anecdotal information, for example – and estimates show considerable variation, allowing some observers to question the seriousness of the situation (see, for example, Lomborg 2001). The World Conservation Union estimates that 24 per cent of mammals and 12 per cent of birds are now threatened, including 180 animal and 182 bird species that are on the critical list (Hilton-Taylor 2000). Some of these may be keystone species, the loss of which might have a knock-on or cascading effect on biodiversity, whereas others might not be obviously missed if

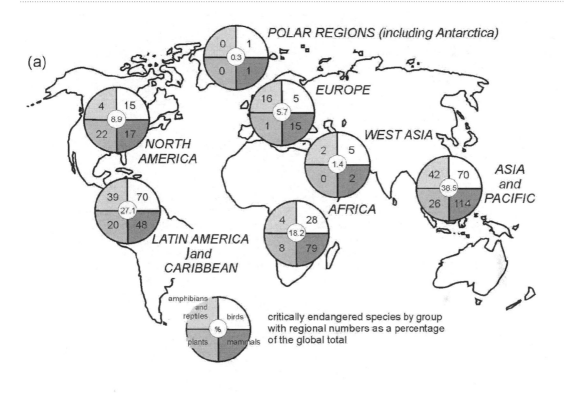

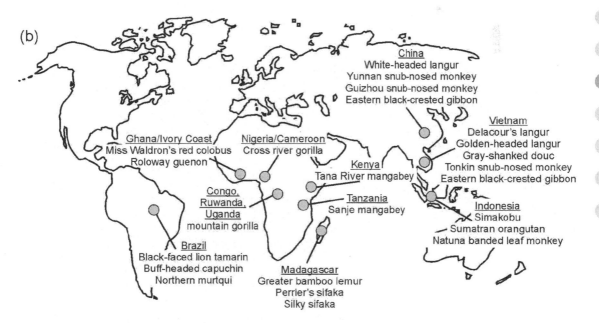

Figure 7.10   (a) Critically endangered species by region, (b) the twenty-five most endangered primates
Source: based on data in (a) UNEP (2002), (b) Hilton-Taylor (2000).

BOX 7.4 CONVENTION ON INTERNATIONAL TRADE IN ENDANGERED SPECIES OF WILD FLORA AND FAUNA

The Convention on International Trade in Endangered Species of Wild Flora and Fauna (CITES) emerged from discussions in the 1960s among members of the World Conservation Union (IUCN) who were concerned that the growing world trade in plant and animal wildlife was threatening the survival of some species. The convention was signed by eighty members of the IUCN in 1972 and on 1 July 1975 it came into force, with the aim of protecting species by restricting trade in live plants and animals and the products derived from them. Since it was first adopted, additional governments have signed up and now more than 160 states are party to the agreement.

CITES operates by assigning species to three groups, listed in appendixes, with the most threatened being in Appendix I. Commercial trade in these species is prohibited. Species in Appendix II are considered to be under threat if trade is not controlled, but some trade is allowed through the use of export permits. Individual countries may list species that are not necessarily threatened but for some reason require protection. They may be traded under permit and are grouped in Appendix III. Through this approach, CITES provides protection to some 5000 animal and 25,000 plant species.

Species may move between appendixes if their situation changes. Elephants, for example, were moved into Appendix I from Appendix II in 1989 when it became clear that they needed additional protection. The move was accompanied by a worldwide ban on ivory trading in 1990. Since then a number of African nations with well developed elephant conservation programmes have lobbied for a relaxation of the regulation and in 1999 Botswana,

Namibia and Zimbabwe were allowed to resume limited trade in ivory. Unfortunately, this was followed by an increase in ivory poaching and smuggling.

CITES claims that no species has become extinct as a result of trade since the convention entered into force, but it has not been problem-free. With so many countries involved, operating at different economic or bureaucratic levels and with variations in their capacity to regulate the trade, provisions of the agreement are not always enforced effectively. Some of the states involved have yet to pass laws to enforce all elements of the convention and many transactions continue to go unreported. Black market trade thrives in some species – those perceived to provide health benefits, for example – and although illegal import and export operations are regularly exposed, the underground trade is estimated to be large.

Despite these problems, CITES is generally seen as having a positive impact on species survival and attempts to improve it continue, with regular reviews of the situation, through standing committees and biannual meetings of the parties to the agreement.

For more information see:

CITES Web page, viewed at http://www.cites.org/eng/disc/what.shtml (accessed 10 June 2003).

DeSombre, E.R. (2002) 'Convention on International Trade in Endangered Species', in A.S. Goudie (ed.) *The Encyclopedia of Global Change*, New York: Oxford University Press.

Smith, Z.A. (2000) *The Environmental Policy Paradox* (3rd edn), Upper Saddle River NJ: Prentice Hall.

they were to become extinct. In the broadest sense, however, all species are integral components of the ecosystem in which they live, and the loss of even one species would have some effect – be it great or small – on the integrity of the system.

SPECIES EXTINCTION

With endangered species and those lost to an area through extirpation, there is at least some hope of recovery, but with extinction there is none. When a species becomes extinct it is effectively removed from the biosphere and cannot be replaced. It is

a natural process, brought about, for example, by the inability of a species to cope with changing environmental conditions or with increased competition from another species. It is also part of the Darwinian concept of 'the survival of the fittest', in which organisms that adapt to change survive, whereas those that do not adapt become extinct (see Chapter 1). Perhaps as many as 98 per cent of the species that ever existed on earth are now extinct, although many are still represented by their descendants. The modern horse (*Equus*), for example, is a descendant of the 'dawn horse' (*Eohippus*), which became extinct some 40 million years ago (Stiling 1992). Mass extinctions punctuate

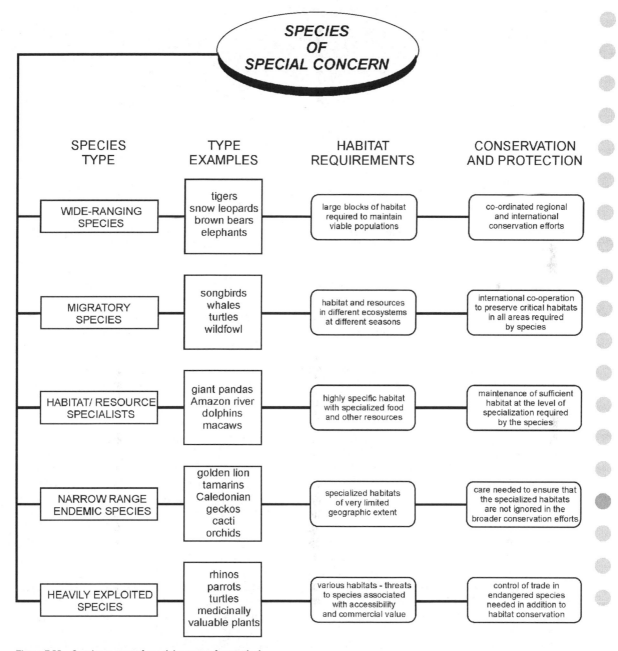

Figure 7.11   Species groups of special concern for survival
Source: based on WWF data. Viewed at http://www.worldwildife.org (accessed 2 July 2003).

the environmental history of the earth (Table 7.2), with some, such as the elimination of the dinosaurs at the end of the Cretaceous, giving rise to much speculation (Alvarez *et al.* 1980; Hecht 1997). The most recent of these episodes, in the late Pleistocene mega-fauna extinctions, may well represent the first significant contribution of human beings to the extinction of other animal species (see Chapter 1).

Allowing for such episodes when rates were higher than normal, the fossil record suggests

TABLE 7.2 MAJOR SPECIES EXTINCTIONS

Time	Cause	Results
c. 440 mya[a] end of Ordovician	Climate change? Severe cooling?	25% of species families lost, mainly marine
c. 365 mya near end of Devonian	Climate change?	19% of species lost
c. 250 mya end of Permian	Plate tectonics Volcanic activity Climate change	90% of species lost; 70% of terrestrial vertebrates lost; nearly 30% of insect species lost
c. 215 mya near end of Triassic	Meteorite impact? (Botswana/South Africa) Climate change	20% of all groups of species lost
c. 65 mya late Cretaceous	Meteorite impact (Yucatan peninsula) Volcanic activity? Climate change	17% of species lost, including dinosaurs and ammonites
10,000–12,000 ya[a] Pleistocene	Human hunting activities	Many large mammals – giant camels, woolly mammoths, for example – lost
Present/near future?	Human activities – habitat destruction, reduction in biodiversity, introduction of alien species	3000 mammal and 24,000 bird species threatened with extinction

Source: based on data in D. Challinor, 'Extinction of species', in A.S. Goudie (ed.) *The Encyclopedia of Global Change*, New York: Oxford University Press (2002); E. Eldridge, The Sixth Extinction (2001), viewed at http://www.actionbioscience.org (accessed 11 January 2003).
Note: [a] mya million years ago, ya years ago.

that background extinction rates for mammals and birds probably averaged about one species every 500 to 1000 years (May *et al.* 1995). Modern extinctions are much more frequent, but there is no widely accepted figure for the rate at which they are taking place. At one end of the spectrum, there is the estimate of 40,000 species lost every year, first postulated by Myers in 1979, and followed by others ranging up to between 100,000 and 250,000 per year (see, for example, Wilson 1992). These high estimates drove the demand for increased preservation of the earth's biodiversity in the latter part of the twentieth century, but they seem likely to be excessive. Modern interpretations of previously collected data suggest an extinction rate of 0.75 per cent since 1600, and based on recorded numbers of animals and birds, the current loss for all species has been calculated at about 2300 per annum (Heywood and Stuart 1992). Model calculations suggest an extinction rate of 0.7 per cent over the next fifty years, which is about 1500 times the natural background rate, but much less than the values at the other extreme, which could in theory cause the complete extinction of all the earth's known species over the same time period. While there may be no wide agreement on the rate of extinction, it is generally accepted that it accelerated in the last few decades of the twentieth century, and there is growing support for attempts to conserve species and habitat.

PROTECTING BIODIVERSITY

The protection and preservation of biodiversity can take several forms. There are government and non-government organizations, for example, that have identified species in need of protection, individual species being chosen for reasons that include declining numbers, uniqueness, ecological function and aesthetic appeal, with the last often being emphasized in fund raising. Elephants, mountain gorillas, tigers, whales, polar bears and giant pandas have all been targeted, as have the peregrine falcon, osprey and spotted owl, but less photogenic organisms such as insects and fish, despite being endangered, receive less attention,

Plate 7.3   The breeding of endangered animals, such as the tiger, in captivity has allowed the preservation of the species, but their numbers continue to decline in the wild.

except perhaps when their presence interferes with economic development. Such was the case with the snail darter, a small endangered fish whose presence was used by environmentalists to stop the construction of a dam on the Tennessee River in the 1970s (Stiling 1992). The success of this approach has been mixed. The overall populations of African and Asian elephants have experienced continuing rapid decline since the 1970s, because of loss of habitat through the expansion of agriculture and, particularly in Africa, as a result of hunting and poaching for ivory (Cohn 1990). The worldwide ban on trade in ivory put in place by the Convention on International Trade in Endangered Species (CITES), plus greater emphasis on conservation, reduced the pressure on the species and in some protected areas in Africa local elephant populations have stabilized or risen, requiring occasional culling to keep numbers within the carrying capacity of the habitat.

Despite protection programmes set up as early as the 1970s throughout Southeast Asia, tiger populations are under continuing threat from habitat loss and poaching for skins and body parts used in oriental medicine. The tiger population in Malaysia has dropped from 3000 to fewer than 600 in the past fifty years despite being a protected species for a large part of that time. Recent attacks on humans have further threatened the tiger's status by raising suggestions that they should be shot on sight (Pokar 2002). The South China tiger may well be close to extinction, with as few as twenty to thirty animals thought to be surviving in the wild (WWF 1998). Despite the best of intentions, attempts to protect the giant panda, the mountain gorilla and other species have met with similarly limited success. In some cases the population of threatened species has become so low that active programmes to increase the numbers have been set in place. These have met with some success. The Arabian oryx, thought to have become extinct in the wild in 1972, was successfully reintroduced into Oman through the breeding and release of animals in captivity in the 1980s (Middleton 1999).

The reintroduction of threatened bird species has been accomplished in North America. By the early 1940s the whooping crane population, which had once been widespread in North America, had dropped to only fourteen or fifteen individuals. In an attempt to save the species, refuges were created in the nesting and wintering areas used by the birds, and a captive nesting colony was established. Most whooping cranes lay two eggs, with only one chick usually being raised successfully. To increase that rate, a cross-fostering programme was developed in which one of the two eggs was placed in a sandhill crane colony, where it was hatched and the chick raised as part of the colony. Despite some setbacks, the whooping crane population is now about 150 birds and the programme to help them increase continues (Cox 1993). A similar programme was developed for the California condor. By the early 1980s the species was almost extinct, numbering only twenty-seven individuals. All were taken into protective captivity and an active breeding programme was initiated in an attempt to save them, special care being taken to ensure that they did not become socially over-dependent on humans. There are now 200 condors, half of them released into the wild, where some have begun to breed naturally

again. The aim is to have 450 birds by 2020, 300 of them in the wild and the remainder to act as a controlled breeding population (Kaplan 2002). Elsewhere around the world threatened species are being bred successfully at zoos and animal sanctuaries, but their reintroduction into the environment is not always as successful, often because the environment has changed or they have become so habituated to humans that they can no longer cope with the wild or interact with members of their own species.

PARKS, PROTECTED AREAS AND PRESERVES

The problems faced in the preservation of individual species may reflect the impact of continued habitat reduction along with poor management of some programmes, but they may also point to a basic flaw in the approach to conservation. Since, in theory, the various elements in an ecosystem work together for the mutual benefit of all, there may

Plate 7.4 Wildlife sanctuaries such as this example in Australia provide a means of preserving species and also serve an educational function. (Courtesy of Karen Armstrong.)

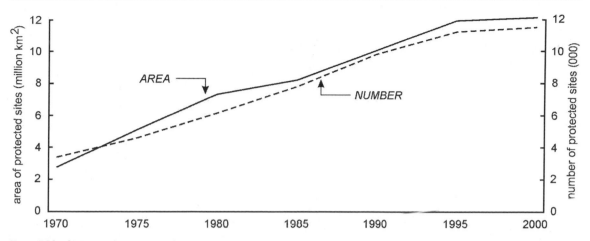

Figure 7.12 Global number and area of protected sites
Source: based on data in UNEP (2002).

be advantages in attempting to maintain the quality of the entire ecosystem rather than concentrating on a specific part of it. Indeed, in a well preserved ecosystem there should be no need for a particular species to be targeted since the development of the natural balance characteristic of such systems should allow all species to survive at a level appropriate to the carrying capacity of the system.

The creation of parks and protected areas is an example of this approach and at first sight it appears to have been very successful, at least in terms of the total area protected (Figure 7.12). Between 1970 and the end of the twentieth century, areas under some form of protection increased from 3 million km^2 to 12 million km^2, including 167 natural heritage sites preserved under the World Heritage Convention (UNEP 2002). There is considerable variation at the national level, however, with New Zealand, Austria, Bhutan and the United Kingdom having more than 20 per cent of their land area under some form of protection, for example, whereas others, such as Uruguay, Fiji and Greece, protect less than 2 per cent and some nations in Africa, such as Equatorial Guinea, protect none (Middleton 1999). Even these figures are a bit misleading and give a false impression of the level at which biodiversity is being protected. Many wilderness areas, nature reserves and parks have been created for aesthetic, political and socio-economic reasons, with conservation factors secondary (Wood 2000). Inaccessible mountainous areas, high-latitude locations with abundant rocks and ice or isolated deserts have often been altered

little by human activities and appear to be ideal for preservation, but, in terms of biodiversity, such areas are not particularly rich and certainly do not compare with the diversity offered by temperate grassland or tropical rain forest, which tend to be underrepresented in any listing of protected areas.

Lost in the figures for the total protected area worldwide is the size of individual sites. The world's largest protected area, at 972,000 km^2, is the Greenland National Park, but most are much smaller. With a total of about 11,500 protected sites covering an area of 12 million km^2, the average size is only 1059 km^2 (UNEP 2002), and sites that size, no matter how well managed, are often too small to maintain viable populations of some species, or the biodiversity expected in protected areas. The optimum size for reserves in the Amazon basin, for example, has been estimated at 5000 km^2 and for the African savanna grassland perhaps as much as 10,000 km^2. In temperate areas, viability may be possible with an area of only 500 km^2, but at present even areas that appear quite extensive may meet only 20–30 per cent of these size requirements (Cox 1993). Other factors, including the shape and fragmentation of protected areas, as well as their proximity to each other, can help to determine their viability (Figure 7.13). Management and enforcement practices as well as the original motivation for the establishment of the reserve – to conserve maximum biodiversity or protect a threatened species, for example – can also have an impact on the success of the site.

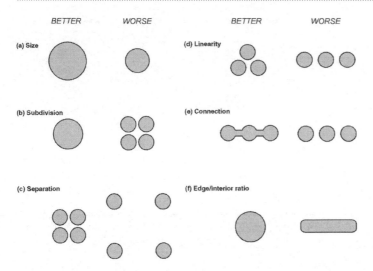

Figure 7.13   A comparison of the factors involved in the design of protected areas

Source: G.W. Cox, *Conservation Ecology: Biosphere and Biosurvival,* Dubuque IA: Wm C. Brown (1993), with permission of the McGraw-Hill Companies.

Theoretical studies have indicated that the likelihood of developing a viable reserve is increased if the site is large and as continuous as possible rather than small and fragmented. Round or blocky reserves appear to be more successful than those that are long or irregularly shaped, since the edge effects, which diminish the quality of the habitat along the boundaries, are reduced. If individual reserves are small and fragmented, distances between them should be small, and if possible some form of corridor should be in place to allow animals to move easily from area to area (Brown and Lomilino 1998). All of that is fine in theory, but in practice conservationists have to work with what is available, and many reserves do not meet these theoretical conditions. That does not mean that they are necessarily unsuccessful. Willingness to invest time and money in promoting a site, coupled with good management and enforcement once the reserve has been established, can bring success, as some of the small prairie grassland reserves in the Great Plains have shown. Other areas may start out as ideal in terms of size, shape and homogeneity but lose these attributes through pressure from economic development or population growth. In South America, for example, protected areas established in the tropical rain forest regularly succumb to the demand for resource development and governments that are unwilling to enforce the legislation in place to protect them (Park 1992). In parts of Africa the pressure comes from a growing population, inexorably encroaching on supposedly protected land, in search of the food they need, either through hunting or clearing the land for agriculture (UNEP 2002).

If a protected area is to be allowed to remain as natural as possible, then it might be expected that management should be minimal, aimed mainly at allowing the various components of the environment to interact as they would normally do in an undisturbed setting. This might include the prevention of poaching and hunting or the removal of trees and other plants, but in some areas the level of interference has become controversial. Wildfires, insect infestations and floods are all natural disturbances from which ecosystems recover naturally with little long-term damage to biodiversity. The human response, however, is to fight them. Particularly in North America it has become normal to fight wildfires in parks, with the aim of preserving the natural environment. In fact there is evidence that such an approach may well be detrimental to conservation (Woods 2000). During the long, hot, dry summer of 1988, for example, fires burned in Yellowstone National Park in the western United States for more than four months, destroying perhaps as much as 500,000 ha of forest. Fire suppression policies in Yellowstone over the previous thirty to forty years were identified by some observers as contributing to the size and intensity of the fire by allowing the build-up of flammable material in the forest. If the natural wildfires characteristic of the area had been allowed to burn they would have

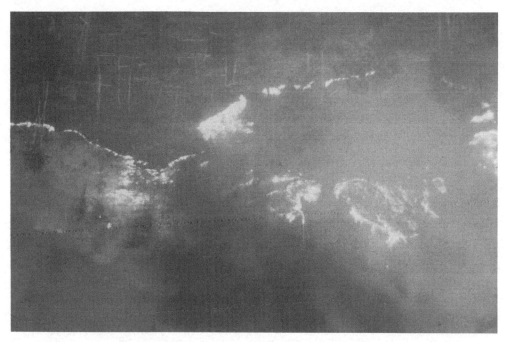

Plate 7.5 Fire is a natural process in many ecosystems, but the modern tendency is to fight all fires and that can lead to problems such as the massive fires that swept through Yellowstone Park in 1988.

removed that fuel periodically, reducing the potential for a fire on the scale of that in 1988, and consequently causing much less environmental damage. The role of fire suppression in conservation became a matter of serious debate following the Yellowstone fires, and initiated a re-examination of the best response to natural disturbances in protected areas (Romme and Despain 1989).

SUMMARY AND CONCLUSION

It is clear that there is worldwide concern over current levels of habitat destruction and the loss of biodiversity. At the same time, there is also evidence of a desire to tackle these problems before they cause irreparable environmental disruption, but the best approach to achieving this has yet to be agreed upon. Disagreement on the level of human intervention that is required or should be allowed, for example, is only one of a number of factors that remain unresolved when consideration is given to the maintenance and protection of the natural environment. In a society driven by a combination of technology and socio-economic issues, weighing the costs and benefits of any activity is important.

In dealing with issues of habitat and biodiversity, it is often difficult to establish the number of organisms involved and their relationship to each other, far less their value to society. That in turn makes it difficult to attract funding for action to conserve and protect them. Environmental organizations are very active in attracting funding through private donations, but the size of the problem is such that it is unlikely to be solved without government involvement. Although there is an abundance of environmental legislation worldwide dealing with issues of habitat and biodiversity (Table 7.3), it is often bypassed or ignored when it appears to block development or economic advance. This is particularly so in the developing world, where nations face the dilemma of trying to develop economically, to meet the needs of populations often deprived of the basic necessities of life, without completely destroying their environment in the process. It is also clear, however, that developed nations are prepared to sacrifice environmental principles if it is economically expedient, as attempts to permit oil exploration in the Arctic National Wildlife Refuge have shown. Such ambivalence is not unusual. Despite many international conferences, commissions and reports by experts on the importance of balancing the needs

TABLE 7.3 EXAMPLES OF INTERNATIONAL ENVIRONMENTAL AGREEMENTS THAT PERTAIN TO HABITAT AND BIODIVERSITY

Year	Agreement
1971	Convention on Wetlands of International Importance (Ramsar Convention)
1973	Convention on International Trade in Endangered Species of Wild Flora and Fauna (CITES)
1979	Convention on the Conservation of Migratory Species of Wild Animals (CMS)
1980	Convention on the Conservation of Antarctic Marine Living Resources (CCAMLR)
1984	Law of the Sea Convention
1985	Tropical Forest Action Plan (TFAP)
1986	Moratorium on commercial whaling
1991	Antarctic Environmental Protocol
1992	Statement of Forest Principles
1993	Convention on Biological Diversity (CBD)
1995	Agreement on Straddling Fish Stocks and Highly Migratory Fish Stocks
1996	Convention to Combat Desertification
2000	Cartagena Protocol on Biosafety
2001	Stockholm Convention on Persistent Organic Pollutants (POPs)

of society with those of the environment, in the final analysis the latter is often ignored. While this applies to all aspects of the environment the effects on the living organisms in the biosphere are often most obvious. Until society can decide on the value it wishes to place on the world's flora and fauna, and chooses to afford the environment the attention that it gives to socio-economic issues, that situation is unlikely to change and the threats to habitat and biodiversity will remain.

SUGGESTED READING

McKee, J.K. (2003) *Sparing Nature: The Conflict between Human Population Growth and the Earth's Biodiversity*, Piscataway NJ: Rutgers University Press. A thought-provoking, sometimes depressing, account of the human impact on the environment, through the numbers and demands of a growing population.

Peterson, D. (2003) *Eating Apes*, Berkeley/Los Angeles/London: University of California Press. A graphic account of the threat to endangered apes in Central Africa as a result of the growth in the bushmeat trade.

Wilson, E.O. (2001) *The Diversity of Life*, London: Penguin. An overview of biodiversity, how it developed and how it is threatened, written by the doyen of modern ecologists.

Goldsmith, B. (ed.) (1998) *Tropical Rain Forest: A Wider Perspective*, London: Chapman and Hall/New York: Kluwer. Comprehensive, interdisciplinary account of tropical rain forests, from the nature of the forests to their inhabitants – human and non-human – and their preservation through sustainable forest management.

QUESTIONS FOR REVISION AND FURTHER STUDY

1   A large area of the tropical rain forest has been incorporated as a protected area. Imagine what your response would be if you were (a) the president of an international lumber company that held cutting rights in the area, (b) a landless peasant who had hoped to clear part of the forest to support his family and (c) a deep ecologist.

2   The distribution of fish depends upon such natural factors as water temperature, the availability of food or the presence of predators. Fish are completely unaware of political or administrative boundaries. What problems does that create when attempts are made to protect stocks from overfishing?

3   The preservation of biodiversity is the goal of environmental groups worldwide. At the same time, the goal of many public health groups is to repeat their success in eradicating smallpox with other insect, bacterial and viral pests, all of which contribute to biodiversity. Is it possible for an environmentalist to reconcile these two approaches?

8

Threats to the Availability and Quality of Water

After reading this chapter you should be familiar with the following concepts and terms:

acid mine drainage	groundwater consumption	phosphates
aquifer depletion	hardness	pulp and paper industry
Aral Sea	intensive livestock farming	salt-water intrusion
aridity	London Dumping Convention	septic systems
BOD	mercury contamination	subsidence
consumptive use	municipal water supply	tanker accidents
continental-scale water	NAWAPA	wastewater treatment
engineering	non-consumption uses	water conflicts
domestic sewage	non-withdrawal uses	water pollution
drought	Ogallala aquifer	water quality indicators
fertilizer contamination	oil pollution	water softeners
fish farming	oxygen sag curve	water transfer
flooding	pathogens	water wastage
grey water	pH	withdrawal uses

Water is one of society's essential resources and always has been, whether society has been primitive or technologically advanced, developed or developing. It is required to meet the basic biological needs of all plants and animals – including the human animal – and is used by society in agriculture, industry, transport and recreation. Since the earth/atmosphere system is closed, the amount of water that it contains is effectively constant, although it changes state and location quite regularly, and the same water must be used again and again. This is made possible by the hydrologic cycle, which cleans and recycles the water constantly (see Chapter 2). Although the hydrologic cycle is very efficient, it is not perfect, and natural variations in availability and quality occur from time to time. Environmental checks and balances normally deal with these variations, but human interference can aggravate them and create serious environmental problems. Society has interfered with all sectors of the hydrologic cycle, damming rivers or changing flow regimes on land, attempting to increase precipitation in the atmospheric sector and causing the oceans to warm and sea level to rise, for example, but it is in the run-off and groundwater sectors that the greatest disruption has taken place, with agricultural, industrial and domestic demands on the system causing local, regional and even continental-scale disruption of the availability and quality of water.

WATER USE

Although water use takes many forms, all can be categorized into withdrawal or non-withdrawal uses. For non-withdrawal uses the water remains in place, usually in its natural location, as in the case of the sea, river or lake, for example. In some cases the location is modified, where rivers have been canalized or dock facilities have been constructed, but the water remains in these locations during its use. Common non-withdrawal uses include recreation – swimming, fishing and boating, for example – and transport by boat or barge, but waste disposal is also a non-withdrawal use of water. Although such activities do not remove water from its location, they may reduce the water quality as a

(a)

(b)

Plate 8.1   Different uses of water. (a) The North American marina with its yachts and powerboats is a recreational facility. (b) The boats in Hong Kong form a floating village where people live year-round.

result of pollution. That is most obvious in the case of waste disposal, but most forms of water transport also pollute, sometimes massively in the case of oil tanker accidents, for example.

Withdrawal of water from its source may be for consumptive or non-consumptive use. In the latter case, water is removed for only a short period, when, for example, it is used to produce hydroelectricity or for cooling purposes in an industrial plant before being returned to the river or lake from which is was taken. With some non-consumptive uses, the water is recycled several times before being returned. Consumption occurs when the water is not returned to its source after it has been used, or returned in a form in which it is no longer usable. This may occur as a result of evaporation or transpiration during use, the absorption of water by plants and animals or the inclusion of the water in finished products such as soft drinks, or household and chemical products. Water that has been contaminated during use is also considered to have been consumed, even if it is returned in its entirety to its source. Some industries can involve both consumptive and non-consumptive use of water. In agriculture, for example, water that is lost through evapotranspiration, or water retained by plants and animals, is consumed, but irrigation water that is returned to the source by way of run-off or by seepage back into the groundwater system is not consumed. The withdrawal of water from the hydrologic cycle increased sixfold during the twentieth century, driven by population growth, industrial development and the increasing use of irrigation in agriculture. Industrial development was an important component of the growth, and in the United States in that time period, combined to some extent with changes in lifestyle, it caused a tenfold increase in demand during a period when total population increased only fourfold. Irrigation was also instrumental in raising the demand for water, with the irrigated area at the end of the century being five times what it had been at the beginning (Gleick 1998). At the end of the twentieth century society was using about 54 per cent of the accessible run-off in the hydrologic cycle and by 2025 it is expected that the figure may be as high as 70 per cent (Postel *et al.* 1996).

Much of the increased demand will come from the developing world, driven in large part by the need to deal with the absence of clean drinking water and poor sanitation, which together cause millions of deaths every year (Uitto and Biswas 2000). In many industrial nations, demand has already peaked. The increasing costs of constructing water supply infrastructure, the environmental costs of diverting and consuming more and more of the world's water and the development of new technologies that have allowed more and more water to be recycled have combined to improve efficiency in the use of water and as a result demand has declined (Gleick 1998). Despite this, total water use is still dominated by the developed world, with 12 per cent of the population responsible for 85 per cent of the world's water consumption (Barlow and Clarke 2002).

WATER AVAILABILITY

Although the earth has an abundance of water, the freshwater sector on which society depends is only 1 per cent of the total (see Chapter 2). Furthermore, the distribution of fresh water in and upon the earth's surface is very uneven, with the amount available subject to seasonal, annual and longer-term variations. The different moisture regimes that have been created as a result influence the distribution of plants and animals and contribute to the development of local, regional and continental-scale landscapes. Like other animals dependent upon water, early human beings tended to congregate in these areas where it was readily available. In areas where the supply was reduced seasonally, early humans led a nomadic existence, following the rains, which supported the plants and animals on which they depended for food. When societies ultimately became more sedentary, the availability of water remained important and it is no accident that the earliest civilizations were water-based, dependent upon the great rivers of the Middle East, India and China (see Chapter 5). The need for water grew as human society became technically more sophisticated. Agricultural communities gravitated to the flat, well watered flood plains of river valleys – despite the potential dangers of flooding – fishing villages grew up along rivers and lakes, industrial centres that used the power of flowing water were attracted to the rapidly flowing rivers and streams in the hills. Water also provided a means of communication and transport where movement overland was difficult.

The uneven natural distribution of water sometimes restricted development where it could not meet the demands of an ever-growing population and to counter these restrictions modern society has spent – and continues to spend – massive amounts of time, effort and money altering that distribution to make water more uniformly available. Thus any consideration of water availability and its environmental impact must include not only natural elements, but also the role of human interference in the system. The scarcity of fresh water is a growing problem in many areas, despite ongoing improvements in the technology of water supply and distribution. The issues involved vary from region to region, but include drought, supply enhancement, urbanization and political conflict, which are often interlinked and often driven by progress in economic development. At times, water availability problems involve excess water rather than scarcity and flooding occurs. Flooding is a natural part of the aquatic environment, which has been aggravated by human interference in the system and causes serious problems for society in both the developing and the developed world.

DROUGHT

Certain parts of the world suffer from aridity, a permanent lack of water, caused by low average rainfall, often (but not always) in combination with high temperatures. The deserts of the world are permanently arid, with rainfall amounts of less than 100 mm per year and evapotranspiration rates well in excess of that amount. Aridity is therefore a permanent feature of the climatology of these areas. In contrast, drought is more often a temporary feature, occurring when precipitation falls below normal or when near normal rainfall is made less effective by other weather conditions such as high temperature, low humidity and strong winds. Although it is temporary, its onset is often unexpected, its duration and intensity are difficult to predict, and organisms often find it more difficult to deal with drought than with aridity. Having plagued mankind for hundreds of years, drought is an environmental issue of long standing and some of its more serious impacts are considered in greater detail in Chapter 9.

WATER SUPPLY ENHANCEMENT

With water being an essential resource for human development, many early settlements grew up in locations where water was readily available. Despite the apparently abundant water supply available when the settlements were founded, with time and growth the demand for water began to exceed supply in some of them, and it became necessary to augment the supply from sources other than precipitation or the existing surface run-off systems. In many cases, water scarcity did not grow with demand, but was there from the beginning. The pull of a resource other than water – precious metals such as gold, for example – was sometimes so strong that concern for the water supply arose only after the settlements were established. Elsewhere, settlers such as the homesteaders who opened up the Great Plains in the nineteenth and twentieth centuries were deluded into thinking that the water supply was adequate to meet their needs, only to find that climate variability in subsequent years left them open to serious water shortages (see Chapter 9). In all of these situations, extra water had to be found to make up the deficit, and in most cases it was, being sought in the transfer of water from areas of plenty via aqueducts and pipes or in the groundwater supply. Successful as these approaches were and continue to be, they often cause acute environmental disruption, plus a mix of economic and political problems.

The transfer of water over greater and greater distances has long been a feature of human development. The Mesopotamians led water from the Tigris and Euphrates to irrigate their fields, but the Romans were the first to engineer aqueducts that could transport large volumes of water over long distances to meet the needs of their cities. At least nine served Rome alone and Roman aqueducts still survive in Spain and southern France. By the end of the nineteenth century, most large cities were piping at least some of their water supply over distances of 100 km or more. The power of a valuable resource to generate its own water supply is well illustrated by the gold-mining towns of Kalgoorlie and Coolgardie in Western Australia. Located in the desert with no local water available, each town was supplied with water delivered in a 550 km long pipeline built between 1898 and 1903 from the Helena River near Perth (Beckinsale 1971). The value of the gold covered the cost of building the pipelines and supplying the water. Considered in their day to be masterpieces of water engineering and pipeline construction, the distances covered by these pipelines are no longer unusual. By 1913 the city of Los Angeles in southern California was

Plate 8.2 An early example of water transfer. This is the Devonport Leat, built in the 1790s to channel water from Dartmoor to the docks at Devonport in Plymouth, twenty-four miles away.

importing water from a reservoir in the Sierra Nevada some 400 km away and has gradually added other long-distance sources – such as Lake Mead on the Colorado River, also more than 400 km from the city – as the demand for water has increased. The removal of fresh water from these rivers and lakes has created serious problems of salinization. Lake Owens, one of the first reservoirs for Los Angeles, has lost so much water that it is now mostly salt flat (Miller 1994) and Mono Lake, some 500 km north of the city, has had so much of its inflow diverted that its level is now 12 m lower than it was in 1941 when it became part of the Los Angeles supply system. It is also hypersaline and exhibits a very low level of biological activity. A citizens' group fought the deterioration of the lake for more than fifteen years and eventually was successful in having the lake's inflow increased, although it will take perhaps another twenty years before there is a significant improvement in the ecology of the water body (Nebel and Wright 2000). The demand for water for domestic and agricultural purposes remains high in California, being administered through the California Water Plan, which includes 1100 km of pipelines, tunnels and aqueducts, bringing water from the wetter north and the

adjacent mountains to the drier land of the central valley and the south of the state.

The conditions that afflicted the environment in and around Mono Lake were created for many of the same reasons, but on a much larger scale, in the Aral Sea basin. Occupying a drainage basin in arid Central Asia, on the border of Kazakhstan and Uzbekistan, the Aral Sea was once the fourth largest inland lake in the world, covering 64,000 km^2, but it is now much depleted. In the 1960s, when the area was still part of the Soviet Union, major irrigation schemes were developed to increase cotton production. This involved diverting most of the water entering the Sea through the Amu Dar'ya and Syr Dar'ya river systems. By the end of the twentieth century, the declining amount of water entering the Aral Sea had caused it to shrink by more than 30 per cent and created what is considered by many environmentalists to be one of the world's greatest environmental disasters (Brown 1991) (Figure 8.1). A fishery that once provided large catches of carp, perch and sturgeon has ended because the waters are now too salty and shallow. Salts, along with toxic pesticides and herbicides that drained off the irrigated land, blow from the exposed sea bed to contaminate the surrounding

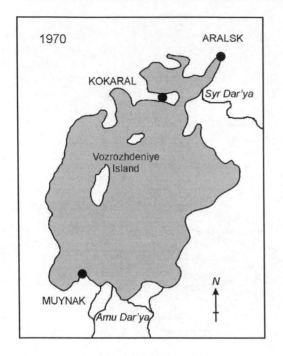

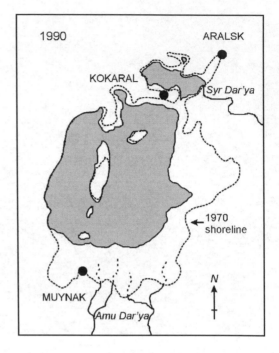

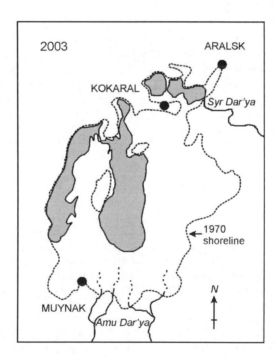

Figure 8.1 The shrinking Aral Sea, 1970–2003

Source: drawn from satellite photography at the Earth Observatory Web page, courtesy of NASA. Viewed at http://earthobservatory.nasa.gov/ (accessed 20 May 2003).

land, and cause respiratory and other health problems for the local inhabitants. Proposed rehabilitation schemes include reduction of the amount of water being diverted from the rivers flowing into the basin or diversion of water from rivers such as the Irtysh that flow into the Arctic. The task is a mammoth one, and given the economic climate in the states of the former Soviet Union, neither approach seems likely to be implemented in the near future (Stone 1999).

Continental-scale water transfer

The proposed diversion of the Siberian rivers into the basins of the Aral and Caspian Seas was only one of a number of continental-scale water engineering schemes given serious consideration in the 1960s to try to meet the growing global demand for fresh water (Figure 8.2). All involved the interbasin transfer of large volumes of water over more than a thousand kilometres. In Africa, a combination of drought and increased withdrawal of water for irrigation has brought about the gradual shrinking of Lake Chad to less than a tenth of what it was half a century ago (Coe and Foley 2001). The transfer of water northwards from the Zaire (Congo) as a potential solution to the scarcity

of water in the Chad basin has been considered since the 1960s, but no serious attempts have been made to take it beyond the concept stage (Simons 1971). The scheme with perhaps the greatest potential for success was the North American Water and Power Alliance (NAWAPA), first proposed in 1964 to divert Canadian rivers flowing into the Arctic Ocean, southwards into the United States by means of a continental-scale network of reservoirs, canals, aqueducts and pumping stations (Figure 8.3). The diverted water was to be used to flush pollution out of the Great Lakes – St Lawrence system and the Mississippi River, but the main purpose of the scheme was the provision of water for irrigation and municipal supply in the arid west and south-west of the United States (Bryan 1973). Although local and regional interbasin diversions have taken place – in the James Bay area of Quebec, for example, and between Colorado and California – there has been no large-scale transfer of water. Even the relatively small-scale transfers that have already taken place cause environmental disruption, however, and the potential damage from the continent-wide interbasin transfers envisaged in these schemes is massive. Significant changes in the hydrologic cycle would occur from the Arctic to Mexico, local and micro-climates would be distorted, stream morphology would be altered

Figure 8.2   Continental-scale water diversion proposals: (1) Canadian Arctic to south-western United States, (2) Russian Arctic to Kazakhstan, (3) Tibet and Three Gorges to northern and western China, (4) Zaire to Lake Chad

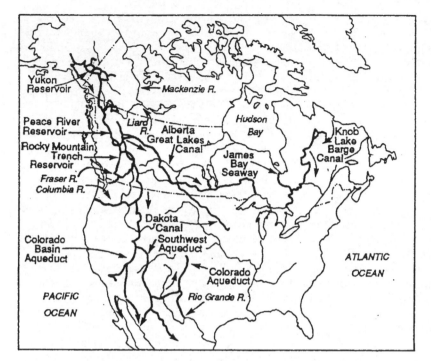

Figure 8.3   The North American Water and Power Alliance
Source: after Healey and Wallace (1987).

because of changes in the rates of erosion and deposition and species transfer and habitat change would have a substantial impact on the aquatic community. Although none of these grandiose schemes has come to pass, future demands for water may see them resurrected. Canadian environmentalists, for example, are concerned that, under the North American Free Trade Agreement (NAFTA), Canada will be unable to prevent the export of fresh water south to the United States and Mexico or the detrimental environmental consequences that would inevitably follow (Crowell 1997).

In China, long-distance interbasin transfer of water has gone beyond the planning stage, with work being implemented in stages to pump water from the middle and lower reaches of the Yangtze River some 1500 km to the north into the Huang He (Yellow River) basin to supply agricultural and industrial needs in the North China Plain. The filling of the lake behind the Three Gorges dam, built across the Yangtze, began in mid-2003 and when it is complete it will allow that diversion to take place. There are also plans to divert water north from tributaries rising on the Tibetan plateau in the upper reaches of the river (Edmonds 1994).

Exploiting groundwater

The groundwater supply is another source of water frequently exploited in those areas where the surface water is unable to meet community needs. Between 1950 and 1990 groundwater consumption in the United States tripled (Owen and Chiras 1995) and the growth in demand is growing globally. Ground water originates as precipitation and percolates down into the pore spaces and cracks in the aquifers that lie beneath the earth's surface. The upper limit of groundwater saturation is the water table (see Chapter 2). Ground water is part of the hydrologic cycle, moving slowly under the influence of gravity and sometimes returning to the surface naturally – through springs, for example. Increasingly it is pumped from wells and boreholes for human use, with modern pumping systems being so efficient that the withdrawal of ground water from aquifers easily exceeds recharge rates and the water table declines (Figure 8.4a). The ground water currently being pumped from aquifers is often the product of precipitation, which fell hundreds or even thousands of years ago in amounts that exceed current values, and modern recharge rates cannot

(a)

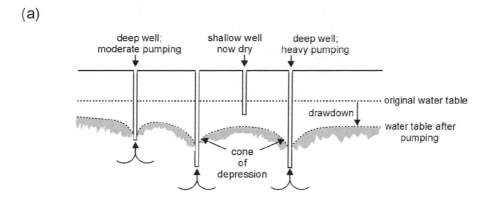

(b)

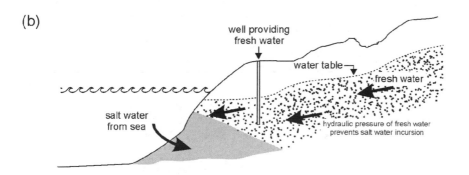

(c)

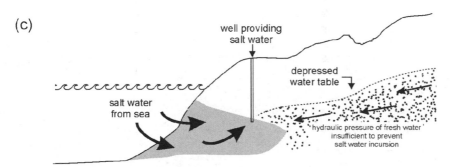

Figure 8.4 (a) Impacts of pumping in the water table. (b) The relationship between fresh water and salt water in the groundwater system in coastal areas. (c) Salt water intrusion associated with excess extraction of fresh water in coastal areas

keep up with the demand on the water supply. In some places the difference between withdrawal and recharge is so great that the ground water is effectively a non-renewable resource and its removal has been referred to as groundwater mining, while the problem may be compounded by the pollution of the groundwater system (Figure 8.5).

Aquifer depletion is a widespread problem, which can have acute local impacts, when a farm or residential well runs dry, for example, or can be felt by millions of individuals when demand draws the water table down over a large area. Currently ground water is being pumped at unsustainable rates in the Great Plains of North America, northern China, North Africa, the Middle East and parts of Mexico. In places the impact is already serious and there is a potential for major water shortages in these areas in the near future if nothing is done to lower the demand or provide water from some source other than the groundwater sector. The world's largest known aquifer, the Ogallala aquifer, which underlies more than 450,000 km^2 of the Great Plains in the United States, is a fossil aquifer containing an estimated 4.01 trillion m^3 of water (Kromm and White 1992) most of which has been in storage since the last Ice Age. It is currently being pumped at a rate that is eight to ten times as fast as it is being recharged, and in the thinner southern parts of the aquifer the withdrawal rate may be as much as 100 times the natural recharge. As a result, the water table has dropped rapidly – an average of 3 m between 1940 and 1980, but less than a metre since then – and it has been estimated that by 2020, one-quarter of the aquifer's original storage will be depleted (Miller 1994). Huge volumes of water remain in the aquifer, but as the water table falls wells have to be deepened, pumping time increases and the cost of water increases. This has caused a reduction in withdrawal rates in some areas, but during the droughts that are common on the Plains the normal response is to pump more ground water and the rates increase again. More than 90 per cent of the water withdrawn from the Ogallala aquifer is used for irrigation, and wastage tends to be high. Government policies contribute to the waste through the subsidization of crops that consume

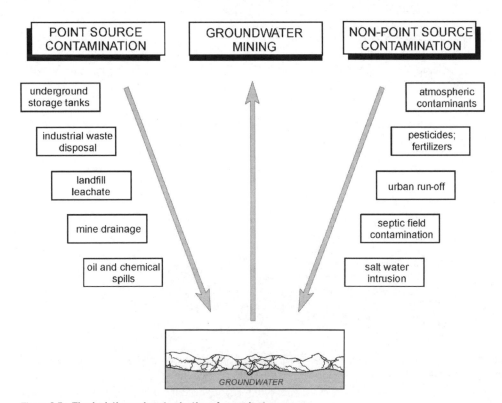

Figure 8.5   The depletion and contamination of groundwater resource

large amounts of water and through water depletion allowances that compensate for the costs of the
falling water table (Miller 1994). If the depletion
of the aquifer is to be slowed, such wastage and the
government policies that tend to encourage it will
have to be reversed.

Similar problems are experienced in other areas
where surface water is in short supply and ground
water is essential to meet local needs. In northern
and western China the declining availability and
quality of surface water have increased pressure
on the groundwater sector. The city of Beijing, for
example, uses 4000 million tonnes of water per year,
but the annual precipitation provides only 800
million tonnes and the difference has to be supplied
from ground water. The result is a declining water
table that causes wells to dry up and emergency
measures have to be introduced to meet essential
needs. In Inner Mongolia, to the north of Beijing,
the water table has dropped by 20 m and in Urumqi,
in the arid west, by 10 m since 1949 (Edmonds
1994).

Subsidence associated with groundwater extraction

In addition to the direct depletion of a resource,
groundwater use is associated with additional
problems. The removal of water from the pore
spaces and cracks in an aquifer with no recharge
creates the potential for subsidence, in which
the land sinks, sometimes gradually, sometimes
catastrophically. Rates vary, but settling rates of
between 10 cm and 15 cm per year are not uncommon, resulting in cumulative subsidence of several
metres in many places. In parts of the San Joaquin
valley in California, for example, the land has sub-
sided by as much as 9 m since it was first settled,
because of groundwater removal (Nebel and Wright
2000). The problem is aggravated in urban areas,
where the weight of the buildings helps to increase
the rate of sinking. In some parts of Mexico City
the land is subsiding at a rate of 40 cm per year
and it is estimated that during the twentieth century
the total subsidence was 7.5 m, causing extensive
damage to buildings, water and sewage pipes
and other elements of the city's infrastructure (Tortajada-Quiroz 2000). Subsidence is common in
most areas where groundwater withdrawal rates
are high, from Arizona, Texas and Florida in the
United States to Tokyo and Osaka in Japan, and
Bangkok in Thailand, which in the early 1980s was

sinking at a rate of 14 cm per year. There is evidence
that a reduction in groundwater removal slows the
rate of subsidence, and in areas where the decline of
the water table has been slowed subsidence has also
decreased. In areas where surface water is limited or
unavailable, however, the demand for ground water
will ensure that subsidence will continue.

Salt-water intrusion

A problem peculiar to coastal areas, in which
ground water is the main source of water, is salt-
water intrusion. Under normal circumstances the
natural gravity flow of fresh water towards the
oceans, plus the lower density of the fresh water,
will allow fresh ground water to extend to the
coast and even beyond. The natural pressure of the
groundwater system prevents the salt water from
intruding into the aquifers. Under strong pumping
the water close to the coast can be removed faster
than it can be replenished from inland and salt
water moves in to fill the void (Figure 8.4b–c); wells
become contaminated and are no longer suitable for
domestic or industrial water supply. In some cases,
if pumping is stopped, the aquifer will recharge
naturally or it may be recharged by pumping fresh
water down into the wells, but in either situation
it will take some time before the system recovers
sufficiently to allow normal use.

URBANIZATION

The rapid growth in population that characterized
the twentieth century (see Chapter 4) brought with
it a significant increase in the demand for water.
Since much of the population growth took place
in urban areas the demand was concentrated and in
some places exceeded supply.

Water supply problems

Attempts at enhancing the supply included the long-
distance transfer of surface water or the increasing
use of ground water, each of which has created
environmental problems. The provision of water
for urban areas in the developed world has been
made relatively easy because of the availability of
financing to provide new infrastructure or upgrade
existing water supply systems, but even there

unexpected increases in demand or the onset of unusually dry conditions can lead to water rationing. Among the developing nations, the inability to finance the necessary improvements in the water supply has contributed to scarcity in urban areas, and large sectors of the world's fastest-growing cities lack an adequate supply of water. In Indian cities such as New Delhi, Mumbai (Bombay) and Calcutta water is supplied for only three and a half to seven and a half hours per day for those connected to the municipal water supply (Biswas 2000) and the situation is similar in parts of north China. In the squatter settlements that are characteristic of the peripheries of most large cities in the developing world there is commonly no piped supply, and water has to be carried from public standpipes or from adjacent surface-water sources, which are often highly polluted. In most of these cities demand easily exceeds supply. The demand for water in Mumbai, for example, is of the order of 3530 million litres per day (mld) whereas the supply system can provide only 2915 mld and the shortfall is likely to continue at least until the end of the first decade of the twenty-first century (Sagane 2000).

Wastage of water

A major contributing factor to water scarcity in some cities is wastage in the supply network. This can include such elements as leakage from broken or corroded pipes or lack of maintenance. Poor management practices such as the widespread use of taps or standpipes also allow higher wastage rates. In Amman, the capital of Jordan, where water is a scarce commodity, leakage from the city's distribution system ensures that losses reach 50 per cent (Salameh and Bannayan 1993) and in Oaxaca, Mexico, the loss rate is estimated to be of the order of 60 per cent, with the norm for large cities in the developing world between 30 per cent and 40 per cent (Biswas 2000). Although losses in the developed nations tend to be less, many cities there are using water supply infrastructure that was constructed in the nineteenth century and is only now being updated. A goal of system losses between 7 per cent and 12 per cent is probably feasible, but beyond that the cost of additional reductions would likely exceed the cost of the water saved (Biswas 2000). The success story of Tokyo, Japan, shows very well what can be done to minimize losses. In 1945 war damage and lack of maintenance meant

that 80 per cent of the water supply was leaking out of the system. Fifty years later losses had been reduced to 9.9 per cent (Takahasi 2000). This was made possible by Japan's rapid economic recovery during that same period, and few cities in the developing world have such resources available. Developed nations have frequently provided financial aid to construct additional water delivery infrastructure, such as dams, pipelines and boreholes, in developing nations, but in many cases the renovation of the existing system to reduce wastage might be more cost-effective than transporting new water from distant sources or from aquifers beneath the city.

REDUCING DEMAND FOR WATER

One of the problems with water scarcity, particularly in urban areas, but also in rural areas where water is used extensively for irrigation, is that solutions are commonly sought on the supply side of the issue. If there is insufficient water, for example, the obvious solution is to look for additional sources to make up the deficit. Another approach is to look at the demand side and to reduce the deficit by reducing demand. In environmental terms, reducing the demand for water is usually more effective than trying to increase supply. Already in many developed countries, a combination of education, regulation and pricing is being used to reduce water consumption, both at the domestic and at the industrial levels. The use of low-volume flush toilets or low-flow shower heads by individuals may be brought about through environmental education or through pricing policies, but the net effect is a reduction in demand, which ultimately impacts positively on the aquatic environment. In industry the incentives are usually economic and can be very effective. Over the last three decades of the twentieth century, for example, when Japanese industrial output was reaching its peak, water use declined by 25 per cent (Postel 1997).

The reduction in demand in the developed world would not have been possible without accurate metering of water use and pricing policies that have forced consumers to give serious consideration to their levels of water consumption. In many developing countries, however, there is little or no metering of water use, the price of water to the consumer remains remarkably low and management is minimal. A study of water utilities in Asia and the

Pacific, carried out for the Asian Development Bank in the late 1990s (McIntosh and Yniguez 1997), indicated that some major cities there have no working system for metering water use, and even where it has been introduced it covers fewer than half the connections. Monthly water bills in the area are commonly less than US$1.00, which along with poor metering encourages excess consumption and high wastage rates. The presence of poorly trained and inexperienced operators or managers plus interference by politicians aggravates the situation. In contrast, well managed water utilities, such as those in Hong Kong, Singapore and Kuala Lumpur, have much higher tariffs, provide 100 per cent coverage, twenty-four hours per day and have much lower rates of wastage and excess consumption (Biswas 2000). Although the costs may initially appear prohibitive for many developing nations, the restructuring of the financing and management systems in urban water utilities has been shown to provide attainable and effective results, and would seem to be an appropriate target for foreign aid, from which consumers, the hydrologic cycle and the wider environment would all benefit.

One approach to dealing with the problems of water supply that is being promoted vigorously in some areas is privatization, which would see water become increasingly a commodity bought and sold for profit by private companies rather than a public good controlled by public utilities. In theory the resulting increase in price would encourage more efficient consumption, thus reducing stress on the water supply system. Where privatization has occurred, however, in both the developed and the developing worlds, it has created additional problems and spawned growing opposition from environmental and development aid groups (see Box 8.1, Globalization and privatization of water).

CONFLICT, GEOPOLITICS AND WATER USE

The natural distribution of resources seldom fits the patterns of human activity that have arisen out of the development of society. The result of this is conflict – sometimes physical, sometimes political – that increases as the commodity in dispute becomes scarce. This is particularly so with water, and access to water has been a source of friction in some areas for hundreds of years (Figure 8.6). As the demand for water has grown, the conflicts have increased. Conflicts can take place at several political levels, including disputes between cities, between states or provinces under federal jurisdiction as in Canada and the United States or on a larger scale between

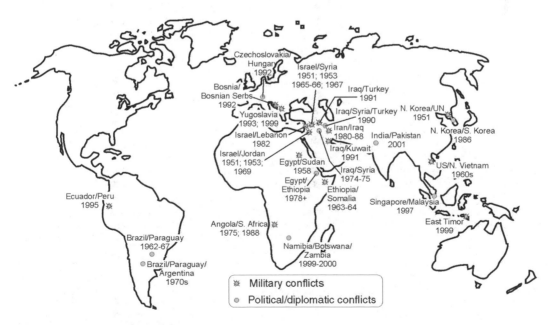

Figure 8.6 International conflicts involving water
Source: based on data in Gleick (1994).

BOX 8.1 GLOBALIZATION AND PRIVATIZATION OF WATER

Water has been regarded traditionally as a common or public good and in the UN Universal Declaration of Human Rights access to clean, safe drinking water is one of the basic human rights. In the past it was a free good, available to anyone who could dip a bucket in a stream or lake, dig a well or collect rainwater, and that may still apply in some areas. In the developed world, however, there is a price for supplying water, normally based on the cost of distributing it rather than on the value of the water itself. The water is provided by a public utility at local government level, at prices governed by the cost of operating and maintaining the system. In the last decade of the twentieth century, however, that structure was increasingly challenged by a move towards privatization, through which water would become an economic good subject to market pressures and international trade, under the management of international companies.

Privatization is an integral part of the globalization of world economics, driven by such international organizations as the World Trade Organization (WTO), the International Monetary Fund (IMF) and the World Bank, through global agreements such as the General Agreement on Tariffs and Trade (GATT) and the General Agreement on Trade and Services (GATS), as well as hemispheric agreements such as the North American Free Trade Agreement signed by the United States, Canada and Mexico. The privatization of water would be considered under GATS.

THE CURRENT STATE OF WATER PRIVATIZATION

Privatization has already taken place in some countries, most notably Britain, where public Water Boards have been replaced by a number of British and international companies. The results have been far from stellar, with the change-over followed by complaints about inadequate supply, poor water quality and price gouging, plus the conviction of several companies on pollution charges. Some of the British companies have become involved in international privatization schemes in the developing world, along with companies from France and the United States, but there too success has been mixed, at best, and the failures have reinforced the concern of those who oppose privatization. In Bolivia, for example, one of these multinationals attempted to force customers to buy permits for collecting rainwater and increased water rates by US$20 per month

(over one-fifth of the minimum wage in the region). After violent public protests the company cancelled its contract. In Buenos Aires, privatization brought a 20 per cent increase in water rates, putting water out of reach of poor families, and the company compounded the problem by failing to renew the sewage system, which had been part of the contract. Similarly, in Manila, rate increases of 50 per cent were introduced following privatization, but little was done to provide the twenty-four-hour water supply promised under the contract.

At a different level, the booming trade in bottled water – often only refiltered tap water – controlled for the most part by the same transnational organizations involved in running municipal supplies, is another indication of privatization. The global aspect of water trading is well illustrated by proposals for the bulk transfer of water on an international scale, from areas such as the Arctic. Water that flows into the ocean unused by society is sometimes considered as wasted, but it is of course intimately involved in the working of the natural environment in the area. As a result of such proposals that indicate lack of concern for the environment fears have risen among nations with an abundance of water that they may be forced to supply fresh water to other areas at the expense of the local environment and society. Canadian environmentalists, for example, have concerns that Canada will lose control over its own water as a result of NAFTA and be forced to sell it southwards to the United States.

WATER AS A SOCIAL OR ENVIRONMENTAL GOOD

Many of the real and perceived problems associated with the privatization of water arise from the fact that too little attention is paid to the fact that water is more than an economic good. It is also a social good that includes important cultural and environmental elements, which are not always represented adequately in market forces. Companies in business to earn a profit for their shareholders cannot afford to be altruistic, and social, cultural and environmental issues that threaten profit margins are unlikely to receive the attention they deserve.

Initial responses to privatization have been mainly negative – sometimes violently negative – but it need not be without benefits. The introduction of a price that exceeds the existing price for water may appear initially negative, but if set at an appropriate level, it

BOX 8.1 – continued

could encourage more efficient use of the water to benefit both people and environment. Infrastructure improvements to cut profit-reducing wastage would reduce supply problems in many areas. The extension of the supply network would give more people access to a safe and reliable water supply, particularly if linked to an appropriate waste disposal system. All these benefits are as naught, however, if people are too poor to afford the cost. If they cannot pay for the water, they do without and have to resort to supplies that are inadequate or polluted. That is a major concern among some groups opposing privatization – multinational companies make a profit from a public good, but large numbers of the public continue to be without access to that good. If privatization is to succeed, it must consider such issues from the viewpoint of the consumer as well as the supplier. It appears that little has been done to ensure that, but, if it is to help deal with the world's water problems, privatization must include not just economic considerations, but also those involving social and environment issues as well. This is unlikely to happen without some form of government regulation to ensure the quality, adequacy and universality of the supply. The dire financial state of many

developing nations leaves them vulnerable to concessionary pressures from multinational companies and it would seem necessary to have an international regulatory body to deal with these issues. The haste with which privatization has been presented has meant that social and environmental issues have received inadequate attention, and until the risks and limitations associated with this omission are adequately investigated, privatization is more likely to aggravate the world's water problems than to alleviate them.

For more information see:

Barlow, M. and Clarke, T. (2002) *Blue Gold: The Fight to Stop the Corporate Theft of the World's Water*, Toronto: Stoddart.

Friends of the Earth (2001) *Stealing our Water: Implications of GATS for Global Water Resources*, viewed at http://www.foe.co.uk/resource/briefings/gats _stealing_water.pdf (accessed 13 June 2003).

Gleick, P.H., Wolff, G., Chalecki, E.L. and Reyes, R. (2002) *The New Economy of Water: The Risks and Benefits of Globalization and Privatization of Fresh Water*, Oakland CA: Pacific Institute for Studies in Development, Environment and Security.

individual nations. They often arise as a result of the occupation of a river basin by a number of separate political units, with those occupying the headwaters of the basin having much more control over water supply than those near the downstream end of the system. The most serious current water disputes in the Middle East, Africa and the Indian subcontinent are associated with such situations in internationally shared river basins. Disagreements can also arise along river systems that are contained within one country, however, as has been the case in the western part of the United States. Although disputes over the sharing of surface water are the most common source of conflict, problems also arise with aquifers that cross the borders between adjacent nations, as in the Middle East between Jordan and Israel.

Water disputes in Africa, the Middle East and Southeast Asia

The most serious problems over water with the potential to contribute to regional and international

conflicts occur in the basins of the Nile, Jordan, Tigris–Euphrates and Indus. All these rivers are in arid areas where water is naturally scarce and the demand is growing as a result of rapidly increasing population, and all are shared by a number of nations with different views on political and economic development. Ironically, these are also the areas in which water played a major role in the development of the first civilizations. The survival of Egypt is utterly dependent upon the river Nile, yet the sources of the water it carries are upstream in Ethiopia and East Africa, and it must pass through Sudan before entering Egypt. Both Ethiopia and Sudan suffer from water scarcity and would benefit from the impoundment and diversion of some of the water that now flows on into the Mediterranean Sea, but that would have an impact on Egypt's water supply. During the second half of the twentieth century disputes over water have flared up between Egypt and Sudan or Egypt and Ethiopia and have been resolved mainly by political or diplomatic means, but tension in the region remains high (Gleick 1998).

In the Jordan valley and adjacent parts of the

Middle East conflict over water is a long-standing issue (Gleick 1994; Kliot 1994), but modern concerns over water scarcity have been overshadowed by the political and religious conflicts that have plagued the area increasingly over the past half-century. Both Israel and Jordan developed plans in the 1950s to divert water for irrigation from the Sea of Galilee and the northern tributaries of the river Jordan, with some success, but with little co-operation, development being accompanied by ongoing political disputes that also involved Syria and led to border clashes or military action on a number of occasions. Since 1967, following the occupation of the Golan Heights, Israel has been in a position to control the headwaters of the river Jordan and the supply of water to irrigation projects in Jordan. In addition, the growth of Israeli settlements in the West Bank, also occupied at that time, has created pressure on the underground water supply. The settlements use much more water than was traditionally consumed by the Palestinian population of the area. Most of the water is being withdrawn from aquifers at rates that exceed recharge, creating the potential for even greater water scarcity in the future. Lack of water, surface and underground, is a severe constraint on agricultural production and economic development, and any resolution of the political disputes in the area will have to include provision for the sharing of the water supply (Hillel 1994).

To the east, in Mesopotamia, the Tigris and Euphrates have met the needs of Syria and Iraq for thousands of years and although there have been disputes between the two states – over the reduction of the flow of water in the Euphrates by Syria, for example – the potential for conflict is centred on Turkey's control of the headwaters of both rivers. Turkey has long-range plans to transport and sell water to Saudi Arabia and the Gulf states, by building pipelines from the upper reaches of the Tigris and Euphrates, and already has numerous dams in place which provide hydro-electricity and irrigate large areas of south-east Anatolia (Miller 1994). Turkey has already shown that it has the ability to restrict the flow of both rivers into Syria and Iraq, and without appropriate agreements such restrictions could have serious political and economic consequences. Environmental impacts are already apparent in southern Iraq, where the reduction in the flow of the Tigris and Euphrates by dam construction in Turkey has contributed to the destruction of large areas of ecologically important marshland in the lower reaches of the rivers (Pearce 2001).

The river Indus and its tributaries were integrated into a major irrigation scheme almost a century ago, when the Indian subcontinent was under British rule. Following independence in 1947, the basin came to be shared by India and Pakistan. Through most of its length the Indus flows through Pakistan, and on the way it provides water for the largest irrigated area in the world, but its headwaters are in India, and along with its tributaries it flows through the disputed territory of Kashmir. This gives India effective control of the water in the river basin and the ability to reduce the supply of water for the irrigation schemes in the Pakistani desert or to create floods by allowing excess water to flow through the system. Both nations signed a treaty in 1960 to share the water in the system, but in response to continuing unrest in Kashmir, India has indicated that it might pull out of the treaty, leaving Pakistan exposed to serious economic loss and the potential for widespread famine (Pearce 2002b). Any action by India would no doubt lead to the escalation of the current conflict between the two nations and since both possess nuclear weapons the consequences could be disastrous.

Water disputes in North America

Water conflicts are also common in politically less volatile parts of the world. In the drier west of the United States, for example, disagreements over the agricultural use of water have arisen since the region was first settled, and in more recent times urban, industrial and environmental issues have contributed to conflict. Disputes also arise from the claims of various indigenous groups to specific water rights. Problems are particularly serious in the Colorado River basin, although agreements among the various states using the basin have allowed for the allocation of water along the course of the river. Despite this, in many years the total amount allocated exceeds the annual river flow and the states at the lower end of the basin suffer shortages and are left with water that has a high level of salinity. The Colorado River water conflicts also have an international element, since the river crosses the Mexican border close to its seaward end, by which time it is reduced to a trickle of highly saline water (Reisner 1993). Water disputes are also common on the Great Plains, despite agreements among the states involved (Owen and Chiras 1995). Again the problem is often one of over-allocation of resources, which is aggravated during drought years.

Along its northern border the United States shares many waterways with Canada. In order to prevent conflict the International Joint Commission (IJC) was established in 1909, with the specific purpose of preventing disputes over the use of waters shared by the two countries. Initial concerns involved changes to the natural flows and levels of boundary waters, but there was also provision for the monitoring and prevention of pollution. The commission has been quite successful in dealing with the former, but less so with the latter, and during the first half-century of its existence water pollution became a major problem, particularly in the Great Lakes basin. Water quality issues are now the major concern of the commission. Despite the generally amicable relationship between Canada and the United States in issues involving water, there are fears in Canada that growing scarcity in the United States will create greater demands on Canadian water and lead to conflict (Heathcote 1997).

FLOODING

Although the scarcity of water creates serious problems in many parts of the world, there are times when the amount of water available in an area exceeds the capacity of the environment to handle it and flooding occurs. Although floods are natural events in many areas, human intervention regularly disrupts the system, with the result that flooding causes damage costing millions and takes hundreds of lives every year (Table 8.1). It can occur as a result of the accumulation of large amounts of precipitation over a major drainage basin over a period of several weeks, as in the case of flooding associated with monsoon rainfall, or it may take place more rapidly as a flash flood following an unusually severe convective storm, for example. In either case, there is usually a time lag between peak precipitation and the peak flood discharge (Figure 8.7). Flooding is most common in river valleys, caused by excessive run-off, but it is also a feature of coastal areas of lakes seas and oceans. A river floods when its channel is incapable of carrying the volume of water added to it and the excess spills over on to the adjacent flood plain.

Heavy and prolonged precipitation, snowmelt, channel constrictions, dam failures and alterations to drainage basins may produce or contribute to flooding, either singly or in combination. Intense, prolonged precipitation will cause flooding, particularly if the ground is already saturated, but the flooding may be aggravated by urban development or deforestation within a drainage basin, both of which lead to an increase in the rate of run-off.

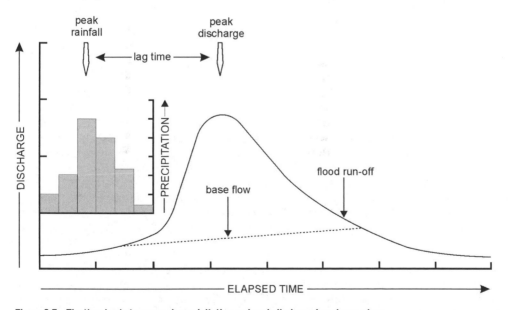

Figure 8.7 The time lag between peak precipitation and peak discharge in a river system

TABLE 8.1 DISTRIBUTION, CAUSES AND COSTS OF FLOODING

Year	Location	Deaths	Damage (US$)	Cause
2002	Eastern Europe and Russia	100+	billions	Heavy rain; dyke failure
2002	China: Yangtze River	1000	?	Monsoon rains
2002	India, Nepal, Bangladesh	800	?	Monsoon rains
1999	India: Orissa	10,000	?	Cyclone
1998	China: Yangtze River	4000	?	Monsoon rains
1997	Southern China, Hong Kong	?	?	Monsoon rains; typhoons
1997	Germany, Poland: river Oder	100?	3 billion	Heavy rain; poor dyke maintenance
1997	USA, Canada: Red River	<50	1 billion approx.	Spring snowmelt
1996	Canada: Sagueney River	10	750 million	Torrential rain; dam and dyke collapse
1993	Western Europe	<10	120 million	Heavy rain
1993	USA: Mississippi River	50	10 billion	Spring snowmelt
1991	Bangladesh	125,000	?	Cyclone: storm surge
1988	Bangladesh	2000	?	Monsoon rains
1982	Peru	2500	?	Torrential rain; El Niño
1973	USA: Mississippi River	11	1.2 billion	Rain; snowmelt
1972	USA: Black Hills, South Dakota	242	163 million	Torrential rains; flash floods
1970	Bangladesh	2,000,000	?	Cyclone; storm surge
1963	Northern Italy	2000+	?	Dam overtopped
1953	Northern Europe	2000+	?	Storm surge; North Sea
1938	China: Huang He	<1,000,000	?	Destruction of dykes by military action
1931	China: Yangtze River	145,000+	?	Prolonged rain
1928	USA: Florida	2400	?	Hurricane
1911	China: Yangtze River	100,000	?	Monsoon rains
1889	USA: Johnstown, Pennsylvania	2000+	?	Dam burst

Source: compiled from various sources.

Similarly, in many areas snowmelt causes annual spring flooding, with the extent and duration of the flooding increased when ice accumulation in the river channel constricts the flow. This is common in Canada and Siberia, where snow in the headwaters of north-flowing rivers begins to melt before the lower reaches are ice-free and able to cope with the increased flow. The Red River flood in North Dakota and Manitoba in 1997 had this origin and caused damage amounting to US$5 billion before it receded (IJC 2000). Because of the time lag involved between the melting and the arrival of the flood peak there was little loss of life, but property damage was high.

Floods are a natural part of the hydrologic cycle and contribute to both erosion and deposition. In human terms, however, they are seen as a serious hazard to life and property, and human responses to flooding reflect that. The most obvious response is to provide protection against it by building embankments or barriers, such as the levees that line major flood-prone rivers like the Mississippi. The diversion of rivers and the straightening or deepening of their channels allow them to carry more water and therefore reduce the amount that spills over on to the flood plain. At the other end of the scale, the response may include acceptance of flooding plus some form of adjustment to minimize the impact. Zoning by-laws, for example, may prevent the use of a flood-prone area for residential development, or require floodproofing of buildings or structures so that flood damaged is minimized. Acceptance of flooding in this way is usually combined with flood frequency analysis and emergency measures and procedures to warn people and secure structures when flooding is imminent.

A modern approach to flood problems is to consider the integrated nature of the entire drainage basin rather than only the area prone to flooding. This recognizes that activities allowed in one part of the basin may have serious consequences elsewhere in the system. The channelization of one section of a river, for example, may cause flooding downstream where the existing channel may be unable to accommodate the new flow regime. Similarly a change in land use near the headwaters of a drainage basin may well have consequences well downstream. Some researchers have blamed deforestation of the upper reaches of the Ganges and Bhramaputra rivers in the foothills of the Himalayas for major flooding problems in Bangladesh, where the rivers combine to flow into the Bay of Bengal, although it may not be the sole cause (Ives and Messerli 1989). Such circumstances, however, suggest that flood prevention requires the management of an entire drainage basin rather than just the areas that are obviously prone to flooding.

The availability of water varies naturally. Where it is not readily available, because of low levels of precipitation or high evaporation rates, for example, and the demand is high, scarcity becomes a problem. There may be times and places, however, when water is physically present, but in a state in which it cannot easily be used by society. Such a situation arises when water becomes polluted and its use depends more on its quality than its availability.

WATER QUALITY

Water quality is under threat everywhere. Even in remote areas the chemical properties of water are being changed by acid precipitation (see Chapter 10), pollution by domestic sewage is ubiquitous, and although the release of industrial effluents into water bodies is subject to increased control it remains a serious problem. As well as deteriorating as a result of the slow but steady addition of contaminants, water quality can also be reduced catastrophically as a result of major chemical spills, oil spills or the accidental release of untreated sewage. These problems and the standards that have been developed to deal with them are usually considered in human terms, but water quality standards are also applied to agriculture, fisheries and certain industries.

In human terms water quality is not an absolute concept but varies according to the proposed use of the water. Water intended for irrigation or for certain industrial purposes would not have to meet the same quality standards as water intended for drinking, for example. There are certain factors that can be considered indicative of water quality, and which can be grouped as physical, chemical or biological indicators (Table 8.2).

PHYSICAL INDICATORS OF WATER QUALITY

Pure water is a colourless, odourless liquid that is a compound of hydrogen and oxygen (H_2O). Natural fresh water is never pure, however, and that is often apparent in its physical properties. Colour is often the most obvious of these properties. Rivers that receive much of their water from areas of peat bog or muskeg are often brown in colour, for example, whereas those originating in meltwater from glaciers may be white or greenish in colour. Abundant iron in the bedrock through which a stream flows may also cause the water to become red or brown. In such cases the quality of the water may not be seriously impaired, but any addition of colour to the water raises aesthetic concerns that may in turn contribute to anxiety about the quality of the water. The colour of many water bodies is a product of the sediments they carry. The numerous 'Red Rivers' around the world and the Yellow River (Huang He) in China, for example, owe their names to the red or yellow colour of the silts and clays that they contain. These rivers have high turbidity levels, and while their water might be used directly for irrigation, it usually requires some form of filtering before it is suitable for industrial or domestic use. The presence of bacteria and other microscopic organisms can also cause surface water to become turbid, or cloudy, and introduce poisons into the system. This is particularly so with algal blooms, which are not only aesthetically unpleasant but cause a serious loss of quality through the toxins they introduce into the water during their growth and decay.

From a human point of view, taste and odour are important indicators of water quality. Taste and odour are caused by the presence of solids and liquids of biological, mineral or human origin dissolved or incorporated in the water, and the sensitivity of human tastebuds and noses to many of these products means that even very small amounts

TABLE 8.2 INDICATORS OF WATER QUALITY

Physical properties	Chemical properties	Biological properties
Colour Not necessarily harmful; aesthetic concern; streams flowing through peat often brown, for example	**Acidity and alkalinity** Fresh water tends to be neutral (pH 6–7); deviation from that may indicate contamination	**Micro-organism content** Presence of algae and bacteria can make water unusable; *E. coli* bacterium indicates faecal contamination
Turbidity Cloudiness caused by sediment or bacteria; may require treatment such as filtering before use	**Hardness** Caused by the presence of dissolved calcium and magnesium compounds; prevents soap from lathering and causes scale build-up in boilers	
Taste and odour Presence of dissolved solids of biological, mineral or human origin; very small amounts of some chemicals make water unpalatable	**Dissolved oxygen** Important for biological and chemical processes; levels indicated by chemical and biochemical oxygen demand	
Temperature Influences the dissolved oxygen content and the rate at which biological and chemical reactions occur		

Source: compiled from various sources.

of some chemicals make water unpalatable. Water that has a taste or smell that makes it unsuitable for human use, however, may still be used for some agricultural or industrial purposes. The taste and odour of water can vary naturally according to local environmental conditions. Groundwater that has been in storage for many years, for example, can taste of iron or smell of sulphur because of the minerals that it has incorporated in its time underground. Seasonal differences in the amount of organic matter incorporated in surface water may make it unpleasant for a period of time, particularly when the by-products of decaying vegetable matter are incorporated in the water. Increasingly, problems with taste and odour are caused by human activities. Even very small amounts of petroleum products can contaminate large volumes of water, remaining detectable even after they have been diluted to levels of less than one part per million (ppm). Ironically, one of the most common methods of ensuring that drinking water is safe to consume – chlorinization – imparts a flavour to water that is unpleasant to some consumers. Temperature is also a property that has a role in water quality, mainly indirectly, however, through its influence on oxygen levels in the water and the

rate at which chemical and organic processes take place.

CHEMICAL INDICATORS OF WATER QUALITY

The chemical composition of fresh water reflects the rock types, sediments or vegetation with which the water has been in contact, the biochemical processes in which it has been involved and sometimes the deposition of materials from the atmosphere above it (atmospheric loading). As a result the chemical content of water varies considerably from place to place and from time to time. There are certain factors that all water bodies have in common, however, and these give an indication of potential problems with its quality.

Acidity

The hydrogen ion concentration or pH (potential hydrogen) of water is a measure of its acidity or alkalinity. pH is measured on a logarithmic scale of 0 to 14 on which 7 is neutral, acidity increases from 7 to 0 and alkalinity increases from 7 to 14. Fresh water is normally only slightly acid (pH 6–7) although it can cover a wider range depending upon local conditions. In areas of acidic bedrock such as granite, for example, or where it is in contact with peat, surface water can be many times more acidic – pH 4–5 perhaps – whereas in areas of limestone or chalk bedrock, rich in calcium or magnesium carbonate, the pH of the water supply may be considerably higher. Once the normal pH of the water in an area has been established, any deviations may indicate contamination and therefore a reduction in quality. The acidification of lakes in North America and Europe in the 1950s, 1960s and 1970s associated with acid precipitation caused a substantial decline in water quality over large areas (see Chapter 10), but the falling pH in these lakes was only one indication of the deterioration. Accompanying it was a change in chemical content, with levels of calcium and magnesium declining, to be replaced by elevated sulphate concentrations and an increase in the amounts of potentially toxic metals such as copper and aluminium (Brakke *et al.* 1988). For aquatic organisms, such as fish, this represented a serious threat to survival and the impact was disastrous in many areas. In human terms, the impact of increasing acidity is less direct and much more difficult to identify. Aluminium or heavy metals such as mercury, copper, cadmium and zinc, liberated from the soil or rocks by acid water, may eventually reach the human body via plants and animals in the food chain or through drinking water. The corrosion of storage tanks or distribution pipes by acidified water can also add metals to drinking water: the liberation of lead from lead piping or from solder on copper piping is a particular concern. Although individual doses in such cases would be small, regular consumption might allow the metals to accumulate to toxic levels.

Hardness

Water in areas where the bedrock is rich in calcium and magnesium compounds are much better equipped to deal with increased levels of acidity, but the presence of these compounds has other implications for water quality. Compounds such as calcium and magnesium carbonate cause hardness in water, which can create problems when it is used for certain purposes. Hard water does not lather well, for example, but creates soap scum, which may be an inconvenience for many domestic uses, but can be costly in an industrial setting. The deposition of the calcium and magnesium salts when the water is boiled produces scale on everything from kettles to industrial boilers, reducing their efficiency and adding to costs. In areas where hard water is the norm, water softeners are often used to remove the calcium and magnesium. The most common method involves cation exchange, which is similar to the process of leaching in soils (see Chapter 3). Water softeners contain sodium, usually in the form of brine, and when hard water is passed through the system the calcium ions are replaced by sodium ions. When the exchange capacity of the softener is full, reversing the process can recharge it. Brine pumped through the system, for example, will cause the captured calcium ions to be replaced by sodium ions and the softener will be ready for use again. The decision to use a softener or not illustrates the way in which the concept of water quality depends upon its end use. Water rich in calcium or magnesium will not be of an appropriate quality for certain uses, but at the same time the sodium-rich water produced by water softeners may not be suitable for other uses.

Dissolved oxygen is also an important chemical indicator of water quality. It makes an essential contribution to the nature and rate of chemical processes and is particularly important in maintaining water quality suitable for the survival of aquatic organisms, from bacteria to fish. Chemical and biochemical oxygen demand values are routinely used as indicators of pollution levels in water bodies.

BIOLOGICAL INDICATORS OF WATER QUALITY

Biological activity in a water body is related to its physical and chemical characteristics, and any change in water quality induced by a change in these characteristics will be reflected in the nature and level of organic activity. In acidified lakes, for example, pH values provide one indication of water quality, but these lakes are also deficient in aquatic organisms as a result of acidification and that is a biological indicator of reduced quality. The presence or absence of specific organisms is often a good indication of water quality Since different organisms have different tolerances of such factors as sediment load, nutrient levels, acid content or oxygen demand, the presence or absence of specific organisms is often a good indication of water quality. Highly polluted streams, for example, may support populations that include little more than sludge worms or communities of anaerobic bacteria, whereas a clear, well oxygenated stream with low acidity will support a vibrant, mixed population of aerobic bacteria, insects and fish.

The quality of a water body changes with time through a process called eutrophication, characterized by increasing levels of nutrients in the water, which encourages high organic productivity. Because of this, eutrophic lakes often contain high levels of algae, bacteria and other microscopic organisms that cause them to be cloudy. So much organic material also means that the water does not meet drinking-quality standards and in extreme cases thick mats of floating algae and the contamination caused when the organic material begins to decay prevent the use of the water for recreation. The biodegradation of the organic matter also causes change in biochemical oxygen demand (BOD), one of the main biological indicators of water quality. Biodegradation is brought about by the action of aerobic bacteria that consume and metabolize the organic material, but to do that they require access to oxygen, which they obtain from the aquatic environment in which they live. As the amount of organic material in a water body increases and the bacteria attempt to increase the rate of biodegradation, the demand for oxygen grows also. If the BOD exceeds the available dissolved oxygen in the water, the water body becomes deficient in oxygen or anoxic, and aquatic organisms suffer. High temperatures add to the problem, since warm water is able to hold less oxygen than cool water, and under conditions of high BOD and high temperature fish kills are not uncommon. Under natural conditions, water bodies recover when temperatures fall and the organic content of the water declines – at the end of the summer growing season, for example – but for many lakes the addition of biodegradable material from human sources continues year round and the BOD is always high.

In rivers, the flow of water and the natural reaeration associated with turbulence help to return the oxygen to the water more rapidly, but rivers too suffer from variations in water quality along their courses and that is often indicated by changing levels of BOD. When organic waste, whether of human or natural origin, is released into a stream, the demand for oxygen from the bacteria and other organisms that will digest it, and from chemical oxidation processes, is met by the oxygen dissolved in the water. Thus, immediately downstream from the source of waste, the BOD increases and the dissolved oxygen content of the stream falls. As the organic material is gradually decomposed and converted into carbon dioxide and water, the demand for oxygen declines. At the same time, natural aeration processes return oxygen to the water and ultimately the addition of oxygen exceeds its use, the dissolved oxygen levels in the stream rise again and its quality improves. If the dissolved oxygen levels along a stream are graphed, there is a characteristic dip in the graph as the oxygen level declines and then recovers, which is referred to as the oxygen sag curve (Figure 8.8). The depth and extent of the sag depend upon such factors as the initial dissolved oxygen content of the stream, the BOD of the waste, the rate at which the waste is added, the flow rate and volume of the stream and its capacity for reaeration. Problems arise if new waste is added to the stream before the dissolved oxygen level has recovered and the sag is extended downstream. Although this can happen under natural conditions, it is most often associated with multiple sources of effluent from human activities, which effectively cause the deterioration of water quality along great lengths of the river or stream.

These physical, chemical and biological indicators point to aspects of water quality that may place constraints on specific uses. Such constraints may have natural origins, but since the middle of the nineteenth century the main problems have arisen as a result of pollution associated with human activities.

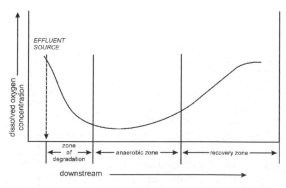

Figure 8.8 The oxygen sag curve associated with the disposal of effluent into a waterway

WATER POLLUTION: SOURCES, IMPACTS AND SOLUTIONS

Within the aquatic environment there are processes capable of dealing with and neutralizing almost any contaminant added to it. They may be physical, chemical or biological in nature, or a combination of all three, and without them the earth's water would quickly become unable to fulfil its various functions in the environment. Their ability to maintain water quality is not unlimited, however, and in many areas they have been overwhelmed, causing normal aquatic processes to be adversely affected. When that point is reached, a water body is considered to be polluted. The number of potential water pollutants is almost infinite, but since current contaminants come overwhelmingly from human sources, they are frequently categorized according to the human activities that produced them. Accordingly, they will be considered here in terms of their origins in urban/domestic, agricultural and industrial activities (Figure 8.9).

URBAN/DOMESTIC SOURCES, IMPACTS AND SOLUTIONS

The main source of pollution in urban areas is waste water or sewage produced as a result of domestic and industrial activities that is allowed to drain into adjacent water bodies. The industrial contribution depends very much upon the nature of the industry involved, but the contents of domestic waste water are remarkably similar in urban areas around the world. They are predominantly organic, including human waste, detergents or other cleansers and grey water, which does not contain the products of bodily functions, being mainly the result of bathing, showering and washing. Domestic sewage is a ubiquitous pollutant, which continues to be released untreated into the world's waterways, disrupting the aquatic environment through its impact on the biochemical oxygen demand, causing health problems and constraining the human use of rivers and lakes. In the past, when sewage volumes remained small, the environment was able to neutralize the contaminants, through oxidation and the activities of a variety of micro-organisms that broke them down into less noxious by-products. As populations grew and lifestyles and technology changed, the volume of sewage increased and the natural approach was no longer

Plate 8.3   Water polluted by sewage or bird droppings keeps this beach empty of bathers.

SOURCES OF WATER POLLUTION

POINT SOURCE CONTAMINATION	NON-POINT SOURCE CONTAMINATION
underground storage tanks	atmospheric contaminants
industrial waste disposal	pesticides; fertilizers
landfill leachate	urban run-off
mine drainage	septic field contamination
oil and chemical spills	salt water intrusion

Figure 8.9   Sources and forms of water pollution

feasible. The direct treatment of sewage became necessary, but since it was not always available where or when most needed serious water pollution resulted.

Pollution of water by sewage

In the nineteenth century, as the economic and technological benefits of the industrial revolution began to spread around the world, they were accompanied by the spread of pollution and one of the most common pollutants was domestic sewage. Existing, primitive sanitation technology could not accommodate the growing populations concentrated in urban areas and the pollution of rivers, lakes and ground water by domestic sewage was a common characteristic of the first industrial cities. A similar pattern has emerged more recently in the developing world, where rapid population growth has outstripped the provision of adequate sanitation and raw sewage is piped directly into local water bodies or carried there by surface run-off (Brown et al. 1999). In mega-cities, such as Cairo, Mexico City, Karachi and Manila, that grew rapidly in the last third of the twentieth century, the infrastructure could not keep up with the population growth and as a result existing sanitation systems were overwhelmed. Millions of cubic metres of mostly untreated sewage is discharged into rivers and lakes or the sea every day from these and other cities in South America, Asia and Africa (Biswas 2000).

Since most of the material in domestic sewage is organic, the initial impact is on the BOD of the water body in which it is deposited. In some cases the demand from the microscopic organisms attempting to break down the organic contaminants is so high that they effectively use up all of the oxygen available in the water and other aquatic life forms such as fish are unable to survive. It was largely for this reason that salmon and trout disappeared from the major rivers of Europe as the industrial revolution progressed, and fish kills caused by oxygen depletion continue to be a feature of rivers that receive large volumes of raw sewage. Suzhou Creek, which flows through Shanghai in China, receives so much sewage that it contains no dissolved oxygen and the adjacent Huangpu River, into which some 450,000 tonnes of human excrement are dumped every year, is little better (Edmonds 1994).

Even in the developed world, where environmental regulations are supposedly more stringent, millions of litres of raw or partially treated sewage are released into waterways every year, either deliberately or accidentally. Since the construction of treatment facilities is both time-consuming and costly, many communities continue to do little more than screen solids from the effluent while small particulates, organisms and materials in solution are carried on to contaminate the environment. Even where treatment facilities have been built, problems can arise where sewer systems and storm drains are combined, an approach not uncommon in the early development of sanitary engineering and therefore present in many older cities in the developed world. In such systems, sewage and urban run-off share the same drainage pipes, which often cannot cope with the increased flow initiated by storms or periods of prolonged precipitation. This causes the areas served by such systems to be flooded with a mixture of sewage and rainwater, and to reduce the flooding or to prevent damage to the treatment plants it may be necessary to release untreated sewage into rivers or lakes. In Osaka, Japan, for example, despite major improvements in wastewater collection and treatment, flooding during the rainy season regularly releases sewage into the environment and will continue to do so until a project to replace the existing combined sewerage system with a separate system is complete (Nakamura 2000).

From a human point of view, sewage pollution is aesthetically unpleasant, but it is much more. It can be a major source of disease when it contaminates sources of drinking water or areas used for swimming and water-based recreation. Sewage contains a wide variety of pathogens responsible for some of the world's most lethal and debilitating diseases. The urbanization that accompanied the industrial revolution was punctuated with deadly outbreaks of cholera, typhoid and dysentery associated with the pollution of the water supply by human sewage. The resulting high mortality rates in Victorian Britain eventually led to increased attention to sanitation and health and legislation to deal with the problems (Mannion 1997). For most of the developed world, major health issues associated with sewage disposal are in the past, largely because of improved sanitation, but also because of better monitoring systems. These, for example, can identify sewage contamination through indicators such as E. coli and other bacteria, allowing boil water advisories to be issued or bathing beaches to be closed before the problems become deadly. The developing world is much less fortunate, however, with cholera prevalent in Afghanistan, India and China (WHO 1999) and about 3 million people dying in Africa every year as a result of waterborne diseases, such as cholera, typhoid, schistosomiasis and various diarrhoeal diseases commonly associated with sewage contamination. Those who escape death may suffer the depredations of intestinal worms or become blind as a result of diseases such as trachoma (UNEP 2002).

Phosphate pollution and eutrophication

A common ingredient of detergents in the middle of the twentieth century was phosphorus in the form of phosphate. After use it was released in the grey water component of sewage into rivers and lakes where it acted as a fertilizer and accelerated the natural aging or eutrophication of the water body. The flow of sewage with high levels of phosphate into the sea also caused rapid increases in the growth of algae – algal blooms – particularly in coastal areas or in enclosed seas such as the Mediterranean (Pearce 1995a). Eutrophication is characterized by increased nutrient levels in the water leading to the rapid growth of aquatic plants, particularly algae. When the plants begin to die, the increase in bacterial activity raises the BOD, deoxygenating the water and creating problems for fish and other aquatic organisms. Depending upon the type of algae, toxic chemicals may also be released in the process. By the late 1960s and early 1970s, Lake Erie in North America was in a state of advanced eutrophication as a result of phosphorus loading from municipal sewage and fertilizer run-off from adjacent farmland. Great mats of algae floated on its surface, rotting plant material contaminated great stretches of shoreline, a combination of high BOD levels and toxins caused a serious decline in fish populations and many environmentalists considered the process to be irreversible (Ashworth 1986). Co-operation between the Canadian and US governments through a series of Remedial Action Plans turned the situation round, however. Municipal phosphorus loading was decreased by 80 per cent by the end of the century, slowing algal growth and reducing oxygen depletion to such an extent that the fishery has recovered (Environment Canada 1999). Phosphorus discharges have also

been lowered by between 50 per cent and 80 per cent in Europe since the early 1980s through the increased use of low-phosphate detergents coupled with better wastewater treatment (UNEP 2002).

Sanitation and wastewater treatment

The problems associated with water pollution from urban/domestic sources can be solved through improved sanitation and wastewater treatment, but only with a massive investment of time and money. In its broadest sense, sanitation includes measures to promote health and prevent disease, but in terms of the reduction in water pollution it can be considered in terms of the appropriate disposal of human waste and liquid household waste. When population numbers are small, the environment can cope with these materials being released into local water bodies or deposited in pits or latrines, where they eventually decompose as a result of natural processes. In some Asian countries, human waste has been spread on fields as fertilizer for hundreds of years – the so-called 'night soil' of China, for example – and while it does provide crop nutrients it is also insanitary, helping to spread pathogens and other problem pollutants such as heavy metals. Latrines, earth closets or pits continue to be used in rural areas even in the developed countries, where they are also common in parks and recreation areas. If sited carefully and regularly maintained they provide an environmentally safe and cost-effective means of disposing of small quantities of human waste. They are dry systems with no additional water being added to the solids during disposal. Most modern large-scale sewage waste disposal systems are wet systems using water to carry the waste from its source to treatment centres. Where the production of sewage is limited or access to the sewer network is not possible, septic systems are a suitable alternative to latrines or earth closets (Figure 8.10). They involve the use of water for solid waste disposal and can cope with household liquids or grey water. Sewage and waste water are discharged into an underground concrete or fibre-glass tank in which sedimentation and bacterial digestion of the waste takes place. The liquid then drains into a leaching bed, consisting of a perforated pipe system that gradually releases it into crushed rock and gravel, where it is filtered before being absorbed by the soil. A well maintained septic system is an effective way of dealing with limited

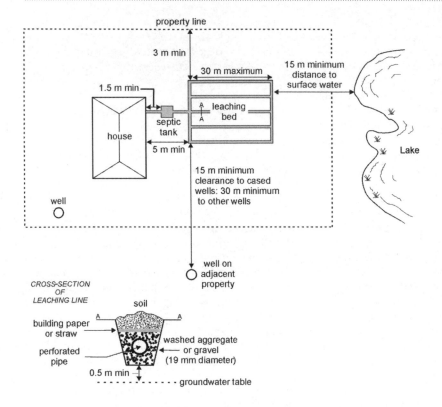

Figure 8.10 The components and structure of a modern septic system

amounts of sewage. If not properly maintained, however, it can lead to contamination of soil and the local groundwater system.

Modern sewage systems consist of networks of waste pipes carrying solids and waste water from flush toilets in homes or other sources to treatment plants. These systems can process waste so well that the effluent produced can be released into rivers and lakes with no chemical or biological impact on the aquatic environment. While this is the ideal situation it is not the norm and the level of treatment varies considerably. Sewage treatment is normally classified as primary, secondary or tertiary (Figure 8.11). Primary treatment involves little more than the removal of solids. Sticks, rags and metal objects are screened out of the effluent, grease and oil are skimmed off the top and suspended sediments are allowed to settle. Undissolved organic material collects in settling ponds as sludge. The effluent remaining after this purely mechanical treatment is commonly released directly into rivers, lakes and the sea, although it is sometimes chlorinated prior to release. Primary treatment does

not remove nutrients, dissolved organic material, bacteria or potentially toxic chemicals. They are released to become a burden on the environment. In plants providing secondary treatment, the primary effluent is subjected to biological purification. Two methods are used – trickling filters or activated sewage sludge. In the former, the primary effluent is allowed to filter slowly through a thick bed of rocks, during which bacteria consume the dissolved organic matter. In plants using the activated sludge method, high oxygen levels maintained in the effluent by aeration encourage the rapid digestion of the sewage by bacteria. In both cases, subsequent sedimentation allows more of the remaining suspended sediments to settle out. Following secondary treatment, the effluent may have regained 85 per cent of its original quality, and direct discharge into the environment would normally cause few problems. In some cases, the secondary effluent may still retain suspended solids, or include dissolved chemicals of various forms. Tertiary treatment using activated carbon filters, reverse osmosis systems and chemical coagulation techniques can

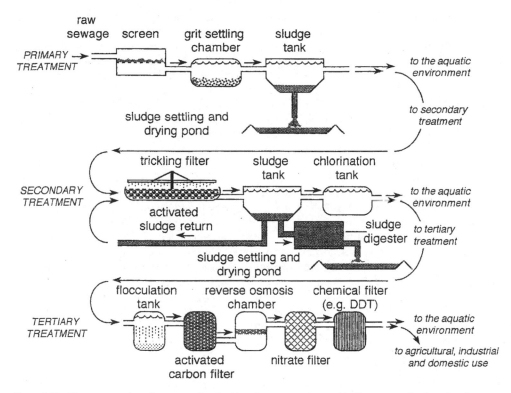

Figure 8.11 The components and processes involved in primary, secondary and tertiary sewage treatment systems

deal with most of the remaining contaminants and provide water that is up to 99 per cent pure (Horan 1990; Reed *et al.* 1995).

Geographical variations in wastewater treatment

Since most of the sewage produced in urban areas is generated at point sources, in theory it should be possible to control its release relatively easily, by collecting it via pipe and tunnel systems, which then deliver it to treatment plants. The infrastructure to provide such control already exists at a high level in the developed world, but is often minimal elsewhere. In Western Europe, for example, more than 90 per cent of the population is served by sewer connections and 70 per cent by treatment plants, whereas in many parts of the developing world a high proportion of the population has no access to proper sanitation at all (UNEP 2002). Worldwide, an estimated 2.4 billion people lack access to improved sanitation, mostly in Africa and Asia, with the 48 per cent access in Asia being the lowest

of any of the continents (WHO/UNICEF 2000). Considerable variations exist between urban and rural areas. In Africa, for example, the overall access to improved sanitation is 60 per cent, varying from as much as 84 per cent in some urban areas to as low as 45 per cent in rural areas, reaching its nadir in Eritrea, where only 1 per cent of the population has sanitation coverage of any kind (WHO/UNICEF 2000).

Even where working sewer systems are present and sanitation has been improved, treatment of the sewage often lags behind. In Eastern Europe, for example, 60–70 per cent of the population has sewer coverage but wastewater treatment is generally inadequate (UNEP 2002), and although in Latin America claims of 10–20 per cent treatment are common, it is more likely that only 2–6 per cent of the sewage collected in major urban centres receives adequate treatment (Biswas 2000). Even in the developed nations, the high participation numbers are often deceptive. Many treatment plants in North America, for example, provide only primary treatment, before releasing effluent back into the environment. In coastal areas the ocean is often seen as capable of absorbing, diluting and degrading larger amounts of sewage without serious damage to the environment, and as a result cities such as Halifax and Victoria in Canada continue to discharge untreated sewage into the sea (Freedman 2001).

Improvements in the developed world have come about as a result of health and anti-pollution legislation combined with the ability to cover the costs. These costs are high and have been met only through the resources made available as a result of the growing economies of the developed nations. Between 1972 and 1992 the government of the United States spent $75 billion on sewage treatment plants and the US Environmental Protection Agency has estimated that an additional $83.5 billion will be required to bring municipal sewage plants up to secondary treatment levels (Smith 2000). Even in a nation as rich as the United States many municipalities will have difficulty meeting their contribution to such costs. In developing nations with little economic growth and growing populations, the costs of providing improved sanitation are overwhelming. The Water Supply and Sanitation Collaborative Council (WSSCC) of the World Health Organization (WHO) has set a target of 2015 to reduce by half the number of people without access to hygienic sanitation facilities (WSSCC 2000). The prospects of that being achieved are dismal, however. Given the rapid growth of populations in the developing world (see Chapter 4), particularly in urban areas, it is estimated that even to maintain the status quo would require a doubling of the rate at which services are being provided in Asia and an increase in the present rate of thirty-three times in Africa (Brown *et al.* 1999).

AGRICULTURAL SOURCES, IMPACTS AND SOLUTIONS

The development of agriculture is one of society's great success stories. Without it the world's population would be much smaller than it is, differently distributed and involved in fewer cultural or socio-economic activities (see Chapter 4). At first sight agriculture appears relatively innocuous in environmental terms, being the extension and management of natural processes such as plant and animal growth. In creating modern agriculture,

however, society has extended these processes beyond their natural limits to cause direct environmental damage in the form of the loss of natural vegetation, soil erosion or water pollution and less direct damage through the disruption of hydrologic and atmospheric processes.

Water pollution is caused by both arable and pastoral agricultural activities on land and by fish farming in lakes or coastal waters (Figure 8.9). Perhaps the most common pollutant in water bodies is soil particles washed off the land as a result of soil erosion (see Chapter 6). Such sediments cause the water to become less transparent, which reduces light penetration and disrupts the ability of aquatic plants to photosynthesize. When the sediments settle out, they may cover food supplies or spawning beds, aquatic food chains are disrupted and the overall productivity of the water body is reduced. Although the products of soil erosion do disrupt the aquatic environment, they are not usually considered as pollutants, perhaps because they are seen as inert or inactive natural elements in the environment. The main contributors to water pollution from agricultural activities are the organic and inorganic chemicals washed off the land by run-off or waste released from animal operations.

Fertilizers, pesticides and herbicides

Much of society's success in providing food for a rapidly growing population, particularly in the second half of the twentieth century, has been possible through the use of artificial or inorganic fertilizers. The Green revolution of the 1960s and 1970s, for example, which increased grain production by 60–100 per cent, and sometimes more, was driven by a combination of new, higher-yielding crops, requiring large amounts of nitrogen-based fertilizer for maximum output. Nitrogen used in this way eventually makes its way back into the atmosphere by way of the nitrogen cycle (see Chapter 3), but before it does so it can contribute to water pollution. Particularly where nitrogen-based fertilizers are used excessively and not completely absorbed by the growing plants, nitrates are frequently leached through the soil into the groundwater system, contaminating the water supply of communities that depend upon ground water (Sampat 2000). Nitrate contamination is a chronic problem in the Canadian prairie provinces and has been detected in varying amounts in wells in forty-nine of the United States

(UNEP 2002). Although it is seldom present at levels that cause serious problems for most consumers, it causes methaemoglobinaemia – blue-baby syndrome – in infants through its ability to bind with haemoglobin and reduce the ability of the blood to carry oxygen. Surface water becomes contaminated with nitrates through surface run-off, which carries fertilizer into rivers and lakes, leading to accelerated eutrophication. The problem is aggravated in areas where excessive amounts of fertilizer are used in an attempt to improve crop yields in soils that are not particularly suited to arable agriculture, or in those which have experienced reduced fertility through changes in structure or organic content. The situation has improved in North America and Europe with decreasing use of fertilizer, but in other areas, such as Latin America, where nitrate levels continue to rise in the Amazon and Orinoco and surpass international guidelines in Costa Rica, continued excessive use of fertilizers contributes to algal growth and eutrophication in many rivers, lakes and reservoirs (UNEP 2002).

Pesticides are often used in conjunction with fertilizers to improve crop yields and as a result pesticide residues are also common water pollutants. Organochloride pesticides such as DDT, aldrin and lindane were frequently used in the past as very effective insecticides, but they accumulate in the fatty tissues of animals and through biomagnification in the food chain may reach toxic levels in predators. Because of side effects such as sterility, birth defects, cancer and damage to the nervous system, they have been banned or had their use severely restricted in most parts of the world (Simonich and Hites 1995). Organophosphorus compounds such as malathion and diazonon have tended to replace organochlorides, since they are also highly effective against insects but break down rapidly in the environment and do not bioaccumulate. There are claims that the threat from pesticides in polluted water has been overstated – because they do not readily dissolve in water, for example, organochlorides can be removed quite effectively by filtration in most modern municipal water supply systems – but they continue to receive widespread attention as a threat to the environment (Baarchers 1996).

Whereas the sources of urban/domestic pollutants tend to be point sources, with the pollutants emanating from pipes or ditches, and therefore relatively easy to trap and collect for treatment, pollutants associated with arable agriculture are associated with non-point or area sources and are therefore much more difficult to deal with. Once they have been released they have an infinite number of paths by which they can enter water bodies. The most obvious way to reduce fertilizer and pesticide pollution is to reduce the amount of the products used. This can be done by using a more precise estimate of crop requirements, using slow-release fertilizer or incorporating nitrogen-fixing crops such as legumes in rotation, a common approach in the past. A reduction in run-off rates is also beneficial, which can be achieved using techniques designed to combat soil erosion, and a buffer zone of natural vegetation between cultivated fields and waterways would help to trap both soil and pollutants (Miller 1994). Pesticide use can be reduced through integrated pest management, which combines biological, chemical and cultivation techniques for pest control (Horn 1988).

Intensive livestock farming

In the past, when mixed agriculture was more common, animal waste provided fertilizer for cultivated land. It had the advantage that it provided not only nutrients but also the organic material required to maintain a good soil structure. As monoculture became more common and agriculture more mechanized, animals disappeared from many farms and artificial fertilizer replaced the natural variety. Where pastoral farming continued, the waste produced by grazing animals was spread naturally on pasture and grassland, and that situation continues to apply in many parts of the world. In the developed countries, however, the old-style pastoral farming has been replaced in many areas by intensive livestock farming in which cattle, pigs and chickens are confined to feedlots or specifically designed buildings in which their growth, health and development can be easily controlled. Often called factory farms, they have major economic advantages, but have caused controversy among animal-rights groups who regard the confined spaces and artificial conditions under which the animals are raised as inhumane (Freedman 2001). Apart from the animals, the most common product of these facilities is solid and liquid livestock waste, which has become increasingly responsible for water pollution. In Western Europe, for example, fertilizer consumption has fallen since the 1980s, but the eutrophication of water bodies has continued as a result of increased run-off from factory farms (UNEP 2002). In addition to the normal organic materials present in the waste, pathogens, resulting from the insanitary conditions in which the animals live, and antibiotics, fed to the animals to manage infections, are also common. Although regulations exist for dealing with the high levels of waste generated, events such as heavy rain followed by increased run-off or flooding, coupled with lax inspection or enforcement, regularly allow these operations to contaminate both surface water and groundwater supplies. All of these factors came into play in Milwaukee, Wisconsin, in 1993, when the contamination of the city's drinking water with the bacterium *Cryptosporidium* commonly present in animal waste caused more than 400,000 people to become sick and 104 to die. The bacterium had been allowed to enter the water system following a combination of heavy precipitation, the illegal release of waste from a local stockyard and the failure of the city water filtration plant (Smith 2000).

In large open-air feedlots, liquid and semi-solid waste from cattle and pigs is collected in holding ponds where the liquids evaporate, leaving behind organic sludge, which can be treated or composted to be used as fertilizer. In the past, untreated waste was spread on fields to supply nutrients, with normal biological processes such as bacterial digestion, oxidation or natural solar radiation ensuring that it degraded and any pathogens were destroyed. The danger of such an approach was tragically illustrated in the town of Walkerton, Ontario, in 2000, when contamination of the town's groundwater supply by *E. coli* bacteria caused the deaths of seven townspeople and made some 2000 others sick. The contamination was the result of pollutants from cattle manure being washed by heavy precipitation into the poorly sited and poorly maintained wells from which the town drew its water supply. Human error and inefficient management of a testing programme already in place ensured that even after the problem was recognized nothing was done until the contaminated water had been widely consumed. Although not always accompanied by fatalities, this type of contamination is not uncommon, but the tragedy at Walkerton has again made it clear that much closer monitoring of the disposal of animal waste is required and regulations must be accompanied by much stricter enforcement to take care of the potential for human inefficiency and error (O'Connor 2002).

Fish farming

Another form of farming that has the potential to cause water pollution is fish farming or aquaculture. It has a long history in some parts of the world – monasteries in Europe had fishponds in medieval times, for example, and in China the raising of carp has long made a significant contribution to the food supply in some areas – but modern fish farming is one of the most rapidly growing methods of food production (Figure 8.12). While the traditional fish harvest appears to have reached its maximum of about 80 million to 90 million tonnes per annum in the mid-1980s, aquaculture production has risen sharply since then, to its current level of 40 million to 45 million tonnes, with Asia and the Pacific dominating, with close to 90 per cent of the total (UNEP 2002). Traditional species such as carp are farmed widely in Asia, but tilapia, an African tropical fish, is gaining popularity. In Europe and North America salmon and trout are the most common species, but freshwater catfish leads production in the United States. In addition to these fin fish, shellfish and crustaceans such as oysters, mussels, scallops and shrimp are important aquaculture products, with the latter particularly important in Southeast Asia, where in Thailand, for example, large areas formerly used for rice production have been converted to shrimp ponds (Holmes 1996).

Aquaculture operations can be land-based or water-based. Land-based farms abstract water from rivers or lake into ponds in which the fish are grown, returning the water to its source after use. Water-based operations can be in fresh or salt water and usually involve cages in which the fish are confined. Shellfish are grown in racks suspended in water in coastal areas. Fish farming uses methods not unlike those of intensive livestock farming and therefore poses similar threats to the aquatic environment. Fish cages or ponds are densely stocked with a single species to allow easy management and harvesting, the fish being provided with the amount of nourishment required for optimum growth. Not all of the food is consumed, however, and some falls to the bottom of the water body, increasing the nutrient supply and the potential for eutrophication. Waste products produced by the fish also contribute to eutrophication and increase the biological oxygen demand (BOD). It has been estimated that the production of one tonne of fish also produces waste which has the same effect on BOD as the

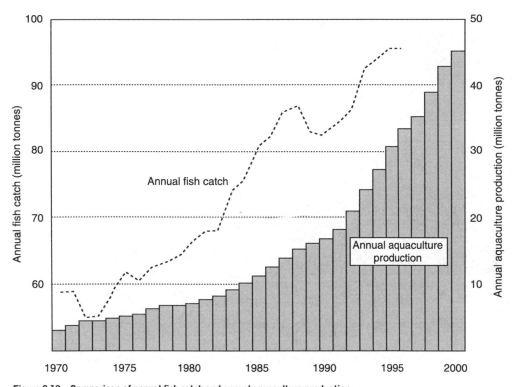

Figure 8.12   Comparison of annual fish catch and annual aquaculture production
Source: based on data in World Resources Institute, *Earthtrends*. Viewed at http://www.earthtrends.wri.org/ (accessed 4 February 2003).

untreated sewage load from twenty people (Smith and Haig 1993). Thus a fish farm producing only 500 tonnes of fish per year would have an impact similar to that of a town of 10,000 people discharging untreated sewage directly into the water, and in Britain in the early 1990s total production had an effect on BOD equivalent to that of 800,000 people (Smith and Haig 1993). Under natural conditions, pollutants such as these would be redistributed through the water by turbulence in stream flow or tidal movements, allowing the natural cleansing processes to convert them more easily into less harmful products. Fish farms tend to be established in sheltered locations, however, in conditions that support the accumulation of pollutants, with consequences ranging from increased BOD to eutrophication and the creation of plankton or algal blooms. Land-based farms using ponds are more susceptible to this than cage operations in lakes or the sea and the shrimp ponds in Thailand, for example, are drained periodically to remove as much as half a metre of rotting shrimp and shrimp faeces (Holmes 1996). If this is not done, the fish begin to suffer from reduced oxygen

levels or succumb to pathogens that thrive in the insanitary conditions and yields decline.

Disease is always a problem when large numbers of a species are confined in close quarters, whether they be cattle or carp. Like factory farmers, fish farmers have to use pesticides and antibiotics to prevent the growth of bacteria and fungi or parasites that would hinder the healthy development of the fish. Apart from the economic consequences of disease outbreaks, if no attempt is made to control them disease may well spread from the farms to cause havoc in the natural environment. The application of chemicals or antibiotics normally controls disease successfully, and these are routinely applied through the water or in food, but may lead to other problems. Regular use of antibiotics as a preventative measure can lead to the development of resistance among certain organisms such as bacteria, and excess antibiotics may be released into the environment, with unknown or unexpected consequences. Pesticides applied to fish cages in Scotland to rid salmon of lice were found to have killed lugworms in the sediments beneath the cages and research is required to find out if the pesticide

Plate 8.4 The fishponds and duck ponds between Hong Kong and Shenzen in southern China provide food for the dense populations of those cities.

is lethal to other aquatic organisms (Edwards 1998).

The initial development of aquaculture was to a large extent uncontrolled, with the introduction of appropriate legislation unable to keep pace with the rapid growth of the industry, and confusion over jurisdiction allowing farms to be established without direct planning control in some cases (Smith and Haig 1993). Although many of the problems have been resolved, the calculation of appropriate discharge rates and the monitoring of effluent release from fish farms continue to present some practical problems. The quantity and nature of effluent from land-based farms can be monitored using conventional methods, but with cage operations, where the effluent diffuses into a larger body of water, measurement is much more difficult. For similar reasons, the collection and treatment of waste from cage farms is not currently feasible. Environmentalists have accused the aquaculture industry of causing serious environmental pollution, and undoubtedly that has been the case in some areas (Holmes 1996). It is in the best interests of the industry, however, to maintain the quality of the environment to allow the production of a superior product that will provide a good return

on investment. That being the case, with proper monitoring of effluents, improved knowledge of the environmental constraints of the water bodies in which they are located and the incentive to maintain water quality, it should be increasingly possible to operate fish farms with minimal environmental deterioration.

INDUSTRIAL SOURCES, IMPACTS AND SOLUTIONS

Of all sources of water pollution, industrial effluent contains the most complex and most dangerous pollutants (Figure 8.9). They can be released during the extraction of raw materials, during processing and during product manufacture and may be organic or chemical in nature. In most cases the pollutants originate in point sources – from manufacturing plant waste pipes into rivers and lakes, for example – but in the extractive industries non-point or area sources are not uncommon and the release of pollutants in flue gases allows the atmospheric loading of water bodies. The acidification of lakes is largely the result of atmospheric loading (see Chapter 10). The impact of industrial effluents is often obvious – through the presence of foam, scum or oil slicks on the surface of the water, for example – but in some cases it may be recognized only following detailed chemical or biological analysis, by which time it may be difficult to halt or reverse deterioration in the aquatic environment. Since many industrial pollutants originate in point sources, monitoring is relatively easy and preventing their release into water bodies is not difficult. The treatment of polluted industrial water, however, presents serious technical problems and considerable costs for some industries, particularly those involved in the production of pesticides, petrochemicals or those using radioactive materials.

Food processing; pulp and paper

Organic pollutants are produced by the food processing and animal product industries and by the pulp and paper industry. The canning of fruit or its processing into jams, preserves or juices can be accompanied by the release of organic fibres and natural sugars into the aquatic environment, leading to increased BOD and nutrient enrichment that encourages eutrophication. Large volumes of water are used for cleaning and processing in abattoirs and fish plants, and if it is allowed to return to local waterways untreated, it takes with it not only organic material, but also pathogens carried by the animals being slaughtered. The production of related non-food items can add chemicals to the waste. The hides of the cattle or other animals, for example, are tanned using chromium salts, formic acid and formaldehyde, which are added to the waste mix. In the developed world, where food processing is carried out on a large scale, abattoir waste is normally piped into municipal sewage systems, with or without pretreatment, although waste from coastal fish processing plants continues to be released directly into the sea. Food processing in the developing world is less centralized and mechanized and less water is used, but there too waste is released into waterways, usually without treatment.

The pulp and paper industry, which uses longfibre softwoods as a raw material, is a significant source of water pollution, particularly in northern Europe and North America, where the boreal forest biome provides the appropriate trees (see Chapter 3). Wood pulp is produced mechanically by grinding down the wood to separate the fibres, or chemically by dissolving the lignin, which holds the cellulose fibres together. Pulp and paper mills require large amounts of water and when that water is returned to the system it is usually polluted (Table 8.3). Ground wood pulp produces large amounts of suspended solids and organic waste, which raise the BOD of the water bodies in which they are dumped. Chemical pulping adds toxic chemicals such as organic mercury, bleaching agents, sulphites and dioxins to the water (Servos 1996), in some cases creating

TABLE 8.3 POLLUTANTS PRODUCED BY PULP AND PAPER MILLS

Effluent	Problem	Solution
Fibre	Fibre mats on stream beds; suspended solids; increased BOD	Recovery and removal of fibre through secondary treatment; aeration of the effluent in ponds
Liquor (chemical by-products of the wood digestion process)	Toxic to aquatic organisms	In-plant recovery and recycling of chemicals
Slime inhibitors (e.g. organo-mercury compounds)	Bioaccumulation of mercury in fish; mercury poisoning	Replacement of organo-mercury compounds
Bleaching by-products	Bioaccumulation of chlorinated organic compounds such as dioxins and furans	Replace chlorine bleach with chlorine dioxide or hydrogen peroxide
Gases (e.g methyl sulphides; methyl mercaptans)	Obnoxious odours	Installation of scrubbers, but difficult to remove completely

Source: compiled from various sources.

serious health problems for communities using the water or eating fish caught in it.

Mercury was routinely used in the industry as a slime inhibitor and was released into the environment in the mill effluent. Once in the water it entered the food chain in the form of methyl mercury, a particularly toxic organic mercury compound, which through bioaccumulation became concentrated in the body tissues of fish at the upper ends of the food chain. Consumption of the fish by humans caused mercury poisoning. Although the effects of mercury poisoning – insomnia, fatigue, blindness, emotional or mental disorientation, paralysis and death – have been known since the nineteenth century, the role of methyl mercury was recognized only in the 1950s, when the inhabitants of the town of Minamata in Japan were poisoned in large numbers through the consumption of fish and shellfish contaminated by methyl mercury released

Plate 8.5 The bioaccumulation of pollutants in fish limits or prevents their consumption from many rivers in the developed world. (Courtesy of Heather Kemp.)

into the water in the effluent from a plastics manufacturing plant (Kurland *et al.* 1960). As a result, this form of mercury poisoning was named Minamata Disease. At about the same time, in north-western Ontario, Canada, the consumption of fish contaminated with mercury from pulp mill effluent caused similar problems on two First Nations reserves, where fish was an important part of the diet (Hutchison and Wallace 1977). In both cases it took more than twenty years before blame was assessed, responsibility for the problem acknowledged and compensation arranged.

To meet the demand for white paper, chlorine was used in the mills to bleach the pulp. Chlorine has a long, successful record as a bleaching agent and is regularly used as a disinfectant in municipal water supplies. In conjunction with organic compounds such as lignin, present in the paper-making process, however, it produces chlorinated organic compounds such as dioxins and furans, which are toxic, highly persistent and bioaccumulate in the environment. In the aquatic environment they accumulate in fish and from there are passed on to human consumers. The health effects of dioxin consumption are complex and varied, ranging from skin problems to cancers, birth defects and serious immunological, neurological and behavioural problems. Although the use of chlorine has been declining, dioxins and furans will persist in the environment for some time because they accumulate and persist in fatty tissue, and fish consumption from waterways that were subject to pollution from pulp mill effluent will continue to be restricted (Ministry of Environment, Lands and Parks, British Columbia/Environment Canada 1993).

Improved effluent control has reduced many of the environmental problems associated with pulp and paper production. Organic material is screened out and recovered, and effluent is aerated in effluent ponds before being released, to reduce the impact on BOD. Used liquor is recycled to recover chemicals, mercury has been banned and chlorine has been replaced by less harmful bleaches such as chlorine dioxide and hydrogen peroxide (Ferguson 1991). The widespread acceptance of non-bleached pulp and paper for uses that do not require bleached stock, along with the increased recycling of paper products, has contributed to the improvements, and financial benefits from the recovery of fibre and chemicals have encouraged pulp and paper companies to implement practices that are less detrimental to the environment.

Mining and manufacturing

The inorganic chemicals in industrial pollution take an almost infinite variety of forms that can be introduced into the environment at all stages of the industrial cycle from extraction to disposal. The extraction of minerals from the earth's crust disrupts its chemical stability. Compounds normally in equilibrium are exposed to processes such as oxidation and reduction and are dissolved in water, to be carried into the aquatic environment to contaminate ground water and surface water. One of the most common expressions of this type of pollution is acid mine drainage, which involves the seepage of water high in sulphuric acid from mining operations into adjacent waterways. It is a universal problem most common in coal-mining areas, but is also associated with nickel and copper mining, where the ores contain sulphur compounds. The most common source of sulphur is iron sulphide (pyrites), which on exposure to oxygen and water, and with the help of bacteria such as *Thiobaccillus thioxidans*, is converted into sulphuric acid. In the process, ferric hydroxide is precipitated, giving acid mine drainage its characteristic yellow-brown colour. The precipitate also builds up as a brown coating on rocks and sediments in stream beds. Acid mine drainage is highly acidic and corrosive, and contains toxic metals such as aluminium, copper, zinc, manganese and beryllium leached from the local bedrock by the acid. This combination of acid and minerals completely disrupts the aquatic ecosystem, and may render the water unsuitable for municipal, industrial or recreational use. In the United States, 2.7 million tonnes of acids are estimated to drain every year from existing and abandoned mine workings, polluting some 12,000 km of streams (Chiras 1998). Methods developed to reduce the acidity of mine drainage include the addition of an alkaline buffer such as limestone to the backfill when strip-mining areas are being rehabilitated and the storage of water from working mines in holding ponds, where it can be neutralized before release. Such methods can be incorporated relatively easily into current operations, but to apply them to abandoned mine workings, which make a major contribution to acid mine drainage, would take many years and would be costly.

Ores containing only small percentages of the mineral being sought can be worked successfully using modern mining techniques, although the volume of rock waste produced will be high (see Chapter 6). Run-off from spoil tips can add to acid mine drainage and carry sediments into the aquatic environment. To save on transport costs, these minerals are usually concentrated on site before being sent on to a smelter, and the resulting waste creates potential environmental problems. A mixture of fine particles and the chemicals used in the beneficiation process, the wastes are usually stored in settling ponds once they leave the plant. In the ponds, the evaporation of water takes place, the sediments settle and some of the chemicals may be recovered for reuse. Seepage from the ponds can allow groundwater contamination, and following heavy precipitation they may overflow or burst, sending large amounts of toxic chemicals into adjacent rivers and lakes. The problems are exacerbated in mountainous regions, where slopes are steep, rainfall is heavy and run-off is rapid (Fox 1997). Cyanide used in the extraction of gold, for example, can cause significant fish kills if allowed to escape into the environment, and heavy metals such as lead, tin and cadmium are considered particularly hazardous. In early 2000, for example, the wall of a tailings pond at a mine in Baia Mare in Romania gave way and released 100,000 m^3 of waste water containing cyanide into a tributary of the river Danube, where it poisoned large numbers of plants and fish (UNEP 2002). Tailings from the uranium industry have their own specific problems. They are radioactive. During the refining process they are stored in ponds in much the same way as other tailings, and have the same potential to cause water pollution. Although the level of radioactivity in the tailings is low, there is some concern that even low-level exposure can have serious cumulative effects (Ritcey 1989).

Plastics and petroleum industries

Because most manufacturing processes use a variety of raw materials to produce a finished product they can create complex chemicals, which need to be disposed of when production is complete. The plastics and petrochemical industries, which use and manufacture complex hydrocarbons, have produced some particularly potent pollutants. The methyl mercury that caused Minamata Disease, for example, originated in a plastics plant, and polychlorinated biphenyls (PCBs), considered by some environmentalists to be among the most toxic pollutants that society has released into the biosphere, is a product of the petrochemical industry. PCBs are a group of highly stable chlorinated hydrocarbons once widely used as liquid insulators in the electricity distribution industry and as plasticizers and synthetic resins in the plastics industry, where they were highly regarded for their stability. That same stability, however, allowed them to pass along food chains and accumulate in the environment so that they are now found in locations far removed from their industrial sources. Seals, fish and humans in the Arctic have traces of PCBs in their bodies. It is possible that PCBs present in polluted water are less hazardous than was once thought, but tests have shown that they are carcinogenic and may impair the immune system, perhaps in part because of impurities, such as dioxins, that they often contain (Baarchers 1996). Because of this, their production and use have been banned in most industrial countries since the late 1970s, but considerable quantities remain in use in the developing world. Even with no additional production, the stability of PCBs will ensure that they remain in the environment for at least several decades.

The pharmaceutical industry is another source of highly complex compounds that often make their way into the aquatic environment. Although the industry itself may not be a direct water polluter, many of its products find their way into rivers and streams. Antibiotics are found in the animal waste from factory farms, for example, but recent studies in Europe have shown that a wide range of drugs consumed by humans is also present in water systems there. These include anti-cholesterol drugs, antibiotics, analgesics, hormones, such as oestrogen, and a variety of drugs used in chemotherapy, passed into the sewage systems through normal human bodily functions – 30–90 per cent of antibiotics administered to people are excreted in urine, for example (Pearce 1999). Drugs are not normally included in water quality monitoring, so there is no base against which current levels can be compared; there is no indication of how many of these chemicals eventually reach the water users or what their impact might be. Excess consumption of antibiotics is known to create resistance among harmful bacteria, and chemicals designed to kill cancerous cells can also damage healthy cells, while the intake of unnecessary hormones might have an impact on endocrine systems. Much more monitoring and testing is necessary to establish the extent of the problem, but dealing with it will not be easy as modern medicine comes to rely more and more on complex pharmaceuticals.

TRENDS IN WATER POLLUTION LEVELS

Although industrial development and water pollution have grown simultaneously since the earliest days of the industrial revolution, and probably even before that, by the second half of the twentieth century pollution had reached levels that could no longer be ignored. Most rivers passing through urban industrial areas or lakes offshore from manufacturing centres contained high levels of chemical pollutants, ranging from metals to fertilizers, petrochemicals and petroleum products. In biological terms, they were effectively dead, containing little oxygen, carrying toxins that poisoned living organisms and bearing little resemblance to the waterways they had once been. How far they had changed was well illustrated in the summer of 1969, when the Cuyahoga River, flowing through Cleveland, Ohio, caught fire and burned for eight days (Smith 2000). Similar, if less spectacular, events occurred in other river basins in the developed world, leading to the realization that major efforts were needed to reduce pollution. In the developed world, the passage of anti-pollution legislation showed that there was a will to deal with the problem and there were also the resources to finance the necessary improvements. As a result, in North America pollution in the Great Lakes was significantly reduced, fish returned to streams from which they had been long absent and the Cuyahoga is once again being used by recreational boaters and anglers (Miller 1994). In rivers such as the Thames and the Rhine in western Europe, water quality has improved appreciably, allowing the aquatic environment to rebound, but in the east pollution levels in the Volga and its tributaries in Russia are still considered to be extremely high and not improving (UNEP 2002).

Except in some parts of Asia, where improved wastewater treatment and clean-up programmes have been successful, water pollution continues to worsen in the developing world. Estimates of deaths from water-related diseases, for example, range between 2 million and 5 million per year and Gleick (2002) has projected that as many as 76 million people will die by 2020 of water-related diseases that could be preventable given an appropriate commitment to improved sanitation. Few developing nations have the finances to deal with the problems that they face, and the developed nations have generally not lived up to their commitments to improve the situation. Although legislation may exist to reduce water pollution, the financial and political reality in most of the developing world means that it is often not enforced. Indeed, as a result of the competition for foreign investment, some countries see themselves forced to relax existing environmental laws on water protection in an attempt to encourage transnational companies to establish new industrial enterprises. In some countries, free-trade zones have been created in which environmental standards are set lower than in the other parts of the country and the disposal of pollutants into the aquatic environment remains unregulated (Brown *et al.* 1999; Barlow and Clarke 2002).

POLLUTION IN THE OCEANS

Because of the natural circulation of the earth's water, many of the pollutants that are released into rivers and lakes eventually make their way to the sea. Although oceans have the ability to dilute and neutralize large amounts of pollution, variations in depth and circulation patterns ensure that some parts of the ocean are subject to serious contamination. This is particularly true of coastal areas, including deltas and estuaries, but it also applies to large enclosed seas such as the Mediterranean and Baltic where the large-scale circulation is restricted. The major rivers of the developed world create pollution plumes when they enter the oceans with their loads of sewage, chemicals, nutrients and other contaminants. In many cases, pollutants are deliberately released into the oceans. Coastal cities routinely dump raw or minimally treated sewage into the oceans, where the nutrients it contains may be responsible for red tides (algal blooms), the heavy metals it contains may contaminate fish, making them unsuitable for human consumption, and the human waste it contains may create aesthetic pollution and cause health hazards to those using adjacent beaches for recreation. Beaches around the North Sea and in the Mediterranean are routinely cited for their hazard to health, and in Australia the famous white sands of Bondi beach suffer contamination from the large volumes

of sewage released by the city of Sydney after little or no treatment (Beder 1990; Pearce 1995a).

Past practice in some areas was to dump municipal and industrial waste into the ocean, where it was expected that it would cause no problems, and garbage scows from New York, for example, routinely sailed out to the waters beyond the edge of the continental shelf to dispose of the city's waste. Waste chemicals in steel drums were also dumped in the oceans with little thought of future contamination when the drums rusted through and the chemicals leaked into the oceans. Obsolete or unwanted military supplies, from complete aircraft to shells, bombs and drums of nerve gas, joined the other garbage from time to time, particularly after World War II, and more recently radioactive materials from nuclear reactors or manufacturing processes have been added to the mix. Growing concern in the early 1970s led to the Convention on the Prevention of Marine Pollution by Dumping of Wastes and other Matter signed in London in 1972. Popularly called the London Dumping Convention, it has been updated regularly so that by 1996 it included a prohibition on all dumping unless explicitly permitted, which has caused industrial nations to adopt alternative approaches to waste disposal that are less harmful to the environment (Stokke 2002). Unfortunately, there are many states that have not signed the convention, particularly coastal states in the developing world, and the convention is unable to deal with accidental dumping – following a shipwreck, for example – or illegal dumping. A matching programme to deal with pollution entering the oceans from the land, the Global Program of Action for the Protection of the Marine Environment from Land-based Activities (GAP), was introduced in 1995, and has been adopted by 108 nations, but it will be some time before its impact can be gauged (UNEP 2002).

OIL POLLUTION

One of the most serious threats to the oceans comes from oil pollution. Shipping activities are responsible for about 50 per cent of the oil entering the world's oceans, largely as a result of tank flushing and bilge pumping, and perhaps as little as 2 per cent on average is contributed by tanker accidents, although these proportions may vary considerably from year to year. The remainder is provided by offshore petroleum production and exploration. Large spills from oil tanker accidents are the major source of sudden, large-volume pollution episodes (Figure 8.13), but offshore oil operations also make major contributions. In 1977, for example, a blow-out from the *Ekofisk Bravo* platform released 14,000 tonnes of crude oil into the North Sea (Jenkins 1980), and the largest accidental oil spill yet recorded – nearly 500,000 tonnes – was the result of a blow-out at a well in the Gulf of Campeche off the coast of Mexico (Cutter *et al.* 1991). Even regular activities in the offshore environment can contribute significant amounts of oil pollution. More than 16,000 tonnes of oil were added to the Norwegian sector of the North Sea between 1984 and 1990 in the form of oil-based drilling muds. Over the same period perhaps as much as eight times that amount was discharged into the British sector (Pearce 1995c). In addition to ongoing shipping and offshore oil production activities, there are other unexpected events, such as the deliberate release of more than 800,000 tonnes of oil into the Persian Gulf during the Gulf War in 1991, which may be few in number, but no less serious.

The size of a particular oil spill is not always a good indication of its ultimate impact. The blow-out of 500,000 tonnes at Campeche in 1979 was the largest offshore oil spill on record. However, a combination of winds and currents prevented the main oil slicks from coming ashore, and most of the oil dispersed in the open ocean. Damage to the aquatic environment did take place, but it was much less than would have occurred in the more complex and productive coastal environment. In contrast, the release of a much smaller amount of oil – 38,000 tonnes – by the *Exxon Valdez* in Alaska killed thousands of seabirds and mammals and caused incalculable environmental and economic damage (Davidson 1990) (Figure 7.4)

The spills that normally receive most attention are those that occur in coastal areas, where they foul great stretches of sandy beach or rocky shoreline. Reports in the media carry pictures of oil-soaked seabirds and mammals, many destined to die because the oil has destroyed their natural insulation and buoyancy or because they have ingested toxic chemicals in the oil. The effects of the spill reach far beyond these more obvious impacts, however. Organisms living in the intertidal zone are poisoned and bottom-dwelling shellfish become contaminated as the effects of the spill descend through the water. Often the food chain is broken,

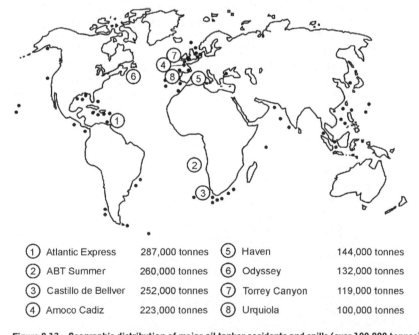

①	Atlantic Express	287,000 tonnes	⑤	Haven	144,000 tonnes
②	ABT Summer	260,000 tonnes	⑥	Odyssey	132,000 tonnes
③	Castillo de Bellver	252,000 tonnes	⑦	Torrey Canyon	119,000 tonnes
④	Amoco Cadiz	223,000 tonnes	⑧	Urquiola	100,000 tonnes

Figure 8.13 Geographic distribution of major oil tanker accidents and spills (over 100,000 tonnes), 1962–2000
Source: based on data in International Tanker Owners Pollution Federation, *Past Spill Statistics.* Viewed at http://www.itopf.com/stats.html (accessed 2 July 2003).

either directly when an entire group of organisms is wiped out by the initial spill or indirectly when predators begin to die after ingesting a contaminated food supply. Thus the entire ecosystem suffers.

Many of the ecological impacts have economic implications. The aesthetic and physical degradation that follows coastal spills, for example, may bring major economic losses through the disruption of the tourist industry. Commercial shellfish beds may have to be closed because oysters or clams have been tainted by the oily water they have ingested. Similarly, local fisheries may be decimated if spills occur near fish-spawning beds or on migration routes and fish farming operations may have to close down.

CLEANING UP OIL SPILLS

Although the pollutants are mainly hydrocarbons, and therefore subject to biodegradation by bacteria and other organisms in the natural environment, the process is slow and the impact of the pollutants may be felt for many years after the initial contamination. As a result, some form of direct clean-up or rehabilitation is required. Oil spills at sea are attacked using floating booms that contain the oil until it can be removed by pumps or skimmers; by spraying the slick with chemical dispersants that break up the oil; or by burning the oil on the surface or in the vessel from which it is escaping. None of these systems is ideal. Booms do not work well when the seas are rough. Chemical dispersants were once popular because they caused the oil to sink and the slicks to disappear quite quickly. Evidence from the areas contaminated by

the *Torrey Canyon* and *Amoco Cadiz* spills (Figure 8.13), where large volumes of dispersants were used, suggests that the contamination of bottom-dwelling organisms by oil and the chemical dispersants persisted some ten to fifteen years after the events (Pearce 1993). Once the oil comes ashore, clean-up becomes difficult, particularly on rocky coasts. Absorbents such as straw or peat are used to soak up the oil, contaminated sand is removed and steam is used to clean coated rocks. Nothing is completely successful, however. Absorbents cannot remove all of the oil and the removal of shoreline material may have significant ecological and morphological consequences. Steam cleaning was used extensively following the *Exxon Valdez* accident, and although it effectively removed the oil, it also tended to scald and kill the organisms that had escaped the initial effects of the spill (Davidson 1990).

Despite ongoing attempts to improve clean-up technology, the success rate is not inspiring. Clean-up costs are high – more than US$2 billion in the case of the *Exxon Valdez* – and the returns are unimpressive. Scientists estimate that even under ideal conditions, with state-of-the-art technology provided rapidly by well trained technicians, no more than 10–15 per cent of the oil from a major spill can be recovered (Miller 1994). The residue remains in some form or other in the environment, its clean-up dependent upon natural processes. Comparing the costs with the returns, it has been suggested that savings could be made by letting nature take its course, since it appears to accomplish most of the clean-up anyway (Lomborg 2001). The money saved could be used to promote prevention programmes or develop appropriate technology (Figure 8.14). Public attitudes are in favour of

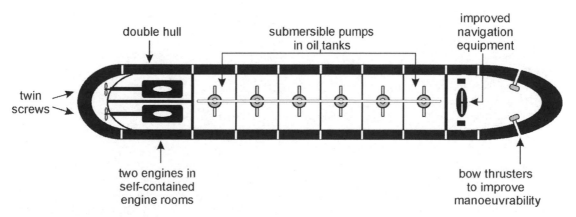

Figure 8.14   Plan view of a safer oil tanker

the clean-up response to oil pollution, however (Holloway 1996), and until that changes, billions of dollars will be spent on the cleaning and rehabilitation of polluted coastlines.

SUMMARY AND CONCLUSION

Water is essential to the working of the earth/atmosphere system and to the survival of human beings, yet it is often treated in a very cavalier fashion, with little apparent thought to its availability and quality. The earth/atmosphere system contains a finite amount of fresh water, which is regularly replaced and cleaned by natural recycling through the hydrologic cycle.

The demand for water is growing so rapidly that in some areas the hydrologic cycle cannot replace it fast enough to meet the needs of domestic, industrial, agricultural, transport and recreational consumers. Nor is it able to deal with the pollution created by them. Local schemes to offset the limited availability and poor quality of water are not uncommon, but the dimensions of the problem are now so great that continental-scale water diversions are being considered. Disputes over water can become so serious that they can lead to conflict or aggravate existing conflicts between states.

With the existing high demand for water in the developed nations and a growing demand from the developing world, problems of water availability will increase and although some progress has been made in reducing pollution that too will get worse in some areas. For some observers, however, the greatest threat to water comes from the privatization of the resource, which would see it converted into a commodity, controlled by profit and loss, rather than as a public good available to all who need it.

SUGGESTED READING

Hessler, P. (2001) *River Town*, New York: HarperCollins. An account of life along the Yangtze river in Sichuan, in an area to be flooded by the Three Gorges dam project. Includes an interesting chapter on attitudes to the flooding that will follow the completion of the project.

Kandel, R. (2003) *Water from Heaven*, New York: Columbia University Press. A comprehensive, non-technical, look at water from basic hydrology to its use and abuse by society.

Ward, D.R. (2002) *Water Wars: Drought, Folly and the Politics of Thirst*, New York: Riverhead Books. Consideration of the physical, social, economic and political aspects of the world's most pressing water problems and the people working to try to solve or reduce them.

QUESTIONS FOR REVISION AND FURTHER STUDY

1   Find out about the water supply and wastewater disposal systems in your community. Ask such questions as:
    (a)   Where does the water come from?
    (b)   What treatment does it receive before it is used, e.g. filtration, chlorination?
    (c)   How is it used, e.g. residential, industrial, commercial?
    (d)   How much is used?
    (e)   Are there any restrictions on use?
    (f)   How much does it cost?
    (g)   Has the cost changed over the years?
    (h)   Has the community done anything to discourage waste or promote conservation?
    (i)   Are there any water supply or quality problems?
    (j)   How does the community dispose of waste water – into river, lake or sea, for example?
    (k)   What treatment does the waste water receive before disposal?

    Using the answers to these questions, create a poster containing a mixture of text, maps, annotated diagrams and photographs that provides a snapshot of water use and wastewater disposal in your community.
2   In the dry south-western United States the limited surface water supplies are being utilized to their maximum and groundwater sources are being rapidly depleted. As the area continues to develop, it is conceivable that demand will ultimately exceed supply. What do you think will happen then?

9

Drought, Famine and Desertification

After reading this chapter you should be familiar with the following concepts and terms:

aridity	famine prediction	rural depopulation
boreholes	FEWS	Sahel
contingent drought	fuelwood	salinization
Depression (The)	Great American Desert	seasonal drought
desertification	Great Plains	shifting sands
desertification – prevention,	Green revolution	teleconnection
reversal	over-grazing	TOGA
drought prediction	Pueblo drought	UNCCD
Dustbowl	rainfall variability	UNCOD
famine		

Drought and famine are problems that have plagued society for thousands of years and have proved intractable even when tackled with modern technology. Desertification is a natural process that has always been part of the normal working of the environment in some areas – along the world's desert margins, for example – but concern over its impact on the environment and society increased only in the second half of the twentieth century.

All three of these issues involve natural environmental activities, but current approaches to them include consideration of the human contribution to their development as well as their impact, individually and in combination, on society. Together, drought, famine and desertification have been implicated in many human catastrophes, past and present, and are likely to continue to contribute to future suffering, particularly in the developing world, where, for example, large parts of Africa south of the Sahara continue to be in the grip of all three (Figure 9.1).

DROUGHT

There is no universally accepted definition of drought, but in the broadest sense, it occurs when there is insufficient moisture to meet the needs of plants and animals. It is sometimes confused with aridity (Figure 9.2), which is permanent dryness – characteristic of the world's deserts, for example – whereas in areas suffering from drought, the dryness will end when moisture conditions return to what is considered normal for the area. Over thousands of years certain plants and animals have adapted to life under conditions of extremely low levels of moisture, which allows them to survive in the earth's arid zones. Their needs are met, therefore there is technically no drought, despite the prevailing aridity. In reality the situation is much more complex and even organisms that have minimal moisture requirements can suffer from drought under certain circumstances. Fluctuations in the normal weather patterns, for example,

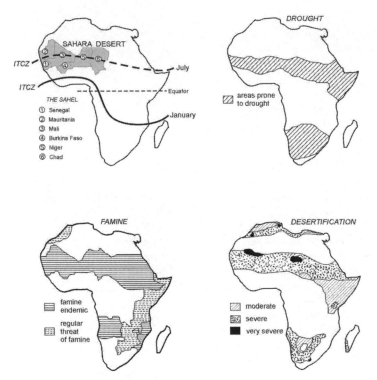

Figure 9.1   Drought, famine and desertification in Africa
Source: after Thomas (1993) with permission.

may reduce even the small amount of precipitation available, thereby changing the whole relationship between the flora and fauna and the other elements of the arid environment. If plants and animals can no longer cope with the reduced water supply, they will suffer the effects of drought despite their adaptations to limited moisture availability. Depending upon the extent of the change, plants may die from lack of moisture, they may be forced out of the area as a result of competition with species more suited to the new conditions, or they may survive, but at a reduced level of productivity. The situation is often more complex for animals, but the response is often easier. In addition to requiring water, they also depend upon plants for food, and their fate will therefore be influenced by that of the plants. They have one major advantage over plants, however. Being capable of movement, they can respond to changing conditions by migrating to areas where their needs can be met. Given time, some degree of balance will be attained again, but in areas where precipitation is both slight and

variable the environment tends to be in a continuing state of flux.

Aridity is not a prerequisite of drought, but some of the worst droughts ever experienced have occurred in areas that are arid or semi-arid (Figure 9.3). The problem lies not just in the small amount of precipitation available, but in its variability or unreliability. Much of the recurring drought in Africa is associated with the inherent unreliability of precipitation over most of the continent (Nicholson 1989). Along the desert margins, for example, annual precipitation is low, ranging between 100 mm and 400 mm per year, but sufficient to provide enough forage to support small herds of cattle, goats and camels. These values are based on the averages of long-term observations, which mask annual totals that range well above and below the mean (Figure 9.4), allowing some years of plenty, but also allowing drought to occur with some regularity. In India there are distinct wet and dry seasons associated with the monsoon circulation, but variability in the timing and amount

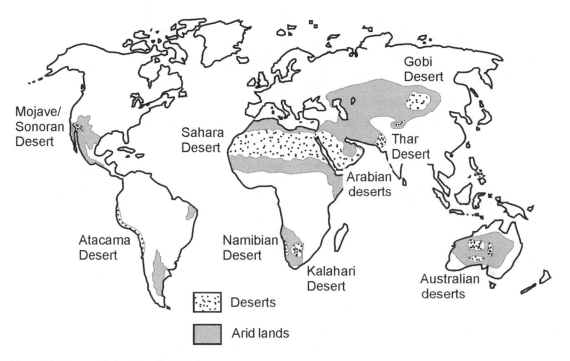

Figure 9.2 The world's deserts and arid lands

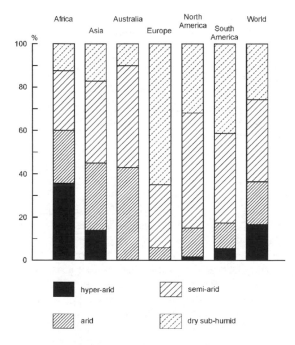

Figure 9.3 Distribution of arid lands by continent

Source: based on data from the World Resources Institute. Viewed at http://ideas.wri.org/pubs_content_c.cfm?ContentID=722 (accessed 2 July 2003).

of rain in the wet season frequently causes an extension of the dry season and the seasonal drought that it brings. The failure of the monsoon is a recurring theme in studies of the meteorology of India (Fagan 1999). Weather records at Beijing, in drought-prone northern China, show that the city receives close to 600 mm of precipitation in an average year. However, the amount falling in wetter years can be six to nine times that of drier years. Only 148 mm were recorded in 1891, for example, and 256 mm in 1921, compared with a maximum of 1405 mm in 1956 (NCGCC 1990). Rainfall variability is now recognized as a major contributor to drought, and the nature of the environment reflects that variability rather than the so-called normal conditions.

The human animal, like other species, is forced to respond to the changing environmental conditions brought on by drought. In earlier times, this often involved migration, which was relatively easy for primitive communities, living by hunting and gathering, in areas where the overall population was small. As societies changed, however, this response was often no longer possible. In areas of permanent or even semi-permanent settlement, with their associated physical and socio-economic structures, migration was no longer an option – indeed, it was

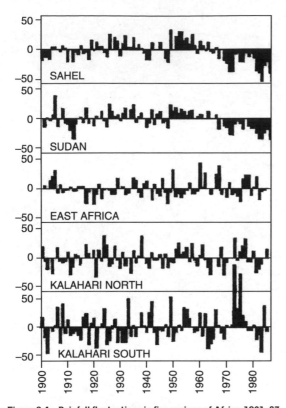

Figure 9.4 Rainfall fluctuations in five regions of Africa, 1901–87, expressed as a percentage departure from the long-term mean

Source: after Nicholson (1989) with the permission of the Royal Meteorological Society.

PERCEPTIONS, IMAGES AND IMPACTS OF DROUGHT

Drought can be defined or measured in a number of ways. To some it is a long dry spell during which lawns turn brown and crops shrivel and die. To others it is a complex combination of meteorological elements, expressed in some form of moisture index, or a moisture deficiency measured against normal or average conditions, established through long-term observation (Katz and Glantz 1977). Drought is very much a human concept, and many current approaches to the study of drought focus on its impact on human activities (see e.g. Mortimore 1989; Glantz 1994). Agriculture is particularly susceptible to moisture deficiency and the retardation of crop growth or development is commonly one of the first effects of drought. This may lead to economic considerations, when, for example, dry conditions reduce yield and therefore income, but if the drought is extended the impacts are often more serious, particularly in the developing world, where crop failure can lead to famine and death. The soil water on which plants depend is often quickly depleted when drought occurs, but it may take some time before other sectors feel the impact. Activities dependent upon surface water or ground water may be unaffected by short-term moisture shortages, for example, but if the drought continues for several months or years, stream flow will be reduced and lake water levels will decline. Ultimately, this may have an impact on the production of hydro-electric power or navigation on waterways. Even large water bodies such as the Great Lakes feel the impact of long dry spells when access to marinas and other boating facilities is restricted by shallow water and commercial lake vessels are unable to carry a full cargo because of the risk of grounding. The last sector of the hydrologic cycle to experience the effects of drought is usually the groundwater system. Water takes a long time to move into and through that system, which dampens the immediate effects of drought. If the shortage of water continues for some time, however, its impact is eventually felt in the groundwater system also. When the drought has lifted, soil moisture levels are replenished relatively quickly, often in only a few weeks or months, but rivers and lakes take much longer to rebound – several years in the case of the Great Lakes, for example – and the groundwater system may continue to show evidence of drought for a considerable time after moisture

almost a last resort. The establishment of political boundaries, which took no account of environmental patterns, also restricted migration in certain areas. As a result, in those areas susceptible to drought, the tendency, perhaps even the necessity, to challenge the environment grew. If sufficient water was not available from precipitation, either it had to be supplied in other ways – by well or aqueduct, for example (see Chapter 8) – or different farming techniques had to be adopted to reduce the moisture need in the first place. The success of these approaches depended very much on such elements as the nature, intensity and duration of the drought, as well as on various human factors, which included the numbers, stage of cultural development and the technological level of the peoples involved. Thus a number of human ingredients have been added to the physical components of drought, and these have been responsible for exacerbating problems of water availability, not just in areas experiencing drought, but in adjacent areas as well.

conditions elsewhere have returned to normal. The overall duration of the recovery depends on such factors as the intensity and duration of the drought, as well as the rate at which precipitation amounts return to normal.

Every generation has had its images of drought. In the early part of the twenty-first century it is southern Africa; in the 1990s it was East Africa; in the 1980s it was Ethiopia; in the 1960s it was the Sahel; in the 1930s it was the Dustbowl on the Great Plains. Although these events have gripped the popular imagination, they are only the more serious examples of a problem that is worldwide and time-less. The inhabitants of Africa, south of the Sahara, have suffered the effects of drought for hundreds of years, and it strikes areas as far apart as Australia and the Canadian prairies, Brazil, India and China with some regularity and often with considerable severity. Drought is expected in such areas, but elsewhere it is highly irregular and, as a result, often more serious. Such was the case in 1976, when the normally humid United Kingdom was sufficiently dry that the government felt it necessary to appoint a Minister of Drought. Attitudes to drought include a strong element of denial, and when drought strikes there is a natural tendency to hope that it will be short and of limited intensity. The traditional Australian approach to drought, for example, has been to denigrate it as a hazard and to react to its onset with surprise (Heathcote 1987). The inhabitants of drought-prone areas, therefore, may not respond immediately to the increased aridity, continuing to cultivate the same crops, perhaps even increasing the area under cultivation to compensate for reduced yields, or retaining the flocks and herds that expanded during periods of abundant moisture. With prolonged drought in arable areas the crops die and the bare earth is exposed to the ravages of soil erosion. If no attempt is made to reduce the animal population in areas of pastoral farming during drought, the land falls victim to over-grazing, plant cover is destroyed, unprotected soil becomes more susceptible to erosion and desertification becomes a possibility.

Although drought is the most persistent climato-logical problem that society has had to face, responsible perhaps for the decline and decay of civilizations in the past (see Box 9.1, Drought and civilization), not all areas suffer equally. Its impact is greatest in areas that suffer either seasonal drought or contingent drought. Seasonal drought is characteristic of the subtropics and associated with

circulation patterns that create distinct wet and dry seasons. Although the dry season returns year after year there is no pattern to its length and intensity while the precipitation in the wet season is often unreliable. The combination of an extended dry season and a wet season that supplies less precipita-tion than normal has been responsible for droughts in Africa, India, China and Australia that have been some of the most catastrophic on record (Morrison 2001; Fagan 1999). The drought of the late 1960s and early 1970s in the Sahel falls into that category. Contingent drought is also associated with irregular and variable precipitation, but in areas that normally have an adequate supply of moisture to meet their needs. The droughts of 1975–76 and 1988–92 in the normally humid United Kingdom were contingent droughts. The Great Plains of North America have suffered contingent drought for thousands of years, from prehistoric times up to the present (Van Royen 1937; Cordell 2001). Arguably the most important of these, in terms of its extent, duration and impact on govern-ment policies and agricultural practices, was the drought of the 1930s, which created the Dustbowl and caused widespread hardship and misery for the inhabitants of the area. Although they were caused by different mechanisms, the droughts in the Sahel and the Great Plains illustrate the importance of human factors such as population size, and cultural or technological development, on the overall impact of drought on society and the environment.

DROUGHT IN THE SAHEL – SEASONAL DROUGHT

The Sahel is a relatively narrow strip of land some 400–500 km wide, stretching from the Atlantic coast of West Africa as far as the Sudan, 5000 km to the east along the southern margins of the Sahara Desert (Figure 9.1). Traditionally, the Sahel has included the nations of Senegal, Mauritania, Mali, Burkina Faso, Niger and Chad, but the name has come to be used more loosely to include adjacent nations such as the Sudan and the northern parts of Nigeria that suffer from similar problems of drought. The Sahel is a transition zone between the desert to the north and the more humid savanna grassland to the south. Being transitional, the vegetation varies from bare soil with bunched or tufted grasses next to the desert to a continuous

BOX 9.1 DROUGHT AND CIVILIZATION

Given the impact that drought has had on populations in different parts of the world in recent times, it is not surprising that it has been linked with the decline and decay of society in the past. Some of the earliest indications of the role of drought in guiding human activities have been obtained from an area stretching from the eastern Mediterranean through Mesopotamia to the Indus valley, which between 2200 and 1900 BC suffered from an abrupt decline in precipitation. These areas included the original agricultural hearths where the first domestication of plants and animals had taken place, and since most of the civilizations that had become established as a result of that agricultural development were dependent upon complex systems of water management and irrigation technology, the climate change had a serious impact. The displacement of the Mediterranean westerlies or a reduction in the moisture they supplied caused drought in the Middle East and changes in the Asiatic monsoon had an impact on the water supply in the Indus basin. Increased aridity in what is now northern Syria from about 2200 BC is thought to have destroyed the agricultural base of the Mesopotamian Subir civilization (Weiss *et al.* 1993) and the general reduction in moisture throughout the area may well have led to the abandonment of sedentary agriculture in favour of a return to pastoralism in Greece, Anatolia, Palestine and northern Mesopotamia (Weiss 2001). In the Indus basin at about the same time, the Harrapan civilization began to decay, with a rapid decline after 1800 BC, apparently as a result of desiccation brought about by changes in the pattern of the Indian monsoon (Calder 1974; Singh *et al.* 1974; Weiss 2001). The irreversible decline of the Harrapan civilization has also been linked with changes in river flow, brought about by geomorphological rather than climatic conditions, when the headwater tributaries of the Sarasvati River, a major waterway in the Indus system, were captured by streams flowing east into the Ganges system and the Bay of Bengal (Possehl 2001). In either case the result was the same. The reduced supply of water meant that less was available for irrigation, the land could no longer produce sufficient grain to meet the needs of the urban populations and the civilization began to decline.

In Egypt the impact of the climate change at this time came about not through the reduction in moisture over the Nile valley, but as a result of the failure of the monsoons in the Indian Ocean to bring the expected heavy rain to the Ethiopian Highlands. Normally the water provided by that rain streamed out of the mountains and moved down the Nile valley, flooding over the river banks to provide the fields with the moisture and fertile silts that made agriculture possible in the desert environment. Between 2200 and 2000 BC rainfall in the Ethiopian Highlands was much less than normal and downstream the harvests failed with some regularity. The resulting famines led to civil disorder, political instability and economic disaster that ultimately led to the break-up of the united Old Kingdom of Egypt into a series of city states and regional units between 2100 and 2050 BC (Fagan 1999; Weiss 2001). When the rains returned – and with them the annual floods – some time after 2000 BC, the kingdoms reunited, illustrating the overwhelming influence of water on the history of the region (Fagan 1999).

Across the Mediterranean Sea from Egypt, about a thousand years later, the flourishing Mycenaean civilization of southern Greece began to disintegrate. The rapidity and extent of the decline was such that it was commonly attributed to the invasion of Mycenae by Greeks from the regions to the north, but a reassessment of the evidence led to the postulation that drought, followed by starvation, social unrest and migration, was the main factor leading to the downfall of the Mycenaeans (Bryson *et al.* 1974; Bryson and Murray 1977; Lamb 1996), although the links are not considered convincing by some researchers (see, for example, Parry 1978).

The abandoned settlements in the south-western United States, at Mesa Verde in Colorado, Chaco Canyon in New Mexico and Canyon de Chelley in Arizona, once supported the Anasazi peoples, a civilization that survived in an arid environment by growing corn (maize), squash and beans using relatively simple gravity-fed irrigation systems (Cordell 2001). Between about 1250 and 1450, however, the Anasazi vacated the area. Their leaving has been blamed on a variety of factors, including warfare, disease, soil depletion and deforestation, but it is becoming increasingly clear that drought probably played a major role. The south-west is an arid area, but the inhabitants had adapted to the dry conditions and were able to sustain a comfortable existence. Analysis of tree ring data from the area has shown that between 1250 and 1450 the normal rainfall patterns were seriously disrupted, which in turn disrupted patterns of socio-economic activity, culminating in the abandonment of the settlements (Cordell 2001). Depending upon a sedentary existence made possible by agriculture, the Anasazi must have found it difficult to adjust to the new climatic reality. Prior to the change, the population

BOX 9.1 – continued

had been growing and it is possible that it had reached the carrying capacity of the land, making it vulnerable to the drought. There is evidence that, even after the rains had begun to fail, the Anasazi tried to maintain their existing lifestyle, but as crop production dwindled, eventually they were forced to move out (Fagan 1999).

In many cases where drought – or some other climatic element – appears to have instigated major societal change in the past, the absence of adequate and appropriate data prevents hypotheses from being developed much beyond the speculative stage. Modern analytical methods involving, for example, the interpretation of tree rings, pollen grains, lake sediments, ice cores and archaeological materials have made it possible to provide a clearer picture of the role of drought in history, however, and its role, both directly and indirectly, in the development of society is increasingly better understood.

For more information see:

Bryson, R.A. and Murray, T.J. (1977) *Climates of Hunger*, Madison WI: University of Wisconsin Press.

Bryson, R.A., Lamb, H.H. and Donley, D.L. (1974) 'Drought and the decline of Mycenae', *Antiquity* 48: 46–50.

Calder, N. (1974) *The Weather Machine and the Threat of Ice*, London: BBC Publications.

Cordell, L. (2001) 'Aftermath of chaos in the pueblo southwest', in G. Bawden and R.M. Reycraft (eds) *Environmental Disaster and the Archaeology of Human Response*, Albuquerque NM: Maxwell Museum of Anthropology.

Fagan, B. (1999) *Floods, Famines and Emperors: El Niño and the Fate of Civilizations*, New York: Basic Books.

Lamb, H.H. (1996) *Climate, History and the Modern World* (2nd edn), London: Routledge.

Parry, M.L. (1978) *Climatic Change, Agriculture and Settlement*, Folkestone: Dawson.

Possehl, G.L. (2001) 'The drying up of the Sarasvati: environmental disruption in South Asian prehistory', in G. Bawden and R.M. Reycraft (eds) *Environmental Disaster and the Archaeology of Human Response*, Albuquerque NM: Maxwell Museum of Anthropology.

Singh, G., Joshi, R.D., Chopra, S.R. and Singh, A.B. (1974) 'Late Quaternary history of the vegetation and climate of the Rajasthan Desert, India', *Philosophical Transactions of the Royal Society* B267: 467–501.

Weiss, H. (2001) 'Beyond the younger Dryas: collapse as adaptation to abrupt climate change in ancient West Asia and the eastern Mediterranean', in G. Bawden and R.M. Reycraft (eds) *Environmental Disaster and the Archaeology of Human Response*, Albuquerque NM: Maxwell Museum of Anthropology.

Weiss, H., Courty, M.A., Wetterstrom, W., Guichard, F., Senior, L., Meadow, R. and Curnow, A. (1993) 'The genesis and collapse of third millennium north Mesopotamian civilization', *Science* 261: 995–1004.

cover of taller grasses and scrubby bush in the south, reflecting the changing availability of precipitation. The normal climate of the Sahel is one of distinct wet and dry seasons associated with the annual migration of the Intertropical Convergence Zone (ITCZ) (Figure 9.1). As the ITCZ moves north in the northern summer, it allows tropical maritime air from the Atlantic Ocean to flood over the area, bringing with it precipitation to cause the wet season, which for most of the Sahel peaks in August. With the return of the ITCZ southwards in the northern winter, the area comes under the influence of the dry tropical continental air from the Sahara Desert and the dry season sets in. Longer periods of drought lasting several years rather than several months are also characteristic of the area. These were long thought to be caused by the failure of the ITCZ to move as far north as normal over an extended period, but this is no longer generally accepted as a sufficient explanation. Extended drought in the Sahel is now seen as part of the broader pattern of continent-wide rainfall vari-

ability, associated with large-scale variations in the atmospheric circulation (Nicholson 1989). Computer modelling suggests that sulphur and other pollutants may have contributed to the recent drought by disrupting the earth's energy budget, which altered atmospheric circulation patterns and led to the displacement of the tropical rain belts to the south of their normal position (Nowak 2002) (Figure 9.5).

Over the centuries, the inhabitants of the region adapted to the seasonal variations, migrating with their herds to follow the rains and the forage they produced, or growing crops such as millet in the short wet season, particularly in the south, where the rainfall was greater. Despite these adaptations to the regular seasonal drought, the population was still vulnerable to spells when the rains were late or less than normal, and over the centuries they suffered many times from these longer periods of drought, usually unnoticed by the rest of the world. It was only in the second half of the twentieth century, with the improvement in global com-

Plate 9.1 Abandoned Pueblo dwellings at Mesa Verde. Once the home of the thriving Anasazi culture which was forced out of the area by increased aridity between 1250 and 1450. (Courtesy of Christine Deschamps.)

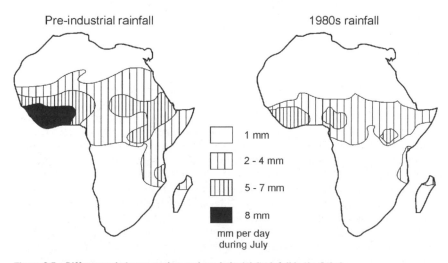

Figure 9.5 Differences between modern and pre-industrial rainfall in the Sahel
Source: after Nowak (2002) with permission.

munication and the power of television that their plight received more universal attention.

Between 1968 and 1974 the Sahel was visited by the third major dry spell to hit the area in the twentieth century. It was by far the most catastrophic. Rainfall amounts were 30–70 per cent below the 1931–60 average and river flows were as much as 60 per cent below average (Nicholson 1989). With the lack of moisture little vegetation grew and there was insufficient forage for the animals. Millions of animals died – perhaps as many as 5 million cattle alone – and that loss along with the failure of the subsistence crops led to famine, malnutrition, disease and the deaths of at least 100,000 people. The latter number would have been higher but for the outside aid that provided for 7 million people at the peak of the drought (Glantz 1977). The years following 1973 were wetter, but by 1980 drier conditions had returned and by 1985 parts of the region were again experiencing fully fledged drought (Cross 1985). It is estimated that the famine of 1984–85 claimed 50,000 lives in Senegal, at least 300,000 in Ethiopia and tens of thousands elsewhere in the Sahel, although the absence of accurate statistics means that the true numbers will never be known (Dando 2002). A return to average rainfall in 1988 provided some respite, but it was short-lived and in 1990 conditions again equalled those during the devastating droughts of 1972 and 1973 (Pearce 1991). Overall the period between 1968 and the mid-1990s was unusually dry, even for a region in which drought is expected.

At that time, also, the direct physical problems of drought were augmented by a number of socio-economic issues. By the mid-twentieth century the introduction of scientific medicine, limited though it was by the standards of the developed world, brought with it previously unknown health services – such as vaccination programmes – that reduced disease and led to improvements in child nutrition and sanitation. Together these helped to lower the infant death rate, and with the birth rate remaining high, populations in the Sahel grew rapidly, doubling between 1950 and 1980 (Crawford 1985). Initially, it was not a major problem, since in the late 1950s and early 1960s a period of heavier, more reliable rainfall produced more fodder, allowing more animals to be kept. Crop yields increased also in the arable areas. The wetter conditions also encouraged cash cropping and even the commercialization of meat production in place of the subsistence agriculture that had existed there

previously. Although this worked well when the rains were plentiful, when drought arrived in the late 1960s the results were catastrophic. Vegetation dried up and died, crop yields fell dramatically and both pasture and cropland became susceptible to soil erosion, which led to the onset of desertification. The high populations of both people and animals could no longer be supported, many died and many migrated into relief camps or into the urban areas in an attempt to survive. Despite attempts to provide food aid, introduce new agricultural techniques and supply additional water from wells or boreholes, the region has not recovered from the intense period of drought that began in 1968. The death of the animals, the destruction of the soil and indeed the destruction of a way of life has meant that the Sahel – along with most of sub-Saharan Africa – remains an impoverished region, dependent upon outside aid and overshadowed by the ever present potential for disaster the next time the rains fail.

DROUGHT ON THE GREAT PLAINS – CONTINGENT DROUGHT

All of the nations in the Sahel are developing nations, and that fact no doubt contributed to the nature and extent of the problems caused by the drought that they faced. It is also clear, however, that economic and technological advance is no guarantee against drought. The net effects will be different, but the environmental processes act in essentially the same way whatever the stage of development.

This is well illustrated by the problems faced by farmers on the Great Plains of North America. Stretching from western Texas in the south, along the flanks of the Rocky Mountains to the Canadian prairie provinces in the north, they form an extensive area of temperate grassland with a semi-arid climate. They owe their aridity in part to low rainfall, but the situation is aggravated by the timing of the precipitation, which falls mainly in the summer months when high temperatures cause it to be rapidly evaporated and therefore less effective in meeting the moisture needs of plants (Figure 9.6). Contingent drought, brought about by the variable and unpredictable nature of the rainfall, is characteristic of the area – consecutive years may have precipitation 50 per cent above normal or 50 per cent below normal – and this has had a major effect on the settlement of the Plains. Averages have little

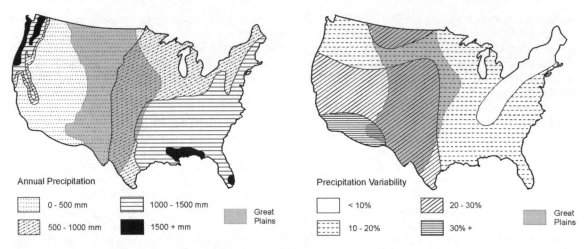

Figure 9.6   Precipitation distribution and variability in the United States
Source: compiled from data in Goudie (1989) and McKnight and Hess (2000).

real meaning under such conditions, and successful agricultural planning is next to impossible. The tendency for wet years and dry years to run in series introduces further complexity (Figure 9.7). Strings of dry years during the exploration of the plains in the nineteenth century, for example, gave rise to the concept of the Great American Desert. Although that view has been criticized for being as much myth as reality, it is now evident that the first Europeans who reached the Plains did so when drought was

rampant, and that coloured their perception of the area (Lawson and Stockton 1981). The conditions they encountered are now known to be quite typical of the area and have been repeated again and again since that time.

The earliest inhabitants of the plains experienced the effects of drought also (Van Royen 1937). Being nomadic hunters, they probably responded in much the same way as the people of the Sahel when drought threatened. They migrated, following the

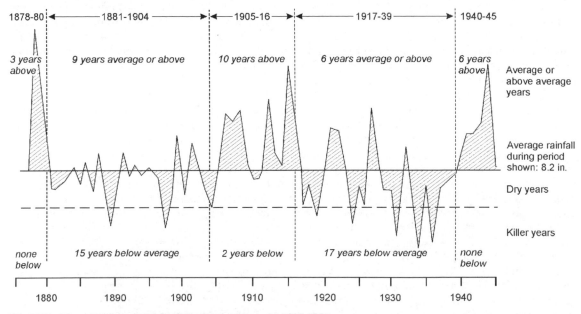

Figure 9.7   Drought years on the Great Plains (eastern Montana), 1878–1945
Source: after Watson (1963) with permission.

animals to moister areas. Communities that had progressed beyond a nomadic lifestyle, however, found that a migratory response to drought was not always easy. Among the Pueblo peoples of the south-west, for example, the development of irrigated agriculture led to higher population densities than elsewhere and the establishment of permanent settlements. Some 200 years of reduced and unreliable rainfall beginning in the mid-twelfth century led to the destabilization of the communities, and despite attempts to resist, the people ultimately had to move, leading to the abandonment of the settlements between 1250 and 1450 (Cordell 2001). Major drought episodes, extensive in both time and place, are also a recurring feature of the historical climatology of the Plains (Bark 1978; Kemp 1982), with the groups of years centred on 1756, 1820 and 1862 being particularly noteworthy (Meko 1992).

The weather on the Plains was wetter than normal when the first agricultural settlers moved to the west in the late 1860s after the Civil War. The image of the Great American Desert had paled, and the settlers farmed just as they had done east of the Mississippi or in the Mid-west, ploughing up the prairie to sow wheat or corn. By the 1880s and 1890s the moist spell had come to an end, and drought once again ravaged the land (Smith 1920), ruining many settlers and forcing them to abandon their farms. As many as half the settlers in Nebraska and Kansas left their farms at that time (Ludlum 1971), like the earlier inhabitants, seeking relief in migration, in this case back to the more humid east. Some of those who stayed experimented with dry farming, while others allowed the land to revert to its natural state, and used it as grazing for cattle, a use to which it was probably much more suited in the first place. The lessons of the drought had not been learned well, however. By the 1920s the rains seemed to have returned to stay and a new generation of farmers moved in. Lured by high wheat prices, they turned most of the Plains over to the plough, seemingly unaware of the previous drought or unconcerned about it. Crops were good as long as precipitation remained above normal, but by 1931 the good years were over and throughout the remainder of the decade the Plains were plagued by dust storms which carried away tens of millions of tonnes of topsoil. In combination with the rundown economic state of the country and the world at that time (a period commonly known as the Depression), these conditions created such disruption of the agricultural and social fabric of the region that the effects have reverberated down through the decades, and the Dustbowl remains the benchmark against which all subsequent droughts have been measured. The only possible response for many of the drought victims of the 1930s was migration, as it had been in the past. The Okies described by Steinbeck in *The Grapes of Wrath*, leaving behind their parched farms in Oklahoma to seek relief in California, had much in common with the indigenous peoples who had experienced the Pueblo drought six centuries before. The societies were quite different, but they felt the same pressures, and responded in much the same way. By migrating, both were making the ultimate adjustment to a hostile environment.

If the drought of the 1930s brought with it hardship and misery, it also produced, finally, the realization that drought on the Plains is an integral part of the environment. It led to the creation of the US Soil Conservation Service in 1935, which, through programmes of erosion control, land restoration and land use planning, contributed to the rehabilitation of the Plains. Intense dry spells have recurred in every decade since then, the most recent being in the first two years of the twenty-first century, when the intensity of the drought on the Canadian prairies matched that of the 1930s. Although such episodes cause significant disruption at the individual farm and local level, the general acceptance of the limitations imposed by aridity has ensured that the overall impact has been less than it would have been in the Dustbowl era. New agricultural techniques – involving dry farming, no-till seeding and irrigation, for example – coupled with a more appropriate use of the land and government aid in the form of crop insurance, help to offset the worst effects of the arid environment, but some problems will always remain. One particularly serious concern is the potential for more frequent and intense periods of drought on the Plains if global warming develops as projected (see Chapter 11). Should that come to pass the southern prairies of Canada and the northern Plains in the United States might once again resemble the Great American Desert visualized by those who explored the western plains in the nineteenth century.

FAMINE

As a result of improved cultivation techniques, the introduction of higher-yielding crop varieties, the increased use of fertilizer and the wider dissemination of these advances to many developing countries, global agriculture has been able to meet the needs of the world's rapidly growing population over the past half-century (DeRose *et al.* 1998). A major increase in production – the Green revolution – was brought about in the late 1950s and 1960s by a combination of increased fertilizer use and the introduction of new higher-yielding varieties of rice and wheat (Figure 9.8). Introduced to developing countries from Mexico to Indonesia, they helped to meet the food needs of the rapidly growing populations in the developing world. As the cost of fertilizers increased in the 1970s, and the vulnerability of some crops to pests and disease required the purchase of pesticides, the Green revolution seemed to falter, leading some observers to doubt the ability of the world to feed itself and prophesy widespread famine (see, for example, Ehrlich 1968).

Such Malthusian predictions continued into the last few years of the twentieth century (see, for example, Brown 1996), despite evidence that the world's farmers were for the most part continuing to meet and even exceed the growing demand for food (Lomborg 2001). Although this picture is bright, it is a general overview and on closer inspection it reveals areas in which the food supply is insufficient to meet the needs of the population. Worldwide, some 800 million people are living with some level of hunger, with more than 500 million in Asia and the Pacific and 200 million in Africa (UNEP 2002). In some cases, the food deficit may represent a slight, short-term dip below the daily dietary energy requirement; in others it may involve extreme food shortages that can result in famine.

THE CAUSES OF FAMINE

Famine is a protracted food shortage that leads to widespread starvation, disease and death. Many different factors contribute to famine and intensify its impact. The role of socio-economic causes such as poverty, civil unrest, war, or inadequate coping mechanisms such as food distribution or transport systems, has received much attention (Middleton 1999; Dando 2002). Drought frequently initiates the process, however, by creating the conditions that allow the other elements to come into play. Drought, for example, may cause crops to die, reducing the local food supply and setting the stage for famine. Famine can be averted, however, by bringing food into the stricken area from elsewhere. Where that is not possible, because of an inadequate transport system, or because political pressure, armed conflict or civil strife makes the distribution of food impossible, famine inevitably intensifies, leading to malnourishment, disease, starvation and ultimately death for those affected by it. In many areas where food continues to be available even after drought strikes, the poverty of the inhabitants is such that they cannot cope with the price increases that commonly accompany food shortages and they feel the full impact of the famine.

Famine in Asia

Links between drought and famine are particularly well marked in India, where records reaching back to the fifteenth century provide evidence that the

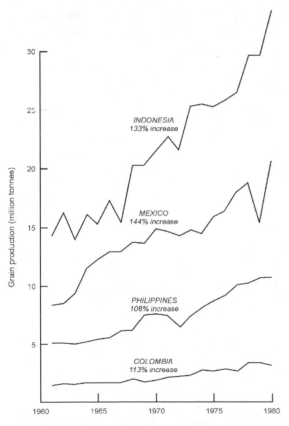

Figure 9.8 Increasing grain production made possible by the Green revolution

Source: based on data from the World Resources Institute, *Earthtrends.* Viewed at http://www.earthtrends.wri.org/ (accessed 15 February 2003).

failure of the monsoons to provide their normal amounts of rain was often a precursor of famine (Morrison 2001). If the length and intensity of the drought were limited, grain stored from a previous harvest or water remaining in tanks and reservoirs might prevent the onset of famine, but throughout the second half of the nineteenth century India was visited by a series of disastrous famines instigated by crop failure following drought. In the worst of these, 4.3 million people died in 1876–78 and 5 million died in 1896–97 (Fieldhouse 2001). Initially these were seen as 'natural' disasters and one of the Malthusian checks on population growth, but it has become clear that although the failure of the monsoon frequently initiated the crop failure that led to famine, socio-economic factors, such as land-holding practices, farming techniques, high grain prices and an inadequate transport network contributed to its intensity (Morrison 2001). The British imperial government set up a commission following the 1876–78 famine to establish means of mitigating future famines, and by the beginning of the twentieth century workable relief measures,

based on increased food production, improvements in rail transport and a more rapid response to the first signs of famine, were in place (Fagan 1999). These did not prevent the last major famine in imperial India, however, which occurred in Bengal in 1943. Initial food shortages were aggravated by an inadequate response on the part of a government beset by wartime communication problems and the presence of large numbers of soldiers and refugees from neighbouring Burma (Fieldhouse 2001). As a result of the Green revolution India was self-sufficient in food grains by the mid-1970s, and although the monsoon continues to fail from time to time and local food shortages occur, there have been no famines to match those of the nineteenth century (UNEP 2002).

Drought has been implicated in other serious famines in Asia, but often in conjunction with other non-natural causes. Perhaps the greatest famine in recent times was that in China between 1958 and 1962, when a combination of drought and problems with the collectivization of agriculture led to widespread harvest failure (Figure 9.9). Over that

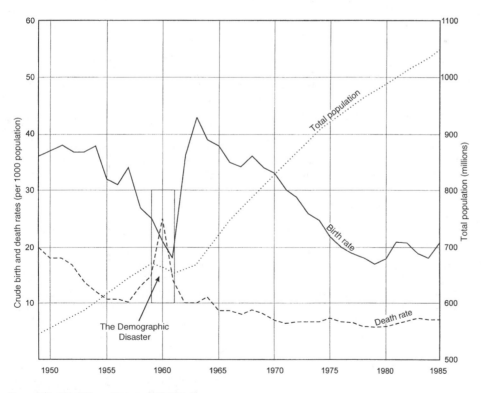

Figure 9.9   The Chinese Demographic Disaster

Source: Jowett (1990) with permission.

four-year period, China suffered 25 million to 30 million more deaths than might have been expected under normal conditions (Jowett 1990). A similar situation developed between 1995 and 1999 in North Korea. A severe famine occurred when existing food supply problems, associated with economic mismanagement and the failure of the collective farming system, were exacerbated by a series of natural disasters, including hailstorms, flooding and drought, which destroyed the already meagre crop production. In addition to killing some 800,000 people, the famine caused chronic malnourishment among millions of others, stunting the growth of perhaps as many as two out of three children under the age of five, the effects of which will continue to be felt many years into the future (Dando 2002).

Famine in Africa

Africa is the continent currently at greatest risk from famine. The areas most frequently affected stretch from the Atlantic coast through the Sahel to Sudan and Ethiopia and from there south in East Africa through Kenya and Tanzania to Zimbabwe and Mozambique (Figure 9.1). The famine in the Sahel between 1968 and 1972 was driven primarily by drought, but its impact was made worse by a number of socio-economic elements. In Ethiopia between 1983 and 1985 perhaps as many as 300,000 people died as a result of famine initiated by drought, although accurate figures were difficult to obtain because of the civil war in the country at that time (Cross 1985). The conflict also made it difficult for aid agencies, which had relief food supplies available, to reach those in need, and that undoubtedly added to the death toll. Drought and locusts also destroyed crops in the northern Ethiopian provinces of Tigray and Eritrea in 1987, putting 3 million people at risk of starvation and death (MacKenzie 1987). Famine struck Ethiopia again in 1993 and 1999, affecting close to 7 million people (UNEP 2002), and although accurate death rates are still difficult to obtain, the rapid supply of food aid kept the rates low, indicating that famine need no longer inevitably follow drought. There is evidence of this from southern Africa also. The area was hit by a massive drought in 1991 and 1992 as a result of rainfall that was significantly below average between 1989 and 1992. Grain production plummeted and 17 million to 20 million people were put at risk of starvation, yet in most areas there

were very few if any famine-related deaths reported (DeRose *et al.* 1998). Widespread co-operation among the countries in the region and the rapid arrival of large amounts of food aid from the United States prevented disaster. Mozambique did not fare as well as the other nations, because of a civil war which discouraged donors from sending food that might be used or stolen by the combatants before it could be distributed to those in need, and it was from Mozambique that the only deaths from the famine of 1991/92 in southern Africa were reported (DeRose *et al.* 1998).

In 2002 famine again threatened southern Africa, with the UN Food and Agriculture Organization (FAO) and the World Food Program (WFP) reporting as many as 10 million people threatened by starvation in Malawi, Zimbabwe, Zambia, Lesotho and Swaziland (FAO/WFP 2002). The region's longest dry spell for twenty years combined with government economic mismanagement and corruption created a demand for 4 million tonnes of food to meet the nutritional needs of the population until the harvest of April 2003. Initial responses to calls for aid were slow, in part because of political problems in some countries, and it is unlikely that southern Africa will fare as well in 2002–03 as it did in the 1991–92 famine. A contributing factor is the severe drought in the northern Great Plains, which has led to a major reduction in yields and caused a decline in the amount of grain available for food aid. The situation is further complicated by the high rates of HIV/AIDS infection in Africa, which make a high proportion of the population more vulnerable to malnutrition and have the potential to elevate the death rates beyond those in 1991–92, when HIV/AIDS had yet to become a serious problem (see Box 4.1, HIV/AIDS).

DROUGHT AND FAMINE PREDICTION

Drought is not the sole cause of famine, but it is certainly a major cause, often initiating the problem only to have it intensified by other factors. Prevention of drought is not feasible – although rain making through cloud seeding to enhance precipitation was once seen as having the potential to reduce its impact – but if it could be predicted, responses could be planned and the consequences much reduced. The simplest approach to prediction is the actuarial forecast, which estimates the probability of drought on the basis of past occurrences. Problems with the length and homogeneity of meteorological records often reduce the reliability of actuarial predictions. Links between meteorological variables and other physical or environmental variables have been assessed as possible predictors. There is, for example, a relationship between sunspot activity and precipitation. Drought on the Great Plains has been correlated with the minimum of the twenty-two-year double sunspot cycle (Schneider 1978). The correlation has no physical theory to explain it, however, and the relationship does not seem to apply outside the western United States. Most modern attempts at drought prediction consider teleconnection, which involves the linking of environmental events in time and place (Glantz *et al.* 1991). Changes in global sea-surface temperatures (SSTs) have been correlated with drought in the Sahel, for example, allowing remarkably accurate rainfall forecasts to be issued for the area up to four months in advance of the rainy season (Owen and Ward 1989; Nicholson 1989). ENSO events have been recognized as precursors of drought in Brazil, Australia, Indonesia and India (Fagan 1999). Drought in north-eastern Brazil commonly occurs in conjunction with an El Niño, and India receives less monsoon rainfall during an El Niño year. The relationship is well marked in India, where monsoon rainfall over most of the country was below normal in all of the twenty-two El Niño years between 1871 and 1978 (Mooley and Parthasarathey 1983). Similarly, in Australia, 74 per cent of the El Niño events between 1885 and 1984 were associated with drought in some part of the interior of the continent (Heathcote 1987) (Figure 9.10). Such figures suggest the possibility of El Niño episodes being used for drought prediction in these parts of the world. Since 1990, the Tropical Ocean and Global Atmosphere programme (TOGA) with its array of fixed and drifting buoys in the Pacific Ocean has provided data on surface and sub-surface ocean temperature, air temperatures and surface winds, which can provide indications of an El Niño event as much as six to nine months before it becomes fully developed (Fagan 1999). As one of the countries that frequently feels the impact of drought during El Niño years, Australia, through its National Climate Centre, issues medium-term (three months ahead) outlooks for rainfall based on the Pacific Ocean data. As well as helping to predict drought, this approach is also seen as contributing to Australia's goal of sustainable development. There is also evidence that the cold phase of an ENSO event – La Niña – can increase precipitation in some areas. The major

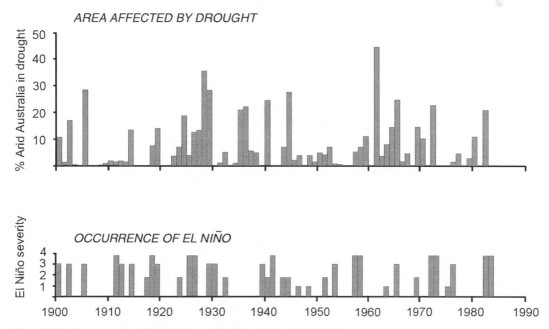

Figure 9.10 The apparent relationship between El Niño and drought in Australia
Source: after Heathcote (1987) with permission.

La Niña in 1988, for example, was associated with increased precipitation and flooding in Bangladesh, Sudan and Nigeria, and may also have prevented drought in the Sahel that year (Pearce 1991). La Niña was identified as a separate entity in the ENSO sequence only in the mid-1980s (see Box 2.7, El Niño–La Niña), and there are as yet insufficient data to establish a firm link between it and rainfall variability, but it is a relationship that merits additional investigation.

Since 1985 the Famine Early Warning System (FEWS) and its successor FEWS NET have provided advance warning of famine in seventeen African nations. Funded by USAID, it uses remotely sensed and ground-based data on such elements as drought and crop health, to provide forecasts of potential food availability, with the aim of reducing the vulnerability of the population to famine (FEWS 2002). FEWS helped to prepare contingency plans to deal with the impact of the 1997 El Niño, provided early warning of an impending food crisis in Ethiopia in 2000 and continues to work with governments and participating aid

organizations to shorten the time between the declaration of a possible famine and the response to it (UNEP 2002). Early warning systems, whether they be for drought or famine, work only if they are heeded and the response is rapid. Unfortunately, for logistical, political and economic reasons, that is not always the case and despite the availability of technology to forecast the famine and the food to deal with it, starvation is still a fact of life in the developing world, particularly in Africa.

DESERTIFICATION

Desertification is the degradation of land in arid, semi-arid and some dry sub-humid areas, initiated by natural environmental change – extended drought, for example – by ecologically inappropriate human activities in marginal environments or by a combination of both (Figure 9.11). In its simplest form desertification is associated with extended periods of drought when land adjacent to the world's tropical deserts becomes more and more

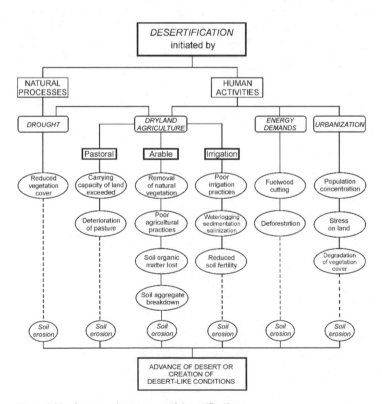

Figure 9.11 Causes and precursors of desertification

arid until eventually desert conditions prevail. With the return of rain, the process is reversed. When considered in this way, desertification is a natural process, which has existed for thousands of years, is reversible and has caused the world's deserts to expand and contract in the past. Most modern approaches to the definition of desertification recognize the combined impact of adverse climatic conditions and the stress created by human activity, while both have been accepted by the United Nations as the elements that must be considered in any working definition of the process (UNEP 2002). The relative importance of each of these elements remains controversial, however (Thomas and Middleton 1994). Some see drought as the primary element, with human intervention aggravating the situation to such an extent that the overall expansion of the desert is increased, and any recovery – following a return to wetter conditions, for example – is lengthier than normal. Others see direct human activity as instigating the process. In reality, there are many complex causes that together bring desert-like conditions to perhaps 3.6 billion km^2 or 70 per cent of the world's drylands (UNEP 2002). The areas directly threatened are those adjacent to the deserts and semi-deserts on all continents (Figure 9.12). Africa has received much attention, but large sections of the Middle East, the Central Asian republics of the former Soviet Union, China adjacent to the Gobi Desert, north-west India and Pakistan, along with parts of Australia, South America and the United States, are also susceptible to desertification. Even areas not normally considered as threatened, such as southern Europe from Spain to Greece, are not immune. At least 50 million people are directly at risk of losing life or livelihood in these regions. In a more graphic illustration of desertification, the United States Agency for International Development (USAID), at the height of the Sahelian drought in 1972, claimed that the Sahara Desert was advancing southwards at a rate of as much as thirty miles (48 km) per year along a 2000 mile (3200 km) front. Although these specific numbers have been treated with a healthy scepticism by some researchers (see, for example, Nelson 1990), they do give an indication of the perceived magnitude of the problem, and remain one of the reasons why there was increasing cause for concern in the decades that followed (van Ypersele and Verstraete 1986).

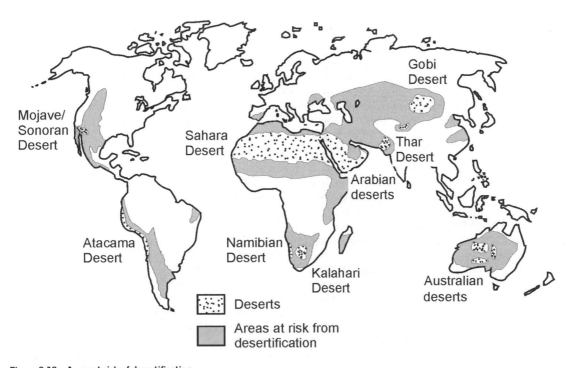

Figure 9.12   Areas at risk of desertification

THE CONTRIBUTION OF DROUGHT TO DESERTIFICATION

In semi-arid areas, where successful farming is never without risk, drought is often the trigger that pushes agricultural activities from marginal to unsustainable and encourages desertification. Human nature being what it is, when drought strikes an area there is a natural tendency to hope that it will be of short duration and of moderate intensity. The inhabitants of drought-prone areas, therefore, may not react immediately to the increased aridity. They may continue to cultivate the same crops, perhaps even increasing the area under cultivation to compensate for reduced yields, or they may try to retain flocks and herds which have expanded during times of plenty. In reality, farmers working marginal land in the developing world have few options but to continue farming as they have always done, as long as they can. If the drought is prolonged, crops die and the bare earth is exposed to wind and water erosion (see Chapter 6). The Dustbowl on the Great Plains developed in this way. Once the available moisture had evaporated and the plants had died, the wind removed the topsoil – the most fertile part of the soil profile – leaving a barren landscape which even the most drought-resistant plants found difficult to colonize (Borchert 1950). In China, in the second half of the twentieth century, the ploughing up of semi-arid grassland in Inner Mongolia, north and west of Beijing, an area of highly unpredictable precipitation, has led to topsoil being blown away to form dunes that are encroaching on adjacent cultivated land (Edmonds 1994).

Prolonged drought in pastoral areas is equally damaging. It reduces the forage supply, and, if no attempt is made to reduce the animal population, the land may fall victim to over-grazing, the most common cause of desertification, affecting some 680 million ha of land (FAO 1996). As the grass cover is lost, the soil is opened up to erosion, allowing fine particles to be blown away and making it easier for any rain that does fall to cause further erosion through rapid run-off, or to percolate into the groundwater system beyond the reach of most plants. Removal of the grass cover also encourages plants that can survive the new conditions to migrate into the area. In North America, over-grazing in the south-western parts of the Great Plains allows unpalatable shrubs such as mesquite and sagebrush to become established, for example (Mannion 1997), and similar developments have

occurred in Australia (Young 1996). Under these conditions there is no guarantee that the return of the rain will be followed by the re-establishment of the grass cover, and productive pasture will have been lost. The retention of larger herds during the early years of the Sahelian drought allowed the vegetation to be over-grazed to such an extent that even the plant roots died. In their desperation for food, the animals also grazed on shrubs or even trees and effectively removed vegetation that had helped to protect the land. The flocks and herds had been depleted by starvation and death by the time that stage was reached, but the damage had been done. When the wind blew, it lifted the exposed, loose soil particles and carried them away, taking with them the ability of the land to support plant and animal life. In combination these human and physical activities seemed to be pushing the boundaries of the Sahara Desert inexorably southwards. Out of this grew the image of the shifting sands, which came to represent desertification in the popular imagination. As an image it was evocative, but the reality of such a representation was increasingly questioned in the 1990s (Nelson 1990; Pearce 1992b).

THE CONTRIBUTION OF HUMAN ACTIVITIES TO DESERTIFICATION

Climate variability clearly made a major contribution to desertification in both the Sahel and the Great Plains, and in concert with human activities created serious environmental problems. An alternative view sees human activity in itself as capable of initiating desertification in the absence of increased aridity (Verstraete 1986). For example, human interference in areas where the environmental balance is delicate may be sufficient to set in motion a train of events leading eventually to desertification. The introduction of arable agriculture into areas more suited to grazing, or the removal of forest cover to open up agricultural land or to provide fuelwood, may disturb the ecological balance to such an extent that the quality of the environment begins to decline, and, if nothing is done, the soil becomes highly susceptible to erosion. In such cases, desertification is initiated by human activities, with little or no contribution from nature.

Soil takes a physical beating during cultivation: its crumb structure is broken down as its individual

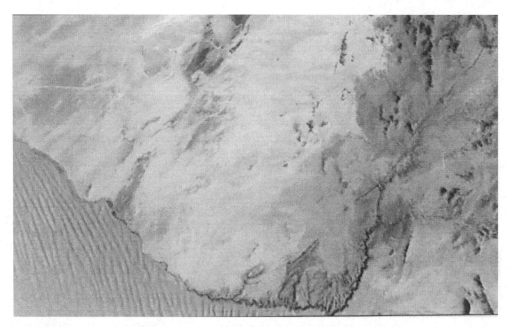

Plate 9.2   Shifting sands along the Kuiseb River in Namibia. The dunes south of the river reach heights of 150 m. Prevailing winds blow sand to the north but it is blocked by the river. The river has become so choked by sand that it reaches the sea only in times of flood. (Courtesy of NASA: Earth Observatory.)

constituents are separated from each other. In addition, cultivation destroys the natural humus in the soil and the growing crops remove the nutrients, both of which help to hold the soil particles together into aggregates (see Chapter 3). The soils in semi-arid areas are notoriously low in both organic matter and nutrients and if nothing is done to replace them – which many farmers cannot afford to do – the soil becomes highly susceptible to erosion and desertification. Modern agricultural techniques, which allow the soil to lie exposed and unprotected by vegetation for a large part of the growing season, also contribute to the problem. When wind and water erode the topsoil it becomes impossible to cultivate the land, and even natural vegetation has difficulty re-establishing itself in the shifting mineral soil that remains.

The removal of trees and shrubs to be used as fuel has had similar effects in parts of the developing world, where the main source of energy is wood. According to the UN Food and Agriculture Organization, about 1.8 billion m³ of fuelwood are cut every year, about a quarter of it in Africa (FAO 1996). Particularly in drier areas, the woodlands that do exist are thin and scrubby and sufficient wood can be obtained only by cutting over large

areas, with the result that fuelwood consumption has been responsible for the degradation of 137 million ha of land (FAO 1996). In Sudan, for example, the growing demand for fuelwood was a major factor in the decline of the total wood resource by 3.6 per cent annually in the 1970s and early 1980s (Callaghan *et al.* 1985). At about the same time, along the desert margins in Africa and the Middle East, the woody steppe was being cut for fuelwood at a rate of about 20 million ha per year, leaving a potential target for desertification (Le Houerou 1977).

In arid areas susceptible to desertification, irrigation is often used to provide the moisture that does not arrive naturally in the form of precipitation. Irrigation has contributed significantly to the increase in the world's food supply since the middle of the twentieth century, but under certain circumstances it can contribute to desertification, by causing salinization of soil. Salinization is the build-up of salts in the upper layers of the soil. It has a number of causes but in irrigated areas it is usually associated with the evaporation of irrigation water. In areas where air temperatures are high, evaporation rates are high also, and this increases the salinity of surface water stored in lakes or

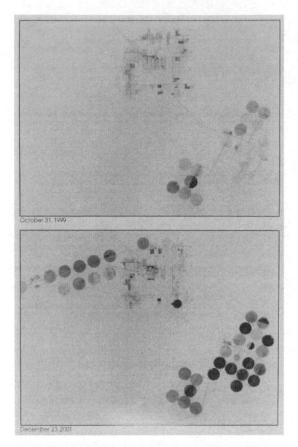

October 31, 1999

December 23, 2001

Plate 9.3 **The expansion of irrigation in southern Egypt. The dark circles are about 1 km in diameter and show the results of central pivot irrigation. In the two years separating the images (31 October 1999, above, to 23 December 2001, lower), the area under irrigation has increased threefold. (Courtesy of NASA: Earth Observatory.)**

reservoirs. Ground water also tends to contain high levels of salts, particularly if it has been in the system for a long time. When surface or ground water is used to irrigate crops, the salts remain in the soil after the water has evaporated and accumulate over time. The process of evaporation also leads to saline water being drawn up through the soil profile by capillary action to contribute to salt accumulation near the surface. At best, the salt build-up can lead to a reduction in crop yields; at worst it makes the land sterile and unsuitable for agriculture. The vegetation cover is lost and the chemical action of the salt helps to break down the soil aggregates, with the result that once-productive soil has become susceptible to erosion and desertification. A combination of good drainage and sufficient water to

flush the excess salt out or to leach it back down through the soil may prevent or even reverse the process, but it is costly and the drainage of salt-laden water may damage the environment in adjacent areas. Salinization is particularly serious in the Middle East, where more than 40 per cent of irrigated desert soils are affected (UNEP 2002). It is less of a problem in Africa, because there is less irrigated agricultural land there, but it stretches from Mexico to Pakistan, to India and into Australia and in the United States some 300,000 ha suffer salinization, reducing crop productivity by as much as 20–30 per cent (Chiras 1998).

RECENT APPROACHES TO IDENTIFYING AND MITIGATING DESERTIFICATION

Current academic and popular attitudes to desertification owe a lot to the findings of a United Nations Conference on Desertification (UNCOD) held in Nairobi, Kenya, in 1977. At the conference, the role of human activities in land degradation was considered to be firmly established, and the contribution of drought was seen as secondary at best. Since human action had caused the problem, it seemed to follow that human action could solve it. In keeping with that philosophy, the United Nations Environment Program (UNEP) was given the responsibility for taking global initiatives to introduce preventative measures that would alleviate the problem of desertification (Grove 1986). When the issue was once again considered in detail at the Earth Summit in Rio, fifteen years and US$6 billion later, few effective counter-measures had been taken, and the UNCOD plan of action was widely seen as a failure (Pearce 1992a).

The data upon which the UNEP responses were based are now considered by many researchers to be unrepresentative of the real situation. Although it is undoubtedly a serious concern in many parts of the world, the extent of irreversible desertification may well have been overestimated (Nelson 1990). Problems arising from the timing and method of collection of the data were aggravated by the UNEP premise that human activity was the main cause of the land degradation that produced desertification. Natural causes such as short-term drought and longer-term climatic change were ignored or given less attention than they deserved, yet both can produce desert-like conditions without input from society. The inclusion of areas suffering from short-term drought may have inflated the final results in the UNEP accounting of land degradation. The basic data, apparently indicating the rapid creation of desert-like conditions, were collected in the early 1970s at a time of severe drought in sub-Saharan Africa. They made no allowance for the characteristic rainfall variability of the area and in consequence overestimated the long-term effects of the drought. A great deal of the information was obtained using remote satellite sensing of vegetation boundaries, which seemed to confirm the steady encroachment of the deserts over the grassland to the south, and the results were incorporated in the UNCOD report of 1977. Since then, however, they have been widely disputed and subsequent studies have found no evidence that the large-scale desertification described in the 1970s continued into the 1980s, when conditions were slightly wetter and the vegetation recovered to some extent (Hulme and Kelly 1993; Tucker *et al.* 1994).

Insufficient attention to the extent of normal annual fluctuations in vegetation boundaries, combined with inadequate ground control, may have contributed to the problem (Nelson 1990). By the mid-1980s a general consensus was forming among those investigating the issue that the approach to defining desertification in terms of vegetation needed to be re-examined. In parts of East Africa, for example, drought and possibly over-grazing have combined to allow the normal grass cover to be replaced by thorn scrub. On satellite photography this appears as an improvement in vegetation cover, yet from a human and ecological viewpoint it is a retrograde step (Warren and Agnew 1988). A more accurate approach to land degradation may be to study the soil. Vegetation responds rapidly to short-term changes in moisture, but damage to soil takes much longer to occur and to reverse. Thus the measurement of soil conditions – nutrient levels, organic content or structure, for example – may give a more accurate indication of the extent of land degradation (Pearce 1992b).

Failure to appreciate the various potential causes of desertification would also limit the response to the problem. Different causes would normally elicit different responses, and UNEP's application of the societal response to all areas, without distinguishing the cause, may in part explain the lack of success in dealing with the problem (Pearce 1992b). The debunking of some of the myths associated with desertification and the realization that, after much

study, its nature and extent are inadequately understood, does not mean that desertification can be ignored (Thomas and Middleton 1994). There are undoubtedly major problems of land degradation in many of the earth's arid or semi-arid lands and the nations occupying the lands affected appeared at the Earth Summit in Rio di Janeiro in 1992 and proposed a Desertification Convention to address their problems. Negotiations continued after the summit, resulting in the United Nations Convention to Combat Desertification (UNCCD), which was signed by 110 governments in 1994 and ultimately ratified by 124 nations in 1998. A much more comprehensive treaty than earlier efforts, the convention is to seek innovative solutions through national action programmes and partnership agreements, with special consideration of regional requirements, such as those of Africa, where the problem is greatest. It recognizes the importance of both local and international activities, often involving non-governmental organizations, in the fight against desertification, and emphasizes the role of education and training in that fight. Along with the provision of expertise and technology, developed nations are expected to mobilize funding for action programmes, which will involve co-ordination of their activities with those of the recipients (Williams *et al.* 1995). Unfortunately the funding nations – mainly in the developed world – have been slow in responding and there is some concern that the many discussions and negotiations that have followed the signing of the convention in 1994 have yet to be linked with the real problems on the ground (UNEP 2002). In the longer term the movement to combat desertification is seen to be part of the broader objective of attaining the sustainable development espoused at the Earth Summit (CSD 1995).

PREVENTING DESERTIFICATION

Often lost in the complexity of attempts to define the issue of desertification more accurately are two questions that remain of supreme importance to the areas suffering land degradation. Can desertification be prevented? Can the desertification that has already taken place be reversed? In the past, the answer to both has always been a qualified yes, and seems likely to remain so, although some observers take a more pessimistic view (see, for example, Nelson 1990). In theory, society could work with the environment by developing a good understanding of environmental relationships in the threatened areas or by assessing the ability of the land to support certain activities, and by working within the constraints that these would provide. In practice, non-environmental elements, such as politics and economics, may prevent the most ecologically appropriate use of the land. The most common approaches to the prevention and reversal of desertification are listed in Table 9.1.

A typical response to variable precipitation in areas prone to desertification is to consider the good years as a bonus and extend production into marginal land at that time. If the opposite move – retreat from the marginal areas – does not take place when the bad years return, the stage is set for

TABLE 9.1 PREVENTION AND REVERSAL OF DESERTIFICATION

Prevention

Good land use planning and management, e.g.:

■ Cultivation only where and when precipitation is adequate
■ Animal population based on the carrying capacity of the land in driest years
■ Maintenance of woodland where possible

Irrigation managed in such a way as to minimize sedimentation, salinization and waterlogging

Plant breeding to increase drought resistance

Improved long-range drought forecasting

Weather modification, e.g.:

■ Rain making
■ Snowpack augmentation

Social, cultural and economic controls, e.g.:

■ Population planning
■ Planned regional economic development
■ Education

Reversal

Prevention of further soil erosion, e.g. by:

■ Contour ploughing
■ Gully infilling
■ Planting or constructing windbreaks

Reforestation

Improved water use, e.g.:

■ Storage of run-off
■ Well managed irrigation

Stabilization of moving sand, e.g. by:

■ Using matting
■ Re-establishment of plant cover
■ Using oil waste mulches or polymer coating

Social, cultural and economic controls, e.g.:

■ Reduction of grazing animal herd size
■ Technology transfer
■ Population resettlement

Source: D.D. Kemp, *Global Environmental Issues: A Climatological Approach* (2nd edn), London/New York: Routledge (1994).

progressive desertification. Experience in the United States has shown that this can be prevented by good land use planning, which includes not only the best use of the land but also the carrying capacity of the land under a particular use (Sanders 1986). To be effective in an area such as the Sahel, this would involve restrictions on grazing and cultivation in many regions, but not only that. Estimates of the carrying capacity of the land would have to be based on conditions in the worst years rather than in the good or even normal years (Kellogg and Schneider 1977; MacKenzie 1987). In theory, that would slow desertification, but at a considerable socio-economic cost to the inhabitants of the area. The nomadic farmers of the Sahel depend upon the herds that they build up in good years as a form of insurance (Stewart and Tiessen 1990), and survive by selling some of the cattle when crops fail. Restricting the growth of flocks and herds during good years would be expected to reduce the potential for desertification, but by effectively removing a traditional safety net, it is unlikely to have popular support. Any planning would have to ensure that such socio-economic attempts to provide a solution would provide real ecological benefits.

One of the stated aims of the UNCCD was to include a 'bottom up' approach to preventing desertification. It would seem therefore that any developments in regions such as the Sahel, would have to consider the traditional herding techniques of the nomadic pastoralists in the area. These techniques involving nomadism and the acceptance of fluctuating herd sizes seem particularly suited to the unpredictable environment of the Sahel, and the nomadic herdsmen may also know more about dealing with pastoral land use problems than they are given credit for by scientists from the developed nations (Warren and Agnew 1988; Pearce 1992b). They represent perhaps the best way to use land in an area where there is no natural ecological equilibrium, only constant flux. As long as the flux remains within certain limits, desertification can be kept at bay, but too much change in one direction, as occurs during a major drought episode, without concomitant change in herding methods could once again encourage desertification. Developments in the Sahelian countries during colonial times, including the establishment of political boundaries that bore no relation to environmental patterns, the introduction of cash cropping and the encouragement of commercial herding, may well have constrained the traditional response to drought and

therefore contributed to the desertification of the second half of the twentieth century in the area.

The problem of the destruction of woodland will also have to be addressed if desertification is to be prevented. Trees and shrubs protect the land against erosion, but they are being cleared rapidly, mainly for fuel or charcoal. Since the scrub woodland of semi-arid areas provides only limited volumes of energy per unit area, large areas have to be cleared to meet energy needs. In northern Africa, which includes the semi-arid desert margins, more than 9 million ha have been stripped of their wood (UNEP 2002), leaving the land susceptible to erosion and desertification. One hundred years ago in Ethiopia 40 per cent of the land could be classified as woodland; by the second half of the twentieth century only 3 per cent could be designated in that way (Mackenzie 1987). A similar situation has developed in arid northern China, where the removal of trees has been responsible for more than 30 per cent of the area considered to be suffering from desertification (Edmonds 1994). Good land use planning would recognize that certain areas are best left as woodland, and would prevent the clearing of that land for cultivation or the provision of fuelwood. The latter problem is particularly serious in sub-Saharan Africa, where wood – in reality often thorny scrub – is the only source of energy for most of the inhabitants. It also has ramifications that reach beyond the fuel supply. Experience has shown that where wood is not available animal dung is burned as fuel, and although that may supply the energy required, it also represents a loss of nutrients and organic material that would normally have been returned to the soil. Any planning involving the conservation of fuelwood must consider these factors and make provision for an alternative supply of energy or another source of fertilizer.

REVERSING DESERTIFICATION

Many of the techniques that could be employed to prevent desertification are also considered capable of reversing it. There are undoubtedly areas that will never be rehabilitated, but there have also been some successes. In North America, land apparently destroyed in the 1930s has been successfully brought back into production through land use planning and direct soil conservation techniques such as contour ploughing, strip cropping and the provision of windbreaks. Irrigation has also become common. Many of these methods can be (and have been) employed with little modification in areas where desertification is rampant. They have worked well in China, for example (Edmonds 1994), and where they have been introduced in Africa they have generally been successful, although there the immensity of the problem is such that their impact has as yet only been local or regional. Dry-farming techniques have been introduced into the Sudan (CIDA 1985); in Ethiopia, new forms of cultivation similar to contour ploughing have helped to conserve water and prevent erosion (Cross 1985); in Mali and other parts of West Africa, reforestation is being attempted to try to stem the southward creep of the desert (CIDA 1985). Throughout the Sahel, boreholes have been drilled to provide access to ground water, but by attracting larger numbers of animals, the vegetation around the borehole tends to be put under greater stress, which has created new problems in a number of places.

There is clearly no universal panacea for desertification. Solutions will have to be specific to the issue and to the location, and will have to include attention not just to the physical symptoms of the problem, but also to socio-economic issues that have often been overlooked in the past (Hekstra and Liverman 1986). Problems arise from the introduction of inappropriate technology, from the unwillingness of farmers and pastoralists to adopt the new methods and from a variety of economic factors, including, for example, fertilizer costs and the availability of labour (Nelson 1990). To be successful, attempts at reversing desertification must have the support of the local population. It is easy to make the scientifically based decision to rehabilitate an area, but reversing land degradation takes time and those who live there are more likely to be supportive if they can see some form of short-term success. African experience in agro-forestry schemes indicates that higher levels of co-operation are achieved when the farmers can see clear and immediate rewards in addition to the less obvious longer-term environmental benefits (Pégorié 1990). In northern Ghana, for example, reforestation has included the planting of cashew trees that will provide farmers with an income from nuts in as little as three years. On the human side, population growth rates and densities must be examined with a view to assessing human pressure on the land. Overpopulation has traditionally been regarded as an integral part of the drought/famine/

desertification relationship, but that too will have to be re-examined (Mortimore 1989). Ironically some areas in the Sahel and parts of East Africa suffer from rural depopulation as a result of the large number of people who sought refuge in urban areas during the droughts, famine and civil strife of the 1960s, 1970s and 1980s and that has led to other environmental issues.

The fight against desertification has been marked by a distinct lack of success. Recent reassessments of the problem, beginning in the late 1980s, suggest that this may be the result of the misinterpretation of the evidence and a poor understanding of the mechanisms that cause and perpetuate the degradation of the land. The additional research required to resolve that situation will further slow direct action against desertification, but it may be the price that has to be paid to ensure future success.

One global environmental issue that has the potential to impact on drought, famine and desertification is global warming (see Chapter 12). According to the latest report of the Intergovernmental Panel on Climate Change (IPCC) – *Climate Change 2001* – warming is likely to be accompanied by the increased occurrence of drought in much of Africa, arid and semi-arid Asia, Australia and North America (IPCC 2001d). Although drought does not always lead to famine or desertification, given the large areas that are already unable to cope with these issues, it seems likely that warming will maintain and exacerbate existing conditions, allowing the images of soil erosion, dying sheep and cattle and emaciated, malnourished children to remain in the media spotlight far into the future.

SUMMARY AND CONCLUSION

Drought and famine are long-standing, related concerns which have plagued society for centuries and to which solutions have been difficult or, in some cases, impossible to find. Desertification is usually seen as a more modern issue, with recent attention emphasizing the human contribution to the problem. It is, however, a natural process in many areas, having much in common with drought and famine. Although relationships among the three issues are not necessarily causal or absolute – famine can be caused by factors other than drought; not all droughts cause famine or desertification – they can be strong. Drought, through its impact on plants and animals, can destroy the food supply and initiate famine. Drought and famine between them contribute to desertification, by allowing productive land to become barren. Desperately attempting to survive, people and animals strip the land of its natural and cultivated vegetation. Without water new plants will not grow, leaving the land bare and open to desertification, which ensures that the famine will persist or at least recur with some frequency. All three problems represent environmental disruption – drought in the hydrosphere, famine in the food chain and desertification in soil and vegetation – but the mechanisms that contribute to these disruptions are not always natural. Poor water management can enhance drought, for example, socio-economic factors can aggravate famine resulting from drought and can also initiate famine even when water is plentiful, over-grazing or poor land management can cause desertification. The complex combinations of these natural and socio-economic drivers create conditions that lead to serious environmental disruption. Society has had limited success in dealing with drought, famine and desertification, and there is no sign that that situation is likely to change in the near future.

SUGGESTED READING

Moeller, S.D. (1999) *Compassion Fatigue: How the Media Sold Disease, Famine, War and Death*, New York/London: Routledge. A broad consideration of the way in which the media report disaster, with the section on famine in Africa particularly pertinent for this chapter.

Middleton, N.J. and Thomas, D.S.G. (eds) (1997) *World Atlas of Desertification* (2nd edn), London: UNEP/Edward Arnold. A cartographic, descriptive analysis of the current state of desertification, covering a wide range of topics beyond the physical aspects of the issue, including poverty, biodiversity and population movements.

Worster, D. (1982) *Dustbowl: The Southern Plains in the 1930s*, New York: Oxford University Press. An environmental history of the Dustbowl era, which considers the physical, socio-economic, political and technological factors that combined to cause the disaster.

QUESTIONS FOR REVISION AND FURTHER STUDY

1   Examine the annual reports of such aid organizations as Oxfam, Save the Children, CARE and UNICEF. How do they contribute to the solution of problems of drought, famine and desertification? Are their approaches to the problems always environmentally sound or is there some validity in the suggestion that they tend to treat the symptoms rather than the disease? Even if they do, is it necessarily the wrong approach in all cases?

2   Although drought may be initiated by natural processes, its impact is often aggravated by human activities. What are they, and how can their harmful effects be reduced?

3   List the human activities and natural processes that contribute to desertification and the actions needed to reduce their impact. Identifying the problem in this way is relatively easy, yet societies in different parts of the world have had very little success in preventing desertification or reversing it. Why is that?

10
Air Pollution and Acid Rain

After reading this chapter you should be familiar with the following concepts and terms:

acid precipitation	fuel desulphurization	radon
acid rain	fuel switching	RAINS
acid rain impacts – aquatic,	HAPs	RSP
health, terrestrial,	indoor air pollution	SCR
urban	lead	scrubbers
aerosols – anthropogenic,	London smog	secondary aerosols
natural	LIMB	secondary pollutants
air pollution	mercaptans	sick building syndrome
Arctic Haze	meteorology and pollution	smog
asbestos	oxides of nitrogen	SPM
Asian Brown Cloud	ozone	spring flush
Bhopal	particulate matter	sulphate aerosols
bronchitis	pH	sulphur dioxide
carbon monoxide	photochemical smog	temperature inversion
carboxyhaemoglobin	pollution sources – area,	tetraethyl lead
catalytic converter	elevated, line, point,	tobacco smoke
criteria pollutants	surface	topography and
emphysema	precipitation scavenging	pollution
FBC	precursors	tree dieback
FGD	primary aerosols	TSP
formaldehyde	primary pollutants	*Waldsterben*

THE CONSTITUENTS OF THE ATMOSPHERE AND AIR POLLUTION

The atmosphere consists of a mix of gases plus liquid and solid particles in proportions that vary in both time and place (see Chapter 2). This variability is characteristic of the atmosphere, brought about by changes in the input of gases and particulate matter from natural sources and the rate at which they are removed by its built-in cleansing systems. When the atmosphere is unable to rid itself of the material being added to it, the build up of gases and aerosols causes it to become polluted. While this may occur as a result of natural processes, pollution is most often considered in human terms, being seen as the result of changes in the natural state of the atmosphere caused by anthropogenic inputs and identified visually or through its impact on human health or the crops upon which people depend or the structures in which they live and work. The impact is felt at a variety of scales and takes different forms. At a global scale, for example, the build-up of particulate matter or aerosols has increased the overall turbidity (dirtiness or dustiness) of the atmosphere and disrupted the earth's energy budget, the inability of the atmosphere to recycle carbon dioxide as rapidly as it is being emitted has led to global warming (see Chapter 11),

(a)

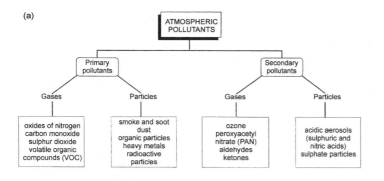

(b)

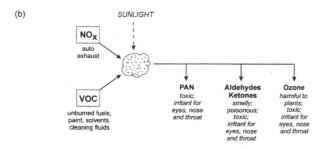

Figure 10.1   (a) Common atmospheric pollutants. (b) The formation and composition of photochemical smog

while the release of chlorofluorocarbons (CFCs) has caused the ozone layer to become thinner and exposed the earth's surface to excess solar radiation (see Chapter 11). At the local or regional level, air pollution is often most obvious in urban areas, where it has caused deterioration in the health and quality of life of urban dwellers for centuries in some cities. Under certain conditions locally produced pollutants can be spread to cause regional problems, as in the case of acid rain, when sulphur dioxide emissions, carried downwind by regional circulation patterns, have led to environmental damage hundreds of kilometres from the pollution source.

Atmospheric pollutants take many forms, but they can be classified according to the way they are formed into primary and secondary groupings (Figure 10.1). Primary pollutants are those emitted directly from a specific source, and while they may be capable of causing environmental problems in their original form, they can also be transformed, through physical and chemical processes in the atmosphere, into more complex secondary pollu-

tants, which are often more disruptive. Through oxidation and combination with water, for example, the primary pollutant sulphur dioxide is converted into sulphuric acid, the main component of acid rain, and the toxic ingredients of photochemical smog are produced by the action of sunlight on primary pollutants such as oxides of nitrogen or volatile organic compounds (Figure 10.1b). All of the processes begin with the emission of a pollutant in the form of an aerosol or a gas, and to understand the development and impact of these processes it is necessary to consider the sources, properties, nature and distribution of the aerosols and gases in the environment.

ATMOSPHERIC AEROSOLS

Variability among the constituents of the atmosphere is greatest with the liquid and solid particles, or aerosols, which range in size from a few molecules to a visible grain of dust. Within that range, the main mass of particulates is concentrated

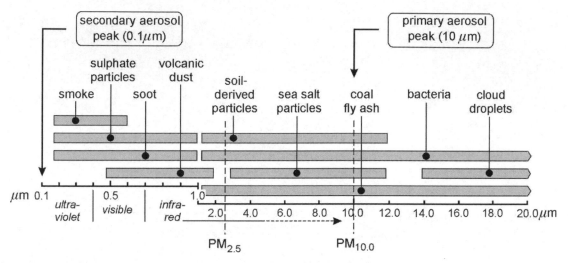

Figure 10.2 A comparison of the size range of common aerosols with radiation wavelengths

in two peaks, one between 0.01 micrometres (µm) and 1 µm, centred at 0.1 µm, and one between 1 µm and 100 µm, centred at 10 µm (Figure 10.2). In dealing with air quality related to particulate matter, standards have been established to limit concentrations of particles with an aerodynamic diameter of less than 10 µm – PM_{10} standards – and a new standard or $PM_{2.5}$ dealing with particles of less than 2.5 µm has been proposed by the United States Environmental Protection Agency (USEPA) (Nazaroff and Alvarez-Cohen 2001). The larger particles between 1 µm and 100 µm are primary aerosols, and include soil, dust and industrial emissions, formed as a result of the physical break-up of material at the earth's surface. The smaller particles are called secondary aerosols, formed as a result of chemical and physical processes in the atmosphere. They include aggregates of gaseous molecules, water droplets and chemical products such as sulphates, hydrocarbons and nitrates. The total global aerosol production is difficult to calculate, but estimates range between 1.0 and 8.5×10^9 tonnes per annum from both natural and human sources (Bach 1979; Bridgman 1994; Penner 2002).

Sources of aerosols

Natural processes such as volcanic activity, forest and grass fires, evaporation, local atmospheric turbulence and biological processes produce large volumes of aerosols, and human activities involving quarrying and mining, industrial processing, energy consumption and agriculture also add to the level of particulate matter in the atmosphere (Figure 2.9). The production of aerosols by individual human events cannot match the rapid and intense contributions of natural activities such as volcanic eruptions. When Mount Tambora erupted in Indonesia in 1815, for example, it released an estimated 150 million tonnes of particulate matter into the atmosphere in a matter of weeks (Rampino and Self 1984) and, more recently, the eruption of Mount Pinatubo in the Philippines produced 30 million tonnes of dust and sulphate aerosols (Brasseur and Garnier 1992). The human contribution, if less spectacular, is steady and ongoing, and in an average year with limited volcanic activity may account for as much as 48 per cent of the finer aerosol fraction (< 2 µm) (Penner 2002). It might be expected that pollution generated from human sources would come mainly from industrial activities in the developed nations of the northern hemisphere, but data from at least some industrialized nations indicate that emissions of particulate matter have been declining since the 1980s (Figure 10.3). In contrast, levels of aerosols and other pollutants have been rising in major cities in the developing world, and in 1999 an extensive area of pollution haze – the Asian Brown Cloud – covered more than 10 million km² of South and Southeast Asia (UNEP 2002) (see Box 10.1, Arctic Haze and the Asian Brown Cloud).

BOX 10.1 ARCTIC HAZE AND THE ASIAN BROWN CLOUD

Both the Arctic Haze and the Asian Brown Cloud are indicators of society's continuing ability to increase atmospheric turbidity through anthropogenic aerosol loading. Although they occupy different geographical locations, they contain similar contaminants and have the potential to threaten the environment at both regional and worldwide scales.

ARCTIC HAZE

Arctic Haze was first recognized in the 1950s from aircraft flying high-latitude routes in the Arctic. Scientific investigation showed that it consisted of dust, soot and sulphate particles, with the sulphate aerosols providing about 90 per cent of the total, and the haze being evident because the particle diameters were of the same order as the wavelength of visible light. The haze is not of local origin. It is created in mid-latitudes by industrial activities in Europe and Asia, before being carried polewards to settle over the Arctic. The haze is most pronounced between December and March, for several reasons, including the increased emission of pollutants at that season, the more rapid and efficient poleward transport in winter and the longer residence time of haze particles in the highly stable Arctic air at that time of year. Arctic air pollution increased from the mid-1950s in parallel with increased aerosol emissions in Eurasia, declining only in the early 1990s when emissions from smelters in Russia decreased as a result of the economic slowdown in that country. The most obvious result of the haze has been a measurable reduction in visibility and the perturbation of the regional radiation budget, but the haze also contains a variety of other substances, including pesticides, persistent organic pollutants (POP), toxic metals and, from time to time, radioactive particles. These often have a more subtle impact on the Arctic environment, being absorbed by the local plants and animals, where they bioaccumulate before being passed on to the indigenous human population. Arctic regions have become a sink for many toxic substances, which accumulate to levels that are often higher than in the source regions to the south. Attempts to mitigate the impact of the haze include the improvement of air quality through such international programmes as the Convention on Long-range Transboundary Air Pollution and the Convention on Persistent Organic Pollutants. These are longer-term measures, however, and in the immediate future Arctic Haze is likely to continue to cause problems for the northern polar and sub-polar environment.

ASIAN BROWN CLOUD

Compared to Arctic Haze, the Asian Brown Cloud (ABC) is a relatively recent discovery. Scientists working on the Indian Ocean Experiment (INDOEX) in early 1999 observed a brown haze covering an area of some 10 million km^2 above South and Southeast Asia as well as the adjacent Indian Ocean, and rising 3 km into the atmosphere. The ABC includes a mixture of dust, soot, sulphates, nitrates, organic particles and fly-ash, emitted by fossil fuel-burning plants, by biomass burning in forested and rural areas, and even by the millions of inefficient cooking stoves used in the region. The cloud may already be reducing the amount of solar radiation reaching the earth's surface by 10–15 per cent, altering atmospheric temperature and circulation patterns sufficiently to disrupt the Asian monsoon and cause a decline in precipitation in India, Pakistan and Afghanistan by as much as 40 per cent. With the support of the United Nations Environment Program (UNEP), Project Asian Brown Cloud has been initiated to study the impact of the haze on regional precipitation patterns, water balance, agriculture and human health. Seen as a consequence of the unsustainable use of energy in Asia, the problem of the ABC was addressed at the World Summit on Sustainable Development held in Johannesburg, South Africa in 2002. Although currently a regional problem, the ABC has the potential to spread sufficiently to have global impacts over the next several decades as population growth and development continue in Asia.

For more information see:

AMAP (1997) *Arctic Pollution Issues: A State of the Arctic Environment Report*, Oslo: Arctic Monitoring and Assessment Programme.

Barrie, L.A. (1986) 'Arctic air pollution: an overview of current knowledge', *Atmospheric Environment* 20: 643–63.

UNEP (2002) *Global Environmental Outlook 3*, London/Sterling VA: Earthscan.

Catania

Plate 10.1 Volcanic activity is an important source of atmospheric particulate matter. This photograph shows the eruption of Mount Etna in Sicily in July 2001, which covered the town of Catania with a layer of ash and sent a plume spreading to the south-east. (Courtesy of NASA: Earth Observatory.)

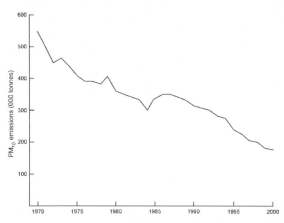

Figure 10.3 Total PM_{10} emissions in the United Kingdom, 1970–2000

Source: based on data in the UK National Air Quality Archive. Viewed at http://www.airquality.UK (accessed 3 March 2003).

Atmospheric distribution of aerosols

Under normal circumstances, almost all of the total weight of the aerosol particles is concentrated in the lower 2 km of the atmosphere, with a mean residence time of between five and nine days, which is sufficiently short that the air can be cleansed in a few days once the emissions of particulate matter have ceased. The equivalent time in the upper troposphere is about one month, and in the stratosphere the residence time increases to two to three years. As a result, anything added to the upper troposphere or stratosphere will remain in circulation for a longer time, and its potential environmental impact will increase.

Much of the particulate matter that reaches the stratosphere arrives as a result of volcanic eruptions sufficiently powerful to penetrate the tropopause. Debris from the explosive eruption of Krakatoa in 1883, for example, reached an altitude of 50 km, well into the stratosphere (Lamb 1970), and the force of the eruption of Mount Pinatubo penetrated the tropopause at about 14 km and carried debris up for another 10–15 km (Gobbi *et al.* 1992). Other natural events such as dust storms seldom carry particles beyond the troposphere, although dust may be carried over considerable distances under favourable wind conditions. During the Pleistocene period, for example, immense amounts of dust were picked up by the wind from land laid bare by retreating ice sheets and glaciers, to be deposited hundreds of kilometres away as loess. Desert dust from the Sahara is regularly carried to South and Central America as well as the United States on the prevailing easterly winds (Perry *et al.* 1997), while dust from the Gobi Desert and adjacent arid lands of northern China is a significant constituent of the atmosphere in Beijing in April and May.

Particulate matter generated by human activities tends to remain closer to the earth's surface than the natural variety, as is often apparent in large urban areas. Occasionally, human and natural events combine to increase the amount and distribution of aerosols into the upper troposphere, however. In the Australian summer of 1983, for example, dust from the desiccated wheatfields and overgrazed pastures of Victoria and New South Wales was picked up by strong winds and the resulting dust storm carried particles as high as 3.6 km into the atmosphere, before depositing 106 kg of dust per hectare on the city of Melbourne (Lourenz and Abe 1983). A much more common source of anthropogenically produced aerosols are the high-flying commercial aircraft that inject soot particles directly into the upper atmosphere. The annual production of elemental carbon from that source is estimated at 16,000 tonnes, of which 10 per cent is added directly into the stratosphere and a large proportion is released into the upper layers of the troposphere (Pueschel *et al.* 1992).

Atmospheric cleansing mechanisms

Whether the aerosols released into it are from natural or human sources, the atmosphere has mechanisms by which it can cleanse itself. Larger particles tend to fall close to their sources as dry sedimentation brought about by the effects of gravity. Smaller particles normally remain in the atmosphere for longer periods, but often they combine together, or coagulate, until they become large enough to fall back to the surface. Individual soot particles, for example, with a diameter of approximately 0.1 μm, tend to link together initially into branches and chains, gradually forming loose spherical aggregates of perhaps 0.4 μm, before falling out of the atmosphere (Appleby and Harrison 1989). One of the most efficient of the atmosphere's self-cleansing mechanisms is precipitation scavenging. It is a process by which rain and snow wash particulate matter out of the atmosphere, and rain

coloured by dust or smoke during the process may be described as 'red rain' or 'black rain'. Precipitation scavenging is known to accompany forest fires, and black, sooty rain fell following the fire-bombing of Hamburg and Dresden during the Second World War, when major conflagrations produced dense smoke clouds. Black rain also fell at Hiroshima and Nagasaki following the nuclear bombing of those cities (Peczkic 1988; Tullett 1984). Effective as these mechanisms can be, cleansing is never complete. There is always a global background level of atmospheric aerosols, which reflects a dynamic balance between the output from natural processes and the efficiency of the cleansing mechanisms. The addition of aerosols from industrial and agricultural activity over the past 200 years or so should be causing the level to rise, and that appears to be so, although the amounts are poorly quantified and the fraction of the background aerosol level that is of anthropogenic origin is not easily estimated (IPCC 2001a).

ATMOSPHERIC GASES

The gaseous composition of the atmosphere is remarkably uniform throughout, particularly in the troposphere, where most of the air is located (see Chapter 2). Oxygen and nitrogen, which together account for 99 per cent of all the gases in the atmosphere, maintain their levels as a result of efficient recycling systems. The relative proportions of minor gases can change quite dramatically or slowly and steadily, however, and it is these changes that lead most often to pollution problems.

One of the most important of the minor gases is carbon dioxide, which makes up only 0.03 per cent of the atmosphere by volume. Since the second half of the eighteenth century, atmospheric carbon dioxide levels have been rising steadily as emissions exceed the rate at which it can be recycled into its individual elements. Carbon dioxide is not normally considered as a pollutant. It is a normal constituent of the atmosphere and, being colourless and odourless, its increase has no direct aesthetic impact on the environment and is therefore easily missed. Over the past several decades, however, it has become clear that it does share some of the attributes of gases normally considered to be pollutants, in that it has the potential to have direct and indirect effects on human health and well-being. Rising levels of carbon dioxide bring about the enhancement of the greenhouse effect to cause global warming, which in turn is expected to impact significantly on the natural flora and fauna and cultivated crops upon which society has come to depend. Being produced from a wide range of sources, carbon dioxide emissions are difficult to control, but it is now widely accepted that if global warming is to be curbed, then some form of control must be attempted (see Chapter 11).

There are many other gases that from time to time become constituents of the atmosphere at levels that allow them to be considered pollutants. These include sulphur dioxide, oxides of nitrogen (NO_x), ozone, hydrogen sulphide and carbon monoxide, along with a variety of exotic hydrocarbons and industrial chemicals that even in small quantities can be harmful to the environment. The presence of sulphur dioxide as a pollutant in urban areas has been recognized for some time, but its full impact was not widely appreciated until it contributed to an estimated 4000 deaths during a major smog episode that hit London, England, in 1952

(Brimblecombe 1987) (Figure 10.4). The role of CFCs went similarly unnoticed for almost forty years. Developed in the 1930s, this group of inert gases was not even considered as a source of potential pollutants until the mid-1970s, when their impact on the stratospheric ozone layer was recognized (see Chapter 11). Given that record, it is quite possible that there are pollutants in the atmosphere, causing damage to the environment but as yet unrecognized.

In sharp contrast there have been times when the emission rate of exotic, toxic industrial gases has increased so rapidly that the results have been disastrous. Such was the case in Bhopal, an industrial city in central India, where in 1984 some forty tonnes of methyl isocyanate, a highly toxic chemical used in the manufacture of pesticides, leaked from a chemical plant and drifted over adjacent residential areas. An estimated 2500 people were killed in a matter of hours through exposure to the noxious gas, and a further 200,000 suffered respiratory problems, temporary blindness and severe vomiting (Shrivastava 1987).

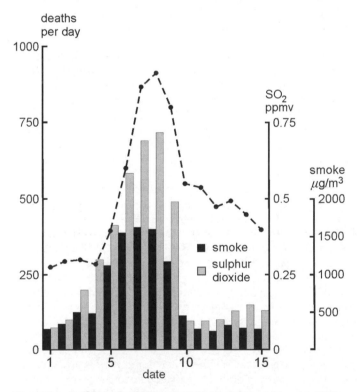

Figure 10.4 Pollution and the death rate in London, England, December 1952
Source: based on data in Bates (1972).

From time to time, exotic gases such as the methyl isocyanate that struck Bhopal cause serious pollution problems, but the most widely monitored gaseous pollutants are gases that are naturally present in the atmosphere. They include carbon monoxide, sulphur dioxide, oxides of nitrogen (NO_x) and ozone (UNEP/WHO 1996), which may be released as a result of biological activity, created during volcanic eruptions or produced by natural forest or grass fires. Increasingly, however, their presence is associated with emissions from industrial or vehicular sources and, since these sources are most frequently found in urban areas, these gases are common constituents of urban air. Also associated with air quality in urban areas is indoor air pollution, caused by the trapping of gaseous and other pollutants within industrial, commercial and residential buildings, creating a threat to the health of employees and residents that may surpass the threat from outdoor urban pollution.

URBAN AIR POLLUTION

Problems of waste disposal arise whenever people congregate in large numbers – as they do in cities. Disposal in urban areas is required for solid waste (see Chapter 6), waste such as sewage released into waterways (see Chapter 8) and also for airborne wastes emitted into the atmosphere in and above cities. When the environment is incapable of dealing with the volume of wastes released into it pollution results. Urban air pollution is not a new issue. Large cities such as London, England, were plagued by it for hundreds of years before measures were taken to deal with it in the second half of the twentieth century (Brimblecombe 1987). Air pollution was one of the elements that elicited a high level of concern during the heyday of the environmental movement in the late 1960s and early 1970s. It was most common in large cities that had high seasonal heating requirements, were heavily industrialized, had large volumes of vehicular traffic or experienced a combination of all three. Even then, however, the serious nature of the problem was already appreciated and air pollution ordinances were beginning to have an effect. In Pittsburgh, for example, the introduction of smokeless fuel, and the establishment of emission controls on the iron and steel industry, brought a steady reduction in air pollution between 1945 and 1965 (Thackrey 1971). Similar improvements were achieved in London, where sunshine levels in the city centre increased significantly following the Clean Air Act of 1956 (Jenkins 1969). The replacement of coal by natural gas as the main heating fuel on the Canadian prairies in the 1950s and 1960s allowed urban sunshine hours to increase there also (Catchpole and Milton 1976). Success was achieved mainly by reducing the level of particulate matter in the atmosphere. Little was done to reduce the gaseous component of pollution, except in California, where, in 1952, gaseous emissions from the state's millions of cars and trucks were scientifically proved to be the main source of photochemical smog (Leighton 1966). Prevention of pollution was far from complete, but the obvious improvements in visibility and sunshine totals, coupled with the publicity that accompanied the introduction of new air quality and emission controls in the 1970s, caused the level of concern over urban pollution to decline markedly by the end of the decade.

MODERN URBAN AIR POLLUTION

By the end of the twentieth century, however, it was clear that urban air pollution was again on the rise, but differing in nature and distribution from the pollution of the earlier period. The air quality in the old, heavy industrial cities of the developed world that had experienced major pollution problems in the 1950s and 1960s improved greatly in the last quarter of the twentieth century, in part because of the introduction of effective emission controls, but also because of the decline in the importance of heavy industry. In Birmingham, for example, once the centre of England's industrial 'Black Country', sulphur dioxide levels declined by over 90 per cent and smoke concentrations by 75 per cent between 1960 and 1992 (UNEP/WHO 1996). The main threat to its air quality now comes from pollutants emitted by vehicles that travel the five motorways that converge on the city. Birmingham's experience – a change from coal to petroleum-based emissions accompanying an overall decline in pollution – has been repeated in most industrial cities in the developed world. This positive trend is being threatened, particularly in North America, by a rapid increase in vehicular traffic in most large cities. The resulting increase in emissions negates the reduction possible from current emission controls; smog alerts have become a regular feature of life in large cities and many jurisdictions are in the process of introducing even more stringent controls.

Overall, however, the focus on urban air pollution has shifted to the large and rapidly growing cities in the developing world, where air quality standards are applied less rigorously than elsewhere, or lag behind what is required to deal with the major increase in emissions generated by a rapidly growing population and the energy use that accompanies it. In India and China, the use of coal to produce electricity or to fuel industry has created pollution containing sulphur dioxide and particulate matter reminiscent of the industrial smog of cities in the developed world several decades ago. During the winter months the large industrial cities of south China disappear from satellite images, hidden beneath a thick blanket of smog for days at a time, and in India there is a significant death toll associated with high pollution levels (Figure 10.5). The most rapidly growing form of air pollution in the developing world is that associated with the burning of petroleum. Cities such as Bangkok, Hong Kong, Lagos, Sao Paulo and Mexico City, for example, all suffer from rapidly growing numbers of cars, trucks, buses and motor bikes, which release high volumes of exhaust fumes containing NO_x and unburned hydrocarbons. Under the high tempera-

tures and abundant sunshine available in these tropical and subtropical locations, the primary pollutants are converted into noxious photo-chemical smog (Figure 10.1b). Although the governments of the developing nations are aware of the environmental, economic and health issues associated with air pollution, and many have attempted to establish and enforce appropriate air quality standards, their efforts have been constrained by limited financial resources that prevent them from advancing the standards quickly enough to match the increase in emissions from the growth of traffic.

It has become increasingly difficult for the atmosphere to cleanse itself in urban areas. Continental-scale air pollution features such as the Arctic Haze and the Asian Brown Cloud suggest that emissions are spilling over from the urban airsheds into the wider environment to contribute to global atmospheric turbidity. Thus, despite significant achievements in combating urban air pollution from the mid-twentieth century on, when the aerosol content of the atmosphere began to decline and levels of gaseous pollutants such as sulphur dioxide were reduced, control is far from complete and effective air quality management strategies continue to be lacking in many areas. To develop appropriate strategies, it is necessary to understand the mechanisms involved, to monitor existing levels of pollutants and to establish standards that will protect human and environmental health. To that end, the UNEP and the WHO have developed a joint programme – GEMS/Air – working along directions recommended at the Earth Summit in 1992, to generate and utilize air quality information in such a way that the problem of urban air pollution can be managed for the benefit of society and the environment (UNEP/WHO 1996).

CAUSES OF URBAN AIR POLLUTION

Urban air pollution, like water pollution and the environmental problems associated with solid waste, is caused by the inability of the local environment to accommodate the level of waste produced by large numbers of people concentrated in relatively small areas. The mechanisms that normally bring about the removal of pollutants from the atmospheric environment are present in and above cities, but the rate at which the pollutants are emitted by urban activities exceeds the rate of removal and they accumulate in the urban atmosphere until

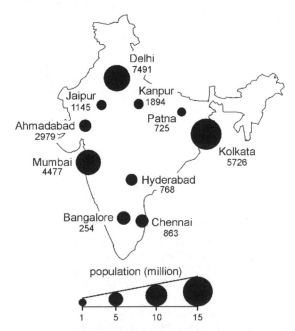

Figure 10.5 Estimated annual premature deaths due to ambient air pollution in Indian cities

Source: based on data from GIS Development. Viewed at http://www.gisdevelopment.net/application/natural.hazards/overview/nho0019.htm (accessed 27 February 2003).

emission rates decline and removal rates can catch up. Although the potential for air pollution is a direct result of the congregation of people and their activities, large cities have their own peculiar climates, some elements of which tend to encourage pollution. With the increase in surface roughness caused by the presence of buildings, for example, average wind speeds are lower in urban areas, which reduces the rate at which pollutants are dispersed. Natural conditions in and around a city can also interfere with dispersal rates and thus increase or diminish pollution potential. Local topography, for example, has an important role in the build-up of pollution levels in some cities. Urban centres in areas with few topographic barriers to air movement – on the Great Plains, for example – are not immune to air pollution, but because they are often windy they are said to be well ventilated, and that helps to disperse emissions relatively quickly, thus preventing pollution levels from rising. In contrast, topography that prevents pollutants from escaping or causes them to become concentrated in specific areas increased the potential for serious pollution. The mountains that curve around the eastern side of Los Angeles to create the basin in which the city lies prevent the dispersal of emissions from the hundreds of thousands of cars and trucks driven in and around the basin every day, allowing the creation of the dangerous photochemical smog that has become synonymous with the name of the city. Mexico City suffers similar (frequently higher) levels of air pollution as a result of its location in a

basin surrounded by high mountains, as does the industrial city of Chongqing in Sichuan, China. Elsewhere, the topography may allow pollutants to be concentrated and moved as a plume away from their points of origin to affect adjacent regions. In southern British Columbia, Canada, for example, contaminated air from the city of Vancouver and its suburbs is funnelled eastwards up the Fraser River to pollute the smaller communities in the valley (Figure 10.6).

In contributing to the distribution and concentration of pollutants, topography often works in combination with meteorological conditions in the layers of the atmosphere closest to the earth's surface – the atmospheric boundary layer. Conditions in the lower atmosphere are particularly important, since they encourage or discourage dispersal and mixing. Turbulent conditions associated with advective (horizontal) and convective (vertical) air movements tend to produce a thicker mixing layer, reducing the concentration of pollutants in the urban atmosphere, whereas calm conditions with descending air will produce a thinner mixing layer in which high levels of pollution can build up. The presence of turbulent or calm conditions depends upon the stability of the atmosphere. An unstable atmosphere is one in which upward motion of the air and its contents is encouraged, which leads to a thicker mixing layer and a reduction in the concentration of pollutants. Instability is also often accompanied by precipitation, which helps to clean the atmosphere by washing particulate

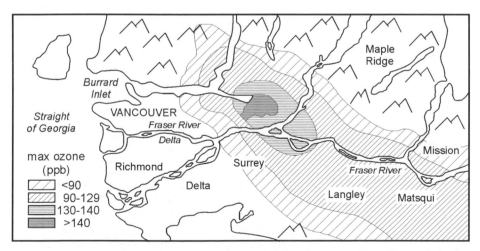

Figure 10.6   Pollution in the Fraser River valley in British Columbia, Canada
Source: after Ministry of Environment, Lands and Parks/Environment Canada (1993) with permission.

matter and gases back to the earth's surface. In contrast, a stable atmosphere is characterized by calm conditions, which limit dispersal, and sometimes by a net downward movement of the air, which narrows the mixing layer and creates a high potential for pollution. The stability of the atmosphere and therefore its potential for pollution can be gauged by comparing atmospheric lapse rates (Figure 10.7).

Natural conditions involving topography and the meteorology of the atmospheric boundary layer are very effective in promoting and sustaining pollution (see Box 10.2, Temperature inversions and air pollution). Under the right conditions – high pressure, clear skies, topographic barriers, for example – atmospheric mixing or the ventilation of an area may be so poor that even small communities

may suffer pollution levels that occasionally equal those in major industrial centres. In settlements such as Fairbanks in Alaska and Whitehorse in the Canadian Yukon during the winter months, the combination of smoke from wood-burning furnaces used for residential heating and gases from car and truck exhausts creates a particularly noxious mix of pollutants that tends to persist in the atmosphere under the cold anticyclonic conditions common in the Arctic in the winter. However, no matter how conducive these conditions may be to the development of high pollution levels, they do not have an effect unless there are sources to provide the pollutants in the first place.

SOURCES OF URBAN AIR POLLUTION

Individual emission sources are many and varied, but can be categorized in several ways. They may be grouped according to their origin in industrial, residential and transport activities, for example, or identified as point, line or area sources, depending upon their location and physical nature (Table 10.1). Point sources include residential chimneys and industrial flues or smokestacks; line sources are located along roads, railways and airport flight paths and involve mobile sources of pollution such as automobiles, locomotives and aircraft; area sources are less common in urban areas, except perhaps in the case of a major fire covering several blocks, but where there is a dense network of point and line sources, the overall effects are such that in reality the entire urban area may be considered as a source of pollution. Sources can also be identified as surface or elevated sources. Pollutants from most line sources are emitted at the surface, although aircraft landing at or taking off from an airport can be both elevated and surface sources. Most point sources are elevated, but with a major difference between residential chimneys, which are of limited height, and industrial smokestacks that range in height from tens of metres to between 300 m and 400 m in some cases.

Among these sources, the easiest to combat are the point sources. They are not difficult to locate and the pollutants released from them can be easily tested for composition, concentration and rate of release. These criteria are much more difficult to establish for individual mobile sources, and attempts at reducing pollution from cars and trucks often involves the use of average emission data,

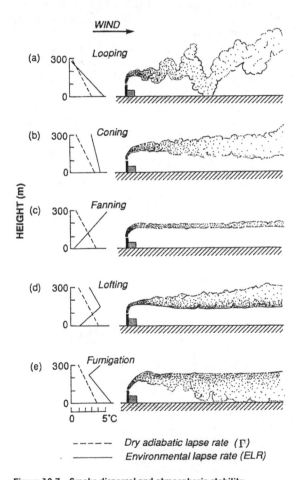

Figure 10.7 Smoke dispersal and atmospheric stability
Source: after Bierly and Hewson (1962) with permission from the American Meteorological Society.

BOX 10.2 TEMPERATURE INVERSIONS AND AIR POLLUTION

One of the most common meteorological conditions promoting pollution is the temperature inversion, a feature that leads to a highly stable atmosphere and has been implicated in many of the world's most serious pollution episodes. An inversion is the reversal of the normal temperature pattern in the troposphere, where temperatures decline with altitude (see Chapter 2). In an inversion the temperature rises with altitude because of the presence of a layer of warm air above the cooler surface air. The warm layer acts to dampen convective activity and mixing. Emissions are trapped beneath the inversion and the surface layer becomes highly polluted, particularly if the inversion remains in place for some time.

Inversions are commonly caused by strong radiation cooling of the ground under clear skies at night (Figure 10.8a). In valleys or depressions, cool air drainage downslope on to lower ground reinforces the radiation cooling (Figure 10.8b). Temperature inversions are also caused by the adiabatic warming of descending air in an anticyclone or high-pressure system. Serious smog episodes in London in 1952 and New York in 1963 were caused by a combination of radiation cooling and stationary high-pressure systems. Local inversions in coastal areas are frequently caused by the incursion of cold air off the sea or off large lakes (Figure 10.8c). Being denser than the warm air over the land, the sea air pushes under it, raising it to form a warm inversion layer aloft. This type of marine inversion layer contributes to the pollution problems of Los Angeles when relatively cool air from offshore moves inland and displaces the warm air over the basin upward. Since the altitude of the inversion layer so formed is normally less than that of the surrounding mountains, the combination of the inversion lid and the mountain barrier acts to contain the pollution in the basin. When air masses cross mountain barriers and begin to descend over the downwind side, the air is subject to adiabatic warming and its temperature rises. This is the origin of the warm Chinook winds of North America and the Foehn winds of Alpine Europe. The presence of very cold air at the surface sometimes prevents the warm air from reaching the surface and it spreads instead across the atmosphere above the cold air (Figure 10.8d). This effectively creates a temperature inversion and raises the potential for the accumulation of pollution over the area.

For more information see:

Oke, T. (1987) *Boundary Layer Climates* (2nd edn), London: Methuen.

Turco, R.P. (1997) *Earth under Siege: From Air Pollution to Global Change*, New York: Oxford University Press.

---

based on levels established during the manufacturing process through the inclusion of pollution control devices in the engine and exhaust system. Since emission levels will change during the life of an engine, some jurisdictions carry out random spot checks or require regular emission testing – tailpipe emission checks – as part of the licensing process in an attempt to reduce pollution. Emissions from elevated sources tend to disperse more widely and rapidly than those from surface sources and this led to the building of higher and higher industrial smokestacks in the 1960s and 1970s – the 'tall stacks

TABLE 10.1 TYPES OF AIR POLLUTION SOURCES

Source	Duration	Height	Example
Point	Continuous	Ground	Camp fire; bonfire
		Elevated	Chimney; smokestack
	Limited duration	Ground	Explosion; idling car exhaust
		Elevated	Shell burst
Line	Continuous	Ground	Busy highway
	Limited duration	Ground	Secondary highway
		Elevated	Airport flight path
Area	Continuous	Variable	Urban fire; forest or grass fire

Source: based on information in T. Oke, *Boundary Layer Climates* (2nd edn), London: Methuen (1987).

(a)

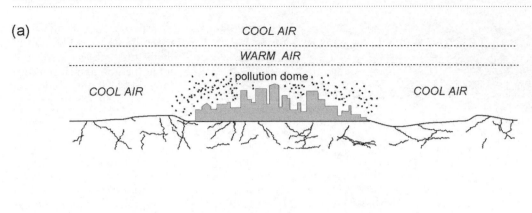

(b)

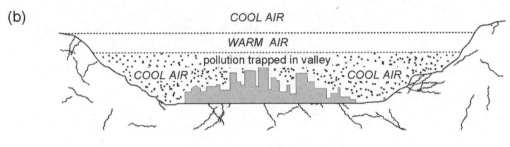

(c)

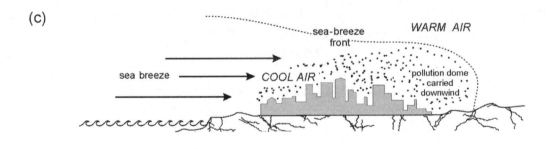

(d)

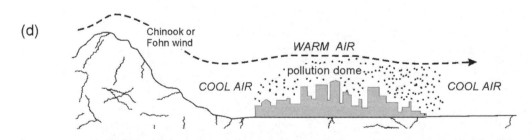

Figure 10.8 Atmospheric pollution associated with temperature inversions: (a) inversion produced by anticyclonic conditions, (b) inversion produced by cold air drainage into valley, (c) inversion produced by a sea-breeze front, (d) inversion produced by the downslope adiabatic warming of air flow

Plate 10.2 The narrow urban canyons in large cities encourage higher pollution levels as a result of reduced ventilation which prevents the dispersal of pollutants emitted from urban vehicular traffic. (Courtesy of Heather Kemp.)

policy'. While this helped to reduce pollution levels in the immediate urban area, it also led to the spread of pollutants downwind to impact on areas with few emissions of their own. The higher smokestacks do nothing to combat surface emissions. The presence of tall buildings along with the reduced air flow in the central areas of most large cities restricts ventilation and allows the build-up of high levels of pollutants at street level.

Industrial sources

Most severe urban pollution episodes are a result of emissions from a combination of industrial, residential and vehicular sources, but the contribution from each sector will vary from city to city. The main pollutants in the air above the new industrial cities that grew out of the industrial revolution in the eighteenth and nineteenth centuries were produced by the use of coal in the iron and steel industries and included high levels of particulate matter as well as acid gases such as sulphur dioxide and the sulphuric acid aerosols derived from it. When electricity became an increasingly important source of energy in the twentieth century it was commonly produced by burning coal, and as a result it contributed smoke and acids to the urban atmosphere. Cities that depended upon the smelting of copper and nickel and other non-ferrous metallic ores also suffered high levels of acidic emissions that devastated the adjacent terrestrial and aquatic environments. Sudbury, Ontario, provides a classic example of the effects of unregulated acid emissions on the environment. When the nickel and copper deposits in the region were first exploited in the nineteenth century, the ore was roasted on open wood-burning fires to get rid of the sulphur it contained. Being released from a ground-level source, the sulphur dioxide produced during the roasting process was poorly dispersed and remained concentrated within a few kilometres of the smelters. As a result the city of Sudbury and the local environment were subject to high levels of acidic pollution, which completely destroyed the vegetation in the area and caused the lakes to become highly acidic (Freedman 2001). The area around Sudbury has begun to recover, but similar problems remain where little or no attempt is made to control the emissions from non-ferrous smelting processes. At Mochegorsk in north-west Russia, sulphur emissions from nickel and copper smelters have destroyed all of the trees within an 18 km radius of the city, exposing bare rock and soil with only sparse patches of shrubs and stunted vegetation (Mackinnon 2002). In the developed world, the relative importance of heavy industry has declined and the control of pollutants such as particulate matter and acid gases has been improved. As a result, the nature of industrial pollution has changed. Petroleum-based industries have introduced a new range of air pollutants, usually organic, released during petroleum refining or the production of petrochemicals that include a range of products from plastics to pesticides. Some of these pollutants may be no more than a nuisance, but others, including some of the solvents produced by the industry, are highly toxic and carcinogenic, while, on occasion – as at Bhopal, for example – their impact can be immediate and deadly.

Certain emissions from petroleum refining are detectable at very low concentrations. The residents of petroleum refining towns are well aware of the obnoxious odours associated with the release of mercaptans, a group of organic compounds containing sulphur. The human nose can detect mercaptans at levels of only a few parts per billion, and while their impact on health at such concentrations may be minimal, they can interfere with the use or enjoyment of public and private space and therefore constitute a nuisance. Odours are a common problem with many urban-based industries, such as meat-packing, food processing, tanning, pulp and paper making, chemical manufacturing and auto body repair and painting. The impact of this type of pollution varies with the odour and with the individual. For some, the odours may be merely disagreeable, but for others they bring on nausea and insomnia. In most cases, the human nose can detect odours long before the vapours that cause them reach life-threatening concentrations, providing an early warning of a potentially more serious situation.

Residential sources

Although pollution from residential sources might at first sight appear to be less likely to cause serious problems than that from industrial sources, some of the earliest recorded urban air pollution episodes, such as those in London from medieval times up to the seventeenth century, were associated with the burning of coal in open fireplaces for domestic heating and cooking (Brimblecombe 1987). Coal continued to be used for residential heating in many industrial countries well into the twentieth century, and the residential component of the aerosols and acid gases that comprised the smog that struck London in 1952 was significant. Clean Air legislation requiring the use of smokeless fuels plus the introduction of natural gas as the preferred source of heating reduced the residential contribution to urban air pollution in most of the developed world shortly thereafter. Coal continues to be used for heating and cooking in some parts of the developing world, producing results reminiscent of the industrial countries in an earlier era. Ankara, Turkey, for example, has been compared with London, England, in the 1960s, because of its dependence on low-grade coal for domestic heating, and as it was in London, the urban atmosphere is polluted with high levels of particulate matter and acid gases (UNEP/WHO 1996). In parts of India and China, coal is also a popular source of energy, but, more often, kerosene, wood, cow dung and other biofuels are burned. In most cases, these fuels are burned inefficiently, leading to high levels of haze containing smoke, carbon particles and unburned fuel, even in small communities. The potential impact is much wider, however. The UNEP (2002) has recognized emissions from millions of inefficient cookers as contributing to the Asian Brown Cloud that has become a persistent feature in the atmosphere over southern Asia (see Box 10.1, Arctic Haze and the Asian Brown Cloud).

Vehicular sources

By far the most ubiquitous source of urban air pollution at the beginning of the twenty-first century is the motor vehicle, powered by the petroleum-burning internal combustion engine. Urban pollution from motor vehicles was initially a feature of large cities in the developed world, with Los Angeles, California, providing the textbook example of the problem. In the second half of the twentieth century, however, global car ownership has grown significantly and increasing proportions of raw materials and manufactured goods are moved using road transport. Because of such changes, pollution from the internal combustion engine is common in all of the world's major cities, and even small or medium-sized communities suffer the effects of motor vehicle exhaust emissions. In larger cities, in both the developed and the developing world, the situation is compounded by increasing volumes of traffic trying to fit inadequate road networks. Slowly moving vehicles and the inevitable traffic jams lead to thousands of vehicles idling in poorly ventilated streets, causing pollution to climb sometimes to dangerous levels.

Most of the pollutants are gaseous, including carbon monoxide, oxides of nitrogen (NO_x), and unburned hydrocarbons, but diesel engines also emit particulate matter. In cities that receive abundant sunshine, the energy available in the solar radiation converts the primary gaseous pollutants from the vehicle exhausts into more complex products that together form photochemical smog (Figure 10.1). Originally identified with Los Angeles in the 1940s and 1950s – and still a problem there – photochemical smog is now common throughout the world in large mid to low-latitude cities where abundant sunshine supports the necessary atmospheric chemical reactions. The city of Cairo in Egypt, for example, suffers from an almost permanent haze caused by exhaust fumes and suspended particulate matter from its 1.2 million automobiles, which combines with fine sand blown into the urban area from the adjacent desert (UNEP 2002). Lack of pollution control legislation or the inadequate enforcement of the standards that do exist, plus poor vehicle maintenance associated with high service costs and limited access to the necessary technology, contribute to high levels of motor vehicle pollution in the developing world. Even with better legislation, more stringently enforced, the continuing increase in the amount of road traffic has created a worldwide problem in which emission levels continue to rise (UNEP/ WHO 1996).

THE MOST COMMON POLLUTANTS AND THEIR EFFECTS

Industrial, residential and vehicular sources of pollution combine in different proportions to create a pollution mix characteristic of a specific urban area. The mix contains a great variety of pollutants, some of which may be no more than a nuisance, whereas others threaten the city infrastructure or compromise the health and well-being of its inhabitants. Although the number of individual pollutants can be large, most of the problems are caused by a small number of gases and aerosols that have a long history and are widely recognized for their ability to disrupt urban air quality. They include particulate matter, carbon monoxide, sulphur dioxide, oxides of nitrogen, ozone and lead, and are the most frequently monitored pollutants. In developing National Ambient Air Quality Standards (NAAQS) in the 1970s, the United States Environmental Protection Agency (USEPA) recognized them as criteria pollutants, and most national jurisdictions in the developed world include them as part of a basic urban air quality monitoring programme. Maximum allowable concentrations of air pollutants in ambient outdoor air have been derived for each of these elements, as well as for as many as 150–60 other toxic chemicals such as volatile hydrocarbons, asbestos, arsenic and a variety of finely divided metals (Nazaroff and Alvarez-Cohen 2001).

Monitoring and air quality assessment are much more sporadic in the developing world, where many government organizations monitor only two or three of the criteria pollutants, and sometimes none. The Global Environmental Monitoring System/Urban Air Pollution and Assessment Programme (GENS/Air), a joint effort of the UNEP and the World Health Organization (WHO), is a co-operative programme designed to improve air pollution monitoring and management worldwide, but most of the assistance it can provide will be needed in the developing world. Air quality standards derived from such programmes are driven by the effects of pollutants on people and other elements in the environment, and must undergo regular revision as knowledge of the impact of the pollutants increases. Once the standards are established, it is important to continue monitoring and to take steps to control emissions so that the standards can be maintained. In the case of the common pollutants, regular monitoring over several decades plus abundant research on the health effects associated with them has allowed the recognition of threshold concentration levels that can be used as targets in improving or maintaining air quality (Table 10.2).

Particulate matter

Suspended particulate matter (SPM) or aerosols make up a high proportion of the pollutants in urban air – perhaps as much as one-third (Nazaroff and Alvarez-Cohen 2001). They include material of various sizes and composition from larger dust particles and water droplets to smaller soot, sulphate and smoke particles plus a variety of mineral, organic and metal particles. The aerosol content of the atmosphere can be represented in a number of ways. The easiest and probably most common is total suspended particulates (TSP), which includes all of the particles in the atmosphere. Observation and testing have shown that health problems arise mainly as a result of the inhalation of smaller particles, and since the 1980s the volume of respirable suspended particulates (RSP) has been measured (Turco 1997). Since RSPs are less than 10 μm in diameter they are commonly referred to as PM_{10}, while there are plans to monitor aerosols of less than 2.5 μm ($PM_{2.5}$), which consist largely of particles produced as a result of combustion and are now considered to represent a significant threat

TABLE 10.2 AMBIENT AIR QUALITY STANDARDS ($\mu g\ m^{-3}$)

Pollutant	Average concentrations	WHO	USA	UK	Australia	India	China
Carbon monoxide	Annual						40,000–60,000
	24 hours	30,000 (1)	40,000 (1)				
	1/3/8 hours	10,000 (8)	10,000 (8)	11,600 (8)	10,000 (8)		
Nitrogen dioxide	Annual	40	100	40	57	60–80	
	24 hours	150				80–120	50–150
	1/3/8 hours	400 (1)		200 (1)	229 (1)		
Ozone	Annual						
	24 hours	120 (8)	235 (1)		195 (1)		120–200 (1)
	1/3/8 hours		157 (8)	100 (8)			
Lead	Annual	0.5	1.5 (quarterly)	0.5	0.5		
SPM	Annual					140–360	
	24 hours					200–500	150–500
PM_{10}	Annual		50	40			
	24 hours	70	150	50	50		
	1/3/8 hours						
$PM_{2.5}$	Annual		15				
	24 hours		65				
Sulphur dioxide	Annual	50	80	20	53	60–80	20–100
	24 hours	125	365	125	213	80–120	50–250
	1/3/8 hours	350 (1)	1300 (3)	350 (1)	530 (1)		

Source: WHO: http://www.who.int/inf-fs/en/fact187.html; United States: http://www.epa.gov/airs/criteria.html; United Kingdom: http://www.aeat.com/netcen/airqual/dailystats/standards.html; Australia: http://www.ea.gov.au/atmosphere/airquality/standards.html; India: http://envfor.nic.in/cpcb/aaq/aaq_std.html; China: R.L. Edmonds, *Patterns of China's Lost Harmony*, London: Routledge (1994).

to human health (WHO 1999). The monitoring observations are commonly expressed as micrograms per cubic metre ($\mu g\ m^{-3}$) and compared against empirically established guidelines to ascertain the level of pollution. In Ankara, Turkey, for example, average PM_{10} values of about 100 $\mu g\ m^{-3}$ are common, which compare with the European Union (EU) guidelines of 75 $\mu g\ m^{-3}$, whereas in Birmingham, England, PM_{10} levels exceed the WHO limits of 70 $\mu g\ m^{-3}$ on at most only a few days in the year (UNEP/WHO 1996). Of the fifteen cities in the world with the highest levels of SPM, twelve are in Asia, with New Delhi, India, having the dubious distinction of recording a maximum TSP level of 420 $\mu g\ m^{-3}$ (UNEP 2002) (Table 10.2).

Aerosols in the urban atmosphere have many sources, but the overall content generally reflects the mix of activities in the urban area. In cities with a heavy industrial base, dust, smoke, soot, sulphate particles, acid droplets and metal particles are likely to be present. Residential heating using coal can also produce most of these, as do thermal electric generating stations, which using pulverized coal also emit fly ash – fine particles representing material in the fuel that is not combustible (Nazaroff and Alvarez-Cohen 2001). The rapid growth in road traffic in the twentieth century added other particulates to the mix. Although many automobile exhaust products are gaseous, they also include smoke and soot particles, with diesel-powered vehicles having particularly high soot emissions. Prior to the introduction of unleaded fuel, gasoline or petrol-burning vehicles emitted fine lead particles, and although lead is no longer added to fuel in most of the developed world, it is still commonly used elsewhere.

Particulate matter creates aesthetic problems, damages the urban infrastructure, threatens the health of urban dwellers and reduces their enjoyment of the community in which they live. Haze caused by aerosols in the atmosphere interferes with visibility and scatters solar radiation so that urban areas receive less sunshine than the surrounding countryside. In many heavy industrial areas the combination of smoke, soot and sulphuric acid aerosols has damaged and discoloured building stone, giving the cities a uniformly black or grey colour. In these cities, dust fall covers trees, shrubs and flowers in parks and open spaces, laundry and dry cleaning costs are high because of the constant soiling by soot and dust, public health is threatened by the chronic pollution and death rates increase remarkably during severe pollution episodes. There is a strong correlation between aerosol levels and mortality. The finer particulate fraction – RSP or PM_{10} – is carried deep into the lungs during respiration and therefore causes the greatest risk to health. Older persons or those suffering from asthma, emphysema, bronchitis or other respiratory conditions, plus those with cardiovascular disease, are most at threat from particulate pollution. Although acute episodes, such as the London smog of 1952, which led to the deaths of some 4000 people within about a week, are now rare, one estimate suggests that 135,000 die annually in the United States alone as a result of particulate pollution in the urban atmosphere (Lomborg 2001) and air pollution has been linked with an alarming rise in asthma among young people in North America (UNEP 2002). In developing countries, perhaps as many as 1.9 million persons die annually as a result of exposure to high concentrations of SPM (WHO 1999).

Trends

There has been a major decline in particulate matter levels in the urban atmosphere since the mid-1960s, mainly as a result of the recognition of the problem, the widespread acceptance that something had to be done to combat it and the availability of the technology to accomplish it (Figure 10.3). Emissions have been reduced directly through the use of electrostatic precipitators, filtration devices and smoke scrubbers to prevent particles from being released into the atmosphere. The introduction of taller smokestacks on thermal electric power stations and smelters helped to reduce particle concentrations in and above cities, by facilitating their dispersal over a wider area – but at the same time creating even more serious problems by causing the spread of acid rain. In addition to these deliberate efforts to reduce emissions, industrial and economic changes also helped to alleviate the problem. Traditional particulate-producing industries declined and coal was increasingly replaced by oil and natural gas as a fuel source for industrial and domestic use. Together, all of these developments helped to reduce airborne particulate pollution significantly, particularly in the developed world, even as populations were growing rapidly, industrial production was increasing and road traffic in urban areas was rising dramatically. The air in most industrial cities is now cleaner than it has been for hundreds of years and that has led to the optimistic outlook that urban air pollution is no longer a serious concern (see, for example, Lomborg 2001). Continuing high mortality rates associated with SPM suggest, however, that optimism is perhaps premature. Certainly in the developing world rapid urbanization has been accompanied by increasing air pollution and air quality SPM guidelines are often not met (Fenger 1999). The presence of particulate matter of urban origin in such larger-scale phenomena as the Arctic Haze and the Asian Brown Cloud also give cause for concern.

Carbon monoxide

When materials containing carbon are burned, the combination of the carbon with atmospheric oxygen typically produced carbon dioxide (CO_2). If the combustion temperature is low, however, or insufficient oxygen is available, carbon monoxide (CO), a colourless, odourless gas, is produced. Although this can happen naturally, the main problems that arise with CO levels in urban areas are caused by human activities. Modern industrial practices usually involve efficient, high-temperature combustion, but the burning of coal in domestic fireplaces, stoves or boilers is an important source of CO in some cities. In Katowice, Poland, for example, residential heating and cooking with coal release some 350,000 tonnes of CO annually (UNEP/WHO 1996). By far the most common source of urban CO, however, is the transport sector, which can account for as much as 90 per cent of emissions in some cities (Bridgman 1994).

City traffic causes elevated levels of CO at street level, usually with a double peak associated with the heavier traffic of the morning and afternoon rush hours. The actual CO concentrations vary from city to city, depending upon the volume and nature of the traffic flow. Slow-moving traffic when temperatures are low tends to produce higher levels of CO, for example, whereas gas emitted from rapidly moving traffic on a warm, windy day will be dispersed rapidly and the concentrations will be lower. Engines that have been poorly maintained or inadequately tuned will have higher emissions also. Exhaust gases may contain more than 1000 μg m^{-3} of CO on exiting the tailpipe, but dispersal reduces that concentration quite rapidly and in many cities typical annual averages based on one-hour maximum levels may vary between about 20 μg m^{-3} and 40 μg m^{-3} with a WHO standard of 25 μg m^{-3}, and eight-hour maximum levels range between 4 μg m^{-3} and 10 μg m^{-3}, with a WHO standard of 10 μg m^{-3} (Table 10.2). The average hourly rate in the city centre of Santiago, Chile, is over 50 μg m^{-3}, twice the WHO standard, and in Taipei, Taiwan, the eight-hour WHO standard is often exceeded (UNEP/WHO 1996). In the United States and Britain the average values have been declining since 1970, with average eight-hour values close to 4 μg m^{-3} in both countries (UK DEFRA 2003; USEPA 1999).

The human impact of CO is brought about by its affinity for haemoglobin, the iron-rich compound found in red blood cells that carries oxygen from the lungs to the body tissues. Together they form a stable compound (carboxyhaemoglobin) which prevents oxygen from being absorbed and carried around the body and impairs the functions of the brain, lungs and heart. High doses of CO are lethal, but the concentrations required are seldom if ever reached in ambient city air. Lower, non-lethal doses of CO cause headaches, drowsiness and reduced brain function, and, as a result, have the potential to contribute to traffic accidents.

Trends

Overall, levels of CO in urban air have been declining for several decades, particularly in the developed world, and the gas is a less serious pollutant than it once was. The decline in urban CO levels has been achieved through improvements in engine performance, new fuel formulations to reduce production of the gas, and the installation of catalytic converters in motor vehicle exhaust systems to remove it before it reaches the atmosphere. In the United States this has made it possible to reduce CO emissions from 84 g per mile travelled in the 1960s to a standard 3.4 g per mile travelled in the 1990s (Nazaroff and Alvarez-Cohen 2001). These technical improvements have been combined with transport infrastructure planning and the control of urban vehicle density and use – as in Singapore, where peak-period traffic is strictly controlled, for example – to further reduce the output of CO (UNEP/WHO 1996). Levels of CO are likely to continue to fall, although in some cities the growth in the number of motor vehicles is making it difficult to attain air quality targets.

Sulphur dioxide

Sulphur dioxide (SO$_2$) is an acid gas produced during the combustion of materials containing sulphur. It is a pollutant in its gaseous form, but it is also a precursor of sulphuric acid, which is created when the SO$_2$ combines with atmospheric water droplets and it is in that form that it often causes the greatest problems. Sulphur dioxide is also the primary ingredient of acid rain. Although it is a naturally occurring gas, its concentration controlled by the sulphur cycle, current concern is with the SO$_2$ produced as a result of human activities. The burning of coal and oil to provide energy for space heating or to fuel thermal electric power plants releases SO$_2$ into the urban atmosphere, as do industrial activities such as the smelting of metallic ores and the refining of petroleum.

Sulphur dioxide was a main component of air pollution in the heavy industrial areas of Europe and North America that grew out of the industrial revolution. With no attempt to control emissions of the gas, concentrations in the air in and above these cities gradually increased to reach maximum levels in the 1950s and early 1960s, when it contributed to most of the extreme urban air pollution episodes of that era. In the London smog of 1952, for example, SO$_2$ concentrations peaked at about 2000 μg m^{-3} (Bates 1972), more than ten times the daily average at the time, and about fifteen times the current WHO twenty-four-hour standard.

The presence of high levels of SO$_2$ in the

urban atmosphere reduced visibility, by scattering incoming light, caused damage to buildings, statues and monuments constructed of limestone and marble, corroded metals and created serious health problems for urban dwellers. When breathed in, some of the gaseous SO_2 is neutralized in the upper respiratory tract, but, particularly when it has combined with water droplets to form sulphuric acid, it can be drawn deep into the lungs, to cause or aggravate respiratory disease. Short exposures – as little as five to ten minutes – to concentrations of about 1 ppmv can increase airway resistance and cause changes in respiration and pulse rates. (The maximum daily average concentration in the London smog of 1952 was 0.75 ppmv.) Exposure to 10 ppmv or more of SO_2 for only five minutes can cause death (Turco 1997). Such responses help to explain the thousands of deaths that accompanied the serious smog episodes of the mid-twentieth century. Since then, and partly as a reaction to such events, legislation and controls have been introduced to curtail SO_2 emissions, with considerable success.

Trends

Worldwide monitoring shows that since the 1960s there has been a widespread decline in the emission and atmospheric concentration of SO_2, particularly in the developed world. In the United States, for example, average concentrations fell by 50 per cent between 1981 and 2001, with emissions falling over the same period by 31 per cent (USEPA 2002). In the United Kingdom between 1990 and 2001 total emissions declined by 70 per cent (UK DEFRA 2003). During the last two decades of the twentieth century, emissions were reduced to one-third of the 1980 levels in Western Europe and even in Central and Eastern Europe, where the problem has been treated less aggressively, SO_2 emissions are only two-thirds of what they were in 1980 (UNEP 2002). In contrast emissions in Asia increased by 32 per cent between 1985 and 1997, in large part because of the continuing growth in the use of coal in India and China, although there was a promising reduction in SO_2 emissions in the latter towards the end of the twentieth century (UNEP 2002). Even where there is no heavy industry, SO_2 levels can be a problem. In Hong Kong, for example, despite having no major coal-burning industries or metal smelters, annual average SO_2 concentrations are

80 μg m^{-3}, or four times the level in the United Kingdom and five times that in the United States (UNEP/WHO 1996). The culprits in Hong Kong are electricity generating stations that burn fossil fuels and the diesel-powered taxis and light buses that are a major component of the city's vehicle mix. Similarly in Madras, India, SO_2 levels are a problem, not because of traditional heavy industries, but because of oil refineries that process high-sulphur crude (UNEP/WHO 1996).

Experience in the developed world indicates that SO_2 emissions can be reduced in those areas where it is currently a problem. Technology is available to cut down sulphur levels in fuel and during the combustion process the addition of limestone can neutralize the acids before they leave the furnace. Flue gas desulphurization (FGD) or the neutralizing of the SO_2 formed during combustion, before it is released into the atmosphere, is one of the most effective ways of reducing acid gas levels. The flue gases are routed through scrubbers in which lime-rich liquid is sprayed into the gas stream, where the chemical reaction between the calcium carbonate in the spray and the SO_2 in the flue gases produces gypsum and neutralizes the SO_2 in the process. As well as reducing urban SO_2 concentrations, these techniques have also had a significant impact on acid rain.

Although SO_2 concentrations in most urban areas continue to decline, there is considerable variation even in those areas where levels are very low. Concentrations climb during specific weather patterns such as calm anticyclonic conditions or persistent inversions, which reduce the rate at which the gas is dispersed. The concentrations reached during such episodes come nowhere near those of the horrendous acid smogs common in the 1950s and 1960s, however, and although SO_2 continues to contribute to urban air pollution, its impact is much reduced.

Lead

As with sulphur dioxide, the reduction in atmospheric lead levels in urban areas is an environmental success story. Lead is a soft, grey heavy metal, which, being malleable and therefore easily worked, has a long tradition of use in areas such as plumbing, printing and ceramics, and more recently it has been used in lead-acid batteries and as an additive in gasoline. Lead in the urban atmosphere is a primary pollutant, usually part of the PM_{10}

fraction of suspended particulates, either in its metallic form or as a compound. It is emitted from lead smelters or from plants manufacturing products containing lead and from the exhaust systems of vehicles burning leaded gasoline. From the mid-twentieth century tetraethyl lead was routinely added to gasoline to improve engine performance, and by the 1970s it was responsible for 90 per cent of the lead present in the urban atmosphere. Its use is now much reduced, but it continues to contribute to serious lead pollution in some cities (UNEP/WHO 1996).

The danger of high lead concentrations in the atmosphere was not appreciated initially, but lead is a highly toxic metal and was undoubtedly responsible for a variety of health problems and premature deaths before its impact was recognized. It has been estimated that in heavily polluted urban environments, city dwellers inhaled as much as 50 $\mu g\ m^{-3}$ of lead every day – one hundred times the maximum daily level allowed by modern health standards – exposing them to potential physical and mental disability (Turco 1997). The ingestion of even small amounts of lead causes loss of appetite, headaches, drowsiness and a variety of behavioural changes. The effects are usually much greater in children than in adults, while pregnant women and their unborn children are also particularly vulnerable (Centers for Disease Control and Prevention 1991). However, because the symptoms of low-level lead poisoning are shared with other ailments, the problem can remain undiagnosed. Since lead accumulates in the body faster than it can be excreted, the continued consumption of even small amounts may ultimately lead to much greater problems, including anaemia, kidney damage, hypertension, brain damage, paralysis and eventually death (Wallace and Cooper 1986).

Trends

Horrific as such symptoms appear to be, initial attempts to reduce lead emissions from burning gasoline were slow and it was not until the large-scale adoption of the catalytic converter that the introduction and use of unleaded gasoline became more widespread. The catalytic converter was designed to reduce pollution from motor vehicles by oxidizing carbon monoxide to carbon dioxide, converting oxides of nitrogen into oxygen and nitrogen and volatile organic compounds (VOCs), such as unburned hydrocarbons, into carbon dioxide and water, and by so doing reducing urban pollution levels. Problems arose because lead deactivates the catalysts, making lead-free gasoline a requirement for vehicles fitted with converters (Baarchers 1996). At about the same time there was increased appreciation of the toxic effects of airborne lead, and from the mid-1970s leaded gasoline began to be phased out in favour of the unleaded variety. Despite initial resistance from gasoline producers and car manufacturers, all of the gasoline sold in North America is now lead-free and although leaded gasoline continues to be sold in Europe, the volume is low and falling. In some countries in the developing world there is little or no progress in converting to lead-free fuel. Unleaded gasoline is not available in Nigeria, for example, and in Lagos, where the lead content of gasoline is highest in the world, atmospheric lead levels at the kerbside have an annual mean of 47 $\mu g\ m^{-3}$, well above the WHO guidelines of 0.5–1.0 $\mu g\ m^{-3}$. Under such conditions lead poisoning is a distinct threat to those living in the urban area. Similar problems with urban lead levels are present in Nairobi, Kenya, on the other side of the African continent. In Asia the picture is brighter, with WHO guidelines being matched or bettered in many of the large cities, despite growing urban traffic (UNEP/WHO 1996).

In North America and Europe the decline in atmospheric concentrations of lead represents a major improvement in air quality. In the United States, for example, levels declined by 94 per cent between 1980 and 1999 and in Britain by 76 per cent between 1970 and 1994, with concentrations now so low that they are difficult to measure (UK DETR 1998; USEPA 1999). It is unlikely that such levels will be reached in the developing world for some time, but there is abundant evidence that the technology exists to bring the lead component of urban air pollution down to negligible levels.

Oxides of nitrogen

The oxides of nitrogen are gases formed by the combination of oxygen and nitrogen, often under high-energy conditions such as the bombardment of the upper atmosphere by cosmic rays, during lightning storms, in high-temperature furnaces and in internal combustion engines. They include nitric oxide (NO), nitrous oxide (N_2O) and nitrogen dioxide (NO_2), which form a group commonly referred to as NO_x. In urban areas the production of oxides of nitrogen is shared relatively evenly between industry and electric utilities (45 per cent) and motor vehicles (47 per cent) although there are variations depending upon the industrial mix in a particular city and the level of motor vehicle ownership (Nazaroff and Alvarez-Cohen 2001). In many large cities in developing countries, for example, the contribution of motor vehicles is higher, but in China until the mid-1990s about 65 per cent of the oxides of nitrogen in the atmosphere originated in industrial emissions (Edmonds 1994). That situation is changing as traffic volumes increase. In the developed world, the introduction of the 'tall stacks' policy reduced the amount of oxides of nitrogen being introduced into the urban atmosphere and the air pollution associated with the gases is primarily a product of motor vehicle exhausts.

The oxides of nitrogen most common in urban air are nitric oxide and nitrogen dioxide. The former is a primary pollutant released directly in exhaust gases, whereas nitrogen dioxide is mainly a secondary pollutant created as a result of the oxidation of nitric oxide. Nitrogen dioxide in turn becomes involved in chemical reactions that produce photochemical smog and is the main precursor of ozone. Oxides of nitrogen can be converted into nitric acid through combination with water vapour in the atmosphere and therefore contribute to acid precipitation. In urban areas, along with sulphuric acid, nitric acid causes damage to buildings and affects the health of the inhabitants. One of the more obvious impacts of oxides of nitrogen is reduced visibility, particularly when nitrogen dioxide levels are high. Being a reddish-brown gas, it is in part responsible for the brown haze that is a common feature of urban smog.

Overall, oxides of nitrogen may be less harmful to city dwellers than fine particulate matter, sulphur dioxide or lead, but exposure to the gases, especially nitrogen dioxide, does produce significant health effects. Like many air pollutants, oxides of nitrogen irritate the respiratory system. Nitrogen dioxide is only slowly soluble in water and as a result it is not neutralized in the upper respiratory tract, but is inhaled into the lungs. When it is eventually converted into nitric acid it damages delicate lung tissue, leading to breathing difficulties and longer-term problems such as bronchitis (Turco 1997). There is also evidence that long-term exposure to even relatively low levels of nitrogen dioxide can lead to reduced resistance to respiratory infections (Miller 1994).

Trends

The results of the monitoring of oxides of nitrogen are ambiguous. Annual averages in many cities have declined or stabilized (Figure 10.9) at or below the WHO guidelines of 50 μg m^{-3}, but in some countries they are beginning to rise again. Levels in the United States rose in the 1990s, for example, and in Japan, where overall concentrations had begun to stabilize in the late 1970s, they are again higher in Tokyo and Osaka (UNEP 2002). In Hong Kong, China, where serious efforts have been made to reduce pollution levels, and nitrogen dioxide levels average less than the WHO guidelines, the growth in the number of motor vehicles and the poor ventilation in canyon-like streets has produced locally high levels of pollution (UNEP/WHO 1996).

The major problem, worldwide, is the proliferation of motor vehicle use. Attempts have been made to reduce emission levels of oxides of nitrogen by preventing their formation in the first place or reducing the amounts released into the atmosphere. Lowering the combustion temperature will reduce the rate at which the oxides are formed and this has been accomplished in automobiles by recirculating cooler exhaust gases back into the engine. The most successful means of reducing emissions, however, has been the development of the catalytic converter, which breaks down the oxides of nitrogen into their component parts – oxygen and nitrogen – before releasing them into the atmosphere as harmless nitrogen, carbon dioxide and water vapour. In the United States, the incorporation of improved combustion technology and catalytic converters into new vehicles has reduced the output of nitrogen dioxide from an average of 5 g per mile travelled in the 1960s to 0.4–0.6 g per mile travelled in 2000 (Nazaroff and Alvarez-Cohen 2001). The 1999 Protocol to abate Acidification, Eutrophication and Ground-level Ozone, from the UN Economic Commission for Europe (UNECE 2000), recognized the need for new reduction commitments for oxides of nitrogen, but, while technology may help to reduce emissions, the effects are continuing to be offset by growing levels of urban traffic, and conditions are unlikely to improve until something is done to counter that aspect of the problem.

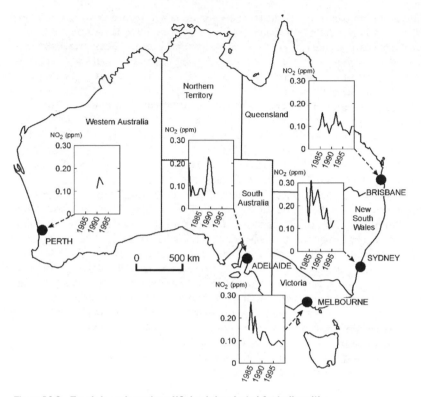

Figure 10.9   Trends in peak one-hour NO$_2$ levels in selected Australian cities

Source: based on data viewed at http://www.dotars.gov.au/mve/urban_air_quality.htm and http://www.apec-vc.org.au/local/air/image1ET. JPG.

Ozone

A blue gas with a pungent odour, ozone is an allo-trope of oxygen in which each molecule contains three atoms rather than the two of normal atmospheric oxygen. It is a very powerful oxidizing agent. Ozone is present in both the troposphere and the stratosphere (see Chapter 2), with tropospheric ozone accounting for about 10 per cent of the total ozone column. Such average values are not particularly representative of reality, however, since ozone concentrations are quite variable in time and place. Ozone is a secondary pollutant, formed as a result of the complex chemical processes that create photochemical smog. The action of sunlight on NO$_x$ released from automobile exhausts provides a starting point (Figure 10.1b). The absorption of solar energy causes nitrogen dioxide to dissociate into nitric oxygen and atomic oxygen. The latter in turn combines with molecular oxygen in the atmosphere to form ozone. Being a strong oxidizing agent, ozone is highly reactive, combining readily with nitric oxide, nitrogen dioxide and certain hydrocarbons and being destroyed in the process. Tropospheric ozone therefore has a short life span, and as soon as the elements necessary for its formation are no longer available its atmospheric concentration declines (Nazaroff and Alvarez-Cohen 2001). A reduction in the supply of NO$_x$, for example, or the absence of sunlight, will reduce the supply of new ozone, and as the remaining gas is lost it is not replaced. As a result, in cities prone to ozone pollution, concentrations peak during the day – motor vehicle emissions high; sunshine available, and decline to zero overnight – emissions low; no sunshine. For similar reasons, ozone concentrations tend to be higher in summer than in winter. Ozone intensity normally peaks four to six hours after concentrations of its precursors have peaked. Thus the increased emissions of NO$_x$ during the morning rush hour are translated into ozone increases that peak in the early afternoon (Bridgman 1994). Where meteorological conditions or topography discourage the dispersal of pollutants, precursors may survive overnight and contribute to ozone production when the sun rises the following day.

Since tropospheric ozone has a relatively short life span it does not usually travel far. Traces of ozone have been measured as much as 800 km downwind from their source, however. Between 30 per cent and 90 per cent of the ozone over eastern Canada originates some 200–800 km away in the United States, while the north-eastern United States receives NO_x, the main precursor of ozone, from sources in the province of Ontario (UNEP 2002).

Ozone is not usually considered to be life-threatening in the concentrations measured in most cities, but it may well have a greater impact than was once thought. Recent studies have shown a strong correlation between ozone levels and the rates of hospitalization and absenteeism among workers (OMA 2000). Even low concentrations of ozone cause nose and throat irritation, and as levels increase, headaches and airway resistance follow. Because the gas is a strong oxidant, long-term exposure contributes to premature aging of the skin and inhalation damages lung tissue (Turco 1997). Non-health impacts include the oxidation of paint and deterioration of tyres and other rubber objects.

Trends

Because of the short time it survives in the atmosphere, ozone levels in urban areas are usually represented as a one-hour average or as a peak concentration. The WHO has established a one-hour standard of 120 ppbv, but it is not uncommon for that to be surpassed. Although peak concentrations in the United States have fallen by almost 30 per cent since 1977, and the hourly average in Los Angeles declined by more than 70 per cent during the same time period (Bridgman 1994), the one-hour air quality standard of 120 ppbv was exceeded in that city on as many as 100 days in the year during the 1990s (Nazaroff and Alvarez-Cohen 2001). In such a case, the combination of chemical, meteorological and anthropogenic conditions, involving abundant sunshine, high levels of automobile exhaust emissions and a stagnant boundary layer, can cause ozone levels to peak even when pollution controls are in place. Mexico City, lying in a valley with poor atmospheric mixing and a growing level of vehicle ownership, has replaced Los Angeles as the city with the highest ozone levels (UNEP 2002). It frequently experiences peak con-centrations as high as 450–70 ppbv, but even where conditions are less conducive to ozone formation hourly peaks above the air quality standards are not uncommon. Concentrations of 300–400 ppbv are recorded in urban Japan, 150–200 ppbv in the cities of Europe, and in Australia concentrations of more than 150 ppbv occur in Sydney and Melbourne (Bridgman 1994). Since it is not possible to change the meterologic conditions or topographic situations that aggravate the urban ozone problem, any reduction must come about through the reduction of vehicle emissions, particularly those containing NO_x, the precursor of ozone. That has been achieved to a large extent by the use of catalytic converters, but the improvements made in that way have been offset by the continuing growth in the number of cars and trucks in urban areas, and perhaps the only way to bring about serious improvement will be to place greater controls on the ownership of vehicles and the numbers allowed into city centres, as has been done already in Singapore (UNEP/WHO 1996).

Hazardous air pollutants

In addition to these criteria pollutants, which are routinely monitored, there are numerous others that are not. Their concentrations are often unknown and their environmental impacts are imperfectly understood, although they are known to be, or suspected of being, toxic or carcinogenic. Some of the most important of these are the hazardous air pollutants (HAPs). They are by-products of a variety of activities, including metallurgical and chemical industrial production, petroleum refining, energy production and use and solid waste incineration. HAPs emitted from such activities include heavy metals such as cadmium, mercury and beryllium, hydrocarbons such as benzene, a known carcinogen, solvents such as formaldehyde and carbon tetra-chloride and undoubtedly others as yet unidentified as hazardous. In some cities, radioactive particles are present in the air, either as a result of their use in manufacturing or as a result of accidents at nuclear power plants or weapons establishments. Although not all of these HAPs are present in all cities, they represent a problem that has yet to be addressed in detail and in the long term may pose a more serious threat to city dwellers than the so-called traditional pollutants. Among the HAPs are a group of persistent organic pollutants (POPs) that include pesticides, such as DDT, aldrin and dieldrin, industrial chemicals, such as PCBs and hexachlorobenzene, and by-products of industrial processes, such as dioxins and furans. In an attempt to deal with them, the Convention on Persistent Organic Pollutants was signed in Stockholm in 2001, setting out measures to control the twelve most harmful organic pollutants (UNEP 2002).

INDOOR AIR POLLUTION

When a city experiences a smog alert, health officials often recommend that individuals with allergies, asthma or other forms of respiratory disease remain indoors on the assumption that indoor air is less likely to cause them harm. Ironically, the quality of the air that they breathe indoors is often no better than that outside and in many cases it may be worse (Figure 10.10). Modern concerns over indoor air pollution began to increase in the 1960s and 1970s, but the problem is one that is as old as society's use of fire and shelter. Prehistoric groups living in caves and burning wood to cook or keep warm were

exposed to remarkably high levels of particulate matter, carbon monoxide and smoke containing organic emissions, and that situation applied as long as people lived in shelters without chimneys, where the smoke accumulated until it gradually escaped through a smoke hole in the roof. That still applies today in some developing countries, where wood, animal dung and crop residues are burned in open fires or poorly vented stoves. Women and young children are particularly exposed to this type of pollution and often succumb to respiratory disease (Miller 1994).

In the workplace

With the industrial revolution, indoor air pollution became a serious problem in the workplace. Particulate matter was generated by the coal, iron and steel industries. Miners' lung, or silicosis, caused by the inhalation of coal dust was a frequent cause of debilitation and death in most mining communities. The air in textile mills contained high levels of fibres that were inhaled and ingested by mill workers,

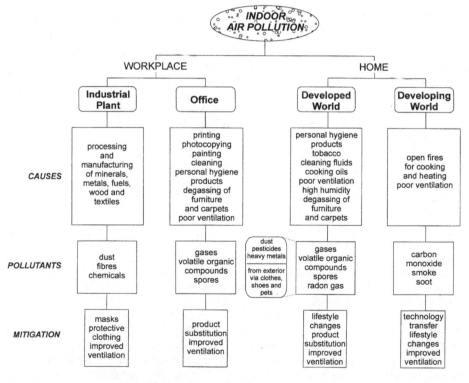

Figure 10.10 The causes and constituents of indoor air pollution

many of whom were young children, and toxic fumes or particles were very much part of the chemical industries that came into being in the nineteenth century. As occupational health standards improved into the twentieth century, the impact of these workplace pollutants was reduced through the use of masks and protective clothing, the provision of ventilation systems that prevented the accumulation of pollutants within a building and the introduction of control systems that reduced the formation or emission of the pollutants in the first place.

'Sick building syndrome'

Whereas safety measures were gradually introduced to protect industrial workers, little or nothing was done initially for office workers, who in comparison appeared to work in a clean, healthy environment. By the second half of the twentieth century, however, that perception changed, when it became clear that many people were working in office buildings with pollution levels that exceeded those of outside air. Such buildings came to be known as 'sick buildings' and those experiencing health problems as a result of working in them were said to be suffering from 'sick building syndrome' (Chiras 1998). Pollutants in these buildings take many forms and have many sources. They may originate with people in the form of tobacco smoke, perfume, deodorant or after-shave; they may result from activities in the building, such as printing, photocopying, painting or the use of cleaning fluids; they may be released from materials in the structure or decoration of the building, such as formaldehyde and other volatile chemicals released from new carpets and paints or the glues and stains used in furniture. In the past, asbestos, used as a fire retardant and to provide insulation around pipes, produced fibres that were a common constituent of indoor pollution, but asbestos has since been banned for such uses in most of the developed world. Together these make up a complex cocktail of pollutants, responsible for a range of health problems experienced by those exposed to them (Baarchers 1996).

These circumstances were aggravated in the mid-1970s, when the energy crisis led to buildings being renovated or retrofitted to reduce heat loss. Cracks and leaks in the structure of buildings were sealed, weather-stripping was added around doors and existing windows were replaced with new energy-efficient versions. Successful as these approaches were in helping to prevent costly heat from escaping from the buildings, they were also responsible for reducing the rate at which pollutants were escaping. Office buildings became increasingly polluted and more and more individuals began to display such health problems as sore throats, sinus infections, headaches, eye irritation, allergies and even skin rashes. One estimate suggests that, in the United States alone, 10 million to 20 million people suffer from these symptoms as a result of indoor air pollution (Chiras 1998). Complaints also include non-specific symptoms, and testing of office air may find no measurable contamination, which can lead to confrontation between management and employees. Whatever the contamination level, indoor air pollution can be controlled relatively easily by improving ventilation in the building – increasing the rate at which indoor air is completely exchanged for new outdoor air, for example – or by banning such activities as smoking, which in the past was one of the most common sources of pollution. Structural changes such as the removal of asbestos or decorative changes involving the replacement of carpeting and furniture that release solvents, such as formaldehyde, into the air also help to reduce concentrations of contaminants. The employment of some or all of these options can reduce indoor air pollution significantly in most office buildings.

In the home

Even if individuals are not exposed to pollutants in the workplace they are not always safe at home. An Australian study, by the Commonwealth Scientific and Industrial Research Organization (CSIRO), found that in Melbourne houses less than a year old contained volatile organic compounds (VOCs) up to twenty times the Australian standard of 500 μg m^{-3} (Edwards 2001). Similar studies in the United States have shown that indoor toxic pollutant levels are often ten to fifty times higher than those outdoors (Renner 2001). Unlike the air in offices and industrial plants, household air is seldom monitored regularly and individual home owners normally have no idea of the forms and levels of pollutants to which they are exposed. Sources of pollution are present in most houses from basement to attic, being produced by heating devices such as furnaces and wood stoves, by lifestyle activities such as smoking or the use of aerosol sprays and by furniture and flooring. Among these sources carpets have gained a reputation for the variety and toxicity of the pollutants they release (Renner 2001). New carpets give off styrene, a suspected carcinogen (Edwards 2001), and act as a reservoir for dust, pesticides, heavy metals and bacteria carried into the house on shoes and clothing or by pets. As they age, the level of pollutants trapped in carpet fibres increases and becomes a particular hazard for young children who spend more time at carpet level than adults. High humidity levels contribute to condensation and dampness in some houses, encouraging mildew and mould, which produce spores, and in some areas, where the soil and rock contain radioactive chemicals, radon gas seeps into buildings through their foundations.

The most common indoor air pollutants

Although the variety of pollutants released from these sources is large – more than 250 VOCs have been detected in indoor air, for example (Turco 1997) – a small number appear to be more dangerous than others, including tobacco smoke, formaldehyde and radioactive radon-222 (Miller 1994). Asbestos has also been included in the list in the past, because of its links with asbestosis, a chronic lung disease, and with lung cancer, both of which were common in workers in asbestos mines or in industries using asbestos products. These workers were exposed to atmospheric fibre levels many times higher than would be experienced in a dwelling with asbestos ceiling tiles or flooring, however. The use of asbestos products has been increasingly restricted over the past several decades and it is not normally considered a threat to health in most houses.

Tobacco smoke

Tobacco smoke is a complex, toxic mix of particles and gases that are capable of causing harm to both the smoker, through direct inhalation from the burning tobacco, and to the non-smoker, through exposure to side-stream and second-hand smoke. The smoke contains fine particles of tars and other organic compounds, many of which are known carcinogens, and the links between tobacco smoke

and the occurrence of lung cancer are well established. A number of respiratory ailments including chronic bronchitis and emphysema are caused or aggravated by tobacco smoke and cardiovascular disease is common among smokers (Baarchers 1996). Other constituents of tobacco smoke include poisonous gases such as carbon monoxide, acetone, formaldehyde and hydrogen cyanide, all of which can cause incremental and potentially permanent damage to the health of those exposed to them. In an average-size room containing even a small number of smokers, short-term air quality deteriorates rapidly, with such pollutants as carbon monoxide and respirable suspended particulates (RSPs) regularly reaching levels that in an outdoor urban setting would trigger smog alerts (Turco 1997). As is the case with urban air pollution, exposure to high levels of carbon monoxide indoors can cause headaches and impair the functioning of the brain for several hours after exposure has ended, while the inhalation of RSPs can have an incremental effect that leads to lung disease. Fortunately, tobacco smoke can be controlled more easily than some other indoor pollutants. Improved ventilation can remove it from a room, a decision can be made not to smoke indoors or individuals may decide to give up smoking completely.

Formaldehyde

One of the constituents of tobacco smoke is formaldehyde, but it is also produced from less obviously hazardous sources. It is used as a solvent in the production of plywood and particle board, carpeting and insulation. When these are used in house furnishings natural evaporation allows the formaldehyde to escape into the building. Exposure to low levels of the gas can cause nausea, dizziness, eye irritation and sore throat, while higher levels can lead to chronic breathing difficulties and possibly cancer. Perhaps as many as 20 million people in the United States suffer from one or more of these problems as a result of exposure to formaldehyde (Miller 2000). To some extent the release of formaldehyde is self-limiting. As the materials containing it age the amount of the gas released decreases, but to remove it from the indoor environment completely, or reduce it to tolerable levels, it may be necessary to replace the items that are emitting it with new products in which formaldehyde is not used. Emissions can also be reduced by the use of sealants or removed by improving the ventilation in the affected building.

Radon

Radon-222 is a radioactive gas produced by the decay of uranium. This occurs naturally in the environment with few if any negative consequences. If the gas is allowed to enter homes, however, it can accumulate to levels that are dangerous for those living there. Not all rocks produce radon, but where it is present it can seep into buildings through cracks in the basement or foundations, through drains and can be carried by water drawn from underground wells drilled in rock containing uranium. Building blocks or concrete containing uranium can also release radon into the house. The inhalation of radon can lead to lung cancer and it has been suggested that exposure to radon indoors is second only to smoking as a cause of lung cancer deaths (Miller 2000). Mortality estimates attribute as much as 14 per cent of lung cancer deaths in the United States to radon exposure, but in Europe the figure is perhaps as low as 1 per cent (Lomborg 2001). Such figures may not be representative of the real situation, however, since few houses have been tested for radon levels and the risks have been extrapolated from the impact of exposure in industrial environments (Baarchers 1996). Exposure levels for residential properties have been suggested in Britain, Canada, the United States and Sweden and these can be met by installing airtight membranes in the foundations of the building or over the basement floor and walls, or by installing relatively simple ventilation systems that prevent the accumulation of the gas in the first place (Nagda 1994).

Indoor air pollution is undoubtedly a serious problem. It causes premature death or disability and leads to a reduction in the productivity of society through lost work time. The numbers involved are difficult to estimate, but it is possible that between 2 million and 3 million premature deaths occur worldwide as a result of indoor pollution, more than four times the number caused by outdoor air pollution (Miller 2000). The impact of indoor pollution can be reduced, and in most developed nations legislation is in place to deal with it. Most of the premature deaths occur in the developing world, however, where charcoal, firewood, agricultural wastes and animal dung are used for cooking and heating, exposing families – particu-larly mothers and young children – to high levels of indoor air pollution (UNEP 2002), and that situation may well worsen before it improves.

ACID PRECIPITATION

Acid precipitation, a product of atmospheric chemical reactions, leads to the deposition of acidic substances on the earth's surface. It falls most commonly as rain, but also includes snow, hail and fog, which together are considered as forms of wet deposition. In addition, it includes dry deposition, which involves the fall-out of oxides of sulphur and nitrogen from the atmosphere, either as dry gases or adsorbed on aerosols such as soot and fly-ash. In contact with moisture in the form of fog dew or surface water, they produce the same effects as the constituents of wet deposition (Mason 1990). The term 'acid rain' is normally used to include all forms of wet and dry deposition.

Although acid rain is most often equated with modern industrial pollution, atmospheric moisture has always contained some level of acidity. Natural processes in the earth/atmosphere system create carbonic, sulphuric and nitric acids, and studies of rainwater chemistry in tropical northern Australia have identified organic acids such as acetic and formic acids as important contributors to atmospheric acidity (Bridgman 1994). When these acids fall out in the rain they become involved in a variety of physical and biological processes at the earth's surface. The return of nitrogen and sulphur to the soil in naturally acid rain helps to maintain nutrient levels, for example. Acid rain also makes a major contribution to the landscape in limestone areas, through the same processes of acid corrosion and solution that are responsible for the deterioration of buildings in urban centres. Current concern over acid rain is not with the natural variety, however, but rather with that produced by the acid gases emitted by modern industrial activities.

Robert Smith, a British chemist, is normally credited with first recognizing the link between air pollution and the acidity of the atmosphere, in his observations of the chemistry of the rain falling in industrial Manchester in the mid-nineteenth century. In 1872 he wrote a book on the subject in which he made the first reference to 'acid rain' (Smith 1872). He went on to become Britain's first Alkali Inspector, responsible for the control of air pollution from industrial sources, but his ideas on acid rain were largely ignored until the mid-twentieth century. By that time the impact of acid rain was becoming increasingly apparent to observers in a number of regions, and studies of the problem by Eville Gorham in the English Lake District in the late 1950s, by Gene Likens and Herbert Borman in New Hampshire and Svete Oden in Sweden in the early 1960s are recognized as laying the foundation for modern interest in acid rain (Park 1987; Middleton 1999).

SOURCES OF ACID RAIN

When Robert Smith described the presence of acid rain in the mid-1850s, the industrial revolution was well under way, and the by-products released into the atmosphere by the industrial activities provided the necessary ingredients for the acid rain. As industrialization continued to expand it was accompanied by an increase in the acidity of the rainfall over more and more of the landscape, but it took almost a century before the seriousness of

the situation was finally recognized. Technological advance in a society often depends upon the availability of metallic ores, which are smelted to produce the great volume and variety of metals needed for industrial and socio-economic development. Considerable volumes of acid gases, such as sulphur dioxide, are released into the atmosphere as a by-product of the smelting process, particularly when non-ferrous ores are involved. The burning of coal and oil to provide energy for space heating or to fuel thermal electric power stations also produces sulphur dioxide. The continuing growth of transport systems using the internal combustion engine – another characteristic of a modern technological society – contributes to acid rain through the release of oxides of nitrogen into the atmosphere (Figure 10.11).

Nature and formation of acid rain

Acid rain produced by human activities differs from natural acid precipitation, not only in its origins,

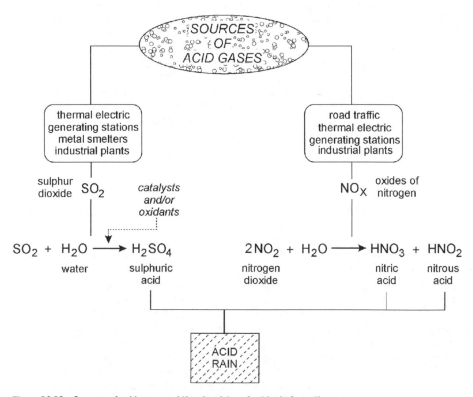

Figure 10.11 Sources of acid gases and the chemistry of acid rain formation

but also in its quality. Anthropologically produced acid rain tends to be many times more acid than the natural variety. The difference is commonly of the order of 1.0 to 1.5 on the pH scale used to measure acidity (Figure 10.12). In North America, for example, naturally acid rain has a pH of about 5.6, whereas measurements of rain falling in southern Ontario, Canada, in the 1970s, when concern over acid rain was at its height, frequently provided values between 4.5 and 4.0 (Ontario: Ministry of the Environment 1980). Similar values for background levels of acidity in rain were recorded in Europe, with the average pH of rain over Britain between 1978 and 1980 being within the range of 4.5 to 4.2 (Mason 1990). To put these values in perspective, vinegar has a pH of 2.7, milk a pH of 6.6 and distilled water a pH of 7.0 (Figure 10.12). Remarkably high levels of acidity have been measured on a number of occasions on both sides of the Atlantic. In April 1974, for example, rain falling in Pitlochry, Scotland, had a pH of 2.4 (Last and Nicholson 1982), and a value of 2.7 was reported from western Norway a few weeks later

(Sage 1980). At Dorset, north of Toronto, Ontario, snow with a pH of 2.97 fell in the winter of 1976–77 (Howard and Perley 1991). Although these values are exceptional, the pattern is not. Acid deposition tends to be episodic, with a large proportion of the acidity at a particular site arriving in only a few rainfall events, during which atmospheric conditions and emission rates have combined to concentrate the acidity. At other times acid levels are much lower and the peaks tend to be masked by the annual averages.

The quality of the rain is determined by a series of chemical reactions set in motion when acidic materials are released into the atmosphere (Figure 10.11). Some of the sulphur dioxide and oxides of nitrogen will return to the surface quite quickly and close to their sources as dry deposition, contributing in many cases to urban air pollution. The remainder will be carried into the atmosphere to be converted into sulphuric and nitric acid, which will eventually return to the earth's surface as acid rain. The processes involved are fundamentally simple. Oxidation converts the gases into acid, in either a

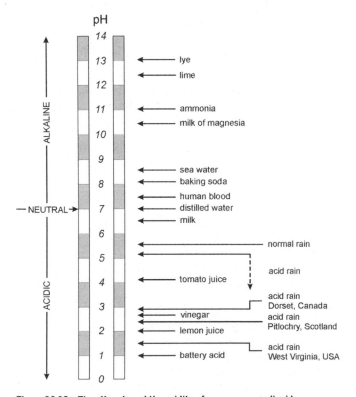

Figure 10.12 The pH scale and the acidity of some common liquids

gas or a liquid phase reaction. The latter is more efficient, producing sulphuric acid from sulphur dioxide six times more rapidly than the gas phase reaction (Mason 1990). The specific rate at which these reactions take place depends upon a number of variables including the concentration of metals such as manganese and iron in airborne particulate matter, the presence of ammonia and the intensity of the sunlight. The metals and ammonia appear to act as catalysts, while the sunlight provides the energy for the production of photo-oxidants – such as ozone and hydrogen peroxide – from other pollutants, and these oxygen-rich products facilitate the oxidation of sulphur dioxide and oxides of nitrogen to sulphuric acid and nitric acid respectively (Cocks and Kallend 1988).

Geographical sources and distribution of acid rain

In the past, acid rain remained very much a local problem, providing its main impact in the vicinity of the emission sources. The gradual increase in the height of smokestacks during the 1960s and 1970s – the so-called 'tall stacks policy' – improved local conditions, but that success was more than offset by the subsequent increase in the geographical extent of the problem. By the mid-1970s stacks ranging in height from 150 m to 300 m were common on smelters and thermal electric generating plants in Europe and North America, with the tallest being the International Nickel Company's 400 m super-stack, built to dispose of exhaust gases from its nickel smelting complex at Sudbury, Ontario. These stacks effectively reduced local pollution by emitting the pollutants outside the boundary layer circulation and into the larger-scale atmospheric circulation with its potential for greater dispersal. Released at a greater height, the pollutants also remained in the atmosphere for longer periods of time, thus increasing the probability that the acid conversion processes would be completed. Overall, local problems were mitigated at the expense of the larger environment.

The traditional sources of acid rain are in the industrialized areas of the northern hemisphere, with sulphur dioxide and oxides of nitrogen the main gases emitted. More than 90 per cent of sulphur dioxide emissions from anthropogenic sources originate in the northern hemisphere, for example (Goudie and Viles 1997). In North America, sulphur compounds accounted for about 65 per

Plate 10.3 The 400 m high 'superstack' at Sudbury, Ontario. Designed to reduce pollution in the immediate area, the release of gases at a higher level in the atmosphere causes serious pollution downwind from the International Nickel smelting complex.

cent of the acid rain in the 1970s, with nitrogen compounds making up the remainder. In Europe, the contribution of sulphur dioxide and oxides of nitrogen at that time split closer to 75 per cent and 25 per cent respectively (Park 1987). Since then, however, declining sulphur dioxide emissions and a growing output of oxides of nitrogen – mainly from an increase in automobile traffic – have combined to bring the relative proportions of these gases closer to the North American values (Mason 1990). Sulphur dioxide emissions in the Western industrialized nations peaked in the mid-1970s or the early 1980s, and have declined significantly since then (Figure 10.13). That trend is likely to continue as new environmental regulations are introduced and enforced, but emissions of oxides of nitrogen may well continue to rise, in part owing to increased vehicular traffic, but also because of the limited

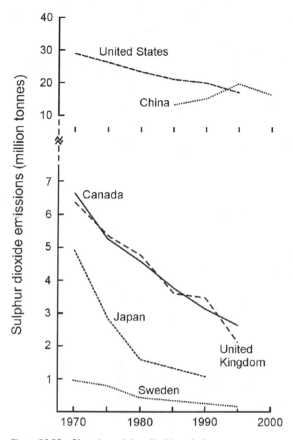

Figure 10.13 Changing sulphur dioxide emissions

Source: based on data from various sources, including World Resources Institute, *Earthtrends*. Viewed at http://www.earthtrends.wri.org/ (accessed 4 April 2003).

attention given to the reduction of oxides of nitrogen in most acid rain control programmes.

The industrial nations of Eastern Europe along with Russia, Ukraine and Kazakhstan – the republics once part of the Soviet Union – continued to emit high levels of sulphur dioxide long after emissions began to be controlled in Western Europe and North America. Even there, however, by 2000 sulphur dioxide emissions were only two-thirds of what they had been in 1980 (UNEP 2002). This was mainly the result of recession caused by the economic disruption that accompanied the break-up of the Soviet Union, and once these have been resolved emissions may increase again. In Asia, Japan has been successful in reducing sulphur dioxide emissions, although rising emissions of oxides of nitrogen continue to cause concern. Elsewhere the trend has been towards increased acidity,

with figures for Asia as a whole indicating an increase of more than 30 per cent in sulphur emissions between 1985 and 1997 (Streets *et al.* 2000), mostly as a result of industrial development in India and China. In recent years, however, China has been able to reverse the trend, achieving a reduction of 15 per cent in sulphur dioxide emissions in the last five years of the twentieth century (UNEP 2002). Most of the emissions are released from coal-fired power plants with outdated control equipment, but across Asia the growing number of motor vehicles is adding to acid deposition by releasing significant amounts of nitrogen oxides.

Once the acid gases are released into the atmosphere they are at the mercy of the prevailing circulation patterns. With almost all of the areas currently producing large amounts of acidic pollution being located in the mid-latitude westerly wind belt, emissions are normally carried eastwards or northeastwards, often for several hundred kilometres, before being redeposited. In this way pollutants originating in the US Midwest cause acid rain in Ontario, Quebec and New England (Figure 10.14). Emissions from the smelters and power stations of the English Midlands and the industrial Ruhr valley in Germany contribute to the acidity of precipitation in Scandinavia. Thus the problem of acid rain transcends national boundaries, introducing political overtones to the problem and creating the need for international co-operation in the search for solutions. Disagreements over acid rain control soured relations between Canada and the United States for more than a decade before the latter passed a comprehensive Clean Air Act in 1990, which included a section dealing with the transboundary transport of acid rain (Howard and Perley 1991). A similar situation prevailed in Europe, where in the early 1980s Britain was the main producer of sulphur dioxide after the USSR. Pumped into the westerly circulation, much of the acid gas made its way to Scandinavia, where Norway and Sweden were among the lowest sulphur dioxide producers on the continent and made little contribution to their own acid rain problems (Park 1987). Considerable animosity was created when, for a time, the British government and the companies emitting the pollutants denied any role in the problem. By 1986, however, their position had become untenable and the British government agreed that sulphur dioxide emissions from factories and power plants in the United Kingdom were responsible for acid rain damage in

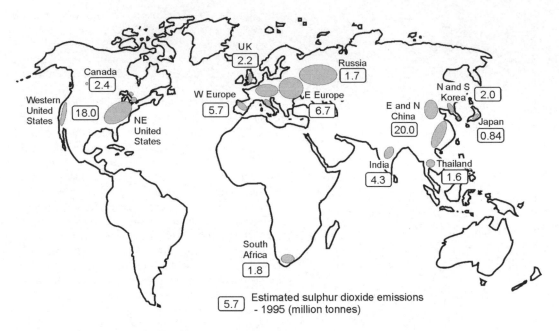

Figure 10.14 Geographic distribution of major sources of acid rain

Norway and Sweden. Albeit with some reluctance, Britain agreed to reduce acid gas emissions, eventually bringing it in line with the control programmes of other members of the European Community.

ENVIRONMENTAL IMPACTS OF ACID RAIN

The impact of acid rain on the environment depends not only on the level of acidity in the rain, but also on the nature of the environment itself. Areas underlain by acidic rocks such as granite or quartzite, for example, are particularly susceptible to damage, lacking the ability to 'buffer' or neutralize additional acidity from the precipitation. In contrast, areas that are geologically basic – with chalk or limestone bedrock, for example – are much less sensitive, and may even benefit from the additional acidity. The high alkalinity of the soils and waters of these areas ensures that the acid added to the environment by the rain is very effectively neutralized. The areas at greatest risk from acid rain in the northern hemisphere are the Precambrian Shield areas of Canada, Scandinavia, Russia and China, where the acidity of the rocks is reflected in highly

acidic soils and water. Many of the folded mountain structures of North America, Europe and Asia are also vulnerable. While areas with similar geological characteristics in the southern hemisphere may be at risk, sources of acid gas emissions there are limited and the increase in acidity to levels common in North America and Europe in the 1970s and 1980s is unlikely.

Aquatic impacts

The impact of acid rain on the environment was first recognized in the rivers and lakes of northern Europe and north-eastern North America. Reduced pH values, which indicated the rising acidity, were accompanied by low levels of calcium and magnesium, elevated sulphate concentrations and an increase in the amounts of potentially toxic metals such as aluminium (Brakke *et al.* 1988). When aquatic communities were examined in areas as far apart as New York State, Nova Scotia, Norway and Sweden there was clear evidence that increased surface water acidity had adverse effects on fish (Baker and Schofield 1985) (Figure 10.15). In some cases the acidity was sufficiently high that mature fish died, but, more commonly, fish populations

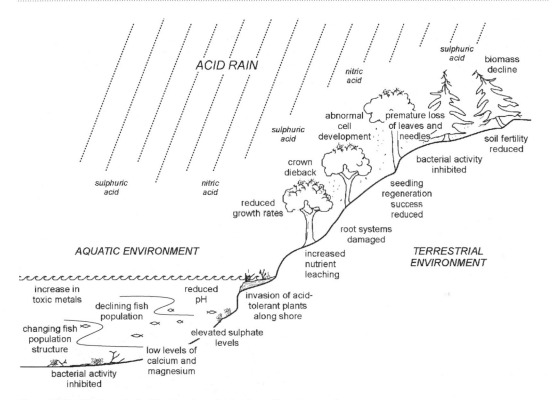

Figure 10.15   The impact of acid rain on terrestrial and aquatic environments

began to decline and their structure to change because of the effects of the increasing acidity on reproduction. Damage to the eggs during spawning and the inability of the young fry to survive the higher acidity, particularly during the spring flush – the rapid release of the winter's accumulation of acidity when the snow melts in the spring – ensured that the proportion of older fish in the population increased and they were not replaced as they died. As a result fish populations in many rivers and lakes in eastern North America, Britain and Scandinavia declined noticeably in the 1970s and 1980s and hundreds of lakes became completely devoid of fish (Harvey 1989) (Figure 10.16). Although the impact on the fish was most obvious, all organisms declined in number and variety during progressive acidification, in part because of the direct impact of the changing pH, but also because of the disruption of the food chain. Although often described as 'dead' or 'dying' some of these water bodies do support life, in the form of remarkably acid-tolerant species of protozoans and bacteria that can survive pH levels as low as 2.0 (Hendrey 1985).

Terrestrial impacts

Terrestrial ecosystems take longer to show the effects of acid rain than aquatic ecosystems. They tend to be more complex and often involve a level of human interference that is greater than in the aquatic environment. The growth, development and decline of plants have always reflected the integrated effects of many variables, including site microclimatology, hydrology, land use change, age and species competition. Acid rain has become part of that long and often complex list, causing damage directly through acid damage to foliage or indirectly through changes in the soil or biological processes controlling plant growth (Figure 10.15). The threat to the terrestrial environment is not universally recognized, however, and there remains a great deal of controversy over the amount of damage directly attributable to acid rain. Reduction in forest growth in Sweden, physical damage to trees in Germany and the death of sugar maples in Quebec and Vermont have all been blamed on the increased acidity of the precipitation in these areas (Kemp

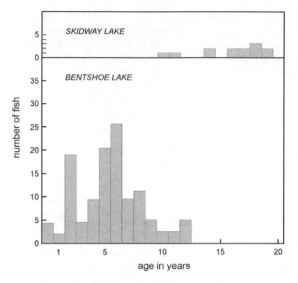

Figure 10.16 The age–class composition of samples of the white sucker population in Lakes Skidway (pH 5.0) and Bentshoe (pH 5.9)
Source: after Harvey (1989) with permission.

1994). Many of the impacts, such as the thinning of annual growth rings, reduction in biomass and damage to fine root systems, are apparent only after detailed examination, but others are more directly obvious and have been described as 'dieback'. This involves the gradual wasting of the tree inwards from the outermost tips of its branches. The process is cumulative over several years until the tree dies. The symptoms of dieback have been recognized in the maple groves and red spruce forests of north-eastern North America, and across fifteen countries and some 70,000 km² of forest in Europe (Park 1991). In Germany dieback, or *Waldsterben*, was particularly extensive and in the 1980s was seen as a major threat to the survival of the German forests (Ulrich 1983). Subsequent research has indicated that forest decline in Europe is a multi-faceted process in which acid precipitation is only one of a series of contributors along with tree harvesting practices, drought and fungal attacks, and its exact role remains controversial (Blank *et al.* 1988; Hauhs and Ulrich 1989; Godbold and Hüttermann 1994). The large-scale death of the forests anticipated in the 1970s and 1980s did not come to pass and the acid rain threat to the terrestrial environment has been described as an environmental myth (Lomborg 2001). While it may have been overstated, acid rain damage to European

and North American forests was a reality, and more recently similar symptoms have been recognized in forests covering a large area of the Sichuan basin and parts of southern China (UNEP 2002; Seip *et al.* 1999).

Impacts on the built environment and human health

In its various forms acid rain has caused damage to buildings and other structures, creating physical and economic costs as well as threatening the world's cultural heritage. Health impacts vary from the direct effects of inhaling acid aerosols to the indirect impacts of the ingestion of metals such as lead, copper and mercury mobilized in soil and water by acid precipitation.

Plate 10.4   A statue damaged by acid rain – a common sight on the historic buildings of Europe.

SOLUTIONS

Solutions to the problem of acid rain are deceptively simple. In theory, a reduction in the emission of acid gases is all that is required to slow down and eventually stop the damage being caused by the acidification of the environment. Translating that concept into reality took some time, however. Initial attempts at dealing with the problem involved adaptation, such as the addition of lime to land or lakes to counteract the increased acidity, and to allow the recovery mechanisms to work more effectively. Experiments involving the liming of lakes in Canada and Sweden have provided encouraging results (Porcella *et al.* 1990), but such artificial buffering wears off in three to five years and re-liming is necessary as long as acid loading continues. The most effective approaches to mitigating the impact of acid rain tackle the problem at its source. They attempt to prevent, or at least reduce, emissions of acid gases into the atmosphere, with most of the attention being given to the abatement of sulphur dioxide pollution. With sulphur dioxide being

formed when coal or oil are burned to release energy, the technology to control it can be applied before, during or after combustion (Figure 10.17). The exact timing will depend upon such factors as the amount of acid reduction required, the type and age of the system and the cost-effectiveness of the particular process (Ellis *et al.* 1990).

Flue gas desulphurization (FGD) is the most common approach to sulphur dioxide reduction. It uses scrubbers to neutralize the acidity of the flue gases before they are released into the atmosphere. With sulphur dioxide removal rates of between 80 per cent and 95 per cent, FGD is more effective than methods such as fuel switching and fuel desulphurization applied prior to combustion and it is technically simpler than fluidized bed combustion (FBC) or lime injection multi-stage burning (LIMB) incorporated in the combustion process itself. It has become the preferred method of dealing with acid rain, and it is likely that future requirements to reduce sulphur dioxide levels will be met in large part by the installation of FGD equipment.

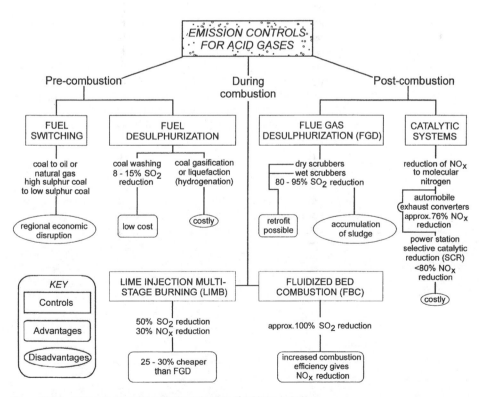

Figure 10.17   Different approaches to the reduction of acid gas emissions

None of the FGD systems works well to remove oxides of nitrogen, however. The best results for the control of oxides of nitrogen – up to 80 per cent reduction – have been obtained using a selective catalytic reduction process (SCR), which decomposes the oxides of nitrogen into the original nitrogen and oxygen, but widespread adoption is unlikely because the costs of installation, maintenance and the replacement of the catalyst are high (Ellis *et al.* 1990). Difficulties also exist with emission reductions from automobile exhausts. Catalytic converters have contributed to some reduction, but their impact is often reduced by the increase in traffic that continues to occur around the world. The development of technology that can produce a cooler-burning engine, or perhaps replace it completely, may be required before acid gas emissions from that source are reduced significantly.

TRENDS

Interest in acid rain has declined remarkably since the 1980s, when many viewed it as the main environmental problem facing the northern hemisphere. It is one issue in which abatement programmes have met with some success (Figure 10.18). Sulphur dioxide levels continue to decline, the rain in many areas is considerably less acid than it was several decades ago and the transboundary disputes that absorbed large amounts of time, money and energy in the 1980s have been resolved. There are indications that lakes and forests are

showing some signs of recovery in the most vulnerable areas of North America and Europe, although that is still a matter of dispute.

Developments such as these have created the perception that the acid rain problem is being solved. It would be more accurate, however, to say that it has changed in nature and in geographical extent. In the developed countries, oxides of nitrogen are gradually making a larger contribution to acidity as levels of sulphur dioxide decline. Among the developing nations of Asia, acid rain is increasing and its full impact may still be in the future. Abatement techniques developed in Europe and North America could be used in Asia. Work is already being done in Asia with the Regional Acidification Information and Simulation (RAINS) programme, developed for use by the European Community, which will allow researchers to alert Asian governments to the extent and intensity of the acid rain problem and to recommend ways of dealing with it (Hunt 1992).

SUMMARY AND CONCLUSION

Air pollution can occur as a result of natural processes, such as volcanic activity, forest fires or dust storms, but it is most often considered in human terms. It has been an environmental issue since the first human beings began to use fire and add the products of combustion to the atmosphere. The atmosphere has the ability to cleanse itself of the solid, liquid or gaseous materials released into it,

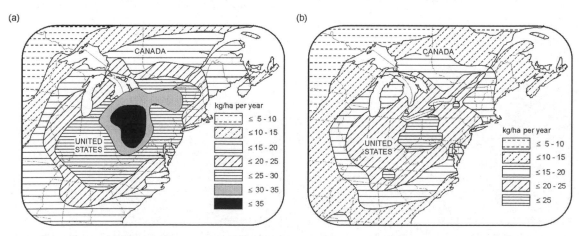

Figure 10.18   Mean wet sulphate deposition in eastern North America (a) 1980–84, (b) 1991–95

Source: Environment Canada, *Indicator: Wet Sulphate Depostion* (2002). Viewed at http://www.ec.gc.ca/soer-ree/English/Indicators/Issues/AcidRain/Bulletin/arind3_e.cfm.

and it is when they are added more rapidly than the atmospheric cleansing processes can remove them that pollution problems arise.

Air pollution increased in the mid-eighteenth century with the advent of the industrial revolution and the introduction of new energy and resource uses. The main concentration of pollutants was in the urban atmosphere of the new industrial towns. Initially the pollutants were particulates and gases produced by the coal-fired heavy industries that dominated early industrial development, but with time the use of oil and gasoline added new pollutants to the mix and together these threatened the health and quality of life of urban dwellers.

By the mid-twentieth century the realization that the rising levels of air pollution were dangerous and unacceptable began to spread and there was a significant improvement of the quality of the urban atmosphere. With the growth in the use of cars and trucks, however, some of the advantages of that improvement have been lost. An additional concern is the problem presented by indoor air pollution, which in some areas presents a greater threat to health than the air in the outdoor environment.

One result of air pollution that has caused serious problems in the aquatic and terrestrial environments is acid rain. Although natural rain is acid, the addition of acid gases such as sulphur dioxide and oxides of nitrogen from industrial activities and the burning of gasoline in the internal combustion engine raised its acidity to unprecedented levels. Before the severity of the problem was appreciated serious environmental damage occurred. Abatement procedures were developed and used successfully in North America and Europe, but acid rain remains a problem in parts of the developing world.

It is now technically possible to reduce acid gas emissions to very low levels. However, as is common with many environmental problems, technology is only one of the elements involved. Economics and politics have retarded the implementation of technical solutions in many cases, and that certainly has been so with acid rain. Nevertheless, progress has been made and, in the developed world at least, interest in acid rain has waned, to be replaced by apparently more pressing issues such as global warming and ozone depletion.

SUGGESTED READING

Davis, D. (2002) *When Smoke Ran like Water*, New York: Basic Books. Urban environmental air pollution, from smog to modern chemical toxins and their effects on health.

May, J.C. (2001) *My House is Killing me! The Home Guide to Families with Allergies and Asthma*, Baltimore MD: Johns Hopkins University Press. A comprehensive account, mainly for a general readership, of the major contaminants that cause indoor air pollution, their effects and ways of dealing with them.

Shah, J.J., Ranankutty, R. and Downing, R.J. (1997) *Rains-Asia: An Assessment Model for Acid Deposition in Asia* Directions in Development, Washington DC: World Bank. Overview of a model developed to investigate energy use and acid rain in Asia and to help identify acid abatement measures.

McCormick, J. (1997) *Acid Earth: The Politics of Acid Pollution* (3rd edn), London: Earthscan. An authoritative global view of the science, economics and politics of acid rain.

QUESTIONS FOR REVISION AND FURTHER STUDY

1   Put together a list of aerosol sources in your community and its region. Include natural and anthropogenic sources. Look beyond the more obvious possibilities such as automobile traffic, power stations or quarries and consider intermittent sources such as forest fires, wind-eroded soils, pollen-producing grassland and woodland in the spring. Which sources provide most aerosols? What role does local climatology play in the dispersal of these aerosols? Do any of the sources make an obvious contribution to global aerosol levels?

2   Under natural conditions the air is never completely clean and human beings have evolved to live with that. What physical characteristics do we possess that allow us to deal with a certain level of atmospheric pollution and what health problems arise when these systems can no longer cope?

3   Having been successful in reducing acid rain in the industrialized world, the developed nations would like to see their success emulated in the developing world for the general good of the global environment. In countries such as India and China, however, the use of cheap energy sources such as coal with its accompanying acid gas emissions is seen as essential for economic development. What can the developed nations do to encourage the reduction of acid rain in such areas without appearing to be attempting to constrain their economic advancement?

11

Ozone Depletion and Global Warming

After reading this chapter you should be familiar with the following concepts and terms:

adaptation	Climatic Optimum	methane
Antarctic ozone hole	global environmental	methyl bromide
business-as-usual scenario	engineering	methyl chloroform
carbon cycle	global warming	mitigation
carbon dioxide	global warming potential	Montreal Protocol
carbon dioxide fertilization	greenhouse effect	nitrous oxide
carbon sequestration	greenhouse gases	ozone
carbon sources and sinks	halocarbons	ozone-depleting substances
carbon taxes	halons	ozone depletion
carbon tetrachloride	heterogeneous chemical	ozone layer
catalyst	reactions	polar stratospheric clouds
catalytic chain reaction	hydrogen oxides	polar vortex
CFCs	hydrogen radical	radiation blindness
chlorine monoxide	IPCC	rising sea level
chlorine monoxide dimer	Kyoto Protocol	skin cancer
chlorine nitrate	Medieval Warm Period	ultraviolet radiation
clathrates	melanoma	

Ozone depletion and global warming became the most widely known global environmental issues in the last couple of decades of the twentieth century. Media coverage generated widespread popular concern over the topics and interest grew rapidly in the scientific community, where international investigators studying them garnered a large proportion of the scientific research budget dedicated to environmental issues. Global warming has probably now surpassed ozone depletion as the most prominent environmental issue, highlighted by the work of the meteorologists, climatologists and other scientists involved in the work of the UNEP/WHO-sponsored Intergovernmental Panel on Climate Change (IPCC), an important source of information on the scientific aspects, impacts and mitigation of global warming (see Box 11.1, IPCC).

Controversy over the implementation of the Kyoto Protocol, designed to limit global warming, has also kept the issue in the public eye. Despite this high profile there has been little success in dealing with the various aspects of global warming. In contrast, international co-operation in controlling the release of ozone-depleting substances (ODS) has allowed considerable progress in diminishing the causes of ozone depletion, and this may be reflected in the reduced general interest in the topic – success often appearing less newsworthy than failure in reporting on environmental issues. Although stratospheric ozone levels continue to reach record lows, particularly over Antarctica, the marked decline in the consumption of ODS is likely to be followed by the recovery of the ozone layer to pre-1980 values by the middle of the twenty-first century.

BOX 11.1 THE INTERGOVERNMENTAL PANEL ON CLIMATE CHANGE

The Intergovernmental Panel on Climate Change (IPCC) is a group of eminent scientists, originally brought together in 1988 by the World Meteorological Organization (WMO) and the United Nations Environment Program (UNEP). It was charged with assessing the overall state of research on climate change, so that the potential environmental and socio-economic impacts might be evaluated, and appropriate response strategies developed. This involved three working groups charged with the following:

- WG I, to assess available scientific information on climate change.
- WG II, to assess the environmental and socio-economic impacts of climate change.
- WG III, to formulate response strategies.

Their reports – a scientific overview, an impact assessment and a group of response strategies – were produced in 1990. A supplementary report was issued in 1992, generally confirming the original assessments, and by 1994 a second, supplementary report focusing on the radiative forcing of climate had been completed.

Because of the rapid accumulation of data in the field of climate change, a second comprehensive report was considered necessary. In addition, the new information to be provided by the report was needed to support work on the United Nations Framework Convention on Climate Change (UNFCCC), signed at the Earth Summit in Rio de Janeiro in 1992. Most of the contributors to the second assessment were university or government scientists, with limited numbers from private research agencies, companies and non-governmental organizations (NGOs). Representatives from environmental advocacy groups such as Greenpeace were among those who reviewed the original documents. Completed in December 1995, it was presented to the signatories of the UNFCCC prior to publication in mid-1996.

Central to the report was the recognition of the human contribution in current climate change. It also recommended that action be taken to halt global warming, and because of the time lags involved it concluded that action could no longer be delayed. As part of its terms of reference the IPCC is charged with providing, on request, scientific, technical and socio-economic advice on climate change to the Committee of the Parties (COP) involved in the ongoing development of the UNFCCC, and in so doing it contributed to the process that led to the signing of the Kyoto Protocol in 1997.

A third assessment report from the IPCC working groups was published in 2001, containing an evaluation of the changes that had taken place since the second assessment. It included new data on current and palaeoclimates and a more rigorous evaluation of the quality of the data sets, from which the IPCC claims to have obtained a better understanding of climate change. The observation that the human imprint on current climate change is recognizable in the data, which first appeared in the second assessment in 1996, is confirmed in the third. The results in the third assessment report have become the basis of most of the public policy that is being considered to deal with global warming, and are central to further progress in the development of the provisions of the Kyoto Protocol.

The work of the IPCC has not met with universal approval. It is seen by some as based on bad science that overstates the case on global warming and ignores the many uncertainties that exist. Other observers, less convinced about the existence of global warming, see the reports as alarmist, with the potential to cause major economic disruption as society attempts to deal with a problem that may not exist. The assessments have also become mired in politics, particularly in the way in which they are being used to advance the requirements of the Kyoto Protocol. The controversy will continue as scientists, environmentalists, economists, politicians and the general public debate the issue. Whatever the perceptions of the work of the panel, there can be no denying that the members of the IPCC, through the scientific data they have assembled and analysed, have made a major contribution to the study of climate change.

For more information see:

Essex, C. and McKitrick, R. (2002) *Taken by Storm: The Troubled Science, Policy and Politics of Global Warming*, Toronto: Key Porter Books.

Kemp, D.D. (1997) 'As the world warms: climate change 1995', *Progress in Physical Geography* 21 (2): 310–14.

Masood, E. (1996) 'Climate report subject to scientific cleansing', *Nature* 381 (6583): 546.

Some examples of IPCC reports:

IPCC (1990a) *Climate Change: The IPCC Scientific Assessment*, Cambridge: Cambridge University Press.

IPCC (1990b) *Climate Change: The IPCC Impacts Assessment*, Canberra: Australian Government Publishing Service.

IPCC (1990c) *Climate Change: The IPCC Response Strategies*, Canberra: Australian Government Publishing Service.

IPCC (1996) *Climate Change 1995: The Science of Climate Change*, Cambridge: Cambridge University Press.

BOX 11.1 – continued

IPCC (2001a) *Climate Change 2001: The Scientific Basis,* Cambridge: Cambridge University Press.
IPCC (2001b) *Climate Change 2001: Impacts, Adaptations and Vulnerability,* Cambridge: Cambridge University Press.

IPCC (2001c) *Climate Change 2001: Mitigation,* Cambridge: Cambridge University Press.
IPCC (2001d) *Climate Change 2001: Synthesis Report,* Cambridge: Cambridge University Press.

OZONE DEPLETION

STRATOSPHERIC OZONE

As indicated in Chapter 10, tropospheric ozone is a constituent of photochemical smog and is normally considered to be a pollutant, irritating to eyes and respiratory tissues and harmful to plants. In contrast, stratospheric ozone is an essential component of the earth/atmosphere system, because of its ability to protect the biosphere from excess ultraviolet radiation. Stratospheric ozone is diffused through a layer that extends between 12 km and 50 km above the earth's surface (see Chapter 2). The total amount of ozone at any given time is small – if brought to normal atmospheric pressure at sea level, for example, it would form a band no more than 3 mm thick – but through a dynamic, reversible process in which the gas is continually broken down and reformed, this relatively minor amount of ozone retains the ability to absorb ultraviolet radiation and prevents it from reaching the earth's surface.

Solar radiation provides the energy that allows the earth/atmosphere system to function. As with many essentials, however, there are optimum levels beyond which a normally beneficial input becomes harmful. This is particularly so with the radiation at the ultraviolet end of the spectrum (Table 11.1). At normal levels, for example, it is an important germicide, and is essential for the synthesis of vitamin D in humans. At elevated levels, it can cause sunburn or skin cancer and produce changes in the genetic make-up of organisms. If unrestricted, ultraviolet radiation would disrupt the earth's biotic systems completely and life as it exists today would not be possible. In addition, since it is an integral part of the earth's energy budget, changes in ultraviolet levels have the potential to contribute to climate change.

Human beings and other organisms have evolved to deal with certain levels of ultraviolet radiation, and under normal circumstances the stratospheric ozone layer keeps it within manageable limits. Close to the equator, ozone allows only 30 per cent of the ultraviolet-B (UV-B) radiation to reach the earth's surface. A comparable level for higher latitudes is about 10 per cent, although during the summer months radiation receipts may approach equatorial levels (Gadd 1992). Specific amounts vary along with natural fluctuations in ozone concentrations and such factors as cloudiness or air pollution in the lower atmosphere. Although only a minor gas in terms of the volume it occupies in the atmosphere, ozone is essential to the survival of life on earth because of its ability to filter out a high proportion of the incoming ultraviolet radiation. It removes or converts all of the extremely hazardous UV-C wavelengths and 70–90 per cent of the UV-B rays (Gadd 1992). It does not remove UV-A rays, which are often considered relatively safe – they are used in tanning salons, for example – although long-term exposure to UV-A is now suspected of having potentially detrimental health consequences (Turco 1997).

Daily and seasonal fluctuations in the stratospheric ozone layer were accepted as normal, characteristic of the dynamic nature of the system. By the early 1970s, however, there were indications that the checks and balances thought to be built into the system were failing to prevent a gradual decline in ozone levels. Inadvertent human interference in the chemistry of the ozone layer was identified as the cause of the decline and there was growing concern over the potentially disastrous consequences of

TABLE 11.1 DIFFERENT FORMS OF ULTRAVIOLET RADIATION

Radiation	Wavelength (nm)[a]
Ultraviolet-A (UV-A)	320–400
Ultraviolet-B (UV-B)	280–320
Ultraviolet-C (UV-C)	200–280

Note: [a] 1 nm (nanometre) $= 1 \times 10^{-9}$ m $= 1 \times 10^{-3}$ μm.

elevated levels of ultraviolet radiation at the earth's surface. The depletion of the ozone layer became a major environmental issue by the middle of the decade. Its technological complexity caused dissension in scientific and political arenas, but it garnered much popular attention because its effects impacted directly on individuals. In common with many environmental concerns at that time, however, interest waned in the late 1970s and early 1980s, only to be revived again in 1985 with the discovery of what came to be called the Antarctic ozone hole.

THE PHYSICAL CHEMISTRY OF THE OZONE LAYER

Stratospheric ozone owes its existence to the impact of ultraviolet radiation on oxygen molecules in the upper atmosphere. Oxygen molecules normally consist of two atoms, and in the lower atmosphere they commonly retain that configuration. At the high energy levels associated with ultraviolet radiation in the upper atmosphere, photodissociation occurs, causing these molecules to split apart and produce atomic oxygen. These free atoms then combine quite readily with the available molecular oxygen to create triatomic oxygen or ozone. The action is reversible. The ozone molecule may break down again into its original components, molecular oxygen and atomic oxygen, as a result of the further absorption of ultraviolet radiation or it may combine with atomic oxygen to be reconverted to the molecular form (Figure 11.1). Most of the ozone is produced in the upper stratosphere, at an altitude of 25 km to 50 km, where energy levels are high, and is stored in the lower stratosphere between 12 km and 25 km. At the lower edge of the layer, ozone is destroyed as it diffuses into the troposphere. The production rate is greatest in the tropical stratosphere, where radiation levels are high, and from there the ozone moves outwards into higher latitudes (Horel and Geisler 1997).

The ozone layer is in a constant state of flux as the molecular structure of its constituents changes, and the total amount of ozone at any given time represents a balance between the rate at which it is being formed and the rate at which it is being destroyed. Because it requires ultraviolet energy to drive the reactions involved, the ozone layer reduces the overall flow of short-wave solar radiation through to the lower atmosphere, and in so doing protects the environment from the potentially

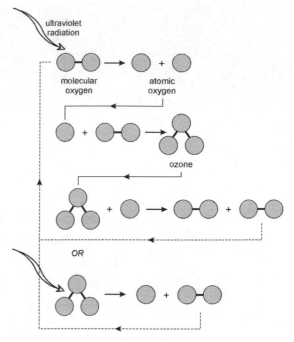

Figure 11.1   Schematic representation of the formation of stratospheric ozone

dangerous consequences of excess ultraviolet radiation.

The role of ultraviolet radiation and molecular oxygen in the formation of the ozone layer was first explained by Chapman in 1930. Later measurements confirmed the basic theory, but observed levels of ozone were much less than expected, given the natural rates of production and decay associated with Chapman's chemistry. Since none of the other normal constituents of the atmosphere – molecular oxygen, nitrogen, water vapour or carbon dioxide – was considered capable of destroying the ozone, attention turned to trace elements in the stratosphere. Initially it seemed that these were present in insufficient quantities to have the necessary effect, but the problem was solved with the discovery that they participated in catalytic chain reactions (Dotto and Schiff 1978). A catalyst is a substance that facilitates a chemical reaction yet is itself unchanged when the reaction is over. Being unchanged, it can go on to promote the same reaction again and again as long as the reagents are available or until the catalyst itself is removed. In this form of chain reaction, a catalyst in the stratosphere can destroy thousands of ozone molecules before it is finally removed. The ozone layer is capable of dealing with the relatively small amounts of naturally occurring catalysts. Recent concern over the depletion of stratospheric ozone has focused on anthropogenically produced catalysts, which were recognized in the stratosphere in the early 1970s, and which have now accumulated in quantities well beyond the system's ability to cope.

Naturally occurring ozone-destroying catalysts

Natural catalysts have always been part of the atmospheric system and many are similar to those now being added to the atmosphere by human activities. The main difference is in production and accumulation, with the natural variety being produced in smaller quantities and remaining in the atmosphere for shorter time periods than their anthropogenic counterparts. The main groups of natural catalysts are hydrogen oxides (HO_x), oxides of nitrogen (NO_x) and chlorine monoxide (ClO) (Table 11.2; Figure 11.2). Of these, the oxides of nitrogen are most efficient, with nitric oxide alone being responsible for 50–70 per cent of the natural destruction of stratospheric ozone (Hammond and Maugh 1974). Although chlorine in conjunction with chlorine monoxide is a very effective catalyst – one chlorine atom can destroy 10^5 molecules of ozone (Rowland 1990) – its production in the natural environment and its diffusion into the stratosphere are insufficient to have much impact on the ozone layer.

Efficient as they are, these catalysts are ultimately limited in the amount of ozone they can destroy. They are naturally cycled into various forms that are relatively inactive and thus lose their ability to destroy ozone. Without this, the ozone layer would be completely eroded, with disastrous consequences (Turco 1997). Hydrogen oxides lose their catalytic capabilities when converted into water vapour, for example, while nitrogen and chlorine oxides are deactivated respectively by reduction into nitric and hydrochloric acid. Nitrogen dioxide (NO_2) readily combines with chlorine monoxide (ClO), one of the most efficient ozone destroyers, to produce chlorine nitrate ($ClONO_2$), a much less reactive compound. This deactivation of the catalysts is often so effective that only a small proportion of them are catalytically active at any given time. In the case of chlorine, for example, as much as 99 per cent may be partitioned into hydrochloric acid and chlorine nitrate, leaving only about 1 per cent active chlorine in the stratosphere (Turco 1997).

THREATS TO THE OZONE LAYER FROM HUMAN ACTIVITIES

Data on stratospheric ozone are available only as far back as the 1950s, but they indicate that ozone levels remained relatively stable, fluctuating only within normal limits, until the mid to late 1970s. By that time it had become clear that human

TABLE 11.2   OZONE DEPLETION POTENTIAL OF VARIOUS HALOCARBONS

Chemical compound	Main use	Life time (years)	Ozone depletion potential (ODP)
CFC-11	Refrigerant, foaming agent, propellant	45	1.0
CFC-12	Refrigerant, foaming agent, propellant	100	1.0
CFC-113	Solvent	85	0.8
Halon-1211	Refrigerant, foaming agent	11	4.0
Halon-1301	Fire extinguisher, fumigant	65	12.0
Methyl chloroform	Solvent, adhesive	4.8	0.1
Carbon tetrachloride	Solvent	35	1.1
Methyl bromide	Fumigant	0.7	0.3–0.9
HCFC-22	Refrigerant, foaming agent	11.8	0.05
HCFC-123	Refrigerant, foaming agent	1.4	0.02
HFC-134a	Refrigerant, foaming agent	13	0.0

Source: based on data in R.P. Turco, *Earth under Siege*, Oxford/New York: Oxford University Press (1997); E.A. Ripley, 'Chlorofluorocarbons', in A.S. Goudie (ed.) *The Encyclopedia of Global Change*, New York: Oxford University Press (2002); USEPA Web site http://www.epa.gov/ozone/ods.html

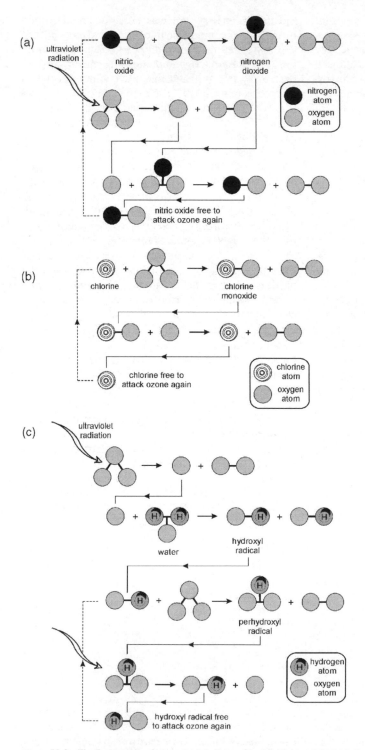

Figure 11.2 **The destruction of ozone by (a) nitric oxide, (b) chlorine, (c) odd hydrogens**
Source: after Kemp (1994).

activities had the potential to bring about sufficient degradation of the ozone layer that it might never recover. The threat was seen to come from four main sources, associated with modern technological developments in warfare, aviation, agriculture and lifestyle and involving a variety of complex chemical compounds, both old and new. Nuclear war, supersonic transports (SSTs) cruising in the atmosphere and agricultural techniques increasingly dependent upon nitrogen-based fertilizers were seen as potential sources of increasing amounts of oxides of nitrogen (NO_x), a group of highly potent destroyers of ozone.

Nuclear war

The analysis of data from atmospheric nuclear tests in the early 1960s was inconclusive about the impact of individual nuclear explosions on atmospheric chemistry, but computer simulations suggested that following a major nuclear conflict 50–70 per cent of the ozone layer might be destroyed, mainly as a result of the synthesis of NO_x from atmospheric oxygen and nitrogen at the high temperatures produced by the thermonuclear explosions (Crutzen 1974; Dotto and Schiff 1978). With the end of the Cold War and agreements to reduce the number of nuclear weapons, the contribution of nuclear war to ozone depletion is no longer given serious consideration.

Supersonic transports

Similarly, the hue and cry that accompanied the development of the original supersonic transports in the 1970s has died away. Like all aircraft, SSTs produce exhaust gases, which include water vapour, carbon dioxide, carbon monoxide, oxides of nitrogen and some unburned hydrocarbons. These are injected directly into the ozone layer, since SSTs commonly cruise at about 20 km above the surface, just below the zone of maximum ozone concentration. Much of the initial concern over the effect of SSTs on the ozone centred on the impact of water vapour, which was considered capable of reducing ozone levels through the creation of hydroxide radicals, known ozone destroyers. So complex is the chemistry of the ozone layer, however, that rather than the expected decrease in ozone, observations indicated a 35 per cent increase in water vapour in

the upper atmosphere was actually accompanied by a 10 per cent increase in ozone (Crutzen 1972). It was suggested that this had been brought about by the ability of the water vapour to convert ozone-destroying NO_x to nitric acid, which does no harm to ozone (Crutzen 1972; Johnson 1972).

Predictions on the potential threat from NO_x included warnings that these gases could reduce ozone levels by as much as 22–50 per cent and lead to thousands of additional cases of skin cancer (Johnston 1971; Crutzen 1972; Hammond and Maugh 1974). Other predicted environmental impacts included damage to vegetation and changes in the nature and growth of some species as a result of mutation. All of these predictions were based on the expectation that the number of SSTs would grow to several hundred by the end of the twentieth century, but that did not happen. The Boeing SST in the United States was cancelled, the Soviet Tupolev-144 and the Anglo-French Concorde were put into production, with the latter the more successful in terms of production numbers and commercial route development. The Tu-144 is no longer flying, only about ten Concordes were in operation at any one time, and their effects on the ozone layer are generally considered to have been negligible. In 2003 both Air France and British Airways, the only airlines flying Concorde, took the aircraft out of service permanently.

Agriculture

The augmentation of levels of nitrous oxide (N_2O) in the atmosphere as a result of the increased use of nitrogen-based fertilizers was also identified as having the potential to cause additional ozone depletion. In the 1960s the so-called Green revolution, which led to a rapid increase in world food production, was made possible by the introduction of genetically improved grains that needed increased amounts of nitrogen fertilizers to allow them to meet their productivity potential. Nitrous oxide is a by-product of that fertilizer use, released into the atmosphere by the denitrification of nitrates in the soil. Being an inert gas, it is not easily removed from the troposphere and it eventually diffuses into the stratosphere, where it becomes part of the ozone-destroying catalytic chain reaction associated with oxides of nitrogen (Figure 11.2a). There is no proof that increased fertilizer use has damaged the ozone layer through the production of

N₂O (Gribbin 1993), but it is a potent ozone-depleting agent with an atmospheric lifetime of over a century, suggesting that N_2O emissions must be treated with caution (Turco 1997).

CFCs and ozone depletion

If there is, or has been, some doubt about the impact of SST exhaust emissions on the ozone layer, or the contribution of nitrogen fertilizers to ozone depletion, the effects of certain other chemicals are now well established. Chlorofluoro-carbons (CFCs) have made a major contribution to ozone depletion, for example, and continue to pose a significant threat to the ozone layer, despite successful attempts at controlling their output and use. CFCs belong to a group of chemicals called halocarbons, which contain carbon and one or more of the halogen elements – chlorine, bromine, fluorine and iodine. CFCs contain carbon, chlorine and fluorine, whereas bromofluorocarbons or halons include carbon, bromine, fluorine and some-times chlorine. They were developed in the 1930s, and after 1960 their production increased rapidly (Figure 11.3). CFCs were used in refrigeration and air conditioning systems, and as propellants in aerosol spray cans dispensing a wide range of products from deodorants to paints and insect repellents. They were also used as foaming agents in the production of insulating foams, polymer foams for upholstery and foam containers for the fast food industry. Halons were found to be ideal for use in fire extinguishers and fire protection systems for aircraft, computer centres and industrial control rooms where conventional fire extinguishing materials such as water or foam would cause damage to delicate instruments.

The popularity of CFCs and halons stemmed from their stability and low toxicity, which meant, for example, that they could be used as propellants in the inhalers required by those suffering respira-tory problems without changing the efficacy of the medication or causing harm to the user. Being inert, they were also ideal solvents for cleaning delicate electronic components such as computer chips. However, the very properties that made CFCs and halons so useful ultimately allowed them to become the major contributors to ozone depletion. Their stability allowed them to accumulate in the environ-ment relatively unchanged. With time they gradually diffused into the upper atmosphere, where they

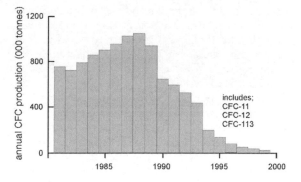

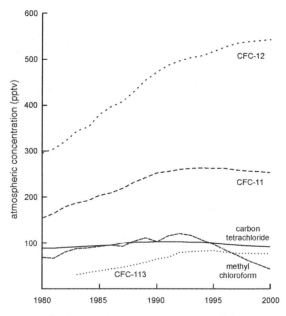

Figure 11.3 CFC production and the concentration of ozone-depleting substances (ODS) in the atmosphere
Source: based on data in UNEP (2002) and World Resources Institute, *Earthtrends*. Viewed at http://www.earthtrends.wri.org/ (accessed 4 April 2003).

encountered conditions under which they were no longer inert, and broke down to release by-products capable of depleting the stratospheric ozone layer. Their capacity to destroy ozone varies and is repre-sented by their ozone depletion potential (ODP). CFC-11, with an ODP of 1, is the standard against which all other ozone-destroying chemicals are measured. The ability of Halon-1301 (ODP 12) to destroy ozone, for example, is twelve times greater than that of CFC-11 (Table 11.2).

In 1974 Mario Molina and Sherwood Row-land, atmospheric scientists at the University of California, Irvine, using the results of their investigations into the photochemistry of the stratosphere, were the first to explain the processes by which the depletion was accomplished. They recognized that the high levels of ultraviolet radiation in the upper atmosphere caused the photochemical degradation of the normally inert CFCs, leading to the release of chlorine. Catalytic chain reactions initiated by the free chlorine then began the process of depleting the ozone layer (Figure 11.2b). The importance of the chlorine catalytic chain lies in its efficiency; it is six times more efficient catalytically than the nitric oxide cycle, for example. The chain is broken only when the chlorine or chlorine monoxide molecules gain a hydrogen atom from one of the hydrogen oxides, or from a hydrocarbon such as methane, and is converted into hydrochloric acid, which diffuses into the lower atmosphere, eventually to be washed out by rain. The combination of chlorine monoxide with nitrogen dioxide to form chlorine nitrate also breaks the chain (Turco 1997). Conclusions similar to those of Molina and Rowland were reached independently at about the same time by other researchers (Crutzen 1974; Cicerone *et al.* 1974; Wofsy *et al.* 1975), and with the knowledge that the use of CFCs had been growing since the late 1950s, the stage seemed set for an increasingly rapid thinning of the earth's ozone shield, followed by a rise in the level of ultraviolet radiation reaching the earth's surface.

The world production of CFCs was less than 50,000 tonnes in 1950, but grew at a rate of about 9 per cent per year in the 1960s and reached 700,000 tonnes in 1973 (Crutzen 1974; Molina and Rowland 1994). Following the development of the theory of ozone depletion in 1974, production levelled off for several years, but by 1986 the total production of CFCs and halons amounted to 1,260,000 tonnes, before falling to 870,000 tonnes in 1990 (Environment Canada 1992). Since then international restrictions on halocarbon production have contributed to a continuing decline, which by 1995 had reached levels common in the early 1960s, the ultimate aim being zero production (Figure 11.3). Even with such a significant reduction in the output of CFCs and halons, however, relief for the ozone layer has not been instantaneous. The inherent stability of these compounds means that they will remain in the atmosphere for periods of forty to 150 years, all the while retaining their ability to destroy ozone. Recent observations by scientists of the IPCC suggest, however, that levels of ozone-depleting gases in the troposphere peaked in 1994 and are slowly declining (IPCC 2001a), which means their diffusion into the stratosphere should also begin to decline and reduce the threat to the ozone.

The net effect of changes in CFC production on long-term ozone depletion has proved difficult to predict. Molina and Rowland's original estimate of a 7–13 per cent reduction in steady state ozone levels was increased to as much as 20 per cent by some studies, but reduced to as little as 5–7 per cent in others (Molina and Rowland 1994). Although all of the researchers acknowledged that the results were at best preliminary, because of inadequate knowledge of the photochemistry of the stratosphere, such conditions were generally ignored as the topic took on a momentum of its own and the dire predictions made during the SST debate were repeated. The spectre of thousands of cases of skin cancer linked with the use of a seemingly innocuous product like hair spray or deodorant was sufficiently different to excite the media and, through them, the general public. Although CFCs were being used as refrigerants and in the production of various types of foam, the problem was usually presented as one in which the convenience of aerosol spray products was being bought at the expense of the global environment. In 1975 there was some justification for this, since at that time 72 per cent of CFCs were used as propellants in aerosol spray cans (Molina and Rowland 1994) and the campaign against that product grew rapidly. The multi-million-dollar aerosol industry, led by major CFC producers such as DuPont, reacted strongly. Through advertising and participation in US National Academy of Science (NAS) hearings, they emphasized the speculative nature of the Molina–Rowland hypothesis, and the lack of hard scientific evidence to support it. The level of concern was high, however, and the anti-aerosol forces met with considerable success. Eventually manufacturers replaced CFCs with less hazardous propellants (Dotto and Schiff 1978; Gribbin 1993). A partial ban on CFCs covering their use in hair and deodorant sprays was introduced in the United States in 1978 and in Canada in 1980. CFC aerosol spray use remained high in Europe, where there was no ban until well into the 1980s.

Response to the concern over ozone depletion at that time is reflected in the stabilization of CFC production for about a decade after 1974, but the CFC controversy had already ceased to make headlines by the late 1970s and the level of public concern had fallen away. Monitoring of the ozone layer showed little change. Ozone levels were not increasing despite the ban on aerosol sprays, which was only to be expected, given the slow rate of decay of the existing CFCs, but the situation did not seem to be worsening. Quite unexpectedly, in 1985, scientists working at the Halley Bay base of the British Antarctic Survey (BAS) announced a major thinning in the ozone layer over Antarctica (Farman *et al.* 1985). This became popularly known as the Antarctic ozone hole, and all the fears that had been raised during the aerosol spray can debate suddenly resurfaced.

THE ANTARCTIC OZONE HOLE

Total ozone levels had been measured by the BAS at the Halley Bay for more than thirty years before the hole was discovered. Seasonal fluctuations were observed for most of that time, and included a thinning of the ozone above the Antarctic during the southern spring, which was considered normal for the area (Schoeberl and Krueger 1986). This regular minimum in the total ozone level began to intensify in the late 1980s, however. The observations indicated that it commonly became evident in late August, and got progressively worse until, by mid-October, as much as 40 per cent of the ozone layer above the Antarctic had been destroyed (Farman *et al.* 1985). Usually the hole would fill again by November, but during the 1980s it began to persist into December (Figure 11.4). The intensity of the thinning and its geographical extent were originally established by ground-based measurements, and later confirmed by overflights from high-flying aircraft and by remote sensing from the Nimbus-7 polar orbiting satellite (Stolarski *et al.* 1986).

The chemistry of the Antarctic ozone hole

When the hole was first discovered, blame fell almost immediately on CFCs and measurements of these chemicals in the stratosphere above the South Pole tended to support that, although the annual fluctuation had been present before CFCs were released into the atmosphere in any quantity. As the investigation of the Antarctic ozone layer intensified, it became clear that the chemistry of the

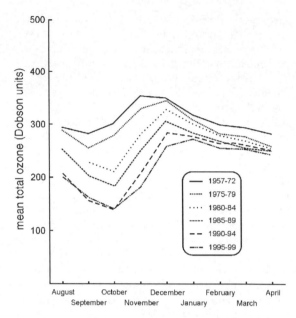

Figure 11.4 Declining ozone levels above Halley Station, Antarctica

Source: based on data on British Antarctic Survey Web page. Viewed at http://www.antarctica.ac.uk/met/jas/ozone/bulletins/bas0201.htm

polar stratosphere is particularly complex and the reactions involved differ from those in which stratospheric ozone is destroyed elsewhere (Figure 11.5). Although the stratosphere is very dry, it becomes saturated at the very low temperatures reached during the darkness of the winter months and clouds form. These polar stratospheric clouds contribute to the thinning of the ozone layer by providing a base, in the form of ice crystals, for a series of heterogeneous chemical reactions that ultimately lead to the release of chlorine (Figure 11.6). They provide the means by which the chlorine normally sequestered in hydrochloric acid and chlorine nitrate is let loose into the atmosphere (Turco 1997). Released in its molecular form (Cl_2) it is initially no threat to the ozone, but as the sunlight returns again to the polar regions photodissociation splits the molecule into active chlorine atoms, which vigorously attack the ozone in a catalytic cycle that includes the formation of chlorine monoxide and the chlorine monoxide dimer (a double molecule – Cl_2O_2 rather than ClO) (Shindell and de Zafra 1995). The process is characterized by the formation of a thick layer of chlorine monoxide that forms rapidly in the lower stratosphere at an altitude of about 20 km and remains as long as chlorine is

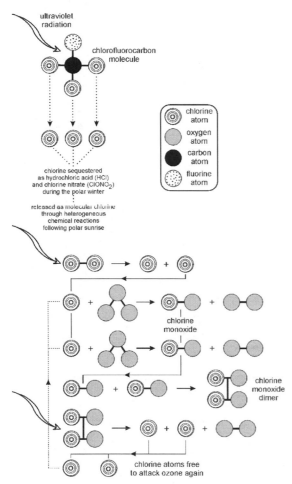

Figure 11.5 The destruction of Antarctic ozone

Source: based on formulae in Turco (1997) and Shindell and de Zafra (1995).

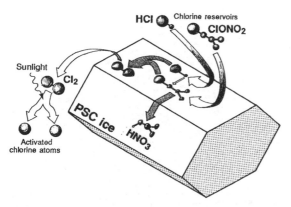

Figure 11.6 Heterogeneous chemical reactions on ice crystals in the Antarctic stratosphere

Source: Turco (1997) with permission.

being released by the heterogeneous chemical reactions. As temperatures rise in the spring and the polar stratospheric clouds evaporate, active chlorine is sequestered once again into hydrochloric acid and chlorine nitrate, allowing the ozone layer to recover (de Zafra *et al.* 1995).

The chemical reactions associated with polar stratospheric clouds take place initially on the surface of ice particles in the clouds. Similar heterogeneous chemical reactions take place on the surface of sulphate particles released into the stratosphere during major volcanic eruptions (Brasseur and Garnier 1992). The eruptions of Mount Pinatubo (Philippines) and Mount Hudson (Chile) in 1991, for example, were followed by the rapid destruction of the ozone layer at levels between 9 km and 13 km (Deshler *et al.* 1992). In the following year, record low ozone levels were reported over the South Pole and over southern Chile and Argentina, while global ozone levels were 4 per cent below normal (Kiernan 1993).

While the thinning of the ozone layer above the Antarctic is a function of the chlorine catalytic cycle, local meteorological conditions contribute to its effectiveness (Bowman 1986). The cooling of the region as winter approaches leads to the creation of the circumpolar vortex, a band of strong winds that circle the pole at about 65°S and extend vertically through the stratosphere (Turco 1997). When it is well established it permits little change of energy or matter across its boundaries and the air inside the vortex is effectively isolated from the rest of the atmosphere (Figure 11.7). Little or no sensible heat is transported into the area from lower latitudes and in combination with the absence of solar radiation during midwinter this ensures that the air within the vortex becomes very cold, allowing the formation of the polar stratospheric clouds essential for the initiation of the ozone depletion cycle. When solar radiation levels first increase in the spring and the rapid destruction of ozone begins, the vortex is still strong enough to prevent the inflow of new ozone from lower latitudes. With no replacement of the lost ozone, concentrations above the pole decrease rapidly – sometimes by as much as 90 per cent in a matter of weeks – but the return of the sun in the spring also causes the decay of the vortex, allowing new ozone into the area along with warmer air, which together end the conditions that encouraged the depletion of the ozone (Turco 1997). As the southern spring and summer progress, the normal chemical composition is re-established and the

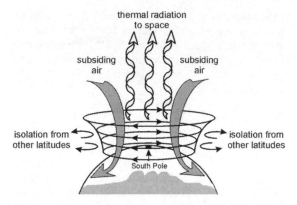

Figure 11.7 Conditions associated with the circumpolar vortex in the southern hemisphere

Source: Turco (1997) with permission.

ozone layer above the Antarctic thickens again. Since the ozone hole was first recognized and the reactions that cause it were identified in the mid-1980s, the spring recovery has never been complete, and the hole has steadily become deeper, larger and persisted longer (Figure 11.8) until in September 2000 it covered an area of 29 million km^2 (WMO 2000).

THE OZONE LAYER ABOVE THE ARCTIC

Following the discovery of the major thinning of the ozone layer over the Antarctic, scientists began to examine the possibility of a similar decline over the Arctic. The European Arctic Stratospheric Ozone Experiment (EASOE) was established in the northern winter of 1991–92 to investigate the nature and extent of ozone depletion over the Arctic (Pyle 1991). Preliminary results indicated the absence of a distinct hole, but decreases of 10–20 per cent in Arctic ozone were detected, particularly in adjacent areas in lower latitudes between 60°N and 30°N (Concar 1992; Turco 1997). Observations showed that the chemical precursors of ozone depletion are probably in place most winters – the ClO dimer, for example – but at levels considerably lower than those in the Antarctic (Shindell *et al.* 1994). Because the Arctic stratosphere is generally warmer than its southern counterpart, polar stratospheric clouds are less ready to form, and ozone destruction is less efficient. The Arctic circumpolar vortex is also less intense, allowing the loss of ozone at the pole to be offset to some extent by the influx of ozone from

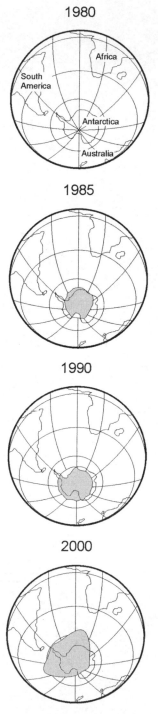

Figure 11.8 The Antarctic ozone hole in the austral spring, 1980–2000

Source: based on data from the NASA Web page. Viewed at http://jwocky.gsfc.nasa.gov/multi/monoct.gif (accessed 16 March 2003).

more southerly latitudes. The relatively free flow of air out of the Arctic in the winter may also contribute to the peculiar patterns of ozone depletion at lower latitudes in the north. Chemicals incapable of destroying ozone because of the lack of solar energy available during the polar night may become energized when carried south into the sunlight of lower latitudes, and cause greater thinning there than at the pole (Pyle 1991). The net result is an average ozone loss of as much as 15 per cent in the Arctic spring, compared with 50 per cent or more in the Antarctic spring (UNEP 2002).

THE ENVIRONMENTAL IMPACT OF OZONE DEPLETION

Declining concentrations of stratospheric ozone allow more ultraviolet light to reach the earth's surface at ratios between 1 : 1 and 1 : 3. A 50 per cent loss in Antarctic ozone, for example, resulted in an increase of 130 per cent in ultraviolet radiation (1 : 2.6), while in the Arctic a 15 per cent loss of

ozone produced a 22 per cent increase in radiation (1 : 1.4) (UNEP 2002). Deciphering the impact of such changes on the environment is difficult, but more and more evidence suggests that even small increases in ultraviolet radiation have the potential to initiate far-reaching effects (Figure 11.9).

Biological effects

Life forms have evolved in such a way that they can cope with normal levels of ultraviolet radiation. They are also quite capable of surviving the increases in radiation caused by short-term fluctuations in ozone levels. The cumulative effects of progressive thinning of the ozone layer would harm most organisms, however.

The most serious concern is over rising levels of biologically active UV-B radiation. It has been known for some time that intense UV-B rays can alter the foundations of life such as the DNA molecule and various proteins (Crutzen 1974). They also inhibit photosynthesis. Although plants

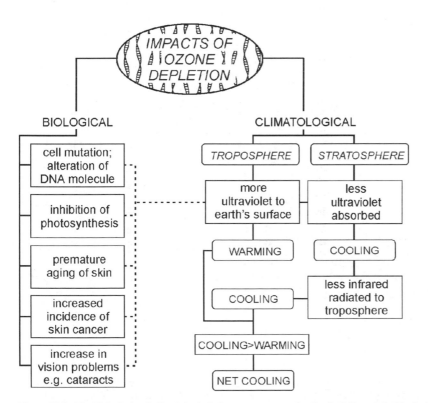

Figure 11.9 **The biological and climatological changes accompanying the depletion of stratospheric ozone**

depend upon radiation to promote photosynthesis, there are threshold levels of UV-B beyond which additional radiation induces stress and lowers productivity (Turco 1997). Experimental exposure of plants to increased UV-B has caused a decrease in yields, with soya bean productivity, for example, suffering a 25 per cent decrease when UV-B exposure was increased by 25 per cent (Gribbin 1993). Other organisms that depend upon photosynthesis, such as phytoplankton, are also vulnerable to overexposure. Any decline in the oceanic phytoplankton population could have severe consequences for the marine food chain. Insects that see in the ultraviolet sector of the spectrum could have their activities disrupted by increased levels of ultraviolet radiation (Crutzen 1974).

Most of the concern over the biological effects of declining ozone levels has been focused on the impact of increased ultraviolet radiation on the human species. The potential effects include the increased incidence of sunburn, premature aging of the skin and possibly a greater frequency of allergic reactions caused by the effects of ultraviolet light on chemicals on contact with the skin. These are relatively minor, however, in comparison with the more serious problems of radiation blindness and skin cancer, both of which would become more frequent with higher levels of UV-B radiation. Long-term exposure to ultraviolet radiation has been linked with cataracts, a clouding of the lens of the eye, and to degeneration of the cornea, the transparent film that protects the lens and the iris. Excess exposure also damages the highly light-sensitive tissue of the retina (Turco 1997).

Skin cancer had a prominent role in the SST and CFC debates of the 1970s (Dotto and Schiff 1978), and it continues to evoke high levels of concern. Skin cancer takes several forms, some of which are non-malignant and curable – basal and squamous cell cancers – but only after painful and sometimes disfiguring treatment. A relatively small proportion – the melanomas – are malignant and commonly fatal (Turco 1997). Although representing only 4 per cent of skin cancers, melanomas account for about 79 per cent of deaths from the disease (ACS 2002). The occurrence of skin cancer tends to be greater among fair-skinned populations, who have low levels of melanin, a dark pigment in the skin that provides protection from ultraviolet radiation and is also responsible for the development of a sun tan. Skin cancer is also more frequent in lower latitudes where ultraviolet levels are high, and

declines polewards, where radiation levels are lower. UV-B levels in Madrid (40°N), for example, are about twice those in Edinburgh (55°N). The number of additional skin cancers to be expected as the ozone layer thins is still a matter of debate, but a commonly accepted estimate is that an increase in UV-B of 1–2 per cent will lead to a 2–4 per cent increase in the incidence of non-melanoma skin cancer (Concar 1992). With the current number of new non-melanoma cases in the United States as estimated by the American Cancer Society (ACS 2002) being in excess of a million, a 3 per cent increase would represent an additional 30,000 cases, of whom perhaps fifty would die. Equivalent figures for malignant melanoma would be in the region of an additional 1600 cases with more than 200 deaths.

The incidence of skin cancer is rising around the world, but this may not be directly or even indirectly related to the thinning of the ozone layer. Rather it may be caused by lifestyle factors, such as the popularity of seaside holidays, or beach vacations, in sunny locations, and fashion trends that encourage a 'healthy tan'. Rising skin cancer levels in northern Europe may reflect the impact of exposure to higher levels of ultraviolet radiation on the beaches of southern Europe, Florida or the Caribbean, which have become popular vacation spots for northerners. In Australia, skin cancer is ten times more prevalent than in northern Europe (Concar 1992), and although summer UV-B levels there increased by 8 per cent during the decade of the 1980s, the real reason for the high rates of skin cancer may be social factors that encourage over-exposure to ultraviolet light, such as fashions that expose arms, legs or other parts of the body, beach sunbathing and other outdoor activities. There is a time lag between exposure and the discovery of skin cancer, and current increases may well represent the results of cell damage initiated ten to twenty years ago. It follows that, despite increasing attempts by health authorities to reduce public exposure to the sun, the rising levels of skin cancer are likely to continue for some time to come.

Climatological effects

The climatological importance of the ozone layer lies in its contribution to the earth's energy budget (see Chapter 2). It has a direct influence on the temperature of the stratosphere through its ability to absorb incoming radiation. Indirectly, this also has an impact on the troposphere. The absorption of short-wave radiation in the stratosphere reduces the amount reaching the lower atmosphere, but the effect of this on tropospheric temperature is limited to some extent by the emission of part of the absorbed short-wave energy downward into the troposphere as infrared radiation.

Natural variations in ozone levels alter the amounts of energy absorbed and emitted, but these changes are an integral part of the earth/atmosphere system, and do little to alter its overall balance. In contrast, chemically induced, anthropogenic ozone depletion could lead to progressive disruption of the energy balance, and ultimately cause climate change. A net decrease in the amount of stratospheric ozone would reduce the amount of ultraviolet energy absorbed in the upper atmosphere, producing cooling in the stratosphere. The radiation no longer absorbed would continue on to the earth's surface, causing the temperature there to rise.

This simple response to declining ozone concentration is complicated by the effects of stratospheric cooling on the system. The lower temperature of the stratosphere would cause less infrared radiation to be emitted to the troposphere, and the temperature of the lower atmosphere would also fall. Since the cooling effect of the reduction in infrared energy would be greater than the warming caused by the extra short-wave radiation, the net result would be a cooling at the earth's surface. The magnitude of the cooling is difficult to assess, but it is likely to be small, with a 20 per cent reduction in ozone concentration producing an estimated decrease in global surface temperatures of only 0.25°C (Enhalt 1980). Any estimate of the impact of ozone depletion on climate must also consider the altitude at which ozone depletion is occurring. Decreases in ozone concentrations immediately above the tropopause cause the greatest reduction in the downward emission of infrared radiation, for example, whereas the depletion of ozone at higher levels would effectively place the zone of greatest concentration closer to the surface and increase the receipt of infrared energy (Harvey 2000).

Changes in the earth's energy budget initiated by declining stratospheric ozone levels are integrated with changes produced by other elements such as atmospheric turbidity and the greenhouse effect. The specific effects of ozone depletion are therefore difficult to quantify, and its contribution to climate change is difficult to assess.

RESPONSES TO OZONE DEPLETION

Growing concern over the health consequences of ozone depletion led the Australian and New Zealand governments to initiate campaigns to discourage exposure to the sun, by advocating the use of sunscreen lotions and wide-brimmed hats and by promoting a variety of behavioural changes that would keep people out of the sun during the high-risk period around solar noon. Given their proximity to the Antarctic ozone hole, the response is not surprising, but as concern spread similar approaches were developed in North America and Europe (Figure 11.10).

Such approaches involve adaptation to the problem of increasing ultraviolet radiation and deal only with the symptoms of the problem rather than the causes. The causes had been recognized since the 1970s, however, and steps were taken to deal with them. The result was the Montreal Protocol, one of the most successful examples of international environmental co-operation, signed in 1987, with the goal of reducing halocarbon use and through that reducing the destruction of the ozone layer. It was the culmination of a series of events which had been initiated two years previously at the Vienna Convention on the Protection of the Ozone Layer. Signatories from thirty-one countries agreed to a 50 per cent cut in the production of CFCs by the end of the century. That figure was deceptive, however, since developing countries were to be allowed to use

---

SUN PROTECTION!

CANCER PROTECTION!

Three quick steps to sun sense

SLIP on a shirt!

SLAP on a hat!

SLOP on some sunscreen!

Figure 11.10 Dealing with ozone depletion: the approach of the Canadian Dermatological Society and the Canadian Cancer Society

CFCs for a decade to allow technological improvements in such areas as refrigeration. The net result turned out to be only a 35 per cent reduction in total CFC production by the end of the century, based on 1986 totals. To deal with that problem, the protocol included provision for an Interim Multilateral Fund to assist developing nations in reducing their dependence on ozone-depleting chemicals. Subsequent meetings in Helsinki (1989), London (1990), Copenhagen (1992), Montreal (1997) and Beijing (1999) expanded and tightened the provisions of the protocol, which had been signed by 182 nations by 2001 (UNEP 2002). By the early 1990s the signatories had agreed to the complete elimination of the production and use of CFCs by the end of the century. It became clear, however, that that was not enough, and for CFCs, carbon tetrachloride and methyl chloroform production bans were brought forward from 2000 to 1996, with a ban being implemented for halons in 1994. Hydrochlorofluorocarbons (HCFCs), widely used as substitutes for CFCs but still containing chlorine, were also to be phased out and methyl bromide, an ozone-destroying fumigant used in the fruit and vegetable industry, was added to the list of banned substances. The use of CFCs in inhalers used by asthma sufferers was discontinued in early 2003. Ozone-destroying chemicals have been replaced by substitutes which are less damaging to the environment, and the ozone layer is expected to begin recovering in the near future, with concentrations returning to pre-1980 levels by the middle of the twenty-first century, if the provisions of the Montreal Protocol continue to be met (UNEP 2002). Given the remarkable level of international co-operation since the protocol was signed, the potential for success is high.

GLOBAL WARMING

Fuelled by a succession of years with record high temperatures, popular interest in global warming grew rapidly in the last two decades of the twentieth century. Local temperature records were broken regularly and analysis of long-term instrumental data indicated that global mean temperatures had risen by more than 0.5°C during the century (Figure 11.11). Exceptional as this current warming is, similar periods of rising temperature are not unknown in the earth's past.

GLOBAL WARMING IN THE CLIMATOLOGICAL RECORD

Since the retreat of the ice from the maximum of the most recent Ice Age some 20,000 years ago, the earth has experienced several spells of significant warming. Proxy data from ice cores and deep oceanic sediments, along with botanical evidence from the northern continents, indicate a period of major warming during the immediate postglacial period between 5000 and 10,000 years ago. This was the Climatic Optimum, which, at its peak between about 5000 and 7000 BP, was accompanied by temperatures perhaps 1–3°C higher than modern normals (Folland *et al.* 1990). Much later, between AD 750 and AD 1200, ameliorating climatic conditions in Europe and North America, and probably

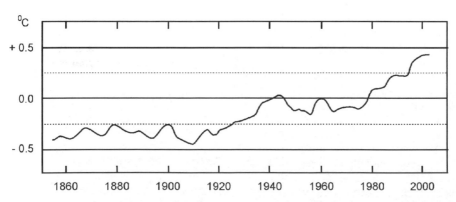

Figure 11.11 Globally averaged surface air temperature changes since 1856 compared with the 1961–90 normals
Source: based on data on the Climate Research Unit Web page. Viewed at http://www.cru.uea.ac.uk (accessed 16 April 2003).

elsewhere, produced the warm spell identified as the Medieval Warm Period – or sometimes the 'Little Optimum' (Lamb 1996; Zhang 1994). Although it appears to have been best developed around the North Atlantic, that perception may be a reflection of the greater availability of proxy data, from such sources as glacier fluctuations, tree ring analysis, pollen analysis and historical documents, in that area (Lamb 1996). Both of these events occurred before the human impact on the climatic environment was globally significant, and were caused by natural variability in the earth/atmosphere system. In contrast, modern global warming appears to have been initiated by human activities that have caused what at first sight seem to be relatively minor changes in the composition of the atmosphere.

THE GREENHOUSE EFFECT

The atmosphere is selective in its response to different types of radiation, allowing as it does most incoming short-wave solar radiation to be transmitted unaltered to the earth's surface, but restricting the return flow of terrestrial radiation out of the atmosphere. Ultraviolet-C radiation and a proportion of the UV-B rays are absorbed by the ozone layer, but UV-A radiation and visible light are unaffected. That radiation is absorbed once it reaches the earth's surface, heating up the surface and allowing it to emit long-wave terrestrial radiation from the infrared end of the spectrum back into the atmosphere. This terrestrial radiation is trapped in the atmosphere, however, where its absorption by a group of minor gases causes the temperature of the troposphere to rise. This phenomenon is called the greenhouse effect, and the gases that cause it are known collectively as greenhouse gases. Without the greenhouse effect, global temperatures would be much lower than they are, averaging −17°C, compared with the existing average of +15°C. The name suggests a similarity to the mechanisms that trap heat in the troposphere and in a greenhouse. The analogy is not perfect – the glass in the greenhouse presents a physical barrier to the flow of energy that is not present in the atmosphere, for example – but it is universally used and accepted in both the media and the scientific community. Although the media sometimes seem to suggest that the greenhouse effect is a recently discovered phenomenon, it is not. It has been characteristic of the atmosphere for millions of years, sometimes more intense than it is now, sometimes less.

There are about twenty greenhouse gases, which account for less than 1 per cent of the total volume of the atmosphere. Water vapour and carbon dioxide have the greatest impact, but methane, nitrous oxide, CFCs and tropospheric ozone also make significant contributions to the greenhouse effect. Although water vapour has received less attention than the other greenhouse gases, it has the greatest impact on the greenhouse effect. Its life span in the atmosphere is relatively short and its distribution in time and place varies considerably, however, more often as a result of the workings of the hydrologic cycle than human activities. Thus when anthropogenic enhancement of the greenhouse effect is being investigated, water vapour receives less consideration than the other gases. Its presence cannot be ignored, however, since it has very positive feedback properties. Any increase in temperature initiated by the enhancement of the greenhouse effect, for example, will raise evaporation rates and release more water vapour into the atmosphere where its greenhouse properties will amplify the original temperature increase.

Given their ability to warm the atmosphere by absorbing terrestrial radiation, any increase in the volume of the greenhouse gases should lead to additional warming. Recent increases appear small, being measured in parts per million, parts per billion and parts per trillion by volume, but in the last decade of the twentieth century it became increasingly clear that even these relatively minor changes were capable of driving the current global warming. According to generally accepted estimates, the earth's surface temperature has increased by between 0.3°C and 0.6°C since 1900, which is well within the range of normal natural variations in global temperatures. The rate of increase is broadly consistent with that expected from the measured rise in greenhouse gas levels over the same period, however, and by 1996 the scientists of the IPCC had come to the conclusion that the global warming signal, initiated by rising greenhouse gas levels, had emerged from the natural background noise. Furthermore, linking the greenhouse gas increases with human activities, they stated that 'the balance of the evidence suggests a discernible human influence on global climate' (IPCC 1996), a position that was reinforced in their third assessment report – *Climate Change 2001* – with new evidence indicating that most of the observed warming in the second half of the twentieth century was attributable to human activities (IPCC 2001d).

GREENHOUSE GASES: CHARACTERISTICS AND SOURCES

Although all greenhouse gases are positive radiative forcing agents – they are capable of disturbing the energy balance in the earth/atmosphere system – they differ in their ability to bring about global warming (Table 11.3). To provide an indication of the contribution of specific gases to global warming, the concept of Global Warming Potential (GWP) was incorporated in the original IPCC Scientific Assessment (IPCC 1990). The GWP of a gas is a measure of the cumulative radiative forcing caused by unit volume of a gas over a given period of time, with the values expressed with reference to the GWP of carbon dioxide. If the GWP of carbon dioxide over a period of 100 years is 1, for example, the comparable GWP for methane would be 21. Equivalent values for nitrous oxide and CFC-11 would be 296 and 3500 respectively (IPCC 1990). The cumulative effects of the greenhouse gases and the relative contribution of specific gases to the overall warming can be estimated from such values by multiplying the particular GWP for each gas by the volume of gas emitted. The GWP approach

TABLE 11.3 GLOBAL WARMING POTENTIAL OF SELECTED GREENHOUSE GASES

Greenhouse gas	Global warming potential (GWP)[a]
Carbon dioxide (CO_2)	1
Methane (CH_4)	21
Nitrous oxide (N_2O)	296
Sulphur hexafluoride (SF_6)	22,200
CFC-11	3500
HFC-23	12,000
HFC-134	1100
Perfluoromethane (CF_4)	570
Perfluoroethane (C_2F_6)	119,000

Source: based on data in IPCC, Climate Change 2001: Synthesis Report, Cambridge: Cambridge University Press (2001d).
Note: [a] GWP represents a combination of the ability of a greenhouse gas to trap heat and its atmospheric life span. These values represent the GWP of the gases over a 100 year period.

aims to provide a simple, direct comparison of the effectiveness of the various greenhouse gases and has been used to estimate the potential impacts of the reduction in emissions of specific gases. In simplifying complex scientific information for use by policy makers, however, the concept of GWP incorporates a number of shortcomings that reduce its effectiveness to such an extent that the accuracy of some of the results obtained from its use are questionable (Harvey 2000). As a result, new methods of assessing the warming potential of greenhouse gases are likely to be required to provide the accurate emission controls that will have to be put in place if global warming is to be managed. Some twenty different gases have been recognized as having greenhouse properties, but attention has been focused on only a few, with carbon dioxide taking centre stage.

Carbon dioxide and global warming

Carbon dioxide was the first of the gases to be recognized as having the potential to contribute to global warming. The link was investigated by a number of researchers, including Tyndall in Britain, Langley and Chamberlin in the United States and Trabert and von Czerny in Europe (Jones and Henderson-Sellers 1990; Mudge 1997), Svante Arrhenius, a Swedish chemist, is usually credited with being the first to provide quantitative predictions of the rise in temperature that might be expected as the concentration of carbon dioxide increased (Hulme 1997). He published his findings in 1896, at a time when the environmental implications of the industrial revolution were just beginning to be appreciated, and although his estimates of carbon dioxide-induced temperature increases were not particularly accurate they were not bettered until 1960 (Bolin 1972). Although little attention was paid to the potential impact of increased levels of carbon dioxide for some time after these initial investigations, by the last quarter of the twentieth century interest in greenhouse gas enhancement had revived and become focused on carbon dioxide, now widely recognized as contributing about 50 per cent of the radiative forcing responsible for current global warming.

Atmospheric concentrations of carbon dioxide have varied considerably in the past (Table 11.4). Analysis of air bubbles trapped in polar ice indicates that the lowest levels of atmospheric carbon

TABLE 11.4 CHANGING ATMOSPHERIC CONCENTRATIONS OF CARBON DIOXIDE AT SELECTED TIMES SINCE THE ICE AGES

Time period	CO_2 level (ppmv)
Glacial periods	180–200
Interglacials	275
Pre-industrial	275
Early industrial (*c.* 1880)	285–290
First measurements at Mauna Loa (1957)	310
IPCC (1990)	353
IPCC (1994)	359
IPCC (2001)	368

Source: various, see text.

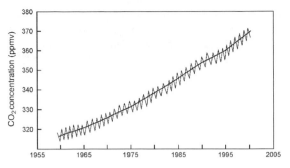

Figure 11.12 Rising levels of atmospheric carbon dioxide measured at Mauna Loa, Hawaii

Source: based on data in Keeling and Whorf (2003). Viewed at http://cdiac.esd.ornl.gov/author/whorf.html.

dioxide occurred during the Quaternary glaciations (Barnola *et al.* 1987). At that time the atmosphere contained only 180–200 parts per million by volume (ppmv), although there is some evidence that short-term fluctuations of as much as 60 ppmv also occurred (Crane and Liss 1985). Levels rose to 275 ppmv during the warm interglacial phases, and that level is also considered representative of the pre-industrial era of the early nineteenth century (Bolin 1986). Carbon dioxide measurements taken by French scientists in the 1880s, just as the effects of the industrial revolution were beginning to be felt, show that levels in the northern hemisphere averaged 285–90 ppmv at that time (Siegenthaler 1984).

When the first measurements were taken at Mauna Loa in 1957, concentrations had risen to 310 ppmv, and they continued to rise by just over 1 ppmv per year to reach 335 by 1980 (Figure 11.12). Since then, increases of about 2–4 ppmv per year brought the levels to 345 ppmv in the mid-1980s (Bolin 1986), 353 ppmv in 1990 (IPCC 1990), 358 ppmv in 1994 (IPCC 1996) and 368 ppmv in 2000 (IPCC 2001d). That represents an increase of about 90 ppmv in less than 200 years. The difference between glacial and interglacial periods was about the same, but then the time interval was measured in tens of thousands of years. The current level of 368 ppmv is about 30 per cent above the pre-industrial value, and without precedent in the past 420,000 years of earth history. Depending upon the success or failure of emission controls, the continuing increase in carbon dioxide emissions could bring atmospheric concentrations of the gas to more than double the pre-industrial level by 2100 (IPCC 2001a).

Under natural conditions the level of carbon dioxide in the earth/atmosphere system is controlled by the flow of carbon in the carbon cycle (Figure 11.13). The carbon cycle is in theory self-regulating and that may have been the case in the past, but current interference by human activities has disrupted the system. The major anthropogenic contribution to the cycle is in the addition of carbon dioxide to the atmosphere as the result of combustion of carbon-rich fossil fuels, industrial processes such as the production of cement and deforestation (Table 11.5). Fossil fuel combustion released carbon dioxide containing an average of 6.3 gigatonnes (1 gigatonne = 1×10^9 tonnes) of carbon per year (Gt C yr^{-1}) in the 1990s – up from 5.4 Gt C yr^{-1} in the 1980s – with coal, the major source, emitting as much as twice the carbon dioxide released by natural gas, per unit of energy provided. Overall, the burning of fossil fuels accounted for about 75 per cent of the rise in carbon dioxide levels in the 1990s (IPCC 2001d). The carbon dioxide released during the chemical changes involved in the production of lime, cement and ammonia augment that released by the burning of the fossil fuels used to provide energy in the manufacturing processes involved.

Deforestation is also a major source of carbon dioxide emissions. The gas is released when the forest is set alight to clear land for agriculture or when the debris left over on the forest floor after harvesting is burned. If it is not burned, the natural decay of the debris also releases carbon dioxide, as does the more rapid decay of carbon in the

TABLE 11.5   MAIN SOURCES OF GREENHOUSE GASES

Sector	Activities	Gases
Energy	Fossil fuel combustion Natural gas leakage Industrial activities Biomass burning	CO_2, CH_4, N_2O, O_3
Forest	Harvesting Clearing Burning	CO_2, CH_4, N_2O
Agriculture	Rice production (paddies) Animal husbandry (ruminants) Fertilizer use	CO_2, CH_4, N_2O
Waste management	Sanitary landfill waste disposal Incineration Biomass decay	CO_2, CH_4, N_2O, O_3 CFCs
Industrial	Metal smelting and processing Cement production Petrochemical production Miscellaneous	CO_2, CH_4, NO_2, CFCs, SF_6, CF_4, C_2F_6

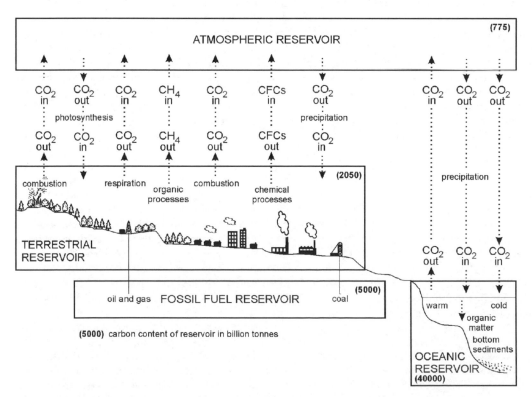

Figure 11.13   Schematic representation of the storage and flow of carbon in the earth/atmosphere system
Source: after Kemp (1994).

Plate 11.1 Burning debris, or 'slash', as part of a forestry operation to prepare the ground for new planting. The burning slash adds carbon dioxide to the atmosphere, but the new trees will replace the carbon sequestrations lost when the original trees were felled.

soils exposed after the forest has been cleared. The situation is complicated by the fact that vegetation will often recolonize the cleared land quite quickly, allowing it to absorb some of the atmospheric carbon dioxide, although never as much as the forest that previously covered the site. As well as contributing carbon dioxide directly to the atmosphere, the destruction of natural vegetation causes carbon dioxide levels to increase by reducing the amount recycled during photosynthesis (see Chapter 3). The role of vegetation in controlling carbon dioxide through photosynthesis is clearly indicated by variations in levels of the gas during the growing season. Measurements at Mauna Loa in Hawaii show patterns in which carbon dioxide concentrations are lower in the northern summer and higher in the northern winter (Figure 11.12). These

variations reflect the effects of photosynthesis in the northern hemisphere, which contains the bulk of the world's vegetation. Plants absorb carbon dioxide during their summer growing phase, but not when they are dormant in the winter, and the difference is sufficient to cause semi-annual fluctuations in global carbon dioxide levels. When the vegetation is removed photosynthesis ceases, less carbon dioxide is recycled and atmospheric levels remain higher than they normally would be. The impact of deforestation is normally considered to be a modern phenomenon, particularly prevalent in the tropical rain forests of South America and Southeast Asia, but Wilson (1978) has suggested that the pioneer agricultural settlement of North America, Australasia and South Africa in the second half of the nineteenth century made an important contribution to rising carbon dioxide levels, providing perhaps twice the amount released by fossil fuel combustion (Stuiver 1987). Tropical deforestation currently emits an estimated 1.6 Gt C yr^{-1} (Harvey 2000), but the contribution of the reduction in photosynthesis to atmospheric carbon dioxide levels remains uncertain.

Although the total annual input of carbon dioxide to the atmosphere is of the order of 7.9 Gt C (emissions + 6.3 Gt C; land use change + 1.6 Gt C), the atmospheric carbon dioxide level is increasing by only about 3.2 Gt C yr^{-1} (IPCC 2001d). The difference is distributed to the oceans and the terrestrial reservoir and probably to other sinks that are as yet undiscovered. The oceans absorb carbon dioxide in a variety of ways – some as a result of photosynthesis in phytoplankton; some through nutritional processes that allow marine organisms to grow calcium carbonate shells or skeletons; some by direct diffusion at the air/ocean interface. In the 1980s, the oceans were commonly considered to be absorbing about 2.5 Gt C yr^{-1}, although there were some indications that the total was much less than that (Harvey 2000) and the figures calculated for the IPCC *Climate Change 2001* report suggest that a more accurate value is closer to 1.7 Gt C yr^{-1}. The terrestrial reservoir absorbs carbon dioxide mainly through photosynthesis. Recent studies have shown that although land use changes such as deforestation continue to release an increasing amount of carbon into the atmosphere, the amount being returned has increased also and the terrestrial reservoir has become a net carbon sink, absorbing some 1.4 Gt C yr^{-1} more than it releases (IPCC 2001d). This may be in part a result of the higher temperatures and higher levels of carbon dioxide, which encourage plants to grow more rapidly and increase the rate of photosynthesis.

The difference between the carbon dioxide emissions into the atmosphere and the amount stored is obviously not accounted for completely by the carbon dioxide sequestered in the oceans or the terrestrial reservoir. The location of the missing carbon is not known. It may represent an accounting error associated with the complexity of the mechanisms involved and the inadequacy of the models used in calculating the flow of carbon dioxide through the system. It may, however, be hidden in the terrestrial biosphere, which has absorbed an increasing amount of carbon dioxide in recent years, or it may be in the oceans, which have the theoretical capacity to absorb all of the additional carbon added to the atmosphere, but appears to account for less than 25 per cent. The location of the lost carbon has important implications for the greenhouse effect and global warming and it remains one of the uncertainties that bedevil attempts to respond to the issue.

Carbon dioxide is the most abundant greenhouse gas, but it is not the most powerful, and the levels of other greenhouse gases are increasing more rapidly. Methane, nitrous oxide and CFCs are the most important of the other gases. Tropospheric ozone is also capable of enhancing the greenhouse effect, but its present concentrations are very variable in both time and place. It is estimated to have increased by 35 per cent since pre-industrial times, but in global terms there have been few observed increases since the mid-1980s, except at some Asian stations (IPCC 2001a). There is no clear indication of future trends.

Methane and global warming

Methane is a natural component of the earth/ atmosphere system, with its origin in the anaerobic decay of organic matter, mainly in the earth's natural wetlands (Figure 11.14). Since the water-logged state of the wetlands precludes the presence of oxygen, the bacteria that accomplish the breakdown of organic matter produce methane as an end product, rather than carbon dioxide. With a GWP of 21, methane is a powerful greenhouse gas and at times its concentration in the atmosphere has been growing at a rate twice that of carbon dioxide. The pre-industrial concentration of methane appears to

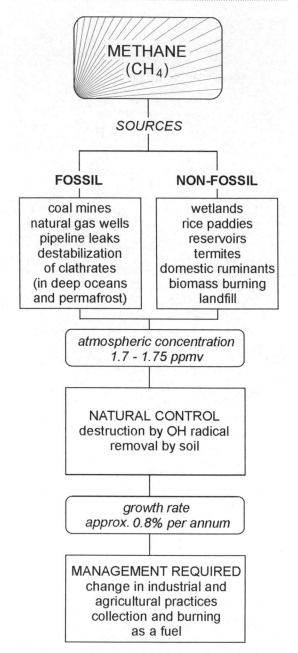

Figure 11.14   Sources, control and management of methane in the environment

have been relatively steady at about 0.8 ppmv, but by the end of the nineteenth century it had risen to 0.9 ppmv, and in 1978, when atmospheric concentrations of methane were first directly measured, the average volume was 1.51 ppmv (Blake and Rowland

1988). Since then it has continued to rise to reach 1.745 ppmv in 1998, although the rate of increase has been quite variable, being near zero in 1992 (IPCC 2001a). The current level is double that of pre-industrial times, but future growth is uncertain, partly because of uncertainty in the assessment of emission rates and partly because the emission and removal rates are both likely to be strongly influenced by future climate change. Methane has a relatively short life span in the atmosphere, being removed through its reaction with the hydroxyl radical (OH) to produce carbon dioxide and water, both of which are greenhouse gases, but less potent than methane.

Slightly more than half of the current methane emissions are from anthropogenic sources, with the most important sources being found in agricultural activities (Figure 11.14). The world's growing population of cattle and pigs release considerable amounts of methane through their digestive processes (Crutzen *et al.* 1986), but by far the largest source of agriculturally produced methane is rice cultivation. Rice paddies, being flooded and therefore providing an anaerobic environment for at least part of the year, act much like natural wetlands. Their total contribution to rising methane levels is difficult to measure, and may have been overestimated in the past, but annual rice production doubled in the second half of the twentieth century and it is likely that methane emissions increased in proportion (IPCC 1990; Schimel 2000). Methane is also likely produced as a result of the flooding of vegetated land for reservoirs, but the amount involved is not known.

The energy industry is another important source of methane. As a by-product of the conversion of vegetable matter to coal it is trapped in coal-bearing strata, to be released into the atmosphere when the coal is mined. It is also one of the main components of natural gas, and escapes during drilling operations or through leaks in pipelines and at pumping stations. Together these sources may account for 15 per cent of global methane emissions (Hengeveld 1991). The disposal of organic waste in landfill sites, where it undergoes anaerobic decay, is also considered to be a potentially significant source of methane, although the absence of appropriate data on the nature and amount of organic waste involved makes it difficult to provide accurate emission values.

The level of atmospheric methane and its impact on global warming will change as the warming progresses and a number of feedback mechanisms begin to operate. The presence of higher levels of the hydroxyl radical, released from the water vapour provided by increased rates of evaporation, should bring about a reduction in methane levels. However, at least some of the additional hydroxyl is likely to be lost in reactions with other anthropogenically produced gases such as carbon monoxide, so that their effects on methane levels may be less than expected. Drier conditions expected to accompany global warming in some areas would lead to a reduction in the area of wetlands and therefore reduce the amount of methane emitted from that source. In contrast to these negative feedbacks, there is one potentially large positive feedback. Large amounts of methane are trapped in high-latitude permafrost and in deep ocean sediments in the form of methane hydrates or clathrates. The volume of methane in these deposits far exceeds that in all of the surface reservoirs, and if it is released will add substantially to methane levels in the atmosphere and positively reinforce global warming. With warming, permafrost is likely to melt and ocean temperatures will rise, both of which would cause the clathrates to decay and release methane. Since it will take considerable time at the present rate of warming before the clathrates are destabilized sufficiently to make a contribution to atmospheric methane levels, there is little concern in the short term (Ruddiman 2001). The clathrates have been identified as potential energy sources, however, and although the technology to exploit them is still in its infancy, any attempts at extracting methane from them for commercial use is likely to be accompanied by the release of some of the gas into the atmosphere.

Methane will continue to contribute to the enhancement of the greenhouse effect, but there is some uncertainty about the rate and timing of that contribution.

Nitrous oxide and global warming

The third highest concentration among the greenhouse gases is nitrous oxide. Its pre-industrial concentration was about 0.27 ppmv, and that grew to 0.314 ppmv by 1998 (IPCC 2001a). It is about a thousand times less common than carbon dioxide, and with a growth rate of 0.2–0.3 per cent per annum it is increasing less rapidly than either carbon dioxide or methane. It is responsible for about 6 per cent of total radiative forcing by green-house gases (IPCC 2001a).

Nitrous oxide is released naturally into the atmosphere through the denitrification of soils, and has a residence time of more than 100 years. It is removed mainly through photochemical decomposition in the stratosphere, in a series of reactions which contribute to the destruction of the ozone layer. It is thought to owe its present growth to the increased use of fossil fuels and the denitrification of agricultural fertilizers, with the latter estimated to provide 50–70 per cent of total emissions (Harvey 2000). The production of nitrous oxide as a result of agricultural activity is difficult to quantify, varying as it does according to such factors as soil moisture levels or the amount and timing of the fertilizer application.

Variations in the growth rate of atmospheric nitrous oxide concentrations have been observed, but they are difficult to explain. The amounts attributable to specific sources cannot be predicted with any accuracy, and although the IPCC reports a better understanding of the overall budget (IPCC 2001a), it still contains many uncertainties, which means that future concentrations of nitrous oxide and its impact remain difficult to predict.

CFCs and global warming

CFCs and other halocarbons released from refrigerators, insulating foams, aerosol spray cans and industrial plants are recognized for their ability to destroy stratospheric ozone, but they are also among the most potent greenhouse gases. HFC-23, for example, has a GWP of 12,000 (i.e. it is 12,000 times more effective than carbon dioxide as a radiative forcing agent). Halocarbons overall are responsible for about 14 per cent of all radiation forcing from greenhouse gases (IPCC 2001d). In the early 1990s, levels of CFCs in the atmosphere were growing at rates of 4–10 per cent, but overall their concentrations were much lower than other greenhouse gases, ranging from CFC-115 at five parts per trillion by volume (pptv) to CFC-12 at 484 pptv. Other halocarbons, such as Halon-1211 and Halon-1301, were growing at rates of as much as 15 per cent at the same time, but from concentrations of only 2 pptv (IPCC 1996).

Apart from methyl bromide and methyl chloride, which have natural sources, the halocarbons are predominantly anthropogenic in origin, and should

therefore be much easier to monitor and control than some of the other gases. According to the IPCC (2001a), atmospheric concentrations of the various halocarbons are consistent with reported emissions. International agreements, beginning with the Montreal Protocol of 1987, were aimed at reducing further damage to the ozone layer by reducing the production and use of CFCs, but they will ultimately have some impact on global warming as well. Already the rate of growth of atmospheric halocarbons has stabilized, and after peaking in 1994 the concentrations of some are beginning to decline (IPCC 2001d). Halocarbons have a long residence time in the atmosphere, however – measured in centuries in some cases – and even after all emissions have ceased they will continue to contribute to global warming for some time to come. As the level of CFCs declines, the abundance of some of the chemicals designed to replace them has been rising. Hydrochlorofluorocarbons (HCFCs) and hydrofluorocarbons (HFCs) were among the earliest replacements for CFCs, and being less stable they remain in the atmosphere for shorter time periods. They are greenhouse gases, however, and therefore can contribute to global warming. Fortunately their present contribution to radiative forcing is small and will decline, since future emissions of these gases are limited by international agreement. Other halogenated compounds, such as perfluorocarbons and sulphur hexafluoride, are currently present in the atmosphere in very small, but growing, amounts, and although they have the potential to make significant contributions to radiative forcing in the future, their main anthropogenic sources are known and can be controlled before their impact is felt (IPCC 2001d).

Future levels of greenhouse gases in the atmosphere

In its Scientific Assessments since 1990 the IPCC has used a variety of scenarios to provide projections of future greenhouse gas levels. These ranged from 'business as usual', in which emissions continued at 1990 levels, through other scenarios in which increasing controls were introduced to reduce the growth of emissions, with projections being made to the year 2100 (IPCC 1990). New, updated emission scenarios were developed for the 2001 assessment, based on four narrative storylines that described the relationships between the forces driving emissions and their evolution. Six scenarios were modelled from a potential forty, using a wide range of the main demographic, economic and technological influences on future greenhouse gas emissions (IPCC 2001b). Projections of the scenarios to 2100 indicate a wide range of possible future emissions (Figure 11.15). Carbon dioxide levels are projected to rise to between 540 ppmv and 970 ppmv from current levels of 365 ppmv, which means that even with the scenario that includes most emission controls the final concentration would be almost double the pre-industrial level. Projections for methane range from 1555 ppbv to 3715 ppbv, and for nitrous oxide from 352 ppbv to 458 ppbv. With such increases the radiative forcing due to greenhouse gases would continue to increase, with that due to carbon dioxide becoming about three-quarters of the total, up from slightly more than half at present, with the net result a continuing rise in global temperatures (IPCC 2001d).

CURRENT AND FUTURE WARMING

The analysis of the instrumental record for the twentieth century shows an increase in global mean temperatures of between 0.3°C and 0.6°C since 1900, but the change has not been even (Figure 11.11). The main increase took place between 1910 and 1940 and again after 1975 (Gadd 1992). Between 1940 and 1975, despite rising greenhouse levels, global temperatures declined, particularly in the northern hemisphere. In addition, analysis of the records suggests that the relatively rapid warming prior to 1940 was probably of natural origin (IPCC 1990). Global surface temperatures have increased by about 0.2°C to 0.3°C since the 1950s (IPCC 1996), with warming particularly evident in the 1980s. The warming continued into the 1990s, which was the warmest decade on record, and 1998 was probably the warmest year since records began. The El Niño event of 1997/98 probably contributed to the high temperature of the latter year, but the decade also experienced relatively cool conditions following the eruption of Mount Pinatubo in 1991. The most recent warming since 1975 has been unevenly distributed, being greater over the land than over the sea, and the largest increases have occurred over the continents in mid to high latitudes in the northern hemisphere (IPCC 2001a).

In addition to the changes recognized at the earth's

(a)

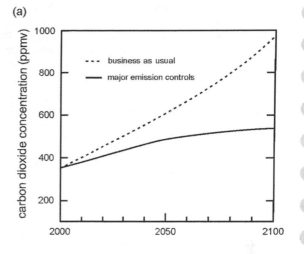

(b)

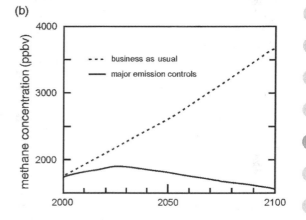

(c)

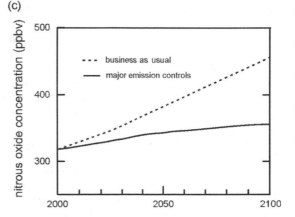

Figure 11.15 Changing greenhouse gas concentrations: comparison of a 'business as usual' scenario with one involving major emission controls

Source: based on data in IPCC (2001d).

surface there have been changes in the troposphere and stratosphere. The records for these areas are much shorter, dating back only to the 1940s, when radiosonde balloon soundings became common. These were augmented in the 1970s by weather satellite observations, which were more precise and provided a more uniform global coverage of upper atmosphere temperatures (Spencer and Christy 1993). In general, the observations indicate that the troposphere has warmed, although not as much as the surface, while the stratosphere has cooled. During the period of the record, however, there have been variations, with the changes in the lower troposphere associated with transient events such as volcanic eruptions or El Niño events, for example (Hansen *et al.* 1995). The cooling of the stratosphere, associated with the depletion of the ozone layer, has been reversed for short periods following volcanic eruptions, such as El Chichon and Mount Pinatubo, as a result of the absorption of radiation by the aerosols injected into the upper atmosphere, but since the mid-1990s global stratospheric temperatures have been at their lowest since radiosonde and satellite observations began (IPCC 2001d).

Given the continued increase in the emissions of most greenhouse gases, global warming will continue, and if society is to cope with the changes that will bring it is important to try to forecast the magnitude and rate of the warming. Estimates of global warming are commonly obtained by employing computer models, which represent physical processes in the earth/atmosphere system through a series of fundamental equations. When solved repeatedly for small but incremental changes during the run of the model they provide a forecast of the future state of the atmosphere. Climate models take various forms and involve various levels of complexity, depending upon the application for which they are designed. Complex, sophisticated General Circulation Models are most often used to simulate global warming and to provide forecasts of future temperatures. They provide full spatial analysis of the earth/atmosphere system through the use of powerful programs capable of processing as many as 200,000 equations at tens of thousands of points in a three-dimensional grid covering the earth's surface, and reaching through two to fifteen levels as high as 30 km into the atmosphere (Hengeveld 1991). To examine the impact of an enhanced greenhouse effect on temperature, for example, the carbon dioxide component in a model

is increased to a specific level. The model is then run until equilibrium is reached among its various components and the new temperatures have been established. In its first Scientific Assessment in 1990 the IPCC estimated that a doubling of carbon dioxide (from its pre-industrial concentration) would produce a temperature rise of 1.9°C–5.2°C, with a best estimate of 2.5°C. By the second assessment in 1995 the range had been reduced to 1°C–3.5°C with a best estimate of 2°C (IPCC 1996). In the third assessment the equivalent range was 1.5°C–4.5°C, with the doubling occurring some time at the beginning of the twenty-second century (IPCC 2001d). The analysis of results from the many simulations that have been run during the 1980s and 1990s have shown similar increases (USGCRP 1995; Kacholia and Reck 1997).

Although the estimated temperature increases are not particularly impressive – mainly because they are global averages – evidence from periods of rising temperature in the past indicates that they are of a magnitude that could lead to significant changes in climate and climate-related activities. During the Climatic Optimum, some 5000 to 7000 years ago, temperatures in North America and Europe were only 2°C–3°C higher than present averages, but they produced major environmental change (Lamb 1996). Evidence from that time period, and from another warm spell in the early Middle Ages, also suggests that the greatest impact of any change will be felt in mid to high latitudes in the northern hemisphere, a situation that is apparent in the model projections (Figure 11.16). The warming would also include regional and seasonal variations. Increases in the Canadian and Russian Arctic, for example, could reach as much as 8°C–12°C in the winter, but less than 1°C in the summer (Hengeveld 1991). In Central Asia summer warming would exceed the average, but elsewhere in Asia warming is expected to be less (IPCC 2001a).

Temperature change and global precipitation

Because the various elements in the atmospheric environment are closely interrelated, it is only to be expected that, if temperature changes, other elements will change also. Moisture patterns are likely to be altered, for example. Climate models tend to be less able to deal with changing moisture levels, however, and forecasts tend to be less robust

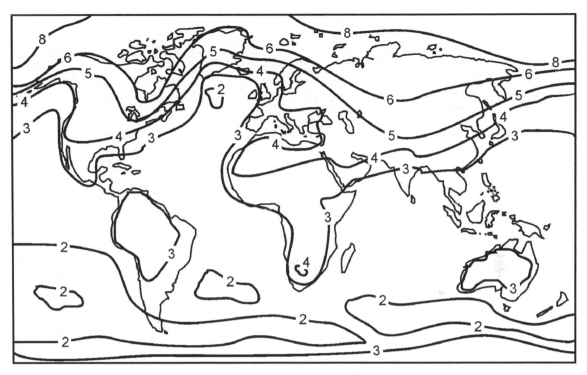

Figure 11.16   Potential temperature change associated with global warming; values represent projected increases over 1961–90 normals by 2100

Source: after IPCC (2001d) with permission.

than those for temperature. Observed changes reported by the IPCC (2001) show that precipitation totals increased towards the end of the twentieth century over land areas in mid to high latitudes in the northern hemisphere, but decreased over the subtropics. Precipitation has also probably increased in the tropics, over both land and sea, but changes over most of the southern hemisphere are uncertain. In keeping with the increased precipitation, cloud cover has increased over the northern hemisphere, and over Australia also. Precipitation projections indicate an increase in global totals, with regional and seasonal variations (Figure 11.17). It is likely to increase in both summer and winter over high latitudes, whereas winter increases are to be expected in northern mid-latitudes, tropical Africa and Antarctica and summer increases in southern and eastern Asia (IPCC 2001d). Warming would cause an increase in the intensity of the Asian monsoon, producing increases of as much as 10–20 per cent during the season, but precipitation totals would be more variable from year to year and the proportion of days with heavy rainfall would increase dramatically (Bhaskaran *et al.* 1995). Similar changes might occur in areas such as northern Australia and West Africa, where monsoon circulation is an integral part of the regional meteorology. In some areas, such as the continental interiors of Eurasia and North America, any increase in precipitation would be more than offset by the increased summer drying that would accompany the warming. That, plus a potential increase in the variability of rainfall, would bring with it an increase in the frequency of drought.

The newest models used to predict climate change are remarkably sophisticated, and confidence in their predictions has increased. Despite this, all models represent a compromise between the complexities of the earth/atmosphere system and the constraints imposed by such factors as data availability, computer size and speed, and the cost of model development and operation, which limit the accuracy of the simulation. While the inadequacies

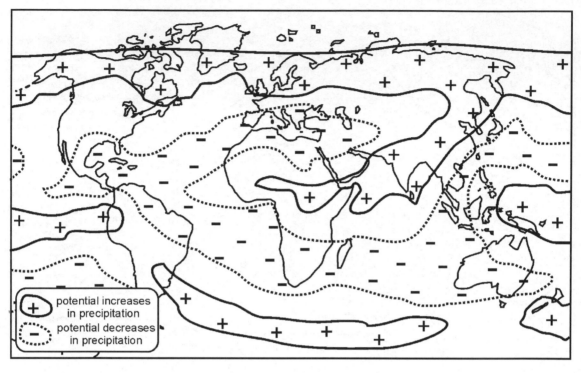

Figure 11.17   Potential changes in precipitation following global warming
Source: after IPCC (2001d) with permission.

of the models continue to be addressed, the real extent of global warming, whether it involves temperature or precipitation, may only become apparent once the changes have taken place.

ENVIRONMENTAL AND SOCIO-ECONOMIC IMPACTS OF GLOBAL WARMING

Given the wide range of possibilities presented in the estimates of future greenhouse gas levels and the associated global warming, plus the great number of regional differences, it is difficult to predict their impacts with any degree of certainty. However, using a combination of investigative techniques – ranging from laboratory experiments with plants to the analysis of physical and economic computer models – researchers have produced results that indicate what the consequences might be in certain key sectors.

It is already clear from the study of weather and climate in the past that both natural and human systems are vulnerable to climate change (see, for example, Lamb 1996). The impact of the current warming is already apparent in the natural environment in the form of physical changes such as shrinking glaciers, changes in the extent of sea ice and the dates of freeze-up and break-up of rivers and lakes or in the form of biological changes such as changing distributions of both plants and animals, the lengthening of the growing season and earlier dates for a range of phenological indicators. The increased frequency of floods and droughts in some areas may be an indication of the impact of warming on socio-economic systems, but both can result from other non-climatic elements, including population and land use change. In a number of important areas, human and physical systems share similar impacts. The effects of warming on plants, for example, are seen not only in natural vegetation, but also in cultivated varieties, with the impacts particularly apparent in agriculture and forestry. Similar links between the impacts on human and natural systems are present in such sectors as hydrology and water resources. In some cases the effects of warming on society are indirect – being felt through changes in plants and animals in the biosphere, for example – but in other areas the effects are more direct – the impact of higher temperatures on the use of energy, for example. Overall, the impacts on both natural and human systems vary considerably with geographic location, time, technology and socio-economic and environmental conditions.

Natural vegetation

Although the impact of increased levels of atmospheric carbon dioxide on plants is commonly considered in terms of rising temperatures, carbon dioxide also participates in photosynthesis (see Chapter 3), and can influence vegetation through that process also. In laboratory experiments under controlled conditions, increased concentrations of the gas (sometimes called carbon dioxide fertilization) enhance growth in most plants, including trees and most commercially important grain crops. The response of natural vegetation or field crops may be less than that of plants in controlled environments, because of a variety of non-climatic variables normally excluded from laboratory experiments, but the biological effects of enhanced carbon dioxide would likely be beneficial to most plants and increase net primary production in most systems (IPCC 2001b).

Higher temperatures working through the lengthening and intensification of the growing season would have an effect on the rates of plant growth and crop yields. The regional distribution of natural vegetation would change, particularly in high latitudes, where the temperature increases are expected to be greatest. Across the northern regions of Canada, Scandinavia and Russia the trees of the boreal forest would begin to colonize the tundra, as they have done during warm spells in the past, at a rate of perhaps 100 km for every 1°C of warming (Bruce and Hengeveld 1985). The southern margin of the boreal forest would also migrate northwards, under pressure from species of the hardwood forest and the grassland, more suited to the new conditions. An expansion of the grassland in western Canada would push the southern boundary of the forest north by 250–90 km (Wheaton *et al.* 1989), and ultimately the forest might disappear completely from central and western Canada (Figure 11.18). These changes will lag behind changes in

climate, however, by decades or even centuries (IPCC 2001b).

Higher temperatures might threaten the forests indirectly through an increase in the frequency of forest fires and insect infestations, and if the change takes place more rapidly than the forest can respond, as seems likely, the rapid die-off of large numbers of trees is a distinct possibility. Whatever the final outcome, it seems likely that global warming would disrupt the northern forest ecosystem, and in turn have a significant effect on those countries, such as Canada, Sweden, Finland and Russia, where national and regional economies depend very much on the harvesting of softwood from the boreal forest.

In lower latitudes, where the temperature element is less dominant and where the warming is likely to be less, the impact of global warming would most likely be felt through changes in the amount and distribution of moisture. In northern Australia, for example, the increased poleward penetration of monsoon rains would encourage the expansion of tropical and subtropical vegetation (Pittock and Salinger 1991; Harvey 2000). In areas experiencing a Mediterranean-type climate – southern Europe, South Africa, parts of South America and Western Australia – the reduced precipitation and higher evaporation rates associated with the warming would cause soil moisture levels to decline. Although the vegetation in these areas has adapted to seasonal drought (see Chapter 3), the extension of the dryness into the normally wet winter season would cause productivity to decrease and ultimately bring about changes in the composition and distribution of the Mediterranean biome (IPCC 1996). Recent studies using observed data and dynamic vegetation models suggest that significant ecosystem disruption is likely to accompany global warming elsewhere also, but the many uncertainties involved make it difficult to assess precisely the intensity and distribution of the changes (IPCC 2001b).

Cultivated vegetation

The conditions likely to alter the regional distribution and productivity of natural vegetation are also likely to change the nature and extent of cultivated vegetation (Table 11.6). Global warming will have a positive impact on the length and intensity of the growing season, and extend the effective season by

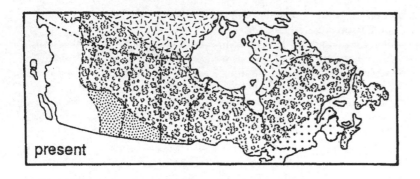

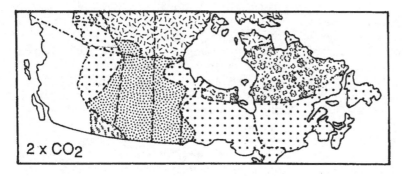

tundra   grassland

boreal forest   semi-arid

temperate forest

Figure 11.18   Projected changes in the distribution of natural vegetation in Canada following global warming
Source: after Hengeveld (1991) with permission.

reducing the risk of frost at the beginning and end of the season. These favourable conditions, however, would also encourage certain crop pests and by altering precipitation and evaporation rates would threaten production in some areas. A significant expansion of agriculture is to be expected in mid to high latitudes, where the greatest warming will be experienced. In the interior of Alaska, for example, a doubling of carbon dioxide levels would raise temperatures sufficiently to lengthen the growing season by three weeks, which would allow land currently under forage crops, or even uncultivated, to produce carrots, cabbage, broccoli, peas and other food crops (Wittwer 1984). Reduced frost risk would be a major benefit in some areas. By 2050, in New Zealand and in the coastal areas of Australia, for example, the frost-free season could be thirty to fifty days longer than at present if the projected warming occurs (Salinger and Pittock 1991).

The greater intensity of the growing season, along with the direct effects of carbon dioxide on photosynthesis, would lead to increased crop yields in some areas (Table 11.6). Model runs for northern Europe provide estimates of increased yields in a variety of crops – cereals, sugar beet, potatoes, hay – from locations as far apart as Germany and Finland, with increases as high as 30 per cent (IPCC 1996). It may not always be possible for growers to take full advantage of the benefits of warming, however, because of the effects on agriculture of non-climatic factors. Warmer climates would allow the northward expansion of agriculture on the Canadian prairies, for example, but the benefits of that would be offset by the inability of the soils

TABLE 11.6 POTENTIAL RESPONSES AND ADJUSTMENTS TO GLOBAL WARMING IN THE AGRICULTURAL SECTOR

Response	Characteristics	Limitations
Change in seasonal activities	Longer growing season and frost-free season Earlier planting and sowing dates Additional harvests possible	Existing season already long Cropping limited by lack of moisture Economic constraints
Introduction of different crop varieties or animal breeds or introduction of new plant and animal species previously unknown in the region	Crop varieties developed to take advantage of new conditions – higher-yielding, longer-season cultivars – drought-tolerant species Animal breeds more suited to the new conditions introduced Possible genetic engineering	Appropriate varieties not available or costly to obtain Lack of tolerance of weeds, pests and diseases Nutritional requirements difficult or costly to meet
Opening of new lands	Extension of cultivated area into regions previously too cold or wet Cropping at higher altitudes	Poor quality of soils Distance from market excessive High introductory costs
Expansion of irrigated lands	Introduction of irrigation and water management systems to areas that have become drier following warming	High costs Limits to viability of water supply Problems with salinization
Miscellaneous adjustments and inputs	Increased fertilizer use Low tillage techniques Abandonment of areas that have become too dry	Fertilizer costs Problems of ground and surface water contamination by fertilizer Increase in the release of nitrous oxide – a greenhouse gas
Greater attention to climate prediction	Improved medium-term or seasonal predictions – along the lines of FEWS, for example	Cost and reliability of modelling Economic, cultural and political constraints on responses to forecasts

Source: compiled from various sources.

in those areas to support anything other than marginal fodder crops, which are unlikely to be profitable (Arthur 1988). In other cases the introduction of crop pests, attracted by the favourable climatic conditions and the availability of the new crops, would prevent yields from reaching their potential.

One of the most serious problems for agriculture is the increasing dryness likely to accompany the rising temperatures in many areas. Moisture problems are expected to be the main constraints on crop yields on the Great Plains, for example, an area in which crops are already vulnerable to drought. Variations in moisture availability would occur from year to year and from place to place – as they

do now – but the increased frequency of drought forecast by the climate models would cause serious declines in production in most parts of the US plains and the Canadian prairies, with model estimates for wheat production being down as much as 20 per cent and for corn (maize) as much as 30 per cent (IPCC 1996). The grain-producing areas of southern Australia might initially experience beneficial effects, but in the longer term yields might decline by as much as 30 per cent. In contrast, models of the impact of warming on agriculture farther north, in Queensland, suggest an increase in grain yields and perennial grasses of between 20 percent and 30 per cent (IPCC 1996, 2001b). Many of the grain-producing areas that would experience lower productivity are areas that currently provide a surplus that helps to meet the needs of those parts of the world where the grain harvest cannot supply what is required. If the surplus is no longer available, and cannot be replaced from alternative sources, the grain-importing countries may face serious food shortages and ultimately face the effects of famine.

Hydrology and water resources

Most of the studies of the impact of global warming on agriculture recognize the importance of an adequate water supply if the benefits of warming are to be maximized. Unfortunately, climate models do not represent precipitation well and the resulting uncertainties mean that there is often little agreement on changes in the hydrologic cycle and water resources. Flooding is projected for some areas – northern Europe, for example – as a result of increased precipitation or changing rainfall patterns such as more frequent heavy precipitation events that encourage greater run-off. Elsewhere, increased snowmelt from mountainous areas and glaciers following warming would increase stream flow. In many areas, however, the impact of warming will come in the form of shortages in the supply of water creating increased pressure on those managing water resources. Shortages are forecast for southern Europe, where both precipitation and run-off are expected to decline (IPCC 2001b). In North America, the world's largest natural aquatic system is threatened. Increased evaporation combined with changes in the timing and distribution of precipitation would reduce mean lake levels in the Great Lakes–St Lawrence system by 30–80 cm, causing the stream flow in the channels connecting the lakes to be reduced by 20 per cent (Sanderson 1987). At present, the Great Lakes water level experiences irregular fluctuations, but the model scenarios suggest that low levels will occur more frequently in the future, perhaps as many as eight years out of ten. This would reduce the system's hydroelectricity output, produced from the flow of water between the lakes, and create major problems for the shipping industry, which would have to limit cargoes to prevent grounding in the shallower water and deal with the subsequent loss in shipping capacity and revenues. Compensating for that to some extent would be the reduced ice cover on the lakes. With a projected winter warming of 6°C, the lakes would become completely free of ice,

allowing a year-round shipping season and a potential increase in shipping volume of 15–30 per cent (Environment Canada 1986).

Changing water levels would also have an effect on ecological resources. Existing wetlands would dry out, reducing habitat for a variety of fish, wildfowl and small mammals. In some areas, replacement wetlands would be created as the waters became shallower, but these would take time to form and reach a reasonable degree of ecological maturity.

One aspect of global hydrology that has received considerable attention is the impact of higher temperatures on sea level. Global warming would cause sea level to rise as a result of the thermal expansion of sea water and the return of additional water to the oceans from melting temperate glaciers. Tide gauge observations show that sea level rose by about 1.5 cm per decade during the twentieth century, which was faster than during the previous 1000 years, but there is no evidence that the rate has accelerated, as might be expected from the increased warming towards the end of the century (Harvey 2000). Estimates of future rises based on model simulation are quite variable. Earlier studies indicated that mean sea level might rise by as much as a metre as early as 2050 (Titus 1986), but the first IPCC assessment in 1990 foresaw a rise of 18 cm by 2030 and 44 cm by 2070. Since then various models have provided figures of between 27 cm and 48 cm for the potential rise in sea level by 2100, and the latest IPCC assessment (2001d) gives a value of 48 cm, within a range of 9–88 cm. With the time lag between temperatures rising and both the thermal expansion of the oceans and the melting of glaciers and ice sheets, the rise in sea level would lag also. Taking that into account, it is possible that the temperature increase following a doubling of atmospheric carbon dioxide could cause enough thermal expansion of the oceans and melt enough ice to raise sea level by 14–17 m, but only over a time span of several thousand years (Harvey 2000). In the extreme case of all of the ice in the Antarctic and Greenland ice sheets melting, sea level would rise by almost 47 m (IPCC 2001d).

Such a rise would be catastrophic, but even the much smaller increases projected for 2100 have the potential to cause serious damage in many areas. Even increases towards the low end of the estimated ranges would be sufficient to cause extensive flooding and erosion in coastal areas, with particular problems for society in urban areas, where the flooding and erosion would cause serious problems for the urban infrastructure, in the form of disruption of sewage and industrial waste disposal facilities, road and rail networks and harbour activities. A sea level rise of only 20 cm would place some 1.1 million ha of land in jeopardy along the east coast of China, from the Pearl River in the south to Bohai Bay in the north (NCGCC 1990). In Bangladesh, a 1 m rise in sea level would inundate 17 per cent of the nation's land in the delta of the Ganges and Brahmaputra rivers. Some 11 million people would be displaced from some of the most intensely cultivated land in the country (Cuff 2002a). Similar effects would be felt in all low-lying coastal plains and deltas around the world, and some of the small island states in the Pacific and Indian Oceans might disappear completely. The highest point on the Maldive Islands in the Indian Ocean is only 6 m above present sea level and the average elevation of the Marshall Islands in the west central Pacific is only 3 m. In all cases, the situation is made more perilous when small rises in sea level are augmented by high tides or storms.

Whether storminess would increase or decrease in a warmer world remains a matter of controversy. Rising temperatures in higher latitudes would reduce the latitudinal temperature that drives mid-latitude extratropical cyclones, and storminess therefore might be expected to decline. It is possible, however, that the storms that do develop will be more intense, fuelled by greater evaporation rates over warmer oceans, and the consequent increase in the energy flux between ocean and atmosphere. This could lead to an increase in the incidence of the intense low-pressure systems that wrought havoc in Britain and western Europe in 1987 and again in 1990 (Simons 1992). Observations of intense storms in the Atlantic and Pacific Oceans between 1900 and 1990 indicate little change in frequency prior to 1970. At that time, however, there was a sharp increase in the Pacific sector, but no significant change in the Atlantic (Lambert 1996). Current models continue to have difficulty dealing with small-scale or regional phenomena and that limits the quality of any forecasts. Until the problems can be resolved, predictions of changes in storminess in mid-latitudes will remain inconclusive.

Similarly, climate models are inconsistent in their predictions of the frequency and intensity of tropical storms – cyclones, typhoons and hurricanes – following global warming. These storms develop only over oceans where sea-surface temperatures

(SSTs) exceed 26°C. With global warming, such temperatures would be exceeded more frequently, and over larger areas, than at present, suggesting that the number of tropical storms should increase. Given the additional energy involved, an increase in the intensity of the storms might be expected also. As yet there is no consistency in the results from investigations into frequency, distribution or intensity of tropical storms (IPCC 2001d). Even with no change in the nature and extent of these storms, rising sea levels will allow them to be more destructive. Aware of the problems they might face in the future, the nations most likely to be affected – in the Indian and Pacific Oceans and the Caribbean – came together as the Alliance of Small Island States (AOSIS), and as such were successful in having their situation addressed as part of the UN Framework Convention on Climate Change at the Earth Summit in Rio de Janeiro in 1992. Without action to address their problems, many of these islands will face serious problems of flooding and erosion and some will become uninhabitable.

MITIGATION

Many aspects of future global warming are difficult to identify and quantify with any degree of certainty. There is, however, enough evidence that the forecast impacts have the potential to cause major physical changes in the natural environment, which in turn would impact adversely on the health and welfare of the world's population. As a result, the search for solutions has intensified. Although there may be individuals and groups who for various reasons are willing to continue with the 'do nothing' or 'business as usual' approach to the problems of global warming, they are becoming a minority, and the urgent need to find solutions is widely accepted. Given the complexity of the problems being addressed, it is not surprising that there is no one approach that satisfies all needs. Most of the options being considered involve either adaptation or prevention, and sometimes a combination of the two.

Adaptation in its simplest form is already part of the earth/atmosphere system, represented by the adjustments that help to maintain some degree of dynamic equilibrium among the various elements in the environment. As part of that environment, human beings have always had the ability to adapt to changing conditions, and in many cases society has been shaped by such adaptations. Adaptation often appears attractive because in the short term it is a relatively simple and low-cost approach. Since the very earliest times in agriculture, for example, farmers have reacted to changing environmental conditions by growing different crops or developing new cultivation techniques. That would continue to be the case as the climate changes with global warming. Other industries would adapt also – the transport industry by changing routes, perhaps, or the construction industry by developing buildings more appropriate to the warmer conditions, for example. With time and continued warming, however, the costs of adaptation may eventually exceed the costs of providing solutions or the cumulative effects of the changes may surpass the ability of society to adjust and preventative measures will be required (Table 11.7).

TABLE 11.7 APPROACHES TO THE MITIGATION OF GLOBAL WARMING

Sector	Approach
Energy	Increase efficiency of energy conversion Switch to low carbon fossil fuels, e.g. natural gas Introduce flue gas decarbonization Increase use of nuclear energy Increase use of renewable energy
Industry	Reduce greenhouse gas emissions – methane from natural gas pipelines and pumping stations, for example Reduce material content of manufactured goods Introduce appropriate technology, e.g. low-carbon or renewable energy forms Transfer technology from developed to developing countries Recycle
Transport	Improve energy efficiency of vehicles Reduce vehicle emissions Reduce vehicle size; improve performance Change land use patterns and lifestyles to reduce transport requirements Integrate transport policies Promote change in travel mode – from car to bus and train
Agriculture	Develop new management techniques based on reduced tillage, recycling of crop residues, restoration of wasteland soils Improve efficiency of energy use Improve nutrition of ruminants to reduce methane production Reduce biomass burning Manage fertilizer use to reduce nitrous oxide production
Forestry	Substitute fuelwood for fossil fuels Improve efficiency of energy use Reduce biomass burning Conserve carbon dioxide in living trees Regenerate existing forests Reforest areas no longer supporting trees
Government	Develop urban and industrial land use planning to minimize energy consumption Plan waste disposal to reduce carbon dioxide and methane production Provide disincentives for excessive energy use – through carbon taxes, for example Provide incentives for reduction in greenhouse gas output – through emissions trading permits, for example Encourage participation at the individual level by offering tax incentives, grants and subsidies for improving energy efficiency, recycling and appropriate waste disposal

Source: compiled from various sources.

Since its inception, the IPCC has included a working group charged with developing response strategies to climate change and formulating approaches by which global warming can be mitigated. Since the early 1990s there have been claims that the technology is available to lower atmospheric carbon dioxide to levels that prevailed in the mid-1970s, mainly through improved efficiency in the use of existing energy sources and the introduction of new and appropriate energy technology (Green 1992). In practical terms such an achievement is unlikely, however. In addition to technology, mitigation will require attention to socio-economic and cultural factors, and since the role of each of these factors will vary from country to country or region to region the provision of a solution that will meet all needs is difficult – perhaps impossible. The reality is that advances in technology offer increasing opportunities for mitigation, but the barriers presented by the non-technical issues continue to slow their implementation. The complexities involved are well illustrated by national responses to the adoption of the Kyoto Protocol, an international agreement on greenhouse gas reduction (see Box 11.2, The Kyoto Protocol).

Approaches to mitigation involve the reduction of greenhouse gas emissions or the provision of sinks to increase the amount of carbon sequestered in the environment. Most attempts at reducing the volume of greenhouse gases released into the atmosphere include improvements in the efficiency of energy use or changes in the sources of energy. Renewable energy, including solar radiation, biomass, wind and water, is seen as having the ability to make a major contribution to the reduction in carbon dioxide, by replacing fossil fuel sources. Proposals to encourage greater efficiency or a more rapid change-over to renewables have involved the consideration of fiscal measures such as taxation. A carbon tax, paid on fuel consumption, has been promoted as a means of reducing fossil fuel use, for example. It would be relatively easy to set up and administer, and the resulting higher fuel prices would in theory lead to improved fuel efficiency. The revenue generated by the tax could be used to aid the transfer of the new technology from developed to developing nations to encourage mitigation in these areas also, or to reverse some of the negative impacts associated with current levels of global warming. Although the energy sector has received most attention in the search for mitigation options, other sectors such as agriculture, construction, transport and manufacturing also contribute greenhouse gases, the output of which can be reduced significantly through technical improvements (IPCC 2001c) (Table 11.7).

All of the greenhouse gases introduced into the atmosphere were kept at appropriate levels in the past by natural recycling processes. Emission levels are now so high, however, that these processes cannot keep up, but if they could be rejuvenated or their capacity increased, perhaps they could contribute to the stabilization and eventual reduction of atmospheric greenhouse gases. The ability of plants to absorb carbon dioxide during photosynthesis has been proposed as a natural method of mitigating global warming, for example. The approach would involve the reforestation of large tracts of land so that carbon could be removed from the atmosphere and sequestered or stored in the growing vegetation. In Canadian studies, after taking such factors as land availability, soil conditions and climate into consideration, it has been estimated that the maximum possible increase in carbon sequestration would be only 9–10 per cent (Van Kooten *et al.* 1992). Although obviously unable to provide a complete or permanent solution, reforestation and other forms of ecosystem management offer a breathing space in which other options can be investigated (IPCC 2001c).

Among suggestions for dealing with global warming are several that can be described as in the realm of global environmental engineering, which combine natural environmental processes with the application of technology on a world scale (IPCC 1996). The oceans, for example, have the capacity to absorb more carbon dioxide than they do at present. Absorption is accomplished in part by phytoplankton during photosynthesis, with the absorbed carbon being moved through the marine food chain into other organisms. When they die they sink to the ocean floor, taking with them the absorbed carbon. In some nutrient-poor parts of the world's oceans the rate of carbon sequestration in this form is low, but could be increased by providing additional nutrients such as iron, spread across the ocean surface in much the same way as farmers fertilize their fields on land (Martin *et al.* 1990). The additional sequestration is likely to be small, however, since a considerable proportion of the carbon would be recycled before it could sink into storage (Turco 1997).

Since the presence of particulate matter in the atmosphere is known to cause temperatures to

BOX 11.2 THE KYOTO PROTOCOL

THE GENESIS OF THE PROTOCOL

One of the agreements signed at the Earth Summit in Rio in 1992 was the Framework Convention on Climate Change (FCCC). The FCCC grew out of the concern over global warming, but was signed only after much controversy and ended up as a relatively weak document, lacking specific emission reduction targets and deadlines. Since it was signed, however, there has been considerable progress towards the achievement of its main aim of bringing about the stabilization of greenhouse gas emissions at a level that would prevent dangerous anthropogenic interference with the climate system. Among the mechanisms set up as part of the process was a Conference of the Parties (COP), designed as an ongoing process to deal with the transfer of scientific and technical knowledge among the parties and to oversee the implementation of the provisions of the convention. It was at one of the COP meetings that the Kyoto Protocol was signed.

The FCCC recognized that economic, technological and socio-political conditions differed among the participants in the convention and, initially at least, that prevented all from being treated equally. It identified so-called Annex 1 nations, which were in theory more able to deal with the requirements of the convention. These included the developed nations – all the members of the Organization for Economic Co-operation and Development (OECD) plus the nations of the former Soviet Union, considered as economies in transition (EIT). The developing nations were included in a non-Annex 1 group, with no commitment initially to reduce greenhouse gas (GHG) emissions, although a number of them – India and China, for example – were and continue to be major emitters of greenhouse gases.

The first step towards providing the FCCC with some regulatory teeth was taken at the initial COP meeting in Berlin in 1995. It produced the Berlin Mandate, in which the parties involved were given the task of negotiating international commitments that would provide specific values and timelines for the reduction of greenhouse gas emissions. After a second COP in Geneva in 1996, at which the findings of the second assessment of the Intergovernmental Panel on Climate Change (IPCC) were considered, the stage was set for COP 3 in Kyoto, Japan, in late 1997. There, negotiations were centred on the creation of a legal framework and the establishment of goals for the mitigation of global warming through the reduction in greenhouse gas emissions identified in the Berlin Mandate. Two weeks of intense and lengthy negotiation produced the Kyoto Protocol, an agreement that is seen by some as the most important international environmental agreement yet signed and by others as the precursor of major economic disaster.

PROVISIONS OF THE PROTOCOL

The Kyoto Protocol established country-by-country greenhouse gas emission reductions based on levels that had existed in 1990. It was to come into effect when ratified by governments of no less than fifty-five Annex 1 nations, which together accounted for at least 55 per cent of the total CO_2 emissions for 1990. Six greenhouse gases were targeted for reduction – carbon dioxide (CO_2), methane (CH_4), nitrous oxide (N_2O), hydrofluorocarbons (HFCs), perfluorocarbons (PFCs) and sulphur hexafluoride (SF_6) – bundled in the form of 'carbon dioxide equivalent' emissions. The average commitment of the Annex 1 nations was to a reduction of 5.2 per cent from 1990 levels, to be achieved by 2012. For many, this value was deceptive because of emission increases that had taken place since 1990. A country in which greenhouse gas emissions had increased by 10 per cent since 1990, for example, would have to reduce emissions by that much as well as by the amount agreed to at Kyoto. Other countries, such as Russia, have already fallen below the 1990 level as a result of economic decline and reduced industrial output over the last decade of the twentieth century. Countries that possess major carbon sinks such as extensive forests, and are able to expand them – by afforestation, for example – can use them to offset the greenhouse gas emissions to which they are committed under the agreement. In reaching their goals individual nations are not constrained by the protocol. It deals only with targets and timing. The mechanisms chosen are entirely at the discretion of national governments, and will vary from place to place depending upon such factors as the mix of emission sources and sinks, the type of energy available and the state of the economy, plus a variety of political and socio-economic issues peculiar to a particular nation.

EMISSIONS TRADING PERMITS

Although the instruments used to reduce greenhouse gases under the Kyoto Protocol are employed at the national level, the efforts of Annex 1 nations have been linked internationally through the creation of

BOX 11.2 – continued

emissions trading permits. This has brought about the development of a market for carbon, through which countries and companies can use differences in international conditions to meet their Kyoto commitments. Permits can be used directly for International Emissions Trading (IET). If a country achieves reductions that exceed its target, for example, it creates emission permits that can be traded at market value to nations that are unable to meet their targets through conventional means. By purchasing these permits, they are allowed to emit greenhouse gases to equivalent levels beyond their committed targets. For some countries, this may be the most cost-effective means of achieving their targets, and at first sight does little to contribute to overall emission reductions. Past experience with acid gas emissions trading suggests that benefits accrue only if the trading involves an offset ratio, in which the buyer of the permit is not allowed the full emissions value of the permit. An offset ratio of 1 : 2, for example would allow the buyer to emit only half the emissions reduction that the vendor achieved to obtain the permit in the first place. The success of the system will also depend upon the value that the open market places on carbon.

A second mechanism by which emission permits might be generated is Joint Implementation (JI), in which a country or company invests in emission reduction or the provision of sinks in another country. Investing in a scheme to increase the use of natural gas, as an energy source to replace coal, for example, or financing the reforestation of a large area to create a sink would generate permits for the investing nation. Both IET and JI schemes are allowed only between Annex 1 nations, but a third scheme – the Clean Development Mechanism (CDM) – is allowed in developing nations that have ratified the protocol. An Annex 1 nation would receive emissions credits for investing in schemes for reducing emissions or creating sinks in the developing world, usually through the transfer of technology. The introduction of renewable energy systems, the improvement of energy efficiency or the increase in the size of existing sinks through improved forestry or agricultural techniques would qualify as CDMs, for example.

DIFFERING RESPONSES TO THE PROTOCOL

The Kyoto Protocol has had a widely mixed reception. It has been embraced by most environmental groups and tends to be seen as positive by the general public. Even these groups do not see it as a final solution, however. At best it is seen as a necessary first step that will have to be repeated with continuing emission reductions in the future. As it stands, although the targets are legally binding, sanctions for failing to meet the targets are weak, with little incentive for the offending nation to comply. There are no direct financial penalties, for example. Countries that fail to meet their targets are assessed an additional 30 per cent reduction penalty, which they may be unable to meet or not even attempt to meet. As of March 2003, fifty-nine of the 186 signatories to the protocol have ratified it, but only twenty-three of these are Annex 1 nations, far below the fifty-five required. The most notable absentee is the United States, the world's leading producer of greenhouse gases, which saw it as detrimental to the nation's development. Ironically, some state governments in the United States, such as those in New England and California, are working quite effectively towards meeting their Kyoto targets. In contrast, in Canada, the federal government has ratified the protocol but a number of provincial governments remain opposed. With its huge stake in the oil, natural gas and coal industries the government of Alberta has been particularly vocal in its opposition. The response from industry has been mixed, with some such as those producing and using fossil fuels seeing the protocol as detrimental, whereas others have embraced its requirements and by improving operational efficiency have not only reduced greenhouse gas output but have also achieved significant economic benefits. Industries involved in the renewable energy field can expect direct gains as the demand for their products rises.

The debate continues on the necessity for the Kyoto Protocol. Some do not accept the existence of human-induced global warming and see no need for the protocol. Even those that accept global warming question the benefits that Kyoto might bring. They see the costs of implementing the protocol as greater than the costs of adjusting and adapting to the changes likely to accompany the warming. In reality the total costs are probably impossible to calculate, including as they do environmental costs that are incapable of being assessed in monetary terms. Problems remain with the estimation and auditing of emission rates, with the amounts of greenhouse gases that are absorbed by particular sinks, with the amount of the burden that should be assumed by the developing nations, and when. The crunch will come in 2012, when the protocol will

BOX 11.2 – continued

have to be revisited and new, potentially costly solutions will have to be sought to the problems that have been revealed in the first attempt to deal with greenhouse gases. Even if the response to the protocol increases before then and all the goals are met, the overall impact on the growth of greenhouse gas emissions will still be limited and the exercise will have to be repeated in the future – probably many times – if warming is to be slowed.

For more information see:

Essex, C. and McKitrick, R. (2002) *Taken by Storm*, Toronto: Key Porter Books.
Oberthner, S. and Ott, H. (1999) *The Kyoto Protocol: International Climate Policy for the Twenty-first Century*, Berlin: Springer.
UNFCCC (1997) *Kyoto Protocol to the United Nations Framework Convention on Climate Change*, viewed at http://unfccc.int/resource/docs/convkp/kpeng.pdf (accessed 1 July 2003).

decline – following major volcanic eruptions, for example – an increase in atmospheric turbidity could compensate for the enhancement of the greenhouse effect. Sulphate particles are particularly effective at intercepting solar radiation, and their injection into the upper atmosphere could provide a layer in which solar radiation is scattered, reducing the amount reaching the earth's surface. Although the concept appears simple, it includes a variety of serious disadvantages, ranging from the potential increase in acid rain as the sulphur is ultimately returned to the earth's surface to the increased thinning of the ozone layer associated with increased sulphate levels in the stratosphere (Turco 1997).

Such esoteric approaches to mitigating global warming make good media copy, but they are unlikely to make a serious contribution to solving the problem. In addition to the technological factors, effective solutions will have to consider the economic, political, cultural and behavioural aspects of the issue. Many of these include barriers to mitigation, which will have to be removed if the problems of global warming are to be successfully addressed. That will require co-operation at regional, national and international levels. Unfortunately, attempts to encourage that co-operation – through the Kyoto Protocol, for example – have met with only limited success and solutions to global warming remain some considerable distance into the future.

SUMMARY AND CONCLUSION

The vulnerability of stratospheric ozone to human interference was discovered almost by accident in the mid-1970s, but in the three decades that followed, depletion of the ozone layer became a major environmental issue, with a very high public profile. By the mid-1980s, spurred on by the discovery of the Antarctic ozone hole, the implications of a much diminished ozone layer for people and the environment were seen as sufficiently serious to merit immediate action. The result was the Montreal Protocol, a landmark international environmental agreement on the protection of the ozone layer, which was signed in 1987. Together with subsequent amendments, it set a timetable for a reduction in the production and use of ozone-destroying chemicals and their replacement by ozone-friendly products. The ozone layer may already be on the road to recovery, with a distinct possibility that it will reach pre-1980 levels by the middle of the twenty-first century.

The outlook is much less bright with the other leading environmental issue – global warming. The earth has experienced warming at various times in the past, but interest in global warming in its present form dates from the creation of the Intergovernmental Panel on Climate Change (IPCC) and the development of the Framework Convention on Climate Change following the Earth Summit in Rio de Janeiro in 1992. Unlike previous warming, the current rise in global temperatures is the result of the enhancement of the greenhouse effect by an increase in the emission of greenhouse gases from anthropogenic sources.

There are many variables involved and it is difficult to predict not only the range of potential impacts, but also their magnitude. Despite the uncertainties, plus a certain degree of dissension among scientists, politicians and businessmen, global warming is widely viewed as a serious threat to the environment and society, which must be dealt with now. Attempts to arrest continued warming have centred mainly on the control and reduction of greenhouse gas emissions, but success is as yet limited. The Kyoto Protocol, signed in 1997, was an attempt to provide a means of achieving success through international co-operation. Unlike the Montreal Protocol, however, it has contributed little to resolving the issues that brought it into being, and may even be abandoned unless the meagre co-operation achieved so far, both nationally and internationally, can be increased. The scientific background to global warming is well established for the most part – although some uncertainties remain – but there is evident reluctance to move to the decision-making stage among those who will have to be involved if mitigation is to succeed.

SUGGESTED READING

Jaccard, M., Nyboer, J. and Sadownik, B. (2002) *The Cost of Climate Policy*, Vancouver: University of British Columbia Press. An analysis of the impact of emissions reduction policies on energy prices, technical options and lifestyle choices with proposals for overcoming the constraints of environmental policy making and the high initial costs that are commonly involved.

Fagan, B. (2001) *The Little Ice Age: How Climate Made History 1300–1850*, New York: Basic Books. A readable account of the last major cold spell in Europe and North America. It helps to put the current warming in perspective.

Harvey, L.D.D. (2000) *Global Warming: The Hard Science*, Harlow: Pearson. A comprehensive, technical, rigorous and critical discussion of global warming. Requires a good scientific background and an understanding of quantitative methods.

Dauncy, G. and Mazza, P. (2001) *Stormy Weather: 101 Solutions to Global Climate Change*, Gabriola Island BC: New Society Publishers. Self-help suggestions on how to deal with global climate change for individuals, companies, local and national governments, with a strong emphasis on appropriate energy use.

QUESTIONS FOR REVISION AND FURTHER STUDY

1 What environmental problems were the Montreal Protocol and Kyoto Protocol designed to alleviate? Why has the former been very successful whereas implementation of the Kyoto Protocol is mired in controversy?

2 In the upper atmosphere the destruction of ozone is considered harmful, yet at the earth's surface it is the production of ozone that is considered harmful. Why does this apparent contradiction exist? Why does the ozone produced in the lower atmosphere not replace that destroyed in the upper atmosphere?

3 Debate the proposition that 'Concentration on negative aspects of global warming has obscured the existence of effects that could be potentially beneficial, given appropriate management and planning'.

12

Problems, Prospects and Solutions

After reading this chapter you should be familiar with the following concepts and terms:

adaptation	Markets First	prevention
anticipatory adaptation	'no regrets' policies	Security First
'business as usual'	Policy First	Sustainability First

CURRENT PERCEPTIONS OF THE STATE OF THE ENVIRONMENT

The human species, which once had a role in the environment that differed little from that of other animals, has over the millennia since it first appeared on earth come to dominate nature and in many places has effectively destroyed it or replaced it. This was made possible by the development of a socio-economic and technological system that, particularly over the past 200–300 years, allowed society to initiate major changes in its relationship with the other elements in the environmental system. Indeed, in many ways, human beings no longer see themselves as part of the environment, but almost as observers on the outside looking in, seeing now a system that is in trouble and needs help, in much the same way, perhaps, as they might look at some animal in trouble that needs to be rescued. Such detachment has undoubtedly contributed to the problems that have arisen in the environment, and there is a perception among some environmentalists, such as the deep ecologists, for example (see Chapter 1) that the situation will be improved only when humans once again begin to consider themselves as an integral part of the natural environment.

Interest in environmental issues grew rapidly in the last two decades of the twentieth century, fostered by individuals and a variety of groups ranging from those trying to ensure the survival of a particular plant or animal species, through those

concerned to preserve a specific, local ecosystem, to those that adopted an international world view. These international environmental groups, such as Greenpeace, Friends of the Earth or the Worldwide Fund for Nature (WWF), have become very much the public face of the environmental movement, through extensive fund-raising efforts, media advertising and continued participation in protest action. In the process they have evolved into international environmental corporations, with teams of managers, statisticians, lawyers and scientists, supported by multi-million-dollar annual budgets, organized professionally in much the same way as the multinational companies whose activities they often criticize. By lobbying governments they have also attempted to ensure that the legislation necessary to mitigate existing environmental problems is put in place.

Although governments were initially slow to adopt a direct approach to environmental problems, they inevitably became involved in the upsurge of interest in the latter part of the twentieth century, often as a result of participation in the international conferences that characterized that time (see Chapter 1). Government departments and agencies at all jurisdictional levels are now major participants in attempts to preserve existing natural environments, to rehabilitate those areas that have been damaged and to reduce or remove the elements that threaten both. This is not unexpected, since governments are in an ideal position to collect and analyse the data needed to make decisions on environmental issues, either directly through government departments or through the provision of external funding for consultants or university research groups. Governments are also in a position to educate the public about environmental issues, pass the necessary legislation and provide the sanctions required to ensure that the provisions of the legislation are followed. Since environmental issues do not recognize national boundaries, the response to environmental problems requires international co-operation, and that has been promoted by the United Nations and its various agencies, with considerable success.

As a result of the activities of environmental groups and governments around the world, knowledge of the environment, the abuse that it has received at the hands of society and the steps that need to be taken to remedy the situation is greater than it has ever been. The picture that emerges from all of that knowledge is not a pretty one. The reduction in acid rain in the developed world, the ban on the production and use of CFCs, the reintroduction of species to the wild, the rehabilitation of landscapes ravaged by mining or the reforestation of clear cuts are all success stories, but they are far outweighed by the increasing levels of pollution in the air and water of the developing world, by the ongoing destruction of habitat and loss of biodiversity worldwide and the seemingly unstoppable threats presented by global warming.

The severity of the threat to the environment is not universally accepted, however, and in the past several years environmental groups, government agencies and international bodies such as the IPCC have come under attack in a number of publications (see e.g. Lomborg 2001; Essex and McKitrick 2002). The authors include political scientists, economists and statisticians as well as atmospheric and environmental scientists ready to challenge the doomsday scenario they see as associated with many environmental forecasts. Although none denies that serious and detrimental environmental change has occurred, the central theme of most of the publications is that the situation is not as grave as it has been represented. Environmental organizations are accused of misrepresenting or overstating data, scientists of basing conclusions on insufficient research and the media of oversimplifying or sensationalizing environmental problems, sometimes in collusion with environmental groups or scientists. Even well respected scientific journals, such as *Nature*, have come under attack for being one-sided

in their representation of environmental issues (Essex and McKitrick 2002). Publications that challenge the existing environmental consensus have been condemned variously as the work of authors who lack the training to appreciate the broad environmental picture or who represent right-wing think tanks and industries, such as those producing or using fossil fuels, which would benefit from the retention of the status quo. Reviewers have demonstrated that they commit the very sins they accuse the environmental scientists of committing – the misuse, misinterpretation and misrepresentation of data – and ignore the crucial role of science in achieving environmental improvements in such areas as air and water pollution, acid rain and the recovery of the ozone hole (Rennie 2002). There is no doubt that environmental groups have on occasion overstated the impact of human activities on the environment – the rates of species extinction claimed by some groups remain controversial, for example (see Chapter 7) – just as scientists have on occasion ignored the uncertainties that they have been trained to expect and reported conclusions with insufficient verification. Similarly, it is true that the media tend to oversimplify environmental issues to meet the needs of their audience or choose the sensational over the mundane. Nevertheless, the progress that has been made in dealing with environmental problems has come about through the application of science to the issues and an unprecedented level of public education, much of it through the media. Success has come about where science and education have been backed up by appropriate levels of funding and coupled with the political will to make decisions that are not always popular with all elements of society. Despite the flaws that may exist in the investigation and analysis of the issues, there can be no denying the existence of an almost endless list of environmental problems, ranging through the absence of clean water for billions of the world's people, the destruction of habitat and the extinction of species, growing levels of air pollution in the developing world, declining fish stocks, destruction of forests, stress on the land from continuing urban sprawl, changing patterns of weather and climate . . . Although the issues are not equally serious, if nothing is done to mitigate them their impact on the functioning of the environment can only increase, leading to a deterioration in the quality of life for most of society. Ignoring them now, when many may still be manageable, has been likened to piling up a debt that will ultimately have to be paid, but at a cost that will exceed by far the cost of an immediate response (Smith 2000).

BUSINESS AS USUAL, ADAPTATION OR PREVENTION?

Much of the difficulty associated with attempts to solve environmental problems arises from the socio-economic and political consequences of the required changes, which are perceived by some to be even more detrimental to society than the continued existence of the problem. In short, the disease is considered less harmful than the cure. The implementation of the requirements of the Kyoto Protocol, for example, fits that category for some observers, with the impact of the changes in energy use on corporate profitability, employment rates and national economic growth being seen as greater than continued global warming. This implies acceptance of a 'do nothing' or 'business as usual' approach in which some level of inconvenience – air pollution, reduced water quality or more frequent flooding, for example – is accepted as the price that has to be paid for the maintenance of a standard of living that is sufficiently high to make the inconvenience worth while. Unfortunately, just as the attributes of a higher standards of living are not shared equally by all members of society, the inconvenience is unlikely to be apportioned equally either. The quality of life in the developing world is generally well below that in the developed countries, and it is clear that if certain environmental issues are not addressed – the provision of clean water, for example – that quality of life is unlikely to improve. Although there may be individuals or groups who for various reasons are willing to continue with the 'business as usual' approach, they are a minority, if not without influence in some cases. The urgent need to provide solutions is widely accepted.

Given the complexity of the environmental problems being addressed, there is no one approach that satisfies all needs. Solutions will have to be flexible enough, for example, to deal with issues that have arisen as a result of deliberate human interference in the environment as well as those that have arisen inadvertently, perhaps as a by-product of some activity designed to improve the quality of life. The development and use of CFCs and their subsequent impact on the ozone layer provide an illustration of the latter situation (see Chapter 11).

There are also problems associated with the synergistic effects created when human and natural changes combine to magnify problems. Water supply issues in areas such as the Great Plains will fall into this category, when the over-extraction of ground water and the increased drought expected to accompany global warming combine (Middleton 1999). Associated with many environmental issues are problems of feedbacks and lagtimes, which are not always well understood and will add to the difficulties of estimating the magnitude and timing of any proposed solutions. Time scales are also crucial, with choices of short-term or long-term approaches requiring consideration in many cases. With such complexity, finding appropriate solutions for any environmental issue may appear almost impossible, but successes at both the local – recycling, for example – and global scales – reduction of acid rain or the banning of CFCs – indicate that some problems at least can be reduced to manageable levels. Whether that success can be replicated in other areas remains to be seen, but there are many options being considered, often involving adaptation or prevention, and sometimes a combination of both.

Adaptation in is simplest form is already part of the environment, represented by the adjustments required to maintain equilibrium among its many ecosystems and the various elements they contain. As part of the environment, human beings have always had the ability to adapt to changing conditions, and in many cases society has been shaped by such adaptation. Indeed, it might be argued that the current dominant position of the human species is very much the result of its successful adaptation to different and changing conditions in the earth/atmosphere system. Can that ability to adapt allow society to deal with the major environmental changes now facing it? In the short term the answer is probably a qualified yes. Adjustment is a continuing response to existing environmental problems, and takes many forms. Changing land use in response to drought is one form, for example, as is the addition of lime to acidified lakes. At the personal level, lifestyle changes such as wearing a hat or using sunscreen to reduce the impact of higher UV-B levels, or drinking bottled water rather than that from a possibly contaminated municipal supply, are also adjustments to a changing environment.

Adaptation often appears attractive because in the short term it is a relatively simple and low-cost approach. With time and continuing environmental change, however, the cost of adaptation may eventually exceed the costs of providing solutions. In theory, adaptation and the development of preventative measures should take place in phase so that solutions can be in place before the cost-effectiveness of adaptation is lost. Adaptation is a reactive approach, often involving little planning, and the long-term impacts of adaptive policies are difficult to predict. Considering the magnitude of current environmental problems, it seems likely that adaptation can be only a temporary measure. Anticipatory adaptation, in which steps are taken in advance to minimize the potential negative effects of environmental change, can extend the utility of the adaptation approach (Goudie and Viles 1997), but ultimately the cumulative effects of the changes are still likely to surpass the ability of society to adjust, and solutions involving preventative measures will have to be found. Given the complexity of many environmental problems, and the difficulties of forecasting their ultimate effects, there is concern that measures will be introduced to deal with projected problems that in reality do not come to pass, resulting in unnecessary economic expense and wasting resources allocated to the issue. In an attempt to deal with such situations, so-called 'no regrets' policies have been developed. These are designed to contribute to the overall health of the environment whether the projected changes take place or not. In short, policy makers should have no reason to regret the adoption of such policies (Goudie and Viles 1997). Policies involving improvements in energy efficiency, for example, aimed initially at mitigating global warming, would bring improvements in urban air quality, atmospheric turbidity, landscape disruption and resource conservation, all of which would be beneficial for the environment even if the projected climate change did not occur.

As research into global change continues, it becomes increasingly clear that the only way to ensure that environmental problems will not become progressively worse is to reduce and ultimately halt the processes that cause them. The technology exists to tackle most environmental problems by amending or replacing the elements involved, but the gap between theory and practice is immense. It has been maintained in large part by a combination of political intransigence and financial constraints in both the public and private sectors of the international economy. Since the 1970s public pressure has had some success in

forcing the political and industrial establishment in the developed world to reassess its position on environmental quality. Oil companies, the forest products industry and automobile manufacturers regularly express concern about pollution abatement and resource conservation and have introduced measures that have had a positive impact in both areas. The growing number of sports utility vehicles on the road, with their high fuel consumption, and the fossil fuel industry's response to the requirements of the Kyoto Protocol, however, indicate that there are limits to industry's commitment to solving environmental issues. The development of appropriate environmental policies and practices is an integral part of government platforms at all administrative levels, and the amount of environmental legislation is climbing steadily, with results that are often encouraging. At times of fiscal restraint, or to meet party political agendas, however, environment ministries and departments suffer cutbacks, and during election campaigns environmentally inappropriate decisions – the opening of a depleted fishery or a reprieve for a polluting industrial plant, for example – for the sake of the votes they can provide are not unknown. Thus there are times when the apparently narrowing gap between what is required to deal with environmental problems and what is being done begins to widen again.

INTERNATIONAL CO-OPERATION

One of the promising signs for the future is the way in which environmental issues have become topics of international concern. Being global in nature and extent, and paying no heed to national borders, they can only benefit from international co-operation. The teamwork that allowed the Montreal Protocol to have such a significant impact on reducing the production and use of CFCs and other ozone-destroying substances is probably the best example of what can be achieved through co-operation. Other conventions and protocols designed to deal with a wide range of issues, including ocean dumping, trade in endangered species, the protection of the tropical forest, preservation of the Antarctic environment, the reduction in persistent organic pollutants and the protection of wetlands, for example (see Table 7.4), have met with mixed success. Co-operation on finding ways of slowing global warming, perhaps the most serious environmental problem facing the world, is less than satis-factory and the likelihood that the provisions of the Kyoto Protocol will have a significant impact on climate warming in the near future is receding as national and industrial interests – in the United States, for example – again appear to be taking precedence over the common good of the earth as a whole.

A comparison of the Montreal and Kyoto proto-cols provides an insight into the nature and com-plexity of the elements that can promote success or retard it. Why has the Montreal Protocol been successfully implemented, while the Kyoto Protocol languishes, unratified by many who initially signed it? The reasons are complex, involving a range of technical, economic, social and political issues that individually and in combination affect each agree-ment differently. Despite major advances in tech-nology, society's knowledge of the working of the various elements in the environment remains incomplete. Scientists are used to dealing with some degree of uncertainty, because they have been trained to accept different levels of confidence in their work. Planners, politicians and policy makers, however, must have reliable estimates of the impact of environmental change if they are to make the decisions necessary to minimize its negative effects and maximize its positive effects. As long as uncertainty exists, it is all too easy to delay action.

This situation applies particularly to global warming and therefore impacts on the Kyoto Protocol. The basic physical chemistry of the role of greenhouse gases in global warming is known, but knowledge of important factors, such as the flow of carbon in the earth/atmosphere system and the contribution of feedback systems to the degree and direction of the warming, remains incomplete. The global warming potentials of the many gases involved are not known with complete accuracy and the contribution of other radiative forcing agents to the total picture is unclear. In contrast, the chemistry of ozone depletion is well established and subject to less uncertainty. In addition the number of ozone-depleting substances is small and they are manufactured by a limited number of know producers in the developed world. The chemicals involved in global warming – carbon dioxide and methane, for example – are ubiquitous, being pro-duced from a myriad of natural and human sources, which makes their complete quantification difficult. Thus setting and achieving objectives have tended to be easier in the case of the Montreal Protocol.

Public perception has also been important. Life-threatening issues such as melanoma associated with ozone depletion produced a level of concern sufficiently high that the public was willing to accept the changes and restrictions required to make the protocol work. The problems that the Kyoto Proto-col is trying to address, however, are less obviously and immediately serious, and although there is considerable public support for the Kyoto agree-ment, it has not always appeared sufficient to cause politicians and administrators to react positively. There are major economic differences also. The ozone-depleting substances were produced by only a small number of companies which after initial resistance worked to provide appropriate substi-tutes and ultimately suffered a limited economic impact. The economic impact likely to follow the reduction of greenhouse gas emissions would be much greater, involving the major energy and manu-facturing industries, and perhaps leading to reduced economic activity and an increase in unemploy-ment. For many the risk is too great, and support for the Kyoto Protocol has suffered.

Thus, in international agreements designed to deal with serious global environmental issues, the interplay of science, technology and socio-economic and political factors has produced sig-nificantly different results. The interrelationships among these various elements and their relative strengths vary with time and place and there is no guarantee that a solution that worked for one problem in one area will be successful elsewhere. That too has to be taken into account in the search for workable solutions to environmental problems.

All attempts at international co-operation, for example, have to consider world socio-economic and technological patterns, which have an impact on the nature and intensity of environmental problems and the solutions that might be applied to them. The most common division of the world on these grounds is into developing and developed countries, often identified as a North–South split, the developed nations being in the North and the developing in the South. In addition there are those nations of the East such as those in Eastern Europe, Russia, other parts of the former Soviet Union and perhaps even China, which have their own peculiar environmental issues arising out of the workings of the centralized, controlled economies associated with communism (Middleton 1999). Each of these areas has different environmental priorities. In the North, it is over-consumption of resources and the environmental damage associated with it that need to be addressed. The nations of the North have also pushed that impact beyond their boundaries by exporting environmentally damaging technologies. While the transfer of technology has been recognized as essential in the search for solutions, the technology introduced to developing nations has often had a detrimental rather than a beneficial effect on the environment. Toxic waste banned from dumps and landfill sites in the North has found its way to the South and pesticides and herbicides banned in the developed world continue to be exported to developing nations to be used in agriculture (Smith 2000). The ongoing environmental problems in the South, however, stem from continued high population growth rates and burgeoning, uncontrolled urban growth (see Chapter 4). In Africa, the implications of the HIV/AIDS epidemic extend beyond health, social development and population structure to the environment. The people in these areas are among the poorest in the world, and although they consume much less of the world's resources than the rich, poverty can force them to use the environment inappropriately. Poverty also means that they are less able to adapt to change, and as a result they are more vulnerable to environmental threats such as drought, soil erosion or water pollution and a variety of natural disasters (UNEP 2002).

Environmental problems of the countries in the East stem from outdated technology, which is wasteful of resources and provides high levels of environmentally damaging pollutants. If global environmental issues are to be dealt with successfully, such differences will have to be considered when reaching solutions. The economic disparity between the North and the other parts of the world has implications in the search for solutions, but the difficulties have been recognized and allowances have been made to reduce the economic impact of the implementation of such agreements as the Montreal Protocol, the Kyoto Protocol and the Statement of Forest Principles on the developing nations. Given the interconnected nature of global ecosystems, the negative impacts of environmental problems are commonly felt outside the areas in which they originate. Solving the problems in one area is not enough. Even if the countries of the North addressed all of the issues that face them, they would still feel the effects of environmental deterioration elsewhere. It would seem therefore that it would be in the best interests of the North to use the economic clout that it has to promote the transfer of appropriate technology and provide financial aid to the South and the East with a view to improving the global environment.

IS SUSTAINABLE DEVELOPMENT THE ANSWER?

Underlying all of this is the concept of sustainable development, which arose in its present form out of the Brundtland Commission and gained prominence through the Earth Summit and its Agenda 21 (see Chapter 5). With its potential to allow development without compromising the environment, it has been widely embraced and applied to issues that range from biodiversity to urbanization. The proposals set out in Agenda 21 are ambitious, requiring changes in previously accepted socio-economic, technological and environmental practices. They will require, for example, greater implementation of the 'polluter pays' principle, the phasing out of subsidies that encourage over-exploitation or inefficient use of resources and the linking of development funding with environmentally appropriate use. They will have to be built into international trade in such a way as to encourage greater social equity and better balanced import/export economies in the developed nations. They will require the North to transfer new technologies to developing nations, rather than the currently common practice of providing obsolescent plant and machinery no longer economically or environmentally viable in the developed world. Socio-economic issues of poverty, education and empowerment – particularly of women – will also have to be addressed (Cuff 2002c). The UN Commission on Sustainable Development (CSD) has assumed the task of promoting the implementation of such issues and monitoring progress towards their achievement, but without the authority to make binding decisions on the issues. Through a series of meetings since 1993 it has successfully encouraged international participation in policy making and in 1997 reviewed the progress that had been made in implementing the provisions of Agenda 21. It concluded that some progress had been made in reducing poverty, curbing pollution and slowing the rate of environmental degradation, but overall the global environment continued to deteriorate and many nations had not lived up to the commitments they had made at the Earth Summit (Chasek 2002). In short, although the policies and principles of sustainable development are now well established, there is a considerable lag in putting them into practice.

In its third Global Environmental Outlook Report (*GEO3*), published in 2002, the United Nations Environment Program (UNEP) included a Sustainability First scenario in its consideration of the changes in the global environment that might take place in the next thirty years. It was one of four scenarios considered, the other three being Markets First, Policy First and Security First (Table 12.1), referring to the different directions that development might take and the impact that they would have on the global environment. The Markets First scenario is little more than a continuation of current practices in which any environmental improvements will come about as a result of market-based forces. The growing scarcity of fossil fuels, for example, and their rising costs would bring about an increasing use of renewable resources, which would ultimately help the environment. Increasing wealth, at least in some areas, would allow people and communities to adapt to environmental change or to cover the costs of dealing with the environmental crises that may arise. This is little different from the 'business as usual' approach espoused by those who see traditional development as having the necessary ingredients to handle the world's socio-economic problems and the environmental deterioration associated with them (see e.g. Lomborg 2001). The UNEP projection indicates that although some areas would benefit, in the

TABLE 12.1 UNEP SCENARIOS FOR ACTION ON GLOBAL CHANGE

Scenario	Characteristics	Results
Markets First	Current values and expectations of developed world prevail Market forces dominate Economic globalization and liberalization continues	Environmental standards decline Pressure on resources remains high Poverty and environmental degradation particularly severe in the developing world
Security First	Striking regional inequalities; high potential for conflict and terrorism Developed nations focus on self-protection, both physically and economically	Increase in socio-economic and environmental stress Little co-operation in solving global environmental problems Overall increase in environmental deterioration Issues of poverty, health and food supply receive less attention
Policy First	Top down approach through government policy and regulatory measures Mix of incentives and disincentives attempt to improve environmental conditions Considerable dependence on technological innovation	Significant improvement in some areas, but at cost of high taxes and increased bureaucracy Unpopular, expensive and drastic action sometimes required might ultimately become unacceptable Potential for continued improvement questionable
Sustainability First	New environment and development paradigm Support for sustainable policy measures and accountable corporate behaviour Collaboration among governments, stakeholder groups, indigenous peoples and individual citizens	Improvements in all areas, usually exceeding those of other scenarios Rate of change slow because of time required to achieve necessary co-operation Continuing and longer-term improvements in environmental issues potentially high

Source: based on information in UNEP, *Global Environmental Outlook 3 (GEO3)*, London/Sterling VA: Earthscan/United Nations Environment Program (2002).

Markets First scenario environmental standards would fall and pressure on resources would remain severe. Poverty and environmental degradation would continue to be characteristic of most of the developing world and global inequalities would continue, with serious implications for global security.

Current socio-economic disparities already create concerns over global security, and the Security First scenario assumes that little will be done to change that situation. The potential for political and economic unrest, brought on by socio-economic and environmental stress, will be high and seen as a threat by economically powerful regions and groups. For self-protection and to retain their wealth and quality of life they will focus inwards, creating an increasingly divided world, lacking the co-operation necessary to solve environmental problems. Isolation increases the disparities and therefore the need for greater security. As more and more is spent on security, there is less to be spent on such issues as poverty, health, food supply and a whole range of environmental issues. In short, the Security First scenario is in many ways a direct extension of the current state of the world with its economic disparities, global terrorism, civil strife, steady flow of refugees, growing emphasis on security in the developed world and continued environmental deterioration. It will do nothing to contribute to the solution of the world's pressing environmental issues.

The two other UNEP scenarios – Policy First and Sustainability First – provide more promising results. They also require greater change. Policy First is very much a top-down approach in which governments develop policies that are pro-environment and anti-poverty, in an attempt to combat the problems associated with market-based approaches. It requires the provision of legislation that can be enforced, supported by a mix of incentives and disincentives such as tax breaks for environmentally friendly activities or the imposition of taxes – carbon taxes, for example – to discourage those that would harm the environment. To be successful it also needs international co-operation, not only between nations but also between multi-national companies and industrial enterprises. When followed through three decades the scenario produced improvements in a number of important areas such as deforestation and the degradation of agricultural land. Economic development and continued population growth in the developing world

ensured that total greenhouse gas emissions continued to rise and many areas suffered as a result of the effects of rising temperatures on the environment. On the positive side, the rate at which greenhouse gases were being emitted fell and there were indications that the temperature increases associated with global warming were about to plateau. Achieving that much had required actions that were unpopular, expensive and sometimes drastic and there were concerns that the high levels of taxation and increased bureaucracy required to maintain the improvements would ultimately become unacceptable. Much of what had been achieved came about as a result of technological advance, but by the end of three decades it was no longer clear that technology could continue to provide the necessary innovation to allow the world to keep ahead of the changes continuing to threaten the environment.

What appeared to be lacking in the Policy First scenario was a fundamental change in human behaviour and in society's demands on the environment. In contrast these were provided for in the Sustainability First scenario, which was based on collaboration at all levels, including governments, stakeholder groups, indigenous peoples and individual citizens. This allowed broad acceptance of what needed to be done to allow all elements of global society to meet basic needs and achieve personal goals without compromising the environment further or threatening the viability of future populations. Although it required considerable government participation in the collection and analysis of data, from which policy could be developed, the Sustainability First scenario also included a strong 'grass roots' element. This was made possible by good communications and education about the issues, coupled with greater transparency and accountability. Together, they allowed those affected by global change to understand what was required of them, and made them more ready to participate in solving the problems.

As the Sustainability First scenario ran its course, improvements were achieved in almost all areas, including reductions in water stress, greenhouse gas emissions, pressure on natural ecosystems and hunger. In other areas, such as biodiversity, however, deterioration continued. Initially the rate of change was relatively slow, because of the time needed to achieve the necessary level of co-operation, and varied from place to place, but in almost all cases, following the Sustainability First scenario produced measurable improvements that exceeded those possible using the other scenarios. Even where the improvements were less than those provided by the alternatives, the potential for continued, long-term improvements was greater with sustainability, as was the case with temperature change, the preservation of biodiversity and sea level change. After the thirty-year run of the Sustainability First scenario, sustainability had not been achieved, but all signs suggested that the world was taking positive steps towards that goal, and, of all the scenarios, the approach that emphasized sustainability was the most likely to bring solutions to the world's environmental problems.

All scenarios attempting to model the future incorporate a group of assumptions that vary with such factors as the nature, geographical extent and time line of the projection. If even one of the assumptions is wrong the quality of the results will be compromised. In addition, since the physical environment is a dynamic entity and human activities are sometimes subject to remarkably dramatic change, the likelihood of the basic assumptions remaining valid through the run of any scenario is limited. Models can be designed to deal with such issues, but particularly if projections are being pushed far into the future, feedback mechanisms and unexpected societal changes continue to reduce the validity of the results. To be successful, for example, sustainability requires the development of an unprecedented level of collaboration among governments – national, local, municipal – non-governmental organizations, industry, indigenous peoples, special interest groups and individuals. If there is no success in achieving that, then one of the basic assumptions of the Sustainability First scenario is lost and reality may bear little relationship to the situation projected in the scenario. Unpredictability is also an integral part of the physical environment, and some observers (see e.g. Harvey 2000) warn of 'surprises' in the form of abrupt and unanticipated changes in, for example, the climate system or oceanic circulation that would prevent environmental and socio-economic goals from being reached.

Whether any of these scenarios represents the real future obviously remains to be seen, but it is also clear that society cannot sit back passively and wait for the results. People are an integral part of the environment and, being responsible for the environmental deterioration that has occurred, they must also participate in the search for solutions, working proactively and decisively. Sustainable development offers the greatest promise, but it is also the approach that demands the greatest change in current socio-economic and cultural patterns and the greatest individual sacrifice. Success or failure in slowing and then reversing environmental deterioration will depend on society's willingness to face these challenges and accept the sacrifices or ignore them and suffer the consequences.

SUGGESTED READING

Brower, M. and Leon, W. (1999) *The Consumer's Guide to Effective Environmental Choice*, New York: Three Rivers Press. A clear account of the relationship between consumers and the environment. It provides the reader with information on what the problems are, how to recognize what is important and what is not and what to do about it.

Brown, L.R., Larsen, J. and Fischlowitz-Roberts, B. (2002) *The Earth Policy Reader*, New York/London: Norton. A readable assessment of progress to building an eco-economy. Includes both successes and failures with suggestions on how to decrease the latter.

Suzuki, D. and Dressel, H. (2002) *Good News – for a Change*, Vancouver: Greystone Books. An account of the way in which individuals are making positive environmental and social contributions to the reduction of environmental damage and helping to rehabilitate existing conditions.

Glossary

abiotic   Non-living. Abiotic components of the environment include soil, water and the atmosphere.

ablation   The loss of ice and snow from a glacier by melting or evaporation.

abyssal zone   That part of the ocean deep lying more than 2000 m beneath the surface.

acid   A compound containing hydrogen which on solution in water produces an excess of hydrogen ions.

acid loading   The addition of acid to water bodies by way of deposition from the atmosphere.

acid mine drainage   The seepage of acid from mining operations into adjacent waterways.

acid precipitation   The wet or dry deposition of acidic substances of anthropogenic origin on the earth's surface. Commonly called acid rain, but also includes acid snow and acid fog.

acid rain   See *acid precipitation*.

adaptation   Response to environmental change in which activities are altered to accommodate the change. In reactive adaptation the response takes place after the change occurs. In anticipatory adaptation the response precedes the change.

aerosols   Finely divided solid or liquid particles dispersed in the atmosphere.

age-specific fertility rate (ASFR)   The total number of births per thousand women within the age range of fifteen to forty-nine, calculated at five-year intervals.

Agenda 21   A blueprint for sustainable development into the twenty-first century, produced at the United Nations Conference on Environment and Development in Rio de Janeiro in 1992.

agrarian civilizations   The agriculturally based civilizations that developed in Egypt, Mesopotamia, the Indus valley and the Yellow River basin (Hwang-He) in China between about 7000 and 3000 years ago (*c.* 5000–1000 BC). Also referred to as the world's agricultural hearths.

agricultural revolution   A period of rapid change in agricultural activities in Britain between about 1750 and 1850, involving new land management and cultivation techniques plus the development of new breeds of domestic animals.

air pollution   The contamination of the atmosphere to such an extent that normal atmospheric processes are adversely affected. Natural occurrences such as volcanic eruptions can cause contamination, but air pollution is commonly equated with contamination from human activities.

air quality standards   A measure of the maximum allowable concentrations of air pollutants in ambient outdoor air.

albedo   A measure of the reflectivity of the earth's surface, usually expressed as a fraction or percentage of the total incident radiation reflected.

algal bloom   The rapid growth in the number of algae in an aquatic environment in which nutrient enrichment or eutrophication has taken place.

alkali   A compound, usually a soluble hydroxide of a metal, such as sodium hydroxide, which on solution with water produces an excess of hydroxyl ions.

allelopathy   The ability of some plants to create conditions that prevent or discourage other plants from sharing their space, thus giving them a competitive advantage.

allotropy   The existence of elements in several different physical forms, but with the same chemical properties. Ozone is an allotrope of oxygen, for example.

anaerobic decay   The breakdown of organic material in the absence of oxygen. Brought about by anaerobic bacteria, the process commonly produces methane.

annual allowable cut   The maximum amount of timber that may be cut in an area without exceeding its sustainable yield.

Antarctic ozone hole   An intense thinning of the stratospheric ozone layer above Antarctica during the southern spring. First reported by the scientists of the British Antarctic Survey at Halley Bay in the early 1980s.

anticyclone   A zone of high atmospheric pressure created by the cooling of air close to the earth's surface (cold anticyclone) or the sinking of air from higher levels in the atmosphere (warm anticyclone).

ANWR   The Arctic National Wildlife Refuge, a 7.7 million ha wilderness reserve set aside to protect wildlife – mainly from oil development – on Alaska's North Slope.

appropriate technology   Technology that allows growth without imposing unmanageable stress on resources or the environment.

aquifer   A layer of rock beneath the earth's surface, sufficiently porous and permeable to store significant amounts of water.

aquifer depletion   The pumping of water from an aquifer at a rate that exceeds its replenishment.

Aral Sea   A lake in Central Asia, once the fourth largest in the world, but now much depleted as a result of water diversion and the expansion of irrigation. The result has been one of the world's major environmental disasters.

Arctic Haze   A result of the pollution of the Arctic atmosphere, mainly in winter, by aerosols, such as dust, soot and sulphate particles originating in Eurasia.

aridity   Permanent dryness caused by low average rainfall, often in combination with high temperatures.

artesian well   A well in which hydrostatic pressure is sufficiently high to raise the water above the aquifer in which it is stored. In a free-flowing artesian well water will rise to the surface without pumping.

asbestos   A fibrous silicate mineral with fire-retardant and insulating properties. Once a significant component of indoor air pollution and a known carcinogen.

Asian Brown Cloud (ABC)   A dense, brownish pollution layer, first identified in 1999, covering most of South and Southeast Asia and the tropical region of the Indian Ocean.

atmosphere   The blanket of air that envelops the earth. It consists of a mixture of gases and aerosols.

atmospheric circulation   The large-scale movement of air around and above the earth, associated with complex but distinct patterns of pressure cells and wind belts.

atmospheric turbidity   A measure of the dirtiness or dustiness of the atmosphere as measured by the reduction in solar radiation passing through it.

azonal soils   Immature soils not yet in balance with their environment.

baby boom   The popular name given to a sharp rise in the birth rate, such as the increase that followed the Second World War in the mid-twentieth century.

baseflow   That part of a river's discharge that is provided by input from the groundwater system.

beneficiation   The concentration of low-quality ores prior to smelting, usually at the mine site. The large volumes of waste rock produced can lead to environmental problems.

benthic ecosystems   Ecosystems located at the bottom of freshwater and marine water bodies.

Bhopal   An industrial city of more than 800,000 people which was the site of a major industrial accident in 1984, when methyl isocyanate leaked from a pesticides plant. The highly toxic gas killed more than 2500 people and caused serious health problems for 200,000 more.

biochemical oxygen demand (BOD)   A measure of pollution in a body of water, based on the organic material it contains. It represents the amount of oxygen that aerobic bacteria would require to biodegrade that organic material. Thus the greater the BOD, the greater the amount of organic matter and the greater the pollution.

biodiversity   The variety of life forms that inhabit the earth. It includes habitat diversity, plant and animal species diversity within the various habitats and the genetic diversity of individual species.

biological diversity   See *biodiversity*.

biological weathering   The disintegration of rock as a result of the activities of plants and animals or the physical and chemical processes associated with these activities.

biomass   The total weight of living organic matter in a given area.

biomass energy   Energy available from organic matter in the environment. It includes the energy in wood, crops, crop residues, industrial and municipal organic waste, food processing waste and animal wastes.

biome   A community of plants and animals in equilibrium with the environmental characteristics – climate, soil, hydrology – of a major geographical area. Usually considered to represent a climax community.

biosphere   The zone of terrestrial life, sometimes called the ecosphere. It is an interactive layer incorporating life on the earth's land and water surfaces plus organisms in the lowest part of the atmosphere and the upper part of the soil and water layers.

biotic   Living. Biotic components of the environment include plants and animals as well as all other living organisms.

birth control   Any actions taken to reduce births, including various types of contraception and abortion.

black earth   See *chernozems*.

boreal forest   The cold, coniferous forest of North America, stretching in an arc from Alaska in the north-west, through central Canada to Newfoundland in the east. Characterized by large stands of few species.

borehole   A hole drilled or dug into the earth's surface through which ground water can be withdrawn. A well.

bronchitis   Inflammation of the bronchial tubes through which air passes into the lungs. It has various causes among which are exposure to air pollution and smoking.

Bronze Age   The first phase of human technological development in which a metal was used. Bronze is an alloy of copper and tin, widely used for tools and weapons some 5000 years ago.

Brundtland Commision   See *World Commission on Environment and Development*.

buffering agent Alkaline or basic material capable of reducing or neutralizing acidity. Calcium carbonate is a very effective natural buffering agent used to prevent and counter the effects of acid rain.

bushmeat Meat harvested from the tropical rain forest, mainly in Africa, but also in Southeast Asia, often at unsustainable rates that have placed gorillas, chimpanzees and other primates in danger of extinction.

business as usual One of the scenarios used to predict the future status of environmental issues such as global warming. It is based on the maintenance of the status quo, with little or no attempt to reduce the output of the agents of change or to mitigate their impact.

calcium carbonate The most common compound of calcium; a white insoluble solid that occurs naturally as chalk, limestone, marble and calcite. Frequently used as a buffering agent to reduce the acidity of soils and water.

carbon cycle A natural biogeochemical cycle through which the flow of carbon in the earth/atmosphere system is regulated. Carbon circulates through the system between several major reservoirs, such as the atmosphere, earth, oceans and fossils fuels, mainly in the form of carbon dioxide.

carbon dioxide One of the variable atmospheric gases, and a greenhouse gas. It is produced by the combustion of carbonaceous substances, by the aerobic decay of organic material by fermentation and the action of acid on calcium carbonate. Its level in the atmosphere is 368 ppmv, but growing.

carbon dioxide fertilization The benefits provided to green plants through more effective photosynthesis as the level of atmospheric carbon dioxide increases.

carbon monoxide A colourless, odourless, tasteless gas produced by the incomplete combustion of carbonaceous material. Common in automobile exhaust gases and tobacco smoke. At low levels of inhalation it causes the functioning of the brain, lungs and heart to be impaired: at high doses it is lethal.

carbon sequestration The storage of carbon in sinks. Carbon is sequestered in growing plants, for example, when carbon dioxide is absorbed from the atmosphere during photosynthesis.

carbon taxes Taxes raised by assessing fossil fuels according to the carbon dioxide they produce when burned. Coal would be assessed higher taxes than natural gas, for example. The aim is to reduce carbon dioxide emissions and slow the enhancement of the greenhouse effect by a realignment of energy use away from high carbon fuels.

carbon tetrachloride An industrial solvent once widely used as a dry-cleaning agent, but largely replaced by other compounds because of its toxicity and its contribution to the depletion of the stratospheric ozone layer.

carboxyhaemoglobin A stable compound formed when carbon monoxide combines with the haemoglobin in blood, impairing the ability of the blood to carry oxygen around the body.

carnivore Meat-eating organism. Most carnivores are animals, but some plants, such as the Venus fly-trap, are carnivorous.

carrying capacity The maximum number of organisms that can be supported by a particular ecosystem. If that number is exceeded some form of environmental disruption will follow.

catalyst A substance that facilitates a chemical reaction, yet remains unchanged when the reaction is over.

catalytic chain reaction A series of chemical reactions made possible by the survival of the catalyst after each individual reaction, allowing it to promote the same reaction again and again.

catalytic converter A pollution control device attached to automobile exhaust systems. Catalysts such as platinum, palladium and rhodium control the output of pollutants such as carbon monoxide and oxides of nitrogen.

certified forest products Products guaranteed to be from forests in which management standards encourage regeneration, the conservation of biodiversity and the maintenance of soil quality to protect the long-term health and viability of the forest.

chemical weathering The breakdown of rocks at the earth's surface as a result of chemical reactions, which destroy the integrity of the rocks. The reactions are initiated by the exposure of the constituents of the rocks to air, water and organic activity.

chernozems Temperate grassland soils characterized by high organic levels, dark colour – black earths – and high fertility.

chlorine monoxide A compound containing chlorine and oxygen that has been implicated in the destruction of the ozone layer.

chlorine monoxide dimer A combination of two chlorine monoxide molecules, which has had a major role in the formation of the Antarctic ozone hole.

chlorine nitrate A combination of chlorine monoxide and nitrogen dioxide involved in the depletion of Antarctic ozone. It does not deplete ozone, but acts as a reservoir for chlorine, which when it is released acts as an ozone-destroying catalyst.

chlorofluorocarbons (CFCs) A group of chemicals containing chlorine, fluorine and carbon. Inert at surface temperature and pressure, in the stratosphere they become unstable and release chlorine, which initiates a catalytic chain reaction leading to the destruction of ozone. CFCs are also powerful greenhouse gases.

circumpolar vortex A band of strong winds that circles the poles in the upper atmosphere. Particularly marked in the winter.

clathrates Deposits of methane hydrate trapped in deep ocean sediments or high latitude permafrost. They are potential energy sources and contain sufficient methane to cause a major enhancement of the greenhouse effect if released.

clear cutting The complete removal of trees from an area during commercial forestry.

Climatic Optimum A period of major warming during the immediate post-glacial period between 5000 and 7000 years ago. Also referred to as the altithermal, hypsithermal and Holocene maximum.

climax community A mature community that represents the final stage in a natural succession, and which reflects prevailing environmental conditions such as soil type and climate. Climax communities are characterized by a diverse array of species and by an ability to use energy and recycle chemicals more efficiently than immature communities.

common-pool resources Resources that are or once were available to all. Water and air are examples of common pool resources.

compost A soil conditioner and fertilizer produced by the controlled decay of organic matter through the activities of aerobic bacteria and other micro-organisms.

conservation An approach to natural resources that involves planning and management to allow both use and continuity of supply.

construction aggregates Natural resources such as sand and gravel used in the creation of infrastructure such as roads, airports and buildings. Extraction of the aggregates can initiate local environmental problems.

consumptive use The use of a resource in such a way that its form or content is changed and it is no longer available for its original use.

continental drift A precursor of the concept of plate tectonics. Lighter continental rocks were seen as floating on the denser rocks beneath the ocean basins and therefore able to drift or change position over time.

continental-scale water engineering The redistribution of continental water supplies through massive interbasin transfers, usually to meet the needs of irrigation or the production of hydro-electricity.

contingent drought A form of agricultural drought characterized by irregular and variable precipitation in areas that normally have an adequate supply of moisture to meet crop needs.

contour ploughing A form of soil erosion prevention involving ploughing across a slope rather than up and down it. This slows the flow of water down the slope and reduces the potential for erosion.

contraceptive pill A chemical contraceptive taken orally. First introduced in 1963, it had an almost immediate effect on the birth rate, particularly in the developed nations.

convection The vertical transfer of heat through a liquid or gas by the movement of the gas or liquid. The fluid in contact with the heat source warms up, expands, becomes less dense and rises through the surrounding gas or liquid.

Convention on International Trade in Endangered Species (CITES) An agreement drawn up in 1973 to protect a wide range of species of plants and animals, thought to be at risk of extinction, by prohibiting trade in the species or products made from them.

Coriolis effect The effect, caused by the rotation of the earth, that brings about the apparent deflection of objects moving in the atmosphere or across the earth's surface.

craton Part of a continental block that has been stable over a long period of time – ancient continental shield areas, for example.

criteria pollutants A group of pollutants considered characteristic constituents of urban air pollution. They include particulate matter, carbon monoxide, sulphur dioxide, oxides of nitrogen, ozone and lead, and are included in most urban air quality monitoring programmes.

crude birth rate (CBR) The number of births per thousand in a population in a given year.

crude death rate (CDR) The number of deaths per thousand in a population in a given year.

cryosols Cold-region soils that are underlain by permafrost.

debt-for-nature An approach to rain-forest protection in which an environmental group or government agency purchases part of the foreign debt of a developing nation and offers to forgive the debt if the debtor agrees to set aside and manage a section of rain forest.

decommissioning wastes The waste material created when an industrial plant ceases production. Potentially serious disposal problems arise with decommissioning wastes produced when nuclear plants are closed down.

decreasers Natural grassland plant species preferred by domesticated livestock. Being preferred, their numbers tend to decrease over time.

deep ecology A holistic approach to environmentalism, which recognizes the importance of individual elements in the environment and their relationship to each other. It is strongly ecocentric.

deforestation The permanent removal of trees from a forested area – as a result of logging, agricultural expansion, or industrial development, for example.

delta A depositional landform created when a river or stream flows into a body of standing water such as a lake or the sea.

demographic transition model A concept that combines population change – birth rates, death rates and population growth – with socio-economic development and reflects the stages through which most European and North American developed nations have passed.

desert An area of permanent aridity in which precipitation is infrequent or irregular in its occurrence and the resulting low annual rainfall totals are exceeded by high evapotranspiration rates.

desertification The expansion of desert or desert-like conditions, initiated by natural environmental change, by human degradation of marginal environments or a combination of both.

domestication The taming of wild animals and the changing of natural plant characteristics for the benefits of humans.

doubling time The time that it takes for the number of people in a region, nation or the world as a whole to double. For most purposes, the doubling time can be estimated by dividing the number 70 by the rate of natural increase of the population.

drought A period of reduced water availability, occurring when precipitation falls below normal, or when normal rainfall is made less effective by other weather conditions such as high temperature, low humidity or strong winds.

Dustbowl An area of the Great Plains, which suffered the effects of drought and desertification in the 1930s.

dynamic equilibrium A degree of balance among the components of the environment achieved by a series of mutual adjustments to their relationships with each other.

earthquake A series of earth movements brought about by a sudden release of energy during tectonic activity in the earth's crust.

ecocentrism An environmental philosophy that takes the view that the natural elements in any environment are important in their own right and should not be considered secondary to the human species or its needs and wants.

ecology The study of the relationships that develop among living organisms and between these organisms and their environment.

ecosphere See *biosphere*.

ecosystem A community of organisms interacting with each other and their environment.

ecotone A transition zone between two ecosystems. It may contain organisms from each of these ecosystems plus organisms peculiar to the ecotone itself.

edaphic factors The physical, chemical and biological properties of soils that contribute to the characteristics of ecosystems. They include such items as pH, structure, texture and organic content.

El Niño A flow of abnormally warm water across the eastern Pacific Ocean towards the coast of Peru. It is associated with changing pressure patterns and a reversal of air flow in the equatorial Pacific, a phenomenon referred to as the Southern Oscillation.

emigration The movement of people from one country with the aim of settling in another.

emphysema A progressive debilitating lung disease caused by smoking and atmospheric pollution in the workplace.

endangered species Species of plants or animals threatened with imminent extirpation or extinction.

energy The capacity to do work. Energy takes a variety of forms and can be converted from one form to another to meet specific needs. In environmental studies, a split into renewable and non-renewable energy is common.

environment A combination of the various physical and biological elements that affect the life of an organism.

environmental equilibrium The concept of dynamic balance among the components of an environment.

environmental impact assessment The formal assessment and analysis of the potential impact of various forms of human activity on the environment.

environmental lapse rate (ELR) The rate at which temperature declines with increasing altitude in the troposphere. Although it varies with time and place the ELR is normally considered to average $-6.4°C$ per 1000 m.

epiphyte A plant that is not rooted in the ground, but lives on or is attached to another plant. The host plant provides no nutrients, only support.

eutrophic lakes Water bodies that have a high concentration of nutrients and are high in organic productivity. Lakes normally become increasingly eutrophic with age as nutri-ents accumulate, but the process can be accelerated by the addition of pollutants such as sewage, phosphates and agricultural fertilizers.

eutrophication The natural aging of a water body, characterized by increasing levels of dissolved nutrients in the water.

evapotranspiration The transfer of water from the terrestrial environment to the atmosphere through a combination of evaporation from the land surface and transpiration from plants.

exosphere The zone at the outer edge of the thermosphere, some 350–500 km from the earth's surface, where gaseous molecules with sufficient velocity may escape the earth's gravity.

exponential growth Cumulative growth, as illustrated by the concept of compound interest. In environmental studies, exponential growth applies particularly to population. If the offspring in a population, for example, reproduce at the same rate as their parents, population will grow exponentially.

extinction The elimination of all individuals in a particular species. The species is effectively removed from the biosphere and cannot be replaced.

extirpation The elimination of a species from a specific area as a result of such factors as disease, over-predation or environmental change. It is less serious than extinction, since members of the species survive in other areas.

famine Acute food shortage leading to widespread starvation. It is commonly associated with large-scale natural disasters such as drought, flooding or plant disease which produce crop failure and disruption of food supply, but may be aggravated by human activities.

Famine Early Warning System (FEWS) An information system funded by USAID aimed at forecasting famine in sub-Saharan Africa.

family planning The regulation of family size through the use of contraception, sterilization and in some cases induced abortion.

fecundity The maximum reproductive capacity of a population.

fertility The number of live births in a population.

fire climax A plant community in which the mix of vegetation reflects the impact of periodic fires rather than the soil and climate conditions that would create the expected climax community in the area.

fire suppression policy An approach to forest management in which fires are suppressed to protect and preserve the forest. This allows the build-up of critical levels of flammable material, creating the potential for abnormally severe fires such as those in Yellowstone National Park in the United States in 1988.

fish farming The breeding of fish in pens in salt water and freshwater environments, allowing greater control of growth rates and harvesting. Waste disposal and disease have created environmental problems in some areas.

flooding The inundation of normally dry land by water. Flooding causes property damage worth millions and takes hundreds of lives every year.

flood plain An area of limited relief bordering a river, inundated when the river overflows its banks during a flood.

flow resources Renewable resources.

flue gas desulphurization (FGD) The process by which sulphur dioxide is removed from waste gases produced by the burning of coal or smelting of ores.

fluidized bed combustion (FBC) A process designed to prevent the formation of sulphur dioxide during the combustion of coal. Limestone present in an aerated mix of coal and sand neutralizes the acid gas that would normally form.

fold mountains Mountains in which the strata are folded and contorted as a result of the movement of the earth's tectonic plates. The Alps and Himalayas are geologically young fold mountains, but eroded remnants of older systems are incorporated in other continental structures.

food chain A group of organisms linked to each other through their production and consumption of food and energy. Most food chains are short and linear with a producer (green plant) at one end and a consumer (carnivore or omnivore) at the other.

formaldehyde A volatile organic solvent widely used in the production of glues and resins. Now recognized as a significant component of indoor air pollution, and a potential hazard to health.

fossil fuel Fuels such as coal, oil and natural gas that are the residues of organic material, buried under sediments or drowned. In the resulting anaerobic environment, decay was limited and they retained much of their stored energy.

Framework Convention on Climate Change (FCCC) One of the conventions signed at the Earth Summit in Rio de Janeiro in 1992. It grew out of concern for global warming and provided a base on which the Kyoto Protocol was built.

fuel desulphurization The treatment of fuels such as coal and oil to reduce their sulphur content prior to combustion.

fuel switching An attempt to control acid gas emissions through the replacement of high-sulphur fuels with low-sulphur alternatives.

fuelwood Wood used as a source of energy for heating or cooking. Fuelwood harvesting makes a significant contribution to deforestation where demand is high.

Gaia hypothesis Developed by James Lovelock, the hypothesis views the earth as a super-organism in which the living matter is capable of manipulating the earth's environment to meet its own needs.

gangue The waste material in an ore body.

garbage Domestic refuse or municipal solid waste, including both organic and inorganic materials.

general fertility rate (GFR) A measure of the total number of births per thousand woman within the age range of fifteen to forty-nine in a population.

global warming The rise in global mean temperatures of about 0.6°C during the twentieth century as a result of the enhancement of the greenhouse effect by rising levels of anthropogenically produced greenhouse gases.

global warming potential (GWP) An index that attempts to quantify the relative impacts of different greenhouse gases on global warming, using the impact of carbon dioxide as a reference point.

Great American Desert The name given to the western interior plains of North America in the nineteenth century, based on observation of the area during a period of natural desertification.

Great Plains An area of temperate grassland with a semi-arid climate in the interior of North America, stretching for some 2500 km from western Texas in the south to the Canadian prairie provinces in the north.

green belt A zone of undeveloped land, either natural vegetation or agricultural land, maintained around an urban area as a means of limiting urban sprawl and the negative environmental consequences that it brings with it.

Green revolution The rapid increase in crop production brought about in the late 1950s and 1960s by a combination of increased fertilizer use and the introduction of new, high-yielding varieties of wheat and rice.

greenhouse effect The ability of the atmosphere to be selective in its treatment of radiation, allowing incoming short-wave solar radiation through to heat the earth's surface but trapping the long-wave returning terrestrial radiation. The net result is a warming of the atmosphere.

greenhouse gases A group of about twenty gases responsible for the greenhouse effect through their ability to absorb long-wave terrestrial radiation. Water vapour is the most effective greenhouse gas, but carbon dioxide, methane, nitrous oxide, CFCs and tropospheric ozone are also making significant contributions to the greenhouse effect.

grey water Waste water that does not contain the products of bodily functions, being mainly the product of bathing, showering, dishwashing and similar activities.

groundwater The water that accumulates in the pore spaces and cracks in the rocks beneath the earth's surface. It originates as precipitation and percolates down into sub-surface aquifers.

groundwater mining The withdrawal of groundwater in amounts that far exceed replenishment rates, causing aquifers to become depleted.

gully erosion A serious form of soil erosion in which the path taken by fast flowing water is marked by deep channels along which topsoil and the underlying subsoil have been eroded.

gyre The roughly circular patterns assumed by the major surface currents in the world's oceans, brought about by a combination of wind patterns and the shape of the ocean basins.

habitat The specific environment in which an organism lives.

halocarbons Compounds containing chlorine, fluorine and bromine – the halogen elements – combined with carbon. They are strong greenhouse gases and have the potential to deplete stratospheric ozone.

halons Synthetic organic compounds containing bromine, similar to CFCs, but with a greater ozone depletion potential.

hardness A chemical property of water that prevents

lathering and causes the build-up of scale on boilers. Caused by the presence of dissolved calcium and magnesium compounds.

hazardous air pollutants (HAPs)   A group of air pollutants not commonly found in the atmosphere and therefore not always monitored. They include heavy metals, radioactive particles and toxic industrial gases.

hazardous waste   Wastes that are dangerous because they are toxic, biologically active, flammable, corrosive, radioactive or include a combination of these factors. The extent of the hazard posed by the waste will depend upon the amount involved, its durability and the methods used to store or dispose of it.

herbivore   A plant-eating organism. The group includes a range of species from large mammals to a variety of insects. Herbivores are primary consumers, directly dependent upon the plant species at the base of all food chains.

heterogeneous chemical reactions   Chemical reactions that take place on the surface of the ice particles that make up polar stratospheric clouds. These reactions cause the release of chlorine into the stratosphere, where it causes ozone depletion and contributes to the creation of the Antarctic ozone hole.

humus   Partly decomposed organic matter, dark brown in colour, colloidal in structure and rich in bacteria and fungi, which is an essential component of good soil.

hydrogen oxides   A group of naturally occurring compounds derived from water vapour, methane and molecular hydrogen, referred to collectively as odd hydrogens, that are very effective ozone destroyers because of their participation in catalytic chain reactions.

hydrologic cycle   A complex group of processes by which water in its various forms is circulated through the earth/atmosphere system.

hydrology   The scientific study of water in the earth/atmosphere system.

hydrophyte   A plant adapted to growth under conditions of abundant soil moisture.

hydrosphere   That part of the earth's crust covered by water, both salt and fresh.

Ice Ages   Periods in the geological history of the earth when glaciers and ice sheets covered large areas of the earth's surface.

immigration   The movement into a country of people seeking permanent residence.

increasers   Natural grassland plant species that are not preferred by domestic livestock. Not being preferred, they survive grazing and their numbers tend to increase over time.

industrial revolution   A period of rapid transition from an agricultural to an industrial society, beginning in Britain in the mid-eighteenth century and spreading to other parts of the world in the next 200 years.

infrared radiation   Low-energy, long-wave radiation, sometimes referred to as heat radiation, with wavelengths between 0.7 μm and 1000 μm in the electromagnetic spectrum.

intensive livestock farming   The raising of animals in close confinement for greater management efficiency, to minimize energy loss and maximize growth rates.

Intergovernmental Panel on Climate Change (IPCC)   A group of eminent scientists brought together in 1988 by the World Meteorological Organization (WMO) and the United Nations Environment Program (UNEP). It was charged with assessing the overall state of research on climate change.

International Tropical Timber Organization (ITTO)   Formed in 1983 to develop the management, harvesting and marketing of tropical rain-forest products to the benefit of producing and consuming countries.

Intertropical Convergence Zone (ITCZ)   A thermal low-pressure belt that circles the earth in equatorial latitudes, between the tropical Hadley cells that lie north and south of the equator.

intrazonal soils   Soils that are not typical of the climatic zones in which they occur, because of the presence of over-riding local factors such as geology or drainage.

Iron Age   The stage in human technological development in which iron was widely used for tools and weapons. It followed the Bronze Age and by about 500 BC the use of iron was common.

irrigation   The provision of water for crops in areas where the natural precipitation is inadequate for crop growth.

jet stream   A fast-flowing stream of air in the upper atmosphere at about the level of the tropopause, associated with zones in which steep temperature gradients exist and where, in consequence, the pressure gradients are also steep.

K-strategists   Organisms that are usually large, have relatively long lives and produce only a limited number of offspring. Characteristically they spend considerable time and energy providing for the survival of these offspring as a means of ensuring the continuation of the species.

Kyoto Protocol   An agreement signed in Kyoto, Japan, in 1997 in which the developed nations of the world agreed to reduce greenhouse gas emissions in an attempt to slow the rate of global warming.

La Niña   An intermittent cold current flowing from east to west across the equatorial Pacific Ocean in those years when El Niño is absent.

laterite   A surface accumulation of iron and aluminium oxides and hydroxides common in soils in humid tropical regions.

leachate   Liquid containing dissolved solids produced during the process of leaching.

leaching   The process by which percolating water removes soluble solids from soils or waste disposal sites.

Lime Injection Multi-stage Burning (LIMB)   A technique developed to reduce acid gas emissions from coal-burning furnaces. Lime injected into the combustion chamber fixes the sulphur in the sulphur dioxide released from the burning coal and prevents it from being released into the atmosphere.

lithosphere   The outermost layer of the solid earth, consisting of the crust and the upper, rigid part of the mantle.

Little Ice Age A period of global cooling lasting for about 300 years from the mid-sixteenth century to the mid-nineteenth. Also referred to as the neoglacial and Fernau glaciation.

London Dumping Convention An agreement signed in London in 1972 to reduce the amount of industrial waste and sewage sludge dumped in the oceans. Since 1996 a protocol to the convention has brought about an end to the dumping of all industrial and hazardous waste.

London smog (1952) A major air pollution event in London, England caused by a combination of meteorological conditions (low temperatures, high pressure, poor ventilation) and energy use (the burning of high-sulphur coal). An estimated 4000 deaths were attributed to the smog.

Luddites Groups of English textile workers in the early nineteenth century who smashed labour-saving machinery in protests over the impact of technology on employment and working conditions.

Markets First One of the scenarios developed by the United Nations Environment Program (UNEP) in an attempt to forecast environmental change. It assumes that most of the world adopts the values and expectations prevailing in the developed world.

mass extinctions Episodes of catastrophic extinction in the earth's history. At least five such events have occurred. The best known is probably the episode some 65 million years ago that led to the demise of the dinosaurs.

Medieval Warm Period A phase in early medieval times from about AD 750 to 1200, sometimes called the Little (Climatic) Optimum, when climatic conditions ameliorated in Europe and North America and probably elsewhere.

melanoma A malignant, normally fatal form of skin cancer, associated with over-exposure to ultraviolet-B radiation.

mercaptans A group of organic compounds that contain sulphur and are characterized by an offensive odour, sometimes likened to rotting cabbage.

mercury poisoning Poisoning caused by the ingestion, inhalation or absorption of mercury or its compounds. It affects the brain, kidneys and bowel, producing symptoms that include amnesia, insomnia, fatigue, blindness and emotional or mental disorders.

mesopause The boundary between the mesosphere and thermosphere lying some 80 km above the earth's surface.

mesophyte A plant that grows where moisture conditions are moderate, neither too wet nor too dry.

mesosphere The layer of the atmosphere above the stratosphere in which the temperature declines with increasing altitude from close to 0°C at the stratopause to −80°C at the mesopause, its upper limit.

methane A simple hydrocarbon gas (CH_4) produced during the decomposition of organic matter under anaerobic conditions. It is a powerful greenhouse gas.

methyl bromide A fumigant used as a pesticide in the fruit and vegetable industry. It is an ozone-depleting substance (ODS) and its production and use are being phased out.

methyl chloroform A volatile, sweet-smelling liquid, once used as an anaesthetic, but more recently as a solvent in the rubber and plastics industries. Like methyl bromide it is an ODS and its production and use are being phased out.

migration The movement of organisms from one area to another across the earth. Migration may be seasonal, involving outward and return journeys – as with migratory birds – or longer-term and permanent.

Montreal Protocol An agreement reached in 1987 in Montreal, aimed at reducing the destruction of the ozone layer by cutting the production and use of ozone-depleting substances.

National Forestry Action Plan (NFAP) One of a series of attempts in the 1980s to maintain tropical forests by developing training and research in areas such as conservation, agroforestry and reforestation.

nekton A group of swimming organisms – ranging from zooplankton to marine mammals such as seals and whales – that inhabit pelagic (open ocean) marine ecosystems.

neritic zone That part of the ocean that lies over the continental shelf, in which waters are usually less than 200 m deep.

niche The position of an organism within a habitat, as defined by its role in that habitat. It includes not only physical location, but also the functional role of the organism in the community, as determined by its needs and its interrelationships with other components of the environment.

noise abatement The reduction or control of noise levels usually achieved through reduction at source, transmission control or receiver control.

noise pollution Unwanted sound, sound without value or sound that causes sufficient disturbance and annoyance to have social and medical implications.

nomadic herding The regular, often seasonal, movement of pastoralists in accordance with the availability of forage for their flocks and herds.

non-consumptive use An activity that does not result in the consumption or depletion of resources. The use of an area for hiking, camping or bird watching is a non-consumptive use, for example.

non-renewable resource A natural resource that cannot be replaced once it has been used. Fossil fuels are non-renewable, for example.

'no regrets' policies Policies designed to deal with projected environmental change, with the ability to contribute to the overall health of the environment whether the projected change actually takes place or not.

North American Water and Power Alliance (NAWAPA) A scheme, first proposed in 1964, to divert Canadian rivers flowing into the Arctic southwards into the United States to provide water for irrigation and the growing municipal demand in the arid south-western states.

nuclear energy The energy released during nuclear fission. If released in an uncontrolled manner it can cause an explosion, but under the controlled conditions of a nuclear reactor it can be used to generate electricity.

nuclear fission A reaction in which the nucleus of an atom of a heavy metal, such as uranium, splits into two relatively equal parts, emitting neutrons as it does so and releasing large amounts of energy.

oceanic circulation   The organized movement of water in the earth's ocean basins. The surface circulation is driven by the prevailing winds, while the sub-surface circulation is driven by temperature and density differences – the thermohaline circulation.

ocean trenches   Deep troughs in the ocean floor in the abyssal zone, associated with the location and movement of the tectonic plates in the earth's crust.

Ogallala aquifer   A major aquifer beneath the central Great Plains of the United States. It supplies most of the irrigation water used in the area, but is increasingly threatened with depletion from over-pumping.

old-growth forest   Forests that have never been logged and are therefore considered to reflect the natural forest community in a particular area. Areas of secondary growth may also fit the pattern if they have had time to recover from prior forestry activities.

oligotrophic lakes   Water bodies that have a low concentration of nutrients and are therefore low in organic productivity.

omnivore   An organism that eats both plants and animals; human beings, are omnivorous, for example.

Organization of Petroleum Exporting Countries (OPEC)   A group of Middle Eastern, African, Asian and South American nations that includes the world's major oil producing and exporting nations. Being in a position to control petroleum production and prices, it has exerted significant influence on the world economy.

organochlorides   A group of organic compounds containing chlorine. Often used in pesticides – DDT, for example – but serious side effects have led to them being banned or severely restricted in their use.

orogenesis   The process of mountain building, brought about by tectonic activity in and beneath the earth's crust.

osmosis   The diffusion of a solvent such as water through a semi-permeable membrane. Osmosis is central to the flow of water through plants and has applications in water purification.

over-grazing   The grazing of land at a rate that exceeds the growth of the vegetation growing on it. Often caused by excess animal populations and can lead to soil erosion and desertification.

oxides of nitrogen (NO$_x$)   A group of gases formed by the combination of oxygen and nitrogen. They include nitrous oxide, nitrogen dioxide and nitric oxide. All of them have an essential role in the nitrogen cycle, but they also contribute to environmental problems such as acid rain, photochemical smog, global warming and ozone depletion.

oxygen sag curve   The characteristic dip in dissolved oxygen levels immediately downstream from an effluent source as represented on a graph of dissolved oxygen levels against distance from the source.

ozone   An allotrope of oxygen containing three atoms of oxygen rather than the normal two. In the troposphere it is regarded as a pollutant, but its presence in the stratosphere protects the earth's surface from excess ultraviolet radiation.

ozone-depleting substances (ODS)   A group of compounds, including CFCs, hydrogen oxides and oxides of nitrogen capable of destroying stratospheric ozone more rapidly than it can reform, therefore causing a thinning of the ozone layer.

ozone depletion potential (ODP)   A measure of the capacity of a specific chemical to destroy ozone, using CFC-11, with an ODP of 1 as a base.

ozone layer   A diffuse layer of ozone present at heights of 20–50 km in the stratosphere which is capable of filtering out a proportion of the sun's ultraviolet radiation.

pampas   The temperate grassland biome in Argentina, South America.

particulate matter   A collective name for fine solid or liquid particles added to the atmosphere by natural or anthropogenic processes at the earth's surface.

pathogens   Living organisms that cause a transmissible or communicable disease.

pelagic ecosystems   Marine or freshwater systems that develop in open water away from the shore.

permafrost   Popularly known as 'permanently frozen ground', but technically it refers to ground in which the temperature has been below 0°C for two years. The ground in some areas that are frozen in this way may have an upper, active layer in which some degree of melting takes place during the warmer season.

peroxyacetyl nitrate (PAN)   One of a group of highly potent oxidants, present in photochemical smog.

perpetual resources   Resources that are in relatively constant supply no matter how they are used – solar energy, for example.

pH (potential hydrogen)   The representation of the acidity or alkalinity of a substance on a logarithmic scale based on hydrogen ion concentration.

phosphates   Salts of phosphoric acid. A source of the phosphorus essential for plant growth, but capable of causing the eutrophication of lakes when leached from agricultural soils or added to the waterways in sewage.

photochemical smog   Smog produced by the action of sunlight on primary combustion products, particularly those such as hydrocarbons and oxides of nitrogen produced by the internal combustion engine.

photosynthesis   A biochemical process in which green plants, through the chlorophyll in their leaves, absorb solar radiation and convert it into chemical energy.

physical weathering   The breakdown of rocks at the earth's surface caused by physical or mechanical processes – such as frost wedging – which destroy the integrity of the rock surface.

phytoplankton   Microscopic plants – such as diatoms – that live in the upper layers of fresh and salt water, where they are moved around by wind, waves and currents.

plantation forests   Cultivated forests in which trees have been planted to replace indigenous species that have been harvested. They may be planted to provide food or fibre crops and for efficiency and maximum profitability tend to involve single species.

plant succession The gradual and sequential change in the structure and content of a plant community at a particular site. The end point in such a succession is a climax community.

plate tectonics The theory that the lithosphere consists of a number of rigid but movable plates. Their movement causes major tectonic activity, from earthquakes, to volcanic activity, to sea-floor spreading and mountain building.

podzol A soil in which minerals and organic matter have been leached from the upper soil horizons and redeposited deeper in the profile. Podzols tend to have low fertility and moderate to high acidity.

polar stratospheric clouds Clouds of ice particles that form in the stratosphere above the poles in the winter.

Policy First One of the scenarios developed by the United Nations Environment Program (UNEP) in an attempt to forecast environmental change. In this scenario strong action is taken by governments in an attempt to reach specific social and environmental goals.

population A group of individuals, usually of the same species or related species, occupying a specific area.

population explosion A very rapid increase in the numbers of an organism, quite common among r-strategists such as grasshoppers or mice, which are small, have a short life span and produce large numbers of offspring.

potential resources Resources available but utilized only when economic, cultural or technological conditions change to create a demand for them.

prairie Temperate grassland in North America.

precipitation scavenging The process by which rain and snow wash particulate matter out of the atmosphere, thus helping to cleanse it.

preservation An approach to nature that requires the retention of the environment in its original state as far as possible.

primary aerosols Large particles with diameters between 1 μm and 100 μm, formed by the break-up of material at the earth's surface and emitted into the atmosphere.

primary pollutants Contaminants added directly into the environment by natural events or human activities.

r-strategists Organisms that are usually small and have a relatively short life span. They mature quickly and produce large numbers of offspring, which they do little to support.

radiation blindness Loss of sight caused by injury to the eye from exposure to excess solar radiation.

radioactivity Radiation emitted as a result of the natural decay of the atomic nuclei of certain elements such as uranium and thorium. Radioactivity can be induced in many elements by bombarding their nuclei with particles such as neutrons – as is done in a nuclear reactor.

radon A radioactive gas produced by the decay of radium. Suspected of causing lung cancer among uranium miners, it is also a hazardous indoor air pollutant in some areas.

rate of natural increase (RNI) The difference between the crude birth rate and crude death rate of a population, usually expressed as a percentage.

recycling The recovery of waste material for reprocessing into new products, or the reuse of discarded products.

reforestation The return of trees to an area from which they have been removed. Reforestation may be achieved by direct planting or seeding or by providing conditions that encourage natural regeneration.

Regional Acidification Information and Simulation (RAINS) A computer model originally developed to study acid rain in Europe, subsequently amended for use in Asia.

regolith The upper, weathered layer of bedrock, consisting of broken rock particles, grading into soil above and solid rock below.

renewable resources A resource that is replaced at a rate that is faster than, or at least as fast as, it is being used. Oxygen in the air, energy from the sun and many plants and animals can be considered renewable resources.

replacement fertility rate (RFR) A rate of 2.1 children born to a woman in her lifetime.

replacement migration The maintenance of population numbers by immigration when the general fertility rate falls below 2.1.

reserves Resources not currently being exploited, but available and able to be extracted or harvested using existing technology under prevailing economic conditions.

resources Any objects, materials or commodities that are of use to society.

respirable particulate matter (RPM) Particles suspended in the air that are small enough to be carried deep into the lungs. Particles with diameters of less than 10 μm (PM_{10}) are most hazardous to health.

ribbon development Development along major arterial roads leading into and out of cities. Often industrial in nature but also including residential development, it contributes to urban sprawl.

Rossby waves Long waves in the circumpolar westerly air flow in the upper atmosphere. They have a role in energy redistribution and in the spread of pollutants that reach the upper atmosphere.

Sahel A semi-arid to arid area, subject to seasonal and longer-term drought, in Africa south of the Sahara Desert.

salinization The build-up of salts in the soil as a result of evaporation and the capillary flow of saline water towards the surface. May be caused by climate change, but common in irrigated areas.

salt water intrusion The replacement of fresh ground water by salt water in coastal areas as a result of excessive pumping from the groundwater reservoir.

sanitary landfill The disposal of domestic refuse or garbage by controlled dumping or tipping.

scrubbers Structures used to reduce acid gas emissions from industrial plants by exposing them to lime-rich solutions that can neutralize their acidity. Aimed at reducing acid rain.

seasonal drought Drought associated with changing air mass distribution that produces distinct wet and dry seasons over the year. Common in the sub-tropics as a result of the movement of the Intertropical Convergence Zone.

secondary aerosols Aerosols formed as a result of chemical and physical processes in the atmosphere. They are concentrated in the size range 0.1–1 μm.

secondary pollutants Pollutants created in the environment by physical or chemical reactions involving primary pollutants.

secure landfill A development of the sanitary landfill concept for the safe disposal of hazardous waste.

Security First One of the scenarios developed by the United Nations Environment Program (UNEP) in an attempt to forecast environmental change. It assumes a world in which inequality and conflict prevail and the need for security becomes predominant.

selective catalytic reduction (SCR) An efficient, but costly, process for reducing the output of oxides of nitrogen from power plants. The catalyst breaks the oxides of nitrogen down into harmless nitrogen and oxygen.

selective cutting An approach to tree harvesting that targets only commercially viable trees and allows the remainder to continue growing. It can also be used as a management technique in which over-mature, dead and diseased trees are removed to provide a better growing environment for seedlings and young trees.

septic system A sewage disposal system consisting of a collection tank and a drainage field, common in rural areas.

sewage Liquid or semi-solid waste mainly from domestic sources. It is predominantly organic, including human waste and food processing residues, but commonly includes detergents and other cleansers.

sheet wash A form of soil erosion in which a reasonably uniform layer of water flows over the soil surface, removing soil particles as it passes.

shield An area of old, stable continental rock, usually with subdued relief as a result of long exposure to erosion.

shifting cultivation A form of cultivation common in the tropics and subtropics in which a small area of land is cleared, cultivated for several years, and then abandoned when soil fertility declines.

sick building syndrome Health problems brought about by poor indoor air quality. Dust, gases, paint and solvent fumes, and a variety of odours, all contribute to the problem.

site rehabilitation The restoration of the environment at an industrial, extraction or disposal site following its use. Restoration can never completely replicate the original site, but it is cleaned up and left in such a state that natural processes can reintegrate it into the local environment.

skin cancer A disease associated with damage to the genetic make-up of skin cells. Although it can be caused by a number of carcinogens, concern has focused on exposure to ultraviolet radiation as the primary cause.

slash and burn Clearing the forest for cultivation, by felling and burning the trees, particularly in the tropics. The ashes produced provide nutrients for the initial crops, but once the fertility declines it is necessary to move on and repeat the process elsewhere.

smog A combination of smoke and fog which creates air pollution.

soil A mixture of weathered rock particles and organic material on the surface of the land in which plants grow.

soil classification A grouping of soils based on the characteristics displayed by mature soils, including colour, texture, structure, chemical composition, organic content and moisture conditions.

soil conservation The preservation of the quality, quantity and productivity of the soil in an area using techniques that slow the rate of soil erosion and maintain fertility.

soil erosion The removal of soil by wind, water and gravity. Although a natural phenomenon, modern concerns are with the elevated rates of erosion associated with inappropriate human use of the land.

solar radiation Radiant energy given off by the sun. Since the sun is a very hot body, radiating at a temperature above 5700 K, the bulk of the energy is ultraviolet radiation and visible light from the short-wave end of the radiation spectrum.

species introduction The introduction of species into an area to which they are not native. If the environmental conditions allow them to survive and prosper they may threaten indigenous species.

spring flush The rapid run-off of water from melting snow and ice at the end of winter. In areas subject to acid rain the winter's accumulation of acidity is flushed into waterways in a matter of days, rapidly lowering the pH and threatening aquatic organism.

Statement of Forest Principles A product of the 1992 Earth Summit, aimed at the sustainable development of the world's forests.

steady-state system A system in which inputs and outputs are equal and constant and in which the various elements are in equilibrium.

steppe Temperate grasslands in Eurasia.

stock resources Non-renewable resources.

Stone Age A stage in human cultural development characterized by the use of stone tools and weapons. It merged into the Bronze Age some 5000 years ago.

stratopause The boundary between the stratosphere and the mesosphere, located at about 50 km above the earth's surface.

stratosphere That part of the atmosphere lying above the troposphere. Characterized by temperature increasing with altitude, because of the absorption of ultraviolet radiation by the stratospheric ozone layer.

strip cropping The practice of alternating narrow strips of crops with strips of land lying fallow to ensure that most of the land retains a vegetation cover. The net result is a cultivated area that has some protection from soil erosion.

strip mining The recovery of coal or mineral ore by stripping the overburden from the surface to expose the coal or ore body, which can then be removed using conventional excavation techniques.

subsidence – atmospheric The sinking of air in the atmosphere, often associated with anticyclones. The net downward movement reduces the probability of precipitation and also traps pollution.

subsidence – landscape   The sinking of the land surface as a result of the removal of sub-surface materials. Common in coal-mining areas and areas where the pumping of ground water exceeds the replenishment of the aquifer.

suburbanization   The spread of urbanization, by expansion adjacent to the edge of the built-up area. It can take several forms, but all contribute to urban sprawl, cause greater demand for resources and increase stress on the environment.

sulphate particles   Produced by the combustion of fossil fuels, sulphate particles make up the largest group of secondary aerosols in the atmosphere, and contribute to a number of environmental issues, including acid rain and Arctic Haze.

sulphur dioxide   An acid gas that is the product of the combustion of materials containing sulphur. In the atmosphere it combines with water to produce sulphuric acid, the main component of acid rain.

suspended particulate matter (SPM)   Particles small enough to remain suspended in the air, where they contribute to air pollution and may cause health problems when they are inhaled.

Sustainability First   One of the scenarios developed by the United Nations Environment Program (UNEP) in an attempt to forecast environmental change. It involves the development of more equitable values and institutions aimed at achieving sustainablility.

sustainable development   Development judged to be both economically and environmentally sound, so that the needs of the world's current population can be met without jeopardizing those of future generations.

taiga   The Eurasian equivalent of the boreal forest, stretching from Scandinavia to eastern Siberia.

tall stacks policy   An approach to the problem of local air pollution, which involved the building of tall smokestacks to allow the release of pollutants outside the local atmospheric boundary layer.

technocentrism   An approach to (environmental) problems that sees technology as being capable of providing the necessary solutions.

technology   Materials, products and processes designed to improve the quality of life of a population. Often used to give one population an advantage over another.

tectonic activity   Movements in the earth's crust associated with folding, faulting, earthquakes and volcanic eruptions, for example.

teleconnection   A causal link between environmental events. The events may be well separated from each other and involve a time lag. For example, an El Niño event in the eastern Pacific late in one year may be linked with the failure of the Indian monsoon the following year.

temperature inversion   The presence of a layer of warm air above cooler surface air. This reverses the normal decline of temperature with altitude in the troposphere.

terrestrial radiation   Radiation emitted from the earth's surface. Since the earth is a relatively low temperature body, terrestrial radiation is low-energy, long-wave radiation from the infrared sector of the spectrum.

thermal pollution   In theory a temperature increase in any part of the environment caused by human activity, but most often applied to temperature increases in water bodies – following the release of cooling water from thermal electric power plants, for example.

thermosphere   The outermost layer of the atmosphere, lying above the mesopause some 80 km above the earth's surface.

total fertility rate (TFR)   An estimate of the average number of children born to a woman during her lifetime.

transpiration   The loss of water from vegetation to the atmosphere by its evaporation through the leaf pores or stomata in individual plants.

tree dieback   The gradual wasting of a tree from the outermost leaves and twigs inwards. Has been linked with acid rain, but there is no conclusive proof that it is the only factor involved.

trophic chain   See *food chain*.

trophic levels   Energy levels within a trophic chain or food chain.

tropopause   The upper boundary of the troposphere. It varies in height from about 8 km at the poles to about 16 km at the equator.

troposphere   The lowest layer of the atmosphere. It contains as much as 75 per cent of the gaseous mass of the atmosphere and it is the zone in which most weather systems develop.

ultraviolet radiation   High-energy, short-wave radiation lying between visible light and x-rays in the electromagnetic spectrum. Ultraviolet rays are an important component of solar radiation.

United Nations Conference on Desertification (UNCOD)   A conference held in Nairobi, Kenya in 1977, that established the modern approach to the problem of desertification. The contribution of climate change working through drought was seen as secondary to the role of human activities in causing desertification.

United Nations Conference on Environment and Development (UNCED)   An international conference – the Earth Summit – held in Rio de Janeiro, Brazil in 1992, with the theme of sustainable development, based on economically and environmentally sound principles.

United Nations Convention to Combat Desertification (UNCCD)   A convention that grew out of the 1992 Earth Summit, aimed at addressing the problems of areas suffering from desertification. In the longer term it was seen as contributing to the broader objective of sustainable development, espoused at the summit.

United Nations Environment Program (UNEP)   Formed at the UN Conference on the Human Environment in Stockholm, Sweden in 1972, UNEP co-ordinates international measures for monitoring and protecting the environment.

urban renewal   The renewal or redevelopment of run-down urban areas. This leads to an improvement in the urban environment and is seen as more environmentally appropriate than the urban sprawl associated with the creation of new urban infrastructure around the edges of the city.

volcano A vent or fissure in the earth's crust through which magma, gases and solids such as volcanic ash are ejected during a volcanic eruption. Volcanoes cause environmental disruption through their contribution to landscape formation, atmospheric turbidity and the destruction of natural vegetation.

'wait and see' scenario The maintenance of the status quo until the nature and extent of an environmental change can be verified. Has much in common with the 'business as usual' approach to change.

Waldsterben Literally, the destruction of the forests. A term coined in Germany to describe the damage caused to forests by acid rain.

waste disposal The storage or destruction of waste material in such a way that the impact on the environment and on society is minimal. Methods range from incineration to sanitary landfill. Waste minimization and recycling help to reduce the volume of waste that requires disposal.

water quality standards Standards developed at the national and international level in an attempt to ensure that water is suitable for human consumption. Standards are also applied to agriculture, fisheries and certain industries – the food industry, for example.

water table The upper level of the saturated or groundwater zone in the rocks beneath the earth's surface.

water vapour Water in its gaseous state, produced from liquid water by evaporation, respiration from animals and transpiration from plants. A significant greenhouse gas.

wilderness An area still in its natural state that has not been significantly disturbed by humans.

World Commission on Environment and Development A Commission chaired by Gro Harlem Brundtland of Norway in 1987. It prompted the concept of sustainable development and led directly to the UNCED.

xerophytic vegetation Plants adapted for life in arid conditions. Adaptations include some that allow moisture to be stored and others that reduce moisture loss by transpiration.

zero population growth (ZPG) A condition in which the crude birth rate and crude death rate are essentially equal and as a result population numbers ultimately stabilize and no growth takes place.

zonal soils Soils with characteristics that reflect regional climatic conditions.

zooplankton Microscopic animal organisms that live in the upper layers of fresh and salt-water environments. They include crustaceans, shrimp-like forms called krill and the eggs or larvae of a variety of shellfish.

References

ACS (American Cancer Society) (2002) *Skin Cancer*. Viewed at http://www.cancer.org/download/PRO/SkinCancer.pdf (accessed 16 June 2003)

Adamson, D.A. and Fox, M.D. (1982) 'Change in Australian vegetation since European settlement', in J.M.B. Smith (ed.) *A History of Australian Vegetation*, Boston MA: McGraw-Hill

Ahearne, J.F. (1993) 'The future of nuclear power', *American Scientist*, 81: 24–35

Ahrens, C.D. (2002) *Meteorology Today* (7th edn), Minneapolis/St Paul MN: West Publishing

Aiken, S.R. and Leigh, C.H. (1995) *Vanishing Rain Forests: The Ecological Transition in Malaysia*, Oxford: Claredon Press

Alberti, M. (2000) 'Urban form and ecosystem dynamics: empirical evidence and practical implications', in K. Williams, E. Burton and M. Jenks (eds) *Achieving Sustainable Urban Form*, London: Spon

Alvarez, L.W., Alvarez, W., Asaro, F. and Michel, H.V. (1980) 'Extraterrestrial cause for the Cretaceous-Tertiary extinction', *Science* 208: 1095–108

Anderson, A.B. (1993) 'Deforestation in Amazonia: dynamics, causes and alternatives', in S. Rietbergen (ed.) *The Earthscan Reader in Tropical Forestry*, London: Earthscan

Anderson, M. (1996a) 'Population change in north-western Europe 1750–1850', in M. Anderson (ed.) *British Population History*, Cambridge: Cambridge University Press

Anderson, M. (1996b) 'British population history 1911–91', in M. Anderson (ed.) *British Population History*, Cambridge: Cambridge University Press

Anthony, D., Telegin, D.Y. and Brown, D. (1991) 'The origin of horseback riding', *Scientific American* 265: 44–8

APC (American Plastics Council) (2002) 2001 *National Post-consumer Plastics Recycling Report*. Viewed on http://www.plasticsresource.com (accessed 10 June 2003)

Appleby, L. and Harrison, R.M. (1989) 'Environmental effects of nuclear war', *Chemistry in Britain*, 25: 1223–8

Archibold, O.W. (2002) 'Grasslands', in A.S. Goudie (ed.) *Encyclopedia of Global Change*, New York: Oxford University Press

Arthur, L.M. (1988) *The Implications of Climate Change for Agriculture in the Prairie Provinces, CCD88–01*, Ottawa: Atmospheric Environment Service

Ashworth, W. (1986) *The Late, Great Lakes: An Environmental History*, New York: Knopf

Baarchers, W.H. (1996) *Eco-facts and Eco-fiction*, London/New York: Routledge

Bach, W. (1979) 'Short-term climatic alterations caused by human activities: status and outlook', *Progress in Physical Geography* 3: 55–83

Baker, J.P. and Schofield, C.L. (1985) 'Acidification impacts on fish populations: a review', in D.D. Adams and W.P. Page (eds) *Acid Deposition: Environmental, Economic and Political Issues*, New York: Plenum Press

Ball, T. (1986) 'Historical evidence and climatic implications of a shift in the boreal forest–tundra transition in central Canada', *Climatic Change* 8: 121–34

Barbour, A.K. (1994) 'Mining non-ferrous metals', in R.E. Hester and R.M. Harrison (eds) *Mining and its Environmental Impact*, Cambridge: Royal Society of Chemistry

Bark, L.D. (1978) 'History of American droughts', in N.J. Rosenberg (ed.) *North American Droughts*, Boulder CO: Westview Press

Barlow, M. and Clarke, T. (2002) *Blue Gold: The Battle against Corporate Theft of the World's Water*, Toronto: Stoddart

Barnola, J.M., Raynaud, D., Korotkevitch, Y.S. and Lorius, C. (1987) 'Vostok ice core: a 160,000 year record of atmospheric CO_2', *Nature* 329: 408–14

Barnosky, A.D. (1994) 'Defining climate's role in ecosystem evolution: clues from Late Quaternary mammals', *Historical Biology* 8: 173–90

Barrie, L.D. (1986) 'Arctic air pollution: an overview of current knowledge', *Atmospheric Environment* 20: 643–63

Barry, R.G. and Chorley, R.J. (1998) *Atmosphere, Weather and Climate* (7th edn), London: Routledge

Bates, D.W. (1972) *A Citizen's Guide to Pollution*, Montreal: McGill-Queen's University Press

Beattie, K.G. (1983) 'Land stresses associated with federal airport facilities', in Environment Canada (ed.) *Stress on Land*, Ottawa: Canadian Government Publishing Centre

Becker, C.M. and Hemley, D.D. (1998) 'Demographic change in the former Soviet Union during the transition period', *World Development* 26 (11): 1957–75

Beckinsale, R.P. (1971) 'The human use of open channels', in R.J. Chorley (ed.) *Introduction to Geographical Hydrology*, London: Methuen

Beder, S. (1990) 'Sun, surf and sewage', *New Scientist* 127 (1725): 24–9

Berthalet, A. and Chavaillon, J. (eds) (1993) *The Use of Tools by Human and Non-human Primates*, Oxford: Clarendon Press/New York: Oxford University Press

Bhaskaran, B., Mitchell, J.F.B., Lavery, J.R. and Lal, M. (1995) 'Climatic response of the Indian subcontinent to doubled CO_2 concentrations', *International Journal of Climatology* 15: 873–92

Biswas, A.K. (2000) 'Water for urban areas of the developing world in the twenty-first century', in J.I. Uitto and A.K. Biswas (eds) *Water for Urban Areas*, Tokyo/New York: United Nations University Press

Blake, D.R. and Rowland, F.S. (1988) 'Continuing worldwide increase in tropospheric methane', *Science* 239: 1129–31

Blank, L.W., Roberts, T.M. and Skeffington, R.A. (1988) 'New perspectives on forest decline', *Nature* 336: 27–30

Bliss, L.C., Heal, O.W. and Moore, J.J. (1981) *Tundra Ecosystems: A Comparative Analysis*, Cambridge: Cambridge University Press

Bolen, W.P. (2000) 'Sand and gravel construction', in *USGS Minerals Yearbook*, Washington DC: United States Geological Survey

Bolin, B. (1972) 'Atmospheric chemistry and environmental pollution', in D.P. McIntyre (ed.) *Meteorological Challenges: A History*, Ottawa: Information Canada

Bolin, B. (1986) 'How much CO_2 will remain in the atmosphere?', in B. Bolin, B.R. Doos, J. Jager and R.A. Warrick (eds) *The Greenhouse Effect, Climate Change and Ecosystems*, SCOPE 29, New York: Wiley

Boothroyd, P. (2000) 'Integrating economy, society and environment through policy assessment', in R.F. Woollard and A.S. Ostry (eds) *Fatal Consumption: Rethinking Sustainable Development*, Vancouver: University of British Columbia Press

Borchert, J.R. (1950) 'The climate of the central North American grassland', *Annals of the Association of American Geographers* 40: 1–39

Boulding, K.E. (1966) 'The economics of the coming spaceship Earth', in H. Jarrett (ed.) *Environmental Quality in a Growing Economy*, Baltimore MD: Johns Hopkins University Press

Bowman, K.P. (1986) 'Interannual variability of total ozone during the breakdown of the Antarctic circumpolar vortex', *Geophysical Research Letters* 13: 1193–6

Brakke, D.F., Landers, D.H. and Eilers, J.M. (1988) 'Chemical and physical characteristics of lakes in the northeastern United States', *Environmental Science and Technology* 22: 155–63

Branson, N. and Heinemann, M. (1973) *Britain in the Nineteen-thirties*, St Albans: Granada Publishing

Brasseur, G. and Garnier, C. (1992) 'Mount Pinatubo aerosols, chlorofluorocarbons and ozone depletion', *Science* 257: 1239–42

Bridgman, H. (1994) *Global Air Pollution*, Chichester: Wiley

Brimblecombe, P. (1987) *The Big Smoke: A History of Air Pollution in London since Medieval Times*, London/New York: Methuen

Broecker, W.S. (1991) 'The great ocean conveyor', *Oceanography* 4 (2): 79–89

Brown, J.H. and Lomolino, M.V. (1998) *Biogeography* (2nd edn), Sunderland MA: Sinauer Associates

Brown, L.R. (1991) 'The Aral Sea: going, going . . .' *Worldwatch* 4 (1): 20–7

Brown, L.R. (1996) *Tough Choices: Facing the Challenge of Food Scarcity*, New York: Norton

Brown, L.R., Gardner, G. and Halweil, B. (1999) *Beyond Malthus*, New York/London: Norton

Bruce, J. and Hengeveld, H.G. (1985) 'Our changing northern climate', *Geography* 14: 1–6

Bryan, R. (1973) *Much is Taken, Much Remains*, North Scituate MA: Duxbury Press

Buchanan, D.J. and Brenkley, D. (1994) 'Green coal mining', in R.E. Hester and R.M. Harrison (eds) *Mining and its Environmental Impact*, Cambridge: Royal Society of Chemistry

Buell, L. (1995) *The Environmental Imagination*, Cambridge MA: Belknap Press

Cairns, J. and Atkinson, R.B. (1994) 'Constructing ecosystems and determining their connectivity to the larger ecological landscapes', in R.E. Hester and R.M. Harrison (eds) *Mining and its Environmental Impact*, Cambridge: Royal Society of Chemistry

Callaghan, T.V., Bacon, P.J., Lindley, D.K. and el Moghraby, A.L. (1985) 'The energy crisis in the Sudan: alternative supplies of biomas', *Biomass* 8: 217–32

Canadian Senate (1999) *Competing Realities: The Boreal Forest at Risk*, report of the Sub-committee on the Boreal Forest of the Standing Senate Committee on Agriculture and Forestry, Ottawa: Government of Canada

Carson, R. (1962) *Silent Spring*, Boston: Houghton Mifflin

Catchpole, A.J.W. and Milton, D. (1976) 'Sunnier prairie cities – a benefit of natural gas', *Weather* 31: 348–54

Centers for Disease Control and Prevention (1991) *Preventing Lead Poisoning in Young Children*, Atlanta GA: Centers for Disease Control and Prevention

Chapman, S. (1930) 'A theory of upper atmospheric ozone', *Quarterly Journal of the Royal Meteorological Society* 3: 103

Chasek, P.S. (2002) 'Commission on Sustainable Development', in A.S. Goudie (ed.) *Encyclopedia of Global Change*, New York: Oxford University Press

Cheng, G. (1991) 'Family planning: the way out', *Beijing Review*, 11–17 March: 27–31

Chiras, D.D. (1998) *Environmental Science: A Systems Approach to Sustainable Development* (5th edn), Redwood City CA: Benjamin/Cummings

Cicerone, R.J., Stolarski, R.S. and Walters, S. (1974) 'Stratospheric ozone destruction by man-made chlorofluorocarbons', *Science* 185: 1165–7

CIDA (1985) *Food Crisis in Africa*, Hull PQ: Canadian International Development Agency

Clark, K. (1969) *Civilization*, London: BBC/John Murray

Clements, F.E. (1916) *Plant Succession: An Analysis of the Development of Vegetation*, Washington DC: Carnegie Institution of Washington

Cocks, A. and Kallend, T. (1988) 'The chemistry of atmospheric pollution', *Chemistry in Britain* 24: 884–8

Coe, M. and Foley, J. (2001) 'Human and natural impacts on the water resources of the Lake Chad basin', *Journal of Geophysical Research* 106 (D4): 3349–56

Cohn, J.P. (1990) 'Elephants: remarkable and endangered', *BioScience* 40 (1): 10–14

Coleman, D.C. and Crossley, D.A. (eds) (1996) *Fundamentals of Soil Ecology*, London: Academic Press

Concar, D. (1992) 'The resistible rise of skin cancer', *New Scientist* 108 (1821): 23–8

Connell, J.H. and Slayter, R.O. (1977) 'Mechanisms of succession in natural communities and their role in community stability and organization', *American Naturalist* 111: 1119–44

Cordell, L. (2001) 'Aftermath of chaos in the Pueblo Southwest', in G. Bawden and R.M. Reycroft (eds) *Environmental Disaster and the Archaeology of Human Response*, Albuquerque NM: Maxwell Museum of Anthropology

Cousens, S.H. (1960) 'Regional death rates in Ireland during the Great Famine from 1846 to 1851', *Population Studies* 14 (1): 55–74

Cox, G.W. (1993) *Conservation Ecology: Biosphere and Biosurvival*, Dubuque IA: Brown

Craig, J.R. (2002) 'Metals', in A.S. Goudie (ed.) *The Encyclopedia of Global Change*, New York: Oxford University Press

Crane, A. and Liss, P. (1985) 'Carbon dioxide, climate and the sea', *New Scientist* 108 (1483): 50–4

Crawford, M. (1985) 'Sub-Sahara needs quick help to avert disaster', *Science* 230: 788

Cross, M. (1985) 'Africa's drought may last many years', *New Scientist* 105 (1443): 3–4

Crowell, G.H. (1997) 'NAFTA: Potential environmental impact', in T. Fleming (ed.) *The Environment and Canadian Society*, Scarborough ON: International Thomson Publishing

Crutzen, P. (1972) 'SSTs – a threat to the earth's ozone shield', *Ambio* (2): 41–51

Crutzen, P. (1974) 'Estimates of possible variations in total ozone due to natural causes and human activities', *Ambio* 3 (6): 201–10

Crutzen, P., Aselmann, I. and Seiler, W. (1986) 'Methane production by domestic animals, wild ruminants, other herbivorous fauna and humans', *Tellus* 38: 271–84

CSD (Commission on Sustainable Development) (1995) *Review of Sectorial Clusters, Second Phase: Land, Desertification, Forests and Biodiversity*, Geneva: UN Department of Economic and Social Affairs (DESA). Viewed at http://www.un.org/documents/ecosoc/cn17/1995/ecn171995-4.htm (accessed 29 April 2003)

CSIRO (1983) *Soils: An Australian Viewpoint*, Melbourne: Commonwealth Scientific and Industrial Research Organization and Academic Press

CSIRO (1997) 'Environmental damage by wild rabbits'. Viewed at CSIRO Australia web site web@aits.csiro.au (accessed 1 May 2003)

Cuff, D.J. (2002a) 'Delta populations and rising sea level', in A.S. Goudie (ed.) *Encyclopedia of Global Change*, New York: Oxford University Press

Cuff, D.J. (2002b) 'Activists and the wood industry', in A.S. Goudie (ed.) *Encyclopedia of Global Change*, New York: Oxford University Press

Cuff, D.J. (2002c) 'Elements of sustainable development', in A.S. Goudie (ed.) *Encyclopedia of Global Change*, New York: Oxford University Press

Cuff, D.J. and Goudie, A.S. (2002) 'Forests', in A.S. Goudie (ed.) *Encyclopedia of Global Change*, New York: Oxford University Press

Currie, J.C. and Pepper, A.T. (eds) (1993) *Water and the Environment*, London: Ellis Horwood

Cutter, S.L., Renwick, H.L. and Renwick, W.H. (1991) *Exploitation, Conservation, Preservation: A Geographic Perspective on Natural Resource Use*, New York: Wiley

Dando, W.A. (2002) 'Famine', in A.S. Goudie (ed.) *Encyclopedia of Global Change*, New York: Oxford University Press

Darwin, C. (1859) *On the Origin of Species*, London: Murray

Davidson, A. (1990) *In the Wake of the Exxon Valdez: The Devastating Impact of the Alaska Oil Spill*, San Francisco CA: Sierra Club Books

Dayton, L. (2001) 'Mass extinctions pinned on Ice Age hunters', *Science* 292: 1819

Dearden, P. and Mitchell, B. (1998) *Environmental Change and Challenge: A Canadian Perspective*, Toronto: Oxford University Press

DeRose, L., Messer, E. and Millman, S. (1998) *Who's Hungry? And How Do we Know? Food Shortage, Poverty and Deprivation*, Tokyo: United Nations University Press

Deshler, T., Adriani, A., Gobbi, G.P., Hofmann, D.J., Di Donfrancesco, G. and Johnson, B.J. (1992) 'Volcanic aerosol and ozone depletion within the Antarctic polar vortex during the austral spring of 1991', *Geophysical Research Letters* 19: 1819–22

Devine, T.M. (1999) *The Scottish Nation 1700–2000*, New York: Viking

de Zafra, R.L., Reeves, J.M. and Shindell, D.T. (1995) 'Chlorine monoxide in Antarctic spring vortex' 1 'Evolution of midday vertical profiles over McMurdo Station, 1993', *Journal of Geophysical Research* 100 (D7): 13999–4007

Dotto, L. and Schiff, H. (1978) *The Ozone War*, Garden City NY: Doubleday

Draper, D. (2002) *Our Environment: A Canadian Perspective*, Scarborough ON: Nelson Thomson Learning

Duxbury, A.C. and Duxbury, A.D. (1997) *An Introduction to the World's Oceans* (5th edn), Dubuque IA: Brown

Ecologist (1972) Blueprint for Survival, London: Penguin

Edmonds, R.L. (1994) *Patterns of China's Lost Harmony*, London/New York: Routledge

Edwards, R. (1998) 'Infested waters', *New Scientist* 159 (2141): 23

Edwards, R. (2001) 'When a new house is positively sickening', *New Scientist* 169 (2281): 20

Ehrlich, P.R. (1968) *The Population Bomb*, New York: Ballantine Books

Ehrlich, P.R. and Ehrlich, A.H. (1991) *Healing the Planet: Strategies for Resolving the Environmental Crisis*, Reading MA: Addison Wesley

Ellis, E.C., Erbes, R.E. and Grott, J.K. (1990) 'Abatement of atmospheric emissions in North America; progress to date and promise for the future', in S.E. Linberg, A.L. Page and S.A. Norton (eds) *Acidic Precipitation* Volume III, *Sources, Deposition and Canopy Interactions*, New York: Springer

Enger, E.D. and Smith, B.F. (2002) *Environmental Science* (8th edn), Dubuque IA: Brown

Enhalt, D.H. (1980) 'The effects of chlorofluoromethanes on climate', in W. Bach, J. Pankrath and J. Williams (eds), *Interactions of Energy and Climate*, Dordrecht: Reidel

Environment Canada (1983) *Stress on Land*, Ottawa: Canadian Government Publishing Centre

Environment Canada (1986) *Understanding CO_2 and Climate: Annual Report 1985*, Ottawa: Atmospheric Environment Service

Environment Canada (1992) *Stratospheric Ozone Depletion*, SOE Bulletin 92–1, Ottawa: State of the Environment Reporting

Environment Canada (1999) *Rising to the Challenge: Celebrating the Twenty-fifth Anniversary of the Great Lakes Water Quality Agreement*, Ottawa: Environment Canada

Epstein, P.R. (2000) 'Is global warming harmful to health?', *Scientific American* 283 (2): 50–7

Essex, C. and McKitrick, R. (2002) *Taken by Storm: The Troubled Science, Policy and Politics of Global Warming*, Toronto: Key Porter Books

Etherington, J.R. (1982) *Environmental and Plant Ecology* (2nd edn), Chichester: Wiley

Evans, A.M. (1993) *Ore Geology and Industrial Minerals: An Introduction* (3rd edn), Oxford/Boston MA: Blackwell

Fagan, B. (1999) *Floods, Famines and Emperors: El Niño and the Fate of Civilizations*, New York: Basic Books

Falkenmark, M. (1994) 'Population, environment and development: a water perspective', in United Nations (ed.) *Population, Environment and Development*, New York: UN Department of Economic and Social Information and Policy Analysis

FAO (1996) *Our Land, Our Future*, Rome/Nairobi: Food and Agriculture Organization; United Nations Environment Program

FAO (1997) *State of the World's Forests 1997*, Rome: Food and Agriculture Organization

FAO (2001) *State of the World's Forests 2001*, Rome: Food and Agriculture Organization

FAO/UNEP (1982) *Tropical Forest Resources*, Forestry Paper No. 30, Rome: Food and Agriculture Organization

FAO/WFP (2002) 'UN agencies warn of massive southern Africa food crisis.' Viewed at http://www.fao.org/english/newsroom/news/2002/5260-en.html (accessed 29 April 2003)

Farman, J.C., Gardiner, B.G. and Shanklin, J.D. (1985) 'Large losses of total ozone in Antarctica reveal seasonal ClO_x/NO_x interaction', *Nature* 315: 207–10

Fearnside, P.M. (1993) 'Deforestation in Brazilian Amazonia: the effect of population and land tenure', *Ambio* 22 (8): 537–45

Fearnside, P.M. (2002) 'Amazonia, deforestation of', in A.S. Goudie (ed.) *Encyclopedia of Global Change*, New York: Oxford University Press

Fenger, J. (1999) 'Urban air quality', *Atmospheric Environment* 33: 4877–900

Ferguson, K. (ed.) (1991) *Environmental Solutions for the Pulp and Paper Industry*, San Francisco CA: Miller Freeman

Ferkiss, V. (1993) *Nature, Technology and Society*, New York/London: New York University Press

FEWS (2002) *Famine Early Warning System*. Viewed at http://www.fews.net/about/index.cfm (accessed 16 June 2003)

Fieldhouse, D. (2001) 'For richer, for poorer', in P.J. Marshall (ed.) *The Cambridge Illustrated History of the British Empire*, Cambridge: Cambridge University Press

Fleming, T (ed.) (1997) *The Environment and Canadian Society*, Scarborough ON: International Thomson Publishing

Folland, C.K., Karl, T. and Vinnikov, K.Y. (1990) 'Observed climate variations and change', in J.T. Houghton, G.J. Jenkins and J.J. Ephraums (eds) *Climate Change: The IPCC Scientific Assessment*, Cambridge: Cambridge University Press

Foster, S.S.D., Lawrence, A.R. and Morris, B.L. (1998) *Groundwater in Urban Development: Assessing Management Needs and Formulating Policy Strategies,*

World Bank Technical Paper 390, Washington DC: World Bank

Fox, D.J. (1997) 'Mining in mountains', in B. Messerli and J.D. Ives (eds) *Mountains of the World: A Global Priority*, New York: Parthenon

Freedman, W. (2001) *Environmental Science: A Canadian Perspective*, Toronto: Pearson

French, H. (1996) *The Periglacial Environment*, Harlow: Addison Wesley Longman

Gadd, A.J. (1992) 'Scientific statements and the Rio Earth Summit', *Weather* 47: 294–315

Gates, D.M. (1993) *Climate Change and its Biological Consequences*, Sunderland MA: Sinauer

Glantz, M.H. (ed.) (1977) *Desertification: Environmental Degradation in and around Arid Lands*, Boulder CO: Westview Press

Glantz, M.H. (1994) *Drought Follows the Plough: Cultivating Marginal Areas*, Cambridge: Cambridge University Press

Glantz, M.H., Katz, R.W. and Nicholls, N. (eds) (1991) *Teleconnections: Linking Worldwide Climate Anomalies*, Cambridge: Cambridge University Press

Gleason, H.A. (1926) 'The individualistic concept of the plant association', *Bulletin of the Torrey Botanical Club* 53: 1–20

Gleick, P.H. (1994) 'Water, war and peace in the Middle East', *Environment* 36 (3): 6–15 and 35–42

Gleick, P. (1998) *The World's Water 1998–1999: The Biennial Report on Freshwater Resources*, Washington DC: Island Press

Gleick, P. (2002) *Dirty Water: Estimated Deaths from Water-related Diseases 2000–2020*, Oakland CA: Pacific Institute. Viewed at http://www.pacinst.org/reports/ water_related_deaths_report.pdf (accessed 30 June 2003)

Gobbi, G.P., Congeduti, F. and Adriani, A. (1992) 'Early stratospheric effects of the Pinatubo eruption', *Geophysical Research Letters* 19: 997–1000

Godbold, D.L. and Hüttermann, A. (eds) (1994) *Effects of Acid Rain on Forest Processes*, New York: Wiley Liss

Goldfarb, T.D. (2001) *Taking Sides: Clashing Views on Controversial Environmental Issues* (9th edn), Guilford CT: McGraw-Hill Dushkin

Goudie, A.S. (1999) *The Human Impact on the Natural Environment* (5th edn), Cambridge MA: Blackwell

Goudie, A.S. (2001) *The Nature of the Environment* (4th edn), Oxford: Blackwell

Goudie, A.S. (2002) 'Groundwater rise under London, England', in A.S. Goudie (ed.) *Encyclopedia of Global Change*, New York: Oxford University Press

Goudie, A.S. and Viles, H. (1997) *The Earth Transformed*, Oxford: Blackwell

Green, C. (1992) 'Economics and the 'greenhouse effect', *Climatic Change* 22: 265–91

Gregory, R.P.F. (1989) *Photosynthesis*, Glasgow: Blackie

Gribbin, J. (1993) *The Hole in the Sky*, New York: Bantam

Grove, A.T. (1986) 'The state of Africa in the 1980s', *Geographical Journal* 152: 193–203

Hammond, A.L. and Maugh, T.H. (1974) 'Stratospheric pollution: multiple threats to the earth's ozone', *Science* 186: 335–8

Hansen, J., Wilson, H., Sato, M., Ruedy, R., Shah, K. and Hansen, E. (1995) 'Satellite and surface data at odds', *Climatic Change* 30: 103–17

Hanson, S. (ed.) (1995) *The Geography of Urban Transport* (2nd edn), New York: Guilford Press

Hardin, G. (1968) 'The tragedy of the commons', *Science* 162: 1243–8

Harrison, C.G.A. (2002) 'Plate tectonics', in A.S. Goudie (ed.) *Encyclopedia of Global Change*, New York: Oxford University Press

Harrison, P. (1992) *The Third Revolution: Environment, Population and Sustainable Development*, London: Tauris

Harvey, H.H. (1989) 'Effects of acid precipitation on lake ecosystems', in D.C. Adriano and A.H. Johnson (eds) *Acidic Precipitation* Volume II, *Biological and Ecological Effects*, New York: Springer

Harvey, L.D.D. (2000) *Global Warming: The Hard Science*, Harlow: Pearson

Hatcher, J. (1996) 'Plague, population and the English economy', in M. Anderson (ed.) *British Population History*, Cambridge: Cambridge University Press

Hauhs, M. and Ulrich, B. (1989) 'Decline of European forests', *Nature* 339: 265

Hazma, A. (1994) 'Urban settlements and the environment in the developing world: trends and challenges', in United Nations (ed.) *Population, Environment and Development*, New York: UN Department for Economic and Social Information and Policy Analysis

Heathcote, I.W. (1997) 'Canadian water resources and management', in T. Fleming (ed.) *The Environment and Canadian Society*, Scarborough ON: International Thomson Publishing

Heathcote, R.L. (1987) 'Images of a desert? Perceptions of arid Australia', *Australian Geographical Studies* 25: 3–25

Heathcote, R.L. (1994) *Australia* (2nd edn), Harlow: Longman

Hecht, J. (1997) 'Did sea-level fall kill the dinosaurs?', *New Scientist* 154 (2076): 19

Hekstra, G.P. and Liverman, D.M. (1986) 'Global food futures and desertification', *Climatic Change* 9: 59–66

Hendrey, G.R. (1985) 'Acid deposition: a national problem', in D.D. Adams and W.P. Page (eds) *Acid Deposition: Environmental, Economic and Political Issues*, New York: Plenum Press

Hengeveld, H.G. (1991) *Understanding Atmospheric Change*, SOE Report 91–2, Ottawa: Environment Canada

Herring, H. (2000) 'Energy efficiency: does it save energy?', *Civil Engineering* 138: 36–8 (special issue *Sustainable Development: Making it Happen*)

Hessing, M. (1997) 'New directions in environmental concern', in T. Fleming (ed.) *The Environment and Canadian Society*, Scarborough ON: Nelson

Hester, R.E. and Harrison, R.M. (eds) (1994) *Mining and its Environmental Impact*, Cambridge: Royal Society of Chemistry

Heywood, V.H. and Stuart, S.N. (1992) 'Species extinctions in tropical forests', in T.C. Whitmore and J.A. Sayer (eds) *Tropical Deforestation and Species Extinction*, London: Chapman and Hall

Hill, R., O'Keefe, P. and Snape, C. (1995) *The Future of Energy Use*, New York: St Martin's Press

Hillel, D. (1994) *Rivers out of Eden*, New York: Oxford University Press

Hilton-Taylor, C. (2000) *2000 IUCN Red List of Threatened Species*, World Conservation Union. Viewed at http://www.redlist.org/info/tables/table4a.html (accessed 1 May 2003)

Hjerppe, R. and Berghall, P.E. (1996) *The Urban Challenge*, Helsinki: UNU/WIDER

Holloway, M. (1996) 'Sounding out science: Prince William Sound is recovering, seven years after the *Exxon Valdez* disaster, but the spill's scientific legacy remains a mess', *Scientific American* 275 (4): 82–8

Holmes, B. (1996) 'Blue revolutionaries', *New Scientist* 152 (2509): 32

Hong, S.M., Candelone, J.P., Patterson, C.C. and Boutron, C.F. (1994) 'Greenland ice evidence of hemispheric lead pollution 2–3 millennia ago by Greek and Roman civilizations', *Science* 265: 1841–43

Horan, N.J. (1990) *Biological Wastewater Treatment Systems: Theory and Operation*, New York: Wiley

Horel, J. and Geisler, J. (1997) *Global Environmental Change: An Atmospheric Perspective*, New York: Wiley

Horn, D.J. (1988) *Ecological Approach to Pest Management*, New York: Guilford Press

Hornby, W.F. and Jones, H. (1993) *An Introduction to Population Geography* (2nd edn), Cambridge: Cambridge University Press

Howard, R. and Perley, M. (1991) *Poisoned Skies*, Toronto: Stoddart

Hudson, P. (1992) *Industrial Revolution*, London: Edward Arnold

Hugo, G. (2001) *International Migration Transforms Australia*, Population Reference Bureau. Viewed at http://www.prb.org (accessed 16 June 2003)

Huisman, M.I.M. (1994) 'Sustainable land development in the Netherlands: the search for a concept', in H.N. Van Lier, C.F. Jaarsma, C.R. Jurgens and A.J. De Back, (eds) *Sustainable Landuse Planning*, Amsterdam: Elsevier

Hulme, M. (1997) 'Global warming', *Progress in Physical Geography* 21 (3): 446–53

Hulme, M. and Kelly, M. (1993) 'Exploring the links between desertification and climate change', *Environment* 35: 4–11 and 39–45

Hunt, P. (1992) 'Putting Asian acid rain on the map', *New Scientist* 136 (1851): 6

Hutchison, G. and Wallace, D. (1977) *Grassy Narrows*, Scarborough ON: Van Nostrand

IJC (International Joint Commission) (2000) *Living with the Red*. Viewed at http://www.ijc.org/pdf/001590app4e.pdf (accessed 5 May 2003)

IPCC (Intergovernmental Panel on Climate Change) (1990) *Climate Change: The IPCC Scientific Assessment*, Cambridge: Cambridge University Press

IPCC (1996) *Climate Change 1995: The Science of Climate Change*, Cambridge: Cambridge University Press

IPCC (2001a) *Climate Change 2001: The Scientific Basis*, Cambridge: Cambridge University Press

IPCC (2001b) *Climate Change 2001: Impacts, Adaptations and Vulnerability*, Cambridge: Cambridge University Press

IPCC (2001c) *Climate Change 2001: Mitigation*, Cambridge: Cambridge University Press

IPCC (2001d) *Climate Change 2001: Synthesis Report*, Cambridge: Cambridge University Press

IPCQL (Independent Commission on Population and Quality of Life) (1996) *Caring for the Future*, Oxford/New York: Oxford University Press

IUCN (International Union for the Conservation of Nature and Natural Resources) (2003) *The 2002 IUCN Red List of Threatened Species*, Gland: World Conservation Union. Viewed at http://www.redlist.org/info/tables/table1.html

Ives, J.D. and Messerli, B. (1989) *The Himalayan Dilemma: Reconciling Development and Conservation*, London: Routledge

Jacobsen, T. and Adams, R.M. (1971) 'Salt and silt in Ancient Mesopotamian agriculture', in T.R. Detwyler (ed.), *Man's Impact on Environment*, New York: McGraw-Hill

Jacobson, J.L. (1991) *Women's Reproductive Health: The Silent Emergency*, Washington DC: Worldwatch Institute

Jenkins, I. (1969) 'Increases in averages of sunshine in Greater London', *Weather* 24: 52–4

Jenkins, R.H. (1980) 'Oily water discharges from offshore oil developments', in C.R. Upton (ed.) *Ninth Environmental Workshop on Offshore Hydrocarbon Development*, Calgary AB: Arctic Institute of North America

Johansson, T.B., Kelly, J., Reddy, A.K.N. and Williams, R.H. (eds) (1993) *Renewable Energy: Sources for Fuels and Electricity*, Washington DC: Island Press

Johnson, F.S. (1972) 'Ozone and SSTs', *Biological Conservation* 4: 220–2

Johnston, H.S. (1971) 'Reduction of stratospheric ozone by nitrogen dioxide catalysts from SST exhaust', *Science* 173: 517–22

Jones, M.D.H. and Henderson-Sellers, A. (1990) 'History of the greenhouse effect', *Progress in Physical Geography* 14: 1–18

Jowett, J. (1990) 'People: demographic patterns and policies', in T. Cannon and A. Jenkins (eds) *The Geography of Contemporary China*, London/New York: Routledge

Jutikkala, E. (1956) 'The great Finnish famine in 1696–97', *Scandinavian Economic History Review* 3: 48–63

Kacholia, K. and Reck, R.A. (1997) 'Comparison of global climate change simulations for a $2 \times CO_2$-induced warming', *Climatic Change* 35: 53–69

Kaplan, M. (2002) 'Plight of the condor', *New Scientist* 176 (2363): 34–6

Katz, R.W. and Glantz, M.H. (1977) 'Rainfall statistics, droughts and desertification in the Sahel', in M.H. Glantz (ed.) *Desertification: Environmental Degradation in and around Arid Lands*, Boulder CO: Westview Press

Keegan, J. (2000) *The First World War*, Toronto: Vintage Canada

Keeling, C.D. and Whorf, T.P. (2003) 'Atmospheric CO_2 records from sites in the S10 air sampling network', in *Trends: a Compendium of Data on Global Change*, Oak Ridge TN: Carbon Dioxide Information Analysis Center, Oak Ridge National Laboratory, US Department of Energy. Viewed at http://cdiac.esd.ornl.gov/trends/co2/sio-mlo.htm

Kellogg, W.W. and Schneider, S.H. (1977) 'Climate, desertification and human activities', in M.H. Glantz (ed.) *Desertification: Environmental Degradation in and around Arid Lands*, Boulder CO: Westview Press

Kemp, D.D. (1982) 'The drought of 1804–1805 in central North America', *Weather* 37: 34–40

Kemp, D.D. (1991) 'The greenhouse effect: a Canadian perspective', *Geography* 76: 121–31

Kemp, D.D. (1994) *Global Environmental Issues: A Climatological Approach* (2nd edn), London/New York: Routledge

Kiernan, V. (1993) 'Atmospheric ozone hits a new low', *New Scientist* 138 (1871): 8

Kittrick, J.A. (1986) *Soil Mineral Weathering*, New York: Van Nostrand

Klein, E.M. (2002) 'Earth structure and development', in A.S. Goudie (ed.) *The Encyclopedia of Global Change*, New York: Oxford University Press

Kliot, N. (1994) *Water Resources and Conflict in the Middle East*, London: Routledge

Kromm, D.E. and White, S.E. (1992) 'The High Plains Ogallala Region', in D.E. Kromm and S.E. White (eds) *Groundwater Exploitation in the Great Plains*, Lawrence KS: University Press of Kansas

Kupchella, C.E. and Hyland, M.C. (1993) *Environmental Science: Living within the System of Nature*, Englewood Cliffs NJ: Prentice Hall

Kurland, L.T., Faro, S.N. and Seidler, H. (1960) 'Minamata disease: the outbreak of a neurological disorder in Minamata, Japan and its relationship to the ingestion of sea food contaminated by mercury', *World Neurology* 1 (5): 370–91

Lamb, H.H. (1970) 'Volcanic dust in the atmosphere with the chronology and assessment of its meteorological significance', *Philosophical Transactions of the Royal Society* A, 266: 453–533

Lamb, H.H. (1996) *Climate, History and the Modern World* (2nd edn), London: Routledge

Lambert, S.J. (1996) 'Intense extratropical northern hemisphere winter cyclone events 1899–1991', *Journal of Geophysical Research* 101: 21319–25

Last, F.T. and Nicholson, I.A. (1982) 'Acid rain', *Biologist* 29: 250–2

Lawson, M.P. and Stockton, C. (1981) 'The desert myth evaluated in the context of climate change', in C.D. Smith and M. Parry (eds) *Consequences of Climatic Change*, Nottingham: University of Nottingham

Layrisse, M. (1992) 'The "holocaust" of the Amerindians', *Interciencia* 17: 274

Leach, G. and Mearns, R. (1993) 'Beyond the woodfuel crisis', in S. Rietbergen (ed.) *The Earthscan Reader in Tropical Forestry*, London: Earthscan Publications

Le Houerou, H.N. (1977) 'The nature and causes of desertization', in M.H. Glantz (ed.) *Desertification: Environmental Degradation in and around Arid Lands*, Boulder CO: Westview Press

Leighton, P.A. (1966) 'Geographical aspects of air pollution', *Geographical Review* 56: 151–74

Lomborg, B. (2001) *The Skeptical Environmentalist*, Cambridge: Cambridge University Press

Lourenz, R.S. and Abe, K. (1983) 'A dust storm over Melbourne', *Weather* 38: 272–4

Louw, G.N. and Seely, M.K. (1982) *Ecology of Desert Organisms*, London/New York: Longman

Lowenthal, D. (2000) *George Perkins Marsh: Prophet of Conservation*, Seattle WA: University of Washington Press

Lowton, R.M. (1997) *Construction and the Natural Environment*, Oxford: Butterworth Heinemann

Ludlum, D.M. (1971) *Weather Record Book*, Princeton NJ: Weatherwise

Lutz, W., Sanderson, W. and Scherbov, S. (2001) 'The end of world population growth', *Nature* 412: 543

MacKenzie, D. (1987) 'Ethiopia plunges towards another famine', *New Scientist* 115 (1573): 26

Mackinnon, M. (2002) 'What hell is this?', *Globe Focus*, 10 August, Toronto: *Globe and Mail*: F1 and F6

MacNeish, R.S. (1992) *The Origins of Agriculture and Settled Life*, Norman OK/London: University of Oklahoma Press

MacPhee, R.D. and Marx, P.A. (1997) 'The 40,000 year plague: humans, hyperdisease and first contact extinctions', in S.M. Goodman and B.D. Patterson (eds) *Natural Change and Human Impact in Madagascar*, Washington DC: Smithsonian Institution Press

Maguire, L.A. (2002) 'Biodiversity', in A.S. Goudie (ed.) *The Encyclopedia of Global Change*, New York: Oxford University Press

Mak, S-L. (1999) 'Where are construction materials headed?', *Building Innovation and Construction Technology* 8. Viewed at http://www.dbce.csiro.au/inno-web/0899/sustainable.htm (accessed 4 December 2002)

Mannion, A.M. (1997) *Global Environmental Change* (2nd edn), Harlow: Addison Wesley Longman

Mannion, A.M. (2002) *Dynamic World: Land-cover and Land-use Change*, London: Edward Arnold/New York: Oxford University Press

Marsh, G.P. (1864) *Man and Nature or Physical Geography as Modified by Human Action*, New York: Scribner

Martin, J., Gordon, R. and Fitzwater, S. (1990) 'Iron in Antarctic waters', *Nature* 345: 156

Martin, P.S. (1984) 'Prehistoric overkill: the global model', in P.S. Martin and R.G. Klein (eds) *Quaternary Extinctions: A Prehistoric Revolution*, Tucson AZ: University of Arizona Press

Mason, B.J. (1990) 'Acid rain: cause and consequence', *Weather* 45: 70–9

Masters, J. (1995) 'Oil and gas utilization: limits of efficiency and their impact on demand', in D. Rooke, I. Fells and J. Horlock (eds) *Energy for the Future*, London: Spon

May, R.M., Lawton, J.H. and Stork, N.E. (1995) 'Assessing extinction rates', in J.H. Lawton and R.M. May (eds) *Extinction Rates*, Oxford: Oxford University Press

Maybeck, M.H. (2002) 'Water quality', in A.S. Goudie (ed.) *The Encyclopedia of Global Change*, New York: Oxford University Press

McIntosh, A.C. and Yneguez, C.E. (1997) *Second Water Utilities Data Book*, Manila: Asian Development Bank

McKnight, T.L. and Hess, D. (2000) *Physical Geography: A Landscape Appreciation* (7th edn), Englewood Cliffs NJ: Prentice Hall

McLellan, A.G. (1983) 'Pits and quarries: their land impacts and rehabilitation', in Environment Canada (ed.) *Stress on Land*, Ottawa: Canadian Government Publishing Centre

Meadows, D.H., Meadows, D.L., Randers, J. and Behrens, W.W. (1972) *The Limits to Growth*, New York: Universe Books

Meko, D.M. (1992) 'Dendroclimatic evidence from the Great Plains of the United States', in R.S. Bradley and P.D. Jones (eds) *Climate since A.D. 1500*, London: Routledge

Mesarovic, M. and Pestel, E. (1975) *Mankind at the Turning Point*, London: Hutchinson

Middleton, N. (1999) *The Global Casino: An Introduction to Environmental Issues* (2nd edn), London/New York: Edward Arnold/Oxford University Press.

Miller, G.T. (1994) *Living in the Environment* (8th edn), Belmont CA: Wadsworth

Miller, G.T. (2000) *Sustaining the Earth* (4th edn), Pacific Grove CA: Brooks Cole

Ministry of Environment, Lands and Parks/Environment Canada (1993) *State of the Environment Report for British Columbia*, Victoria BC/Ottawa: Ministry of Environment, Lands and Parks/Environment Canada

Mistry, J. (2000) *World Savannas: Ecology and Human Use*, London: Prentice Hall

Mnatsakanian, R.A. (1992) *Environmental Legacy of the Former Soviet Republics*, Edinburgh: Centre for Human Ecology

Molina, M.J. and Rowland, F.S. (1974) 'Stratospheric sink for chlorofluoromethanes: chlorine atom-catalysed destruction of ozone', *Nature* 249: 810–2

Molina, M.J. and Rowland, F.S. (1994) 'Ozone depletion: 20 years after the alarm', *Chemical and Engineering News* 72 (33): 8–13

Molnar, S. and Molnar, I.M. (2000) *Environmental Change and Human Survival*, Upper Saddle River NJ: Prentice Hall

Mooley, D.A. and Parthasarathey, B. (1983) 'Indian summer monsoon and El Niño', *Pure and Applied Geophysics* 121: 339–52

Morgan, R.P.C. (1995) *Soil Erosion and Conservation*, Harlow: Longman

Morrison, K.D. (2001) 'Naturalizing disaster: from drought to famine in southern India', in G. Bawden and R.M. Reycroft (eds) *Environmental Disaster and the Archaeology of Human Response*, Albuquerque NM: Maxwell Museum of Anthropology

Mortimore, M. (1989) *Adapting to Drought: Farmers, Famine and Desertification in West Africa*, Cambridge: Cambridge University Press

MPI (Migration Policy Institute) (2003) *United States: Inflow of Foreign-born Population by Country of Birth and by Percentage of Total Inflow by Year*. Viewed at http://www.migrationinformation.org/GlobalData/country-data/data.cfm (accessed 16 June 2003)

Mudge, F.B. (1997) 'The development of the "greenhouse" theory of global climate change from Victorian times', *Weather* 52: 13–6

Murphy, M.J. (1993) 'The contraceptive pill and female employment as factors in fertility change in Britain 1963–80: a challenge to the conventional view', *Population Studies* 47: 221–44

Myers, N. (1979) *The Sinking Ark*, Oxford: Pergamon Press

Myers, N. (1994) 'Population and the environment: the vital linkages', in United Nations (ed.) *Population, Environment and Development*, New York: UN Department for Economic and Social Information and Policy Analysis

Naess, A. (1973) 'The shallow and the deep long-range ecology movement', *Inquiry* 16: 95–100

Nagda, N.L. (1994) *Radon: Prevalence, Measurements, Health Risks and Control*, Philadelphia PA: ASTM

Nakamura, M. (2000) 'Water quality management issues in the Kansai Metropolitan Region', in J.I. Uitto and A.K. Biswas (eds) *Water for Urban Areas*, Tokyo/New York: United Nations University Press

Nash, R.F. (2002) 'Power of the wild', *New Scientist* 173 (2336): 42–5

Nazaroff, W.W. and Alvarez-Cohen, L. (2001) *Environmental Engineering Science*, New York: Wiley

NCGCC (1990) *An Assessment of the Impact of Climate Change Caused by Human Activities on China's Environment*, Beijing: National Co-ordinating Group on Climate Change

Nebel, B.J. and Wright, R.T. (2000) *Environmental Science: The Way the World Works*, Upper Saddle River NJ: Prentice Hall

Nelson, R. (1990) *Dryland Management: the 'Desertification' Problem*, World Bank Technical Paper 16, Washington DC: World Bank

Nepstad, D.C., Verissimo, A., Alencar, A., Nobre, C., Lima, E., Lefebvre, P., Schlesinger, P., Potter, C., Moutinho, P., Mendoza, E., Cochrane, M. and Brooks, V. (1999) 'Large-scale impoverishment of Amazonian forests by logging and fire', *Nature* 398: 505–8

Nicholson, S.E. (1989) 'Long-term changes in African rainfall', *Weather* 44: 47–56

Nowak, R. (2002) 'How the rich stole the rain', *New Scientist* 174 (2347): 4–5

O'Connor, D. (2002) *Report of the Walkerton Inquiry*, Toronto: Ontario Ministry of the Attorney General. Viewed at http://www.attorneygeneral.jus.gov.on.ca/english/about/pubs/walkerton (accessed 5 May 2003)

Oke, T. (1987) *Boundary Layer Climates* (2nd edn), London: Methuen

OMA (2000) *The Illness Costs of Air Pollution,* Ontario Medical Association Web page: http://www.oma.org/phealth/icap.htm (accessed 30 April 2003)

Ontario: Ministry of the Environment (1980) *The Case against the Rain,* Toronto: Information Services Branch, Ministry of the Environment

Ostrom, E. (1990) *Governing the Commons: The Evolution of Institutions for Collective Action,* New York: Cambridge University Press

Ostrom, E. (2002) 'Commons', in A.S. Goudie (ed.) *The Encyclopedia of Global Change,* New York: Oxford University Press

Otok, S. (1989) 'Wars – mortality – poverty', in J.I. Clarke, P. Carson, S.L. Kayastha and P. Nag (eds) *Population and Disaster,* Oxford/Cambridge MA: Blackwell

Owen, J.A. and Ward, M.N. (1989) 'Forecasting Sahel rainfall', *Weather* 44: 57–64

Owen, O.S. and Chiras, D.D. (1995) *Natural Resource Conservation: Management for a Sustainable Future,* Englewood Cliffs NJ: Prentice Hall

Paehlke, R. (1997) 'Green politics and the rise of the environmental movement', in T. Fleming (ed.) *The Environment and Canadian Society,* Scarborough ON: Nelson

Pain, S. and Kleiner, K. (1994) 'Frustrated West watches as Arctic oil spill grows', *New Scientist* 144 (1950): 8–9

Park, C.C. (1987) *Acid Rain: Rhetoric and Reality,* London: Methuen

Park, C.C. (1991) 'Trans-frontier air pollution', *Geography* 76: 21–35

Park, C.C. (1992) *Tropical Rainforests,* London/New York: Routledge

Park, C.C. (1997) *The Environment: Principles and Applications,* London: Routledge

Paton, T.R., Humphreys, G.S. and Mitchell, P.B. (1995) *Soils: A New Global View,* London: University of London Press

Pearce, F. (1991) 'A sea change in the Sahel', *New Scientist* 130 (1757): 31–2

Pearce, F (1992a) 'Despondency descends on Rio', *New Scientist* 134 (1824): 4

Pearce, F. (1992b) 'Miracle of the shifting sands', *New Scientist* 136 (1851): 38–42

Pearce, F. (1993) 'What makes an oil spill a disaster?', *New Scientist* 137 (1858): 11–15

Pearce, F. (1995a) 'Dead in the water', *New Scientist* 145 (1963): 26–31

Pearce, F. (1995b) 'Devastation in the desert', *New Scientist* 146 (1971): 40–3

Pearce, F. (1995c) 'Dirty rigs choke North Sea to death', *New Scientist* 146 (1976): 4

Pearce, F. (1999) 'Something in the water', *New Scientist* 161 (2176): 18–19

Pearce, F. (2001) 'Iraq wetlands face total destruction', *New Scientist* 170 (2291): 4–5

Pearce, F. (2002a) 'Death in the jungle', *New Scientist* 173 (2333): 14

Pearce, F. (2002b) 'Water wars', *New Scientist* 174 (2343): 18

Peczkic, J. (1988) 'Initial uncertainties in "nuclear winter": a proposed test based on the Dresden firestorm', *Climatic Change* 12: 198–208

Pégorié, J. (1990) 'On-farm agroforestry research: a case study from Kenya's semi-arid zone', *Agroforestry Today* 2: 4–7

Penner, J.E. (2002) 'Aerosols', in A.S. Goudie (ed.) *The Encyclopedia of Global Change,* New York: Oxford University Press

Perry, K.D., Cahill, T.A., Eldred, R.A., Dutcher, D.D. and Gill, T.E. (1997) 'Long-range transport of North African dust to the eastern United States', *Journal of Geophysical Research* 102: 11225–38

Pielou, E.C. (1991) *After the Ice Age: The Return of Life to Glaciated North America,* Chicago: University of Chicago Press

Pimental, D., Wilson, C., McCullum, C., Huang, R., Dwen, P., Flack, J., Tran, Q., Saltman, T. and Cliff, B. (1997) 'Economic and environmental benefits of biodiversity', *BioScience* 47 (11): 747–57

Pittock, A.B. and Salinger, M.J. (1991) 'Southern hemisphere climate scenarios', *Climatic Change* 18: 205–12

Plane, D.A. (1995) 'Urban transport: policy alternatives', in S. Hanson (ed.) *The Geography of Urban Transport* (2nd edn), New York: Guilford Press.

Pokar, M. (2002) 'Attacks put tigers in army's sights', *New Scientist* 175 (2359): 5

Ponting, C. (1991) *A Green History of the World,* London: Penguin

Porcella, D.B., Schefield, C.L., Depinto, J.V., Driscoll, C.T., Bukaveckas, P.A., Gloss, S.P. and Young, T.C. (1990) 'Mitigation of acid conditions in lakes and streams', in S.A. Norton, S.E. Lindberg and A.L. Page (eds) *Acidic Precipitation* Volume IV, *Soils, Aquatic Processes and Lake Acidification,* New York: Springer

Postel, S.L. (1997) *Last Oasis: Facing Water Scarcity* (2nd edn), New York: Norton

Postel, S.L., Daily, G.C. and Ehrlich, P.R. (1996) 'Human appropriation of renewable fresh water', *Science* 271: 785–8

Potts, M., Diggory, P. and Peel, J. (1983) *Textbook of Contraceptive Practice* (2nd edn), Cambridge/New York: Cambridge University Press

Press, F. and Siever, R. (1994) *Understanding Earth,* New York: Freeman

Preston, G.T. (1995) 'Fossil power generation and sustainable development', in D. Rooke, I. Fells and J. Horlock (eds) *Energy for the Future,* London: Spon

Price, M.F., Moss, L.A.G. and Williams, P.W. (1997) 'Tourism and amenity migration', in B. Messerli and J.D. Ives (eds) *Mountains of the World: A Global Priority,* New York: Parthenon

Pueschel, R.F., Blake, D.F., Snetsinger, K.G., Hansen, A.D.A., Verma, S. and Kato, K. (1992) 'Black carbon (soot) aerosol in the lower stratosphere and upper troposphere', *Geophysical Research Letters* 19: 1659–62

Pyle, J. (1991) 'Closing in on Arctic ozone', *New Scientist* 132 (1794): 49–52

Pyne, S. (1997) *World Fire: The Culture of Fire on Earth*, Seattle WA: University of Washington Press

Rampino, M.R. and Self, S. (1984) 'The atmospheric effects of El Chichon', *Scientific American* 250: 48–57

Randerson, J (2002) 'Clue to how powerlines could increase the risk of cancer', *New Scientist* 175 (2355): 7

Reed, S.C., Crites, R.W. and Middlebrooks, E.J. (1995) *Natural Systems for Wastewater Management and Treatment* (2nd edn), New York: McGraw-Hill

Reisner, M. (1993) *Cadillac Desert: The American West and its Disappearing Water* (2nd edn), New York: Penguin

Renner, R. (2001) 'Curse this house', *New Scientist* 170 (2289): 36

Rennie, J. (2002) 'Misleading math about the earth', *Scientific American* 286 (1): 61–71

Retherford, R.D., Ogawa, N. and Sakamoto, S. (1996) 'Values and fertility change in Japan', *Population Studies* 50: 5–26

Ritcey, G.M. (1989) *Tailings Management: Problems and Solutions in the Mining Industry*, Amsterdam/New York: Elsevier

Romme, W.H. and Despain, D.G. (1989) 'The Yellowstone fires', *Scientific American* 261: 21–9

Rostow, W.W. (1998) *The Great Population Spike and After*, Oxford/New York: Oxford University Press

Rowland, F.S. (1990) 'Stratospheric ozone depletion by chlorofluorocarbons', *Ambio* 19 (6): 281–92

Ruddiman, W.F. (2001) *Earth's Climate: Past and Present*, New York: Freeman

Rumney, G.R. (1968) *Climatology and the World's Climates*, New York: Macmillan

Sagane, R. (2000) 'Water supply in mega-cities in India', in J.I. Uitto and A.K. Biswas (eds) *Water for Urban Areas*, Tokyo/New York: United Nations University Press

Sage, B. (1980) 'Acid drops from fossil fuels', *New Scientist* 85 (1197): 743–5

Salameh, E. and Bannayan, H. (1993) *Water Resources of Jordan: Present Status and Future Potentials*, Amman: Friedrich Ebert Stiftung

Salinger, M.J. and Pittock, A.B. (1991) 'Climate scenarios for 2010 and 2050 AD Australia and New Zealand', *Climatic Change* 18: 259–69

Sampat, P. (2000) *Deep Trouble: The Hidden Threat of Groundwater Pollution*, Worldwatch Paper 154, Washington DC: Worldwatch Institute

Sanders, D.W. (1986) 'Desertification processes and impact in rainfed agricultural regions', *Climatic Change* 9: 33–42

Sanderson, M. (1987) *Implications of Climate Change for Navigation and Power Generation in the Great Lakes*, CCD87-03, Ottawa: Environment Canada

Schimel, J. (2000) 'Global change: rice, microbes and methane', *Nature* 403: 375–7

Schneider, S.H. (1978) 'Forecasting future droughts: is it possible?', in N.J. Rosenberg (ed.) *North American Droughts*, Boulder CO: Westview Press

Schoeberl, M.R. and Kreuger, A.J. (1986) 'Overview of the Antarctic ozone depletion issue', *Geophysical Research Letters* 13: 1191–2

Schumacher, E.F. (1973) *Small is Beautiful: Economics as if People Mattered*, New York: Harper and Row

Schumacher, M.M. (ed.) (1982) *Heavy Oil and Tar Sands Recovery and Upgrading: International Technology*, Park Ridge NJ: Noyes

Scott, G.A.J. (1995) *Canada's Vegetation: A World Perspective*, Montreal/Kingston: McGill-Queen's University Press

Scrase, H. and Lindhe, A. (2001) *Developing Forest Stewardship Standards: A Survival Guide* (2nd edn), Jokkmokk, Sweden: Taiga Rescue Network. Viewed at http://www.taigarescue.org/publications/pdf/surv_guide.pdf (accessed 9 May 2003)

Segar, D.A. and Segar, E.S. (1998) *Introduction to Ocean Sciences*, Minneapolis/St Paul MN: West Publishing

Seip, H.M., Aagaard, P., Angell, V., Eilertsen, O., Larssen, T., Lydersen, E., Mulder, J., Maniz, I.P., Semb, A., Tang Dagang, Vogt, R.D., Xiao Jinshong, Xiong Jiling, Zhoa Dawel and Kong Guohui (1999) 'Acidification in China: assessment based on studies at forested sites from Chongqing to Guangzhou', *Ambio* 26 (6): 522–8

Servos, M.R. (ed.) (1996) *Environmental Fate and Effects of Pulp and Paper Mill Effluents*, Delray Beach FL: St Lucie Press

Shenstone, M. (1997) *World Population Growth and Movement: Towards the Twenty-first Century*, E2-170\1997E, Ottawa: Government of Canada

Shindell, D.T. and de Zafra, R.L. (1995) 'The chlorine budget of the lower polar stratosphere: upper limits on ClO and implications of new Cl_2O_2 photolysis cross-sections', *Geophysical Research Letters* 22 (23): 3215–8

Shindell, D.T., Reeves, J.M., Emmons, L.K. and de Zafra, R.L. (1994) 'Arctic chlorine monoxide observations during spring 1993 over Thule, Greenland and implications for ozone depletion', *Journal of Geophysical Research* 99 (D12): 25697–704

Shrivastava, P. (1987) *Bhopal: Anatomy of a Crisis*, San Francisco CA: Harper and Row

Siegenthaler, U. (1984) 'Nineteenth-century measurements of atmospheric CO_2: a comment', *Climatic Change* 6: 409–11

Simmons, I.G. (1996) *Changing the Face of the Earth* (2nd edn), Oxford: Blackwell

Simmons, I.G. (1997) *Humanity and Environment: A Cultural Ecology*, Harlow: Longman

Simonich, S.L. and Hites, R.A. (1995) 'Global distribution of persistent organochloride compounds', *Science* 269: 1851–4

Simons, M. (1971) 'Long-term trends in water use', in R.J. Chorley (ed.) *Introduction to Geographical Hydrology*, London: Methuen

Simons, P. (1992) 'Why global warming could take Britain by storm', *New Scientist* 136 (1846): 35–8

Singer, M.J. and Munns, D.N. (1991) *Soils: An Introduction*, New York/Toronto: Macmillan/Collier Macmillan

Smith, B.D. (1995) *The Emergence of Agriculture*, New York: Freeman

Smith, H. and Haig, A.J.N. (1993) 'The development of a control policy for fish farming', in J.C. Currie and A.T. Pepper (eds) *Water and the Environment*, London: Ellis Horwood

Smith, J.W. (1920) 'Rainfall of the Great Plains in relation to cultivation', *Annals of the Association of American Geographers* 10: 69–74

Smith, R.A. (1872) *Air and Rain: The Beginnings of a Chemical Climatology*, London: Longman

Smith, Z.A. (2000) *The Environmental Paradox* (3rd edn), Upper Saddle River NJ: Prentice Hall

Spencer, R.W. and Christie, J.R. (1993) 'Precision lower stratospheric temperature monitoring with the MSU: validation and results', *Journal of Climate* 6: 1194–204

Spinney, L. (1998) 'Monkey business', *New Scientist* 158 (2132): 18–19

Starke, L. (1990) *Signs of Hope: Working towards our Common Future*, Oxford/New York: Oxford University Press

Statistics Canada (2003) *Immigrant Population by Place of Birth and Period of Immigration*. Viewed at http://www.statcan.ca/english/Pgdb/demo25.htm (accessed 16 June 2003)

Stewart, J. and Tiessen, H. (1990) 'Grasslands into deserts?', in C. Mungall and D.J. McLaren (eds) *Planet under Stress*, Toronto: Oxford University Press

Stiling, P. (1992) *Introductory Ecology*, Englewood Cliffs NJ: Prentice Hall

Stokke, O.S. (2002) 'London Convention of 1972', in A.S. Goudie (ed.) *The Encyclopedia of Global Change*, New York: Oxford University Press

Stolarski, R.S., Kreuger, A.J., Schoeberl, M.R., McPeters, R.D., Newman, P.A. and Alpert, J.C. (1986) 'Nimbus 7 satellite measurements of the springtime Antarctic ozone decrease', *Nature* 322: 808–11

Stone, R. (1999) 'Coming to grips with the Aral Sea's grim legacy', *Science* 284 (5411): 30–3

Strahler, A.H. and Strahler, A.N. (1992) *Modern Physical Geography*, New York: Wiley

Strahler, A.H. and Strahler, A.N. (2003) *Introducing Physical Geography*, New York: Wiley

Streets, D.G., Tsai, N.Y., Akimoto, H. and Oka, K. (2000) 'Sulphur dioxide emissions in Asia in the period 1985–1997', *Atmospheric Environment* 34: 4413–24

Stuiver, M. (1987) 'Atmospheric carbon dioxide and carbon reservoir changes', *Science* 199: 253–8

Stutz, F.P. (1995) 'Environmental impacts', in S. Hanson (ed.) *The Geography of Urban Transport* (2nd edn), New York: Guilford Press

Szabo, M. (1995) 'Australia's marsupials: going, going, gone?', *New Scientist* 145 (1962): 30

Taiga Rescue Network (1998) *The Scandinavian Forest Model*. Viewed at http://www.taigarescue.org/publications/ScandinavianForestry.pdf (accessed 2 May 2003)

Takahasi, Y. (2000) 'Water management in metropolitan Tokyo', in J.I. Uitto and A.K. Biswas (eds) *Water for Urban Areas*, Tokyo/New York: United Nations University Press

Thackrey, T.O. (1971) 'Pittsburgh: how one city did it', in R. Revelle, A. Khosla and M. Vinovskis (eds) *The Survival Equation*, Boston MA: Houghton Mifflin

Thomas, D.S.G. and Middleton, N.J. (1994) *Desertification: Exploding the Myths*, Chichester: Wiley

Thomas, W.L. (1956) *Man's Role in Changing the Face of the Earth*, Chicago: University of Chicago Press

Thoreau, H.D. (1854) *Walden*, Boston MA: Ticknor

Tiffen, M., Mortimore, M. and Gichuki, F. (1994) *More People, Less Erosion: Environmental Recovery in Kenya*, Chichester: Wiley

Tiles, M. and Oberdeik, H. (1995) *Living in a Technological Culture*, London/New York: Routledge

Titus, J.G. (1986) 'Greenhouse effect, sea level rise and coastal zone management', *Coastal Zone Management Journal* 14: 147–72

Tortajada-Quiroz, C. (2000) 'Water supply and distribution in the metropolitan area of Mexico City', in J.I. Uitto and A.K. Biswas (eds) *Water for Urban Areas*, Tokyo/New York: United Nations University Press

Tucker, C.J., Newcomb, W.W. and Dregne, H.E. (1994) 'AVHRR data sets for determination of desert spacial extent', *International Journal of Remote Sensing* 15: 3547–65

Tullett, M.T. (1984) 'Saharan dust-fall in Northern Ireland', *Weather* 39: 151–2

Turco, R.P. (1997) *Earth under Siege*, Oxford/New York: Oxford University Press

Turner II, B.L. and Keys, E.G. (2002) 'Carrying capacity', in A.S. Goudie (ed.) *The Encyclopedia of Global Change*, New York: Oxford University Press

Turner, M.D. (1999) 'Merging local and regional analyses of land-use change: the case of livestock in the Sahel', *Annals of the Association of American Geographers* 89: 191–219

Uitto, J.I. and Biswas, A.K. (eds) (2000) *Water for Urban Areas*, Tokyo/New York: United Nations University Press.

UK DEFRA (Department of the Environment, Food and Rural Affairs) (2003) *Digest of Environment Statistics: Air Quality*. Viewed at http://www.defra.gov.uk/environment/statistics/des/airqual/aq3036.htm (accessed 30 June 2003)

UK DETR (Department of the Environment, Transport and the Regions) (1998) *Lead Emissions*. Viewed at http://www.aeat.com/netcen/airqual/emissions/pb.html (accessed 15 June 2003)

UK DETR (2000) *Planning for the Supply of Aggregates in England*, London: Department of Environment, Transport and the Regions

UK DTI (Department of Trade and Industry) (2002) *An Overview of the Coal Industry in Britain*. Viewed at http://www.dti.gov.uk/energy/coal/uk_industry/index.shtml (accessed 10 June 2003)

Ulrich, B. (1983) 'A concept of forest ecosystem stability and of acid deposition as driving forces for destabilization', in B. Ulrich and J. Pankrath (eds) *Effects of Accumulation of Air Pollutants in Forest Ecosystems*, Dordrecht: Reidel

UNAIDS/WHO (2002) *AIDS Epidemic Update*, Geneva: Joint UN Programme on HIV/AIDS and World Health Organization

UNCOD (1977) *Desertification: its Causes and Consequences*, Oxford: Pergamon Press/Secretariat of the United Nations Conference on Desertification

UNECE (United Nations Economic Commission on Europe) (1999) *Protocol to Abate Acidification, Eutrophication and Ground-level Ozone*. Viewed at http://www.unece.org/env/lrtap/multi-h1.htm (accessed 15 June 2003)

UNECE (2000) *Convention on Long-range Transboundary Air Pollution*. Viewed at http://www.unece.org/env/lrtap

UNEP (2002) *Global Environmental Outlook 3 (GEO3)*, London/Sterling VA: Earthscan/United Nations Environment Program

UNEP-WCMC (2000) *Global Biodiversity: Earth's Living Resources in the Twenty-first Century*, Cambridge: World Conservation Press

UNEP/WHO (1996) *Air Quality Management and Assessment Capabilities in Twenty Major Cities*, London: United Nations Environment Program/World Health Organization, Monitoring and Assessment Unit

United Nations (1993) *The Global Partnership for Environment and Development: A Guide to Agenda 21*, New York: United Nations

United Nations (1994) *Population, Environment and Development*, New York: UN Department of Economic and Social Information and Policy Analysis

UNPD (United Nations Population Division) (1999) *World Contraceptive Use 1998*, New York: United Nations

UNPD (United Nations Population Division) (2001) *World Population Prospects 1950–2050* (the 2000 revision), New York: United Nations

USEPA (US Environmental Protection Agency) (1999) *National Air Quality and Emission Trends Report 1999*. Viewed at http://www.epa.gov/oar/aqtrnd99/toc.html (accessed 15 June 2003)

USEPA (2002) *National Air Quality 2001: Status and Trends*. Viewed at http://www.epa.gov/air/aqtrnd01/sulfur.html (accessed 30 June 2003)

USGCRP (1995) *Forum on Global Change Modelling*, Washington DC: US Global Change Research Program

USGS (1996) *Recycling: Metals*, Washington DC: United States Geological Survey

USGS (2002) *Iron and Steel Scrap: Mineral Commodity Summaries, January 2001*, Washington DC: United States Geological Survey

Van Kooten, G.C., Arthur, L. and Wilson, W.R. (1992) 'Potential to sequester carbon in Canadian forests: some economic considerations', *Canadian Public Policy* 18: 127–38

Van Lier, H.N. (1994) 'Landuse planning in perspective of sustainability: an introduction', in H.N. Van Lier,

C.F. Jaarsma, C.R. Jurgens and A.J. De Back (eds) *Sustainable Land Use Planning*, Amsterdam: Elsevier

Van Lier, H.N., Jaarsma, C.F., Jurgens, C.R. and De Back, A.J. (1994) *Sustainable Land Use Planning*, Amsterdam: Elsevier

Van Royen (1937) 'Prehistoric droughts in the central Great Plains', *Geographical Review* 27: 637–50

van Ypersele, J.P. and Verstraete, M.M. (1986) 'Climate and desertification: editorial', *Climatic Change* 9: 1–4

Vera, F. (2000) *Grazing Ecology and Forest History*, Wallingford: CABI Publishing

Verstraete, M.M. (1986) 'Defining desertification: a review', *Climatic Change* 9: 5–18.

Wadhams, P. (2002) 'Cryosphere', in A.S. Goudie (ed.) *The Encyclopedia of Global Change*, New York: Oxford University Press

Walker, M. (2001) 'Biodiversity', *New Scientist* 170 (2288): S24

Wallace, B. and Cooper, K. (1986) *The Citizen's Guide to Lead: Uncovering a Hidden Health Hazard*, Toronto: NC Press

Ward, R.C. and Robinson, M. (2000) *Principles of Hydrology* (4th edn), Maidenhead: McGraw-Hill

Warhurst, A. (1994) 'Environmental best-practice in metals production', in R.E. Hester and R.M. Harrison (eds) *Mining and its Environmental Impact*, Cambridge: Royal Society of Chemistry

Warren, A. and Agnew, C. (1988) *An Assessment of Desertification and Land Degradation in Arid and Semi-arid Areas*, London: Ecology and Conservation Unit, University College

Warren, D.M. and Pinkston, J. (1998) 'Indigenous African resource management of a tropical rainforest ecosystem: a case study of the Yoruba of Ara, Nigeria', in F. Berkes and C. Folke (eds) *Linking Social and Ecological Systems*, Cambridge: Cambridge University Press

WCED (World Commission on Environment and Development) (1987) *Our Common Future*, Oxford: Oxford University Press

Wheaton, E.E., Singh, T., Dempster, R., Higginbotham, K.O., Thorpe, J.P., Van Kooten, G.C. and Taylor, J.S. (1989) *An Exploration and Assessment of the Implications of Climate Change for the Boreal Forest and Forestry Economics of the Prairie Provinces and Northwest Territories: Phase 1*, CCD89-02, Ottawa: Atmospheric Environment Service

White, G. (1789) *The Natural History of Selborne*, London: Dent (1949).

WHO (1999) *Guidelines for Air Quality*, Geneva: World Health Organization

WHO (2000) *The World Health Report 2000. Health Systems: Improving Performance*, Geneva: World Health Organization. Viewed at http://www.who.int/whr/2000/index.htm (accessed 5 May 2003)

WHO/UNICEF (2000) *Global Water Supply and Sanitation Assessment 2000 Report*, Geneva/New York: World Health Organization/United Nations Children's Fund. Viewed at http://www.who.int/water_sanitation_health/Globalassessment/GlasspdfTOC.htm (accessed 5 May, 2003)

Williams, K., Burton, E. and Jenks, M. (eds) (2000) *Achieving Sustainable Urban Form*, London: Spon

Williams, M., McCarthey, M. and Pickup, G. (1995) 'Desertification, drought and landcare: Australia's role in an international convention to combat desertification', *Australian Geographer* 26 (1): 23–32

Wills, B.A. (1992) *Mineral Processing Technology: An Introduction to the Practical Aspects of Ore Treatment and Mineral Recovery* (5th edn), Oxford/New York: Pergamon

Wilson, A.T. (1978) 'Pioneer agricultural explosion of CO_2 levels in the atmosphere', *Nature* 273: 40–1

Wilson, E.O. (1992) *The Diversity of Life*, London: Allen Lane

Winter, J.M. (1977) 'Britain's "lost generation" of the First World War', *Population Studies* 31 (3): 449–66

Wittwer, S. (1984) 'The rising level of atmospheric carbon dioxide: an agricultural perspective', in J.H. McBeath (ed.) *The Potential Effects of Carbon Dioxide: Induced Climatic Change in Alaska. Conference Proceedings*, Fairbanks AK: School of Agriculture and Land Resources Management, University of Alaska

WMO (2000) *Antarctic Ozone Bulletin 5/2000*, Geneva: World Meteorological Organization

Wofsy, S.C., McElroy, M.B. and Sze, N.D. (1975) 'Freon consumption: implications for atmospheric ozone', *Science* 187: 535–7

Wood, P.M. (2000) *Biodiversity and Democracy*, Vancouver: University of British Columbia Press

Woods, R.I. (1996) 'The population of Britain in the nineteenth century', in M. Anderson (ed.) *British Population History*, Cambridge: Cambridge University Press.

Woollard, R.F. and Ostry A.S. (2000) *Fatal Consumption: Rethinking Sustainable Development*, Vancouver: University of British Columbia Press

WSSCC (Water Supply and Sanitation Collaborative Council) (2000) *A Shared Vision for Water Supply, Sanitation and Hygiene and a Framework for Future Action*, Geneva: World Health Organization

WWF (World Wide Fund for Nature) (1998) 'The South China tiger'. Viewed at http://photos.panda.org/image.jsp?image_id=30 (accessed 15 June 2003)

Yi, Z. (1996) 'Is fertility in China in 1991–92 far below replacement value?', *Population Studies* 50: 27–34

Young, A.R.M. (1996) *Environmental Change in Australia since 1788*, Melbourne: Oxford University Press

Zhang, D. (1994) 'Evidence for the existence of a medieval warm period in China', *Climatic Change* 26: 289–97

Index

Note: Information and illustrations contained within plates, figures and tables are indicated by the use of italics, and text contained in boxes by bold.